Lecture Notes in Artificial Intelligence   9222

Subseries of Lecture Notes in Computer Science

More information about this series at http://www.springer.com/series/1244

Stuart P. Wilson · Paul F.M.J. Verschure
Anna Mura · Tony J. Prescott (Eds.)

Biomimetic
and Biohybrid Systems

4th International Conference, Living Machines 2015
Barcelona, Spain, July 28–31, 2015
Proceedings

 Springer

Editors
Stuart P. Wilson
University of Sheffield
Sheffield
UK

Anna Mura
Universitat Pompeu Fabra
Barcelona
Spain

Paul F.M.J. Verschure
Catalan Institution for Research
   and Advanced Studies
Universitat Pompeu Fabra
Barcelona
Spain

Tony J. Prescott
University of Sheffield
Sheffield
UK

ISSN 0302-9743   ISSN 1611-3349 (electronic)
Lecture Notes in Artificial Intelligence
ISBN 978-3-319-22978-2   ISBN 978-3-319-22979-9 (eBook)
DOI 10.1007/978-3-319-22979-9

Library of Congress Control Number: 2015946074

LNCS Sublibrary: SL7 – Artificial Intelligence

Preface

These proceedings contain the papers presented at Living Machines: The 4th International Conference on Biomimetic and Biohybrid Systems, held in Barcelona, Spain, July 28–31, 2015. This followed the first, second, and third Living Machines conferences, which were held in Barcelona in 2012, London in July 2013, and Milan in July 2014. These international conferences are targeted at the intersection of research on novel lifelike technologies based on the scientific investigation of biological systems, or biomimetics, and research that seeks to interface biological and artificial systems to create biohybrid systems. The aim of the conference is to highlight the most exciting international research in both of these fields united by the theme of "living machines."

This year Living Machines returned to Barcelona, to the venue of the first conference in 2012, to Antoni Gaudi's beautiful biomimetic building La Pedrera. Returning to our origins in Barcelona reminds us of the initial conditions from which our community has developed and continues to evolve. The spirit of the conference has remained about more than biomimetics or biohybridity in isolation for engineering new technologies – it is equally about using natural design principles to extend our understanding of the natural world, and of our place in it. After all, through domestication and agriculture, we have been able to create living machines and biohybrid systems for millennia. To find examples of living machines that optimally trade safe and robust autonomy with a capacity to be programmed and controlled, that can make us feel happy, and engage with us socially, we need look no further than our pet dogs!

But our endeavor to create living machines, as represented at LM 2012–2015, is about more than creating malleable autonomous systems. It is, for many of us, about understanding the natural world through building models of natural processes. The models that we might design, build, test, and refine as engineers represent in many cases the theories of the natural world that we might develop as scientists, and the understanding of our place in the natural world that we might consider as philosophers and explore as artists. Living machines are models that enable us to validate and directly test our theories about natural processes.

A striking aspect of the proceedings herein, like previous Living Machines volumes before them, is the extent to which they reflect the fascinating range of natural systems that have inspired our model-making, and the variety of approaches and interdisciplinary skills that we adopt in bringing them to life. Some of our models are physical, such as robotic models of species ranging from the tadpole (see Philamore, Rossiter, and Ieropoulos) to the mantis (see Szczecinski and colleagues) to the human; others are computational, including neurobiologically inspired and neurobiologically constrained models of sensing, from vision and touch to hearing and biosonar (see, e.g., Yamada and colleagues); other models are theoretical, including neural network models of autobiographical memory (Damianou and colleagues) and of knowledge transfer (Terekhov, Montone, and O'Regan); others still explore novel methods like "crowdseeding" (Wagy and Bongard) that involve humans in new ways in the evolution of living machines.

Major themes of the conference that emerged from the variety of fullpaper and short-paper submissions that we received this year were (a) locomotion (particularly for soft bodies), (b) novel sensing and autonomous control systems, and (c) cognitive architectures, social robots, and human-robot interaction.

The main conference, July 29–31, 2015, took the form of a three-day singletrack program, and the three days self-organized quite naturally around these themes. The main program included 18 oral presentations and 29 poster presentations (with 11 poster spotlights), and six plenary lectures from leading international researchers in biomimetic and biohybrid systems. Roger Quinn of the Mechanical and Aerospace Engineering Department at Case Western Reserve University in Cleveland, Ohio presented "Robots Insect Locomotion." Barbara Mazzolai of the Center for Micro-BioRobotics at the Istituto Italiano di Tecnologia, Genoa, Italy, presented "From Plants and Animals to Robots: Movement, Sensing, and Control." Ryad Benosman of the Vision Institute, Universite Pierre et Marie Curie, Paris, France, presented "Neuromorphic Event-Based Time-Oriented Vision: A Framework to Unify Computational and Biological Vision." Robert Richardson of the Institute of Design Robotics and Optimisation in the School of Mechanical Engineering at the University of Leeds, UK, presented "Exploration Robotics." And Jose Halloy, Professor of Physics at the Universite Paris Didero presented "Collective Intelligence in Natural and Artificial Systems."

The conference was complemented by workshops during July 27–28, 2015, held at the Poblenou Campus of Universitat Pompeu Fabra. A workshop on "The Robot Self" was organized by Tony Prescott of the University of Sheffield, UK, by Paul Verschure of Universitat Pompeu Fabra, Barcelona, Spain, and by Kevin O'Regan of Universite Paris Descartes. A workshop on "Nature-Inspired Manufacturing" was organized by Marc Desmulliez and Eitan Abraham of Heriot-Watt University, Edinburgh, UK. And a workshop on "Bio-Inspired Design: Methods and Practice" was organized by Jeannette Yen and Sabir Khan of the Georgia Institute of Technology, Atlanta, USA.

We wish to thank the many people who were involved in making LM 2015 possible. Tony Prescott and Paul Verschure co-chaired and co-planned the meeting, with Stuart Wilson chairing the Program Committee and editing the proceedings volume. Anna Mura was responsible for the overall organization of the conference and with Nathan Lepora was responsible for communications. We are grateful to Louise Caffrey, Mireia Mora, Carme Buisan, Sytse Wierenga, and Pedro Omedas, for assistance with preparing the conference materials and technical support, and Manuel Geerinck for designing the conference poster this year (see manuelgeerinck.com). We are extremely grateful to those who served on the Program Committee, who provided timely and considerate reviews leading to the evaluation of a range of fascinating papers and abstract submissions from broad disciplinary and interdisciplinary perspectives.

Finally, we wish to thank the sponsors of LM 2015: The Convergence Science Network for Biomimetics and Neurotechnology (CSN II) (ICT-601167), which is funded by the European Union's Framework 7 (FP7) program in the area of Future Emerging Technologies (FET), the University of Sheffield, the University Pompeu Fabra in Barcelona, and the Institucio Catalana de Recerca i Estudis Avancats (ICREA).

July 2015   Stuart P. Wilson
                                                Paul F.M.J. Verschure
                                                          Anna Mura
                                                   Tony J. Prescott

Organization

Conference Chairs

Tony J. Prescott   University of Sheffield, UK
Paul F.M.J. Verschure   Universitat Pompeu Fabra and Catalan Institution for
                                 Research and Advanced Studies, Barcelona, Spain

Program Chair

Stuart P. Wilson   University of Sheffield, UK

Local Organizer

Anna Mura   Universitat Pompeu Fabra, Spain

Communications

Anna Mura   Universitat Pompeu Fabra, Spain
Nathan Lepora   University of Bristol, UK

Conference Website

Anna Mura   Universitat Pompeu Fabra, Spain
Sytse Wierenga   Universitat Pompeu Fabra, Spain

Workshop Organizers

Tony Prescott
Paul Verschure
Kevin O'Regan
Marc Desmulliez
Eitan Abraham
Jeannette Yen
Sabir Khan

Program Committee

Federico Carpi   Cecilia Laschi
Holger G. Krapp   Anders Christensen
Marco Dorigo   Jose Halloy

Contents

A Model of Larval Biomechanics Reveals Exploitable Passive Properties for Efficient Locomotion

Dylan Ross, Konstantinos Lagogiannis$^{(\boxtimes)}$, and Barbara Webb

School of Informatics, University of Edinburgh,
10 Crichton St, Edinburgh EH8 9AB, UK
dylan.martin.ross@gmail.com, {klagogia,bwebb}@inf.ed.ac.uk

Abstract. To better understand the role of natural dynamics in motor control, we have constructed a mathematical model of crawling mechanics in larval *Drosophila*.

The model accounts for key anatomical features such as a segmentally patterned, viscoelastic outer body wall (cuticle); a non-segmented inner cavity (haemocoel) filled with incompressible fluid that enables visceral pistoning; and claw-like protrusions (denticle bands) giving rise to asymmetric friction.

Under conditions of light damping and low forward kinetic friction, and with a single cuticle segment initially compressed, the passive dynamics of this model produce wave-like motion resembling that of real larvae. The presence of a volume-conserving hydrostatic skeleton allows a wave reaching the anterior of the body to initiate a new wave at the posterior, thus recycling energy. Forcing our model with a sinusoidal input reveals conditions under which power transfer from control to body may be maximised. A minimal control scheme using segmentally localised positive feedback is able to exploit these conditions in order to maintain wave-like motion indefinitely. These principles could form the basis of a design for a novel, soft-bodied, crawling robot.

Keywords: Larval Drosophila · Biomechanical model · Positive feedback control · Peristaltic motion

1 Introduction

Felicitous use of mechanics can reduce the computational and energetic burdens faced by artificial and biological agents. As an extreme case, the passive dynamic walkers of McGeer were capable of producing naturalistic walking behaviour in the absence of any active control system and using only the energy provided by moving down a slight incline [1]. More recently, robotics has started to move beyond the confines of rigid body mechanics to exploit characteristically soft or compliant phenomena to produce complex mechanical outputs in response to simple control inputs [2,3]. Biology can provide crucial insights for designing such

© Springer International Publishing Switzerland 2015
S.P. Wilson et al. (Eds.): Living Machines 2015, LNAI 9222, pp. 1–12, 2015.
DOI: 10.1007/978-3-319-22979-9_1

time

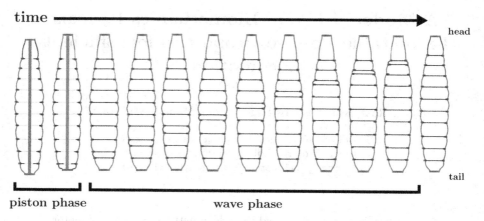

head

tail

piston phase   wave phase

Fig. 1. Crawling in larval *Drosophila*: *Drosophila* larvae crawl using a two-phase visceral pistoning mechanism. During piston phase the head, tail, viscera, and coelomic fluid move forward in tandem. Then, during wave phase, a longitudinal wave of segment compression travels from the posterior to anterior of the body.

systems, as complex biological control problems are often simplified or solved by body mechanics rather than requiring precisely orchestrated neural control [4,5].

Larval *Drosophila melanogaster*, a tiny organism with a nervous system of less than 10,000 neurons, is an excellent example. The larva's primary goal is to acquire and store enough energy to successfully pupate and become a fruitfly; hence its locomotion should be as energy efficient as possible. The *Drosophila* larva possesses a hydrostatic skeleton that runs the entire length of its body and is surrounded by a segmentally patterned cuticle and musculature. Kinematic evidence suggests that it moves using a two-phase *visceral pistoning* mechanism similar to that observed in *Manduca sexta* caterpillars [6,7] (Figure 1). During *piston phase*, the head, tail, and viscera of the organism move forward in a single step. Then, during *wave phase*, a travelling wave of compression propagates from the posterior to the anterior of the animal, moving each segment forwards in sequence. When the travelling wave reaches the head, this process repeats.

Notably, the propagation speed and cycle period of peristaltic waves is highly stereotyped [7,8]. *Drosophila* locomotion differs from that of *Manduca* in that the larva lacks hydrostatic prolegs, and lifts very little from the substrate during crawling [8]. This is reflected by a very low number of circumferential relative to longitudinal muscle fibres in *Drosophila* [9]. Motor neuron recordings show that all muscles within a segment are activated simultaneously during fictive crawling [10], suggesting that the *Drosophila* larva may use an especially simple control scheme to direct crawling behaviour.

We have defined a one-dimensional mechanical model to explore how physical properties of the larval body may simplify its control. We examine why the larva may generate peristaltic waves with a constant period, and how peristalsis may be maintained by a minimal control scheme. Unfortunately, there is little

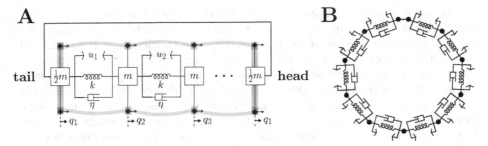

Fig. 2. Model schematic: A we model the segmented cuticle of the *Drosophila* larva as a set of coupled spring-mass-damper systems subject to an asymmetric Coulomb friction force. Volume-conserving coelomic fluid is incorporated as a rigid link connecting the head and tail masses. **B** The presence of this rigid link means that our model has a ring topology.

available experimental data regarding the mechanical properties of larval tissues. We therefore constructed our model in accordance with known anatomy and kinematics.

2   Model Construction

A general practice in modelling soft tissues is to use idealized mass-spring-damper systems [11–14].

Following this approach, we represent the head, tail, and segmental boundaries as 11 point masses, m_i, constrained to move along a single direction parallel to the plane of the substrate. The time varying positions q_i and velocities $\dot{q}_i$ of the masses along this direction of travel describe the state of our model. During peristalsis, the cuticle undergoes reversible viscoelastic strain, modelled here as an ideal spring and damper connected in parallel between neighbouring pairs of masses, characterised by a spring constant, k, and damping constant, η.

Very little change in axial or radial dimensions is observed during larval locomotion [7,8], suggesting that coelomic fluid may be modelled as an incompressible liquid which prevents changes in total body volume. This is enforced in our model by connecting the head and tail masses by a rigid link, which imposes the constraint $\dot{q}_1 = \dot{q}_{11} \; \forall \, t$.

Each body segment in the *Drosophila* larva contacts the substrate via a band of hard, claw-like projections called denticles. Since denticles extend primarily in the posterior direction, we model their interaction with the environment as a directionally asymmetric Coulomb friction force:

$$F_i(\mathbf{q}, \dot{\mathbf{q}}, \mathbf{u}) = \begin{cases} -\mu_{f:b}\mu_{k:s}\mu mg & \text{if } \dot{q}_i > 0 \; \vee \; F_{\text{ext},i} > \mu_{f:b}\mu mg \\ \mu_{k:s}\mu mg & \text{if } \dot{q}_i < 0 \; \vee \; F_{\text{ext},i} < -\mu mg \\ -F_{\text{ext},i} & \text{if } \dot{q}_i = 0 \; \wedge \; -\mu mg \leq F_{\text{ext},i} \leq \mu_{f:b}\mu mg \end{cases}$$

$$(1)$$

where g is standard gravity, μ is a coefficient of static friction specific to the denticle bands and the substrate, $\mu_{k:s}$ is the ratio of kinetic to static friction, and $\mu_{f:b}$ is the ratio of friction in the forward direction to friction in the backward direction. $F_{\text{ext},i}$ is the total non-frictive force being applied to the i-th mass.

As input to our model we allow time-varying tensions, representing muscle forces, to develop between neighbouring masses. We denote the vector of muscle tensions as $\mathbf{u}^{\mathrm{T}} = [u_1, u_2, \cdots, u_N]$ and impose the constraint $0 \leq \mathbf{u} \leq d$ to represent the fact that muscle forces saturate and are purely tensile in nature. A gain parameter $b \geq 0$ allows scaling of $\mathbf{u}$ to an appropriate range for the model's passive forces.

Assembling these elements gives the model shown in Figure 2. Isolating the forces exerted on each mass and applying Newton's second law gives a system of N second-order differential equations which must be solved for q_i and $\dot{q}_i$

$$\mathbf{M}\ddot{\mathbf{q}} = -k\mathbf{D}_2\mathbf{q} - \eta\mathbf{D}_2\dot{\mathbf{q}} + b\mathbf{D}_1\mathbf{u} + \mathbf{F} \tag{2}$$

where $\mathbf{M}$ is the $N \times N$ inertia matrix, $\mathbf{D}_2$ is an $N \times N$ circulant second difference matrix describing the coupling between the head, tail, and segment boundaries, and $\mathbf{D}_1$ is an $N \times N$ circulant backward difference matrix that describes how a particular muscle tension will pull one segment backwards and another forwards

$$\mathbf{M} = \begin{bmatrix} m & & & & \\ & m & & & \\ & & \ddots & & \\ & & & m & \\ & & & & m \end{bmatrix} \quad \mathbf{D}_2 = \begin{bmatrix} 2 & -1 & & & -1 \\ -1 & 2 & -1 & & \\ & \ddots & \ddots & \ddots & \\ & & -1 & 2 & -1 \\ -1 & & & -1 & 2 \end{bmatrix} \quad \mathbf{D}_1 = \begin{bmatrix} 1 & & & -1 \\ -1 & 1 & & \\ & \ddots & \ddots & \\ & & -1 & 1 \end{bmatrix} \tag{3}$$

By introducing spatial and temporal scaling, we can reduce the dimensionality of our parameter space and highlight useful physical relationships.

Letting $\boldsymbol{\chi}$ denote the nondimensionalised position vector, the scaled dynamics may be written

$$\ddot{\boldsymbol{\chi}} = -\mathbf{D}_2\boldsymbol{\chi} - 2\zeta\mathbf{D}_2\dot{\boldsymbol{\chi}} + \mathbf{D}_1\mathbf{u} + \mathbf{G}(\mathbf{x}, \mathbf{u}) \tag{4}$$

where the damping ratio $\zeta = \eta/2\sqrt{km} \geq 0$ specifies the ratio of viscous to elastic and inertial forces. $\mathbf{G}$ denotes the nondimensionalised friction function

$$G_i(\boldsymbol{\chi}, \dot{\boldsymbol{\chi}}, \mathbf{u}) = \begin{cases} -\gamma_{f:b}\gamma_{k:s}\gamma & \text{if } \dot{q}_i > 0 \vee \frac{1}{b}F_{\text{ext},i} > \gamma_{f:b}\gamma \\ \gamma_{k:s}\gamma & \text{if } \dot{q}_i < 0 \vee \frac{1}{b}F_{\text{ext},i} < -\gamma \\ -\frac{1}{b}F_{\text{ext},i} & \text{if } \dot{q}_i = 0 \wedge -\gamma \leq \frac{1}{b}F_{\text{ext},i} \leq \gamma_{f:b}\gamma \end{cases} \tag{5}$$

The parameter $\gamma = \mu mg/b$ determines the magnitude of static friction in the backward direction. We have rewritten $\gamma_{f:b} = \mu_{f:b}$ and $\gamma_{k:s} = \mu_{k:s}$ for notational completeness. The details of this non-dimensionalisation process are provided in the supplementary material [21].

3 Results

We solved (4) using a fixed-step forward Euler method. Integration accuracy was assessed by calculating the summed kinetic, potential, and dissipated energy in our model system over the domain of integration. We tuned the integration timestep until error in total energy fell below 1%, finding a timestep of $10^{-3}t_c$ to be sufficient.

3.1 Passive Dynamics of a Single Segment; Role of Model Parameters

To explore the role of the parameters $\zeta, \gamma, \gamma_{f:b}$, and $\gamma_{k:s}$, we first examined the behaviour of a single segment boundary under the influence of passive mechanical forces. We simplified our analysis by assuming that all other segment boundaries were held fixed to the substrate, enforced by the constraint $\ddot{\chi}_{i \neq j} = 0$, where j specifies the freely moving segment boundary. To reflect the fact that motion of an isolated segment boundary would normally occur within the context of an ongoing peristaltic wave (Figure 1), we set initial conditions such that the free segment boundary was initially at rest away from equilibrium, in an extreme posterior or anterior position (Figure 3A, left panel, $\tau = 0$).

Passive viscoelastic and friction forces alone were able to produce trajectories that qualitatively match movements observed in the real larva (Figure 3A, left panel). Elastic forces exerted by the cuticle initially accelerate the segment boundary towards the cuticle's equilibrium position, which corresponds to the minimum of the elastic potential energy function shown in Figure 3A (right panel). Motion of the segment boundary is opposed by viscous damping and kinetic friction forces, which slow the segment boundary and eventually bring it to rest. Increasing the parameter ζ causes an increase in the magnitude of viscous forces, decreasing segment boundary velocity. When $\zeta < 1$, the moving mass may overshoot the cuticle's equilibrium position before coming to rest. Conversely, $\zeta \geq 1$ implies that the mass will move slowly towards equilibrium without overshoot (Figure 3A, left panel). Examining (5), we see that $\gamma_{k:s}\gamma$ sets the magnitude of kinetic friction when the segment boundary is moving in the posterior direction, and $\gamma_{f:b}$ scales this quantity to specify the magnitude of forward kinetic friction. Accordingly, changing $\gamma_{k:s}$ results in directionally symmetric scaling of segment boundary velocity and final displacement (Figure 3C), while changing $\gamma_{f:b}$ allows directionally asymmetric scaling (Figure 3B).

When the segment boundary comes to rest, static friction may hold it in place. By inspection of (5), γ sets the maximum magnitude of static friction in the posterior direction while $\gamma_{f:b}$ multiplies this quantity to give the maximum magnitude in the anterior direction. The expression $\gamma_{f:b}\gamma \leq \chi \leq \gamma$ thus specifies a stable range of positions in which static friction forces completely oppose elastic forces. A mass at rest within this region will remain at rest (Figure 3A, B, C).

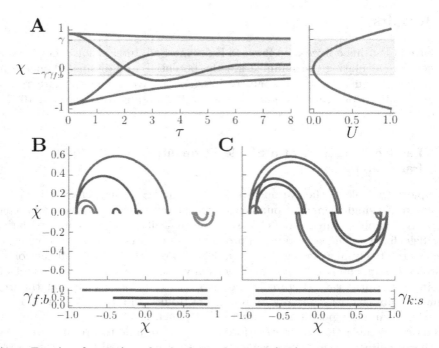

Fig. 3. Passive dynamics of a single segment: A (left) evolution of segment boundary position for $\zeta = 2$ (black) or $\zeta = 0.25$ (green), with $\gamma = 0.9$, $\gamma_{f:b} = 0.4$, $\gamma_{k:s} = 0.1$. (right) elastic energy stored in the cuticle as a function of segment boundary displacement. Shading indicates the range of positions stabilised by static friction. **B** (top) position-velocity phase trajectories for $\gamma_{f:b} = 0.1$ (green), 0.5 (red), or 1.0 (gray). Trajectories with negative velocity are unaffected by change in $\gamma_{f:b}$. (bottom) range of positions stabilised by static friction for these values of $\gamma_{f:b}$. **C** (top) phase trajectories for $\gamma_{k:s} = 0.1$ (green), 0.5 (red), or 1.0 (gray). (bottom) range of positions stabilised by static friction for these values of $\gamma_{k:s}$.

Though the range of possible values for $\zeta, \gamma, \gamma_{f:b}$, and $\gamma_{k:s}$ is very large, we focus on the following cases:

1. $\gamma = 0.9$ and $\gamma_{k:s} = 0.9$ or 0.1, i.e. kinetic friction is either equal to, or far weaker than, maximum static friction. The first case would correspond to the larva "dragging" its denticle bands across the substrate; the latter to lifting the denticles as they move.
2. $\gamma_{f:b} = 1.0$ or 0.4, i.e. either symmetrical friction, or higher backward than forward friction, which could result from denticle orientation [15].
3. $\zeta \in [0, 0.25, 0.5, 1.0, 2.0]$: a representative set of damping ratios which ranges from zero viscous damping to heavily overdamped

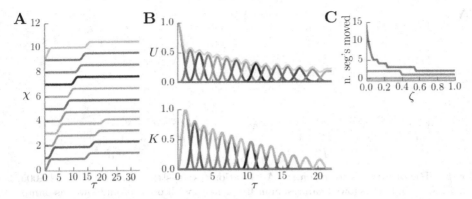

Fig. 4. Passive dynamics of the full body model: A segment boundary displacements in the absence of muscle tension, with $\zeta = 0$, $\gamma = 0.9$, $\gamma_{f:b} = 0.9$, $\gamma_{k:s} = 0.1$. **B** elastic energy stored in each cuticle segment (top, U) and kinetic energy of each segment boundary (bottom, K) during the wave shown in **A**. **C** number of discrete segment boundary movements during a passive peristaltic wave as a function of ζ, for friction conditions $\gamma_{k:s} = 0.1$, $\gamma_{f:b} = 0.4$ (gray), $\gamma_{k:s} = 0.9$, $\gamma_{f:b} = 0.4$ (brown), $\gamma_{k:s} = 0.1$, $\gamma_{f:b} = 1.0$ (orange), $\gamma_{k:s} = 0.9$, $\gamma_{f:b} = 1.0$ (orange).

3.2   Passive Dynamics of the Whole Body

Using the full model (4), we next examined the role of passive mechanics in locomotion. We set initial conditions such that the head segment was almost fully compressed, storing elastic energy. We then integrated our model equations in the absence of active muscle tensions, and observed the passive response of the system.

Under particular parameter choices, our model is capable of producing completely passive peristaltic waves (Figure 4A). In particular, with asymmetric friction forces ($\gamma_{f:b} = 0.4$), low kinetic friction ($\gamma_{k:s} = 0.1$), and low viscous damping ($\zeta < 0.6$), a wave may propagate from posterior to anterior while moving the body across the substrate (Figure 4C). The elastic energy stored in the compressed head segment is converted to kinetic energy as the segment expands. The asymmetry in our friction function means that expansion occurs through forward movement of the head, and due to the rigid link constraint (representing volume conservation of the internal coloemic fluid) this causes compression of the tail segment. The tail segment in turn expands forward, transferring energy to the neighbouring segment boundary, and the process continues. Since friction and viscous damping dissipate an amount of energy each time a segment boundary moves, the travelling wave is gradually attenuated. Nevertheless, if dissipative forces are sufficiently low, the energy supplied by compressing the head segment may propagate through the entire body and even "loop" again from head to tail (Figure 4A). This phenomenon is possible due to hydrostatic coupling between the head and tail, and is prevented by removing the rigid link

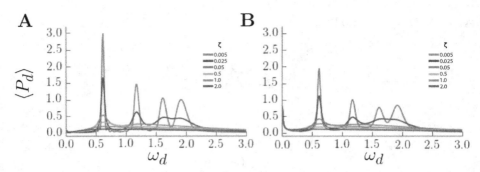

Fig. 5. Resonant frequencies: A Frictionless system with $\zeta \in \{0.005, 0.025, 0.05, 0.5, 1, 2\}$; shows peak resonant frequencies with power going lower as damping ζ increases. Increasing damping ζ drops the peak power and makes the tuning wider. (In the non-dimensionalized system units are $\frac{F_d t_c}{m}$) **B** adding friction does not affect the location of the resonant frequencies but does reduce the amount of power absorbed, without broadening the spectrum of peak responses. Friction also adds an efficient mode close to zero frequency.

constraint in our model. Our analysis suggests a view of the larval body as a ring of energy storage devices with rectified, dissipative connections (Figure 2B).

3.3   Resonance and Preferred Input Timing

We next examined the response of our system to energy input by applying sinusoidal forcing to the tail mass. The work done by a force F_d acting upon a mass to move it a distance dq is given by $dW = F_d dq$. The rate of this process gives the power supplied $P_d = F_d \dot{q}$. Thus, the power provided by applying sinusoidal forcing to the tail mass in the non-dimensionlized system is

$$P_d = \frac{F_d t_c}{m} \cos(\omega_d \tau) \frac{d\chi}{d\tau}. \tag{6}$$

Note that direction and timing of input force are important as power is maximised when force is in phase with velocity $\dot{\chi}$. In general the driving force may add or remove energy during different parts of the cycle and the two may balance, giving on average $\langle P_d(\omega_d) \rangle = 0$ for driving frequency ω_d.

We predicted that in order to maximise energy input, our stimulation frequency would have to relate to the body's passive properties, that define $\dot{\chi}$. Indeed, it is a well known property of n spring-coupled masses to exhibit N (possibly redundant) resonant frequencies at which energy absorption is maximized. Figure 5A shows the average $\langle P_d(\omega_d) \rangle$ power supplied by the driving force, numerically evaluated over 10^2 cycles at frequency ω_d in the absence of friction forces. With low damping coefficient ζ, four peaks are evident. The location of these peaks has been verified by an analysis of the normal modes of our system (not shown).

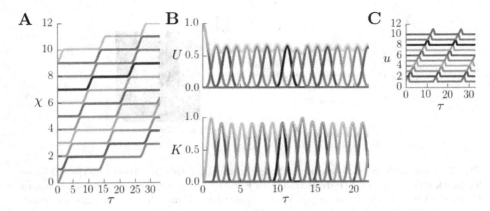

Fig. 6. Local positive feedback control: A segment boundary displacements under positive feedback control (gain $\beta = 0.105$, parameter choices as in Figure 4A). **B** elastic energy stored in each segment (top) and kinetic energy of each segment boundary (bottom) during **A**. **C** muscle tensions produced by the positive feedback control law (7) during **A**. Tensions have been normalized and offset for presentation.

In the frictionless case, viscous damping is the only dissipative force. Increasing damping reduces and broadens the peaks of power absorption. Introducing asymmetric friction decreases the amplitude of the peaks without simultaneous broadening, and also adds a peak towards low frequencies (Figure 5B). The increase in efficiency at low driving frequencies comes due to the reduction of the higher derivatives making the velocity of the driven mass effectively in phase with the driving force as it is dragged against the friction. If the driving force is provided by muscle activation then we can conclude the timing of muscle activation is important in order to achieve efficient locomotion, and this becomes even more evident as damping is decreased.

3.4 Generating Locomotion Through Local Positive Feedback

We constructed a control scheme that would exploit our model's passive dynamics in order to generate forward locomotion. This controller uses segmentally localised positive feedback of cuticle strain rate to produce muscle tensions according to the control law:

$$\mathbf{u} = \beta \mathbf{D}_1^{\mathrm{T}} \dot{\chi} \qquad 0 \leq \mathbf{u} \leq d \qquad (7)$$

which can be interpreted as producing muscle tension across a segment in proportion to the rate at which the segment is shortening, parametrised by the strain rate-tension gain β. Substituting this definition for $\mathbf{u}$ into (4) gives the closed-loop system

$$\ddot{\chi} = -\mathbf{D}_2\chi - 2\zeta\mathbf{D}_2\dot{\chi} + \beta\mathbf{D}_1\mathbf{D}_1^{\mathrm{T}}\dot{\chi} + \mathbf{G} \qquad (8)$$

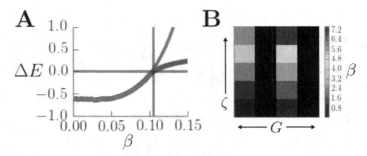

Fig. 7. Tuning feedback gain: A change in total elastic and kinetic energy (ΔE) as feedback gain β is varied (parameters as in Figure 4A). β may be tuned to achieve zero change in total energy (blue lines). With non-saturating muscle tensions, ΔE increases exponentially with β (gray line, $d = \infty$). Decreasing maximum muscle tension d causes ΔE to saturate as β increases (red line, $d = 0.1$). **B** β required to achieve continuous locomotion under various damping conditions. Friction varies left to right as ($\gamma_{k:s} = 0.1$, $\gamma_{f:b} = 0.4$), ($\gamma_{k:s} = 0.1$, $\gamma_{f:b} = 1.0$), ($\gamma_{k:s} = 0.9$, $\gamma_{f:b} = 0.4$), ($\gamma_{k:s} = 0.9$, $\gamma_{f:b} = 1.0$). ζ increases from bottom to top as $[0.0, 0.25, 0.5, 1.0, 2.0]$. Conditions are coloured black if no value of β was able to produce continuous locomotion.

Noting that $\mathbf{D}_1\mathbf{D}_1^{\mathrm{T}} = \mathbf{D}_2$ lets us rewrite this as

$$\ddot{\chi} = -\mathbf{D}_2\chi + (\beta - 2\zeta)\mathbf{D}_2\dot{\chi} + \mathbf{G} \tag{9}$$

This control can be thought of as actively amplifying any passively occurring segment strain, mitigating the effects of dissipative forces and counteracting the attenuation of passive waves described above. Inspection of the closed-loop system (9) shows that as β increases, local positive feedback first acts to cancel viscous damping forces. Increasing β further effectively introduces a positive damping term, which can offset frictional losses. Tuning β allowed us to produce continuous, naturalistic, wave-like locomotion with near-constant total body energy across a range of damping and friction conditions (Figures 6, 7).

Note, however, that this controller cannot produce waves in the absence of passive body motion. If the environment is too frictive, or the head and tail are decoupled, positive feedback fails to produce continuous locomotion.

4   Discussion

We have constructed the first model of crawling mechanics in larval *Drosophila*. The model contains key anatomical features such as a segmentally patterned, viscoelastic cuticle; a non-segmented hemocoel filled with an incompressible fluid that enables visceral pistoning; and asymmetrically frictive denticle bands. Under conditions of light damping and low forward kinetic friction, the passive dynamics of this model naturally produce wave-like motion resembling that of real larvae. Using localised positive feedback of strain rate to produce muscle tensions results in a control that is matched to the passive dynamics of the body and permits an elegant, distributed implementation.

Our model provides insight into the generation of behaviour in larval *Drosophila*. For instance, the propagation speed of peristaltic waves and thus the overall speed of locomotion are highly stereotyped within a given experimental setup [7,8]. Our model suggests that this is due to the existence of resonant modes within the larval body that may be exploited to minimize the energetic costs of locomotion. Locomotion speed has been observed to vary with substrate composition as well as denticle structure [16,17]. Increase in speed over hard substrates has been suggested to represent an escape behaviour in response to undesirable conditions [16], but our model suggests that it may simply represent a change in the frictive forces experienced by the cuticle and denticle bands.

We stress that it is unlikely that crawling behaviour in the real larva is entirely controlled by decoupled, local positive feedback, since propagating waves of motor neuron activity persist in completely isolated nervous system preparations [10]. Positive feedback may still be used to align ongoing neural motor control signals with the mechanics of the body and environment. This is consistent with the observation that waves of muscle activation travel slower in larvae which have been experimentally deprived of mechanosensory input [18,19].

Future work may investigate locomotion in two or three dimensions. This could be accomplished by adding revolute joints and torsional springs at each of the masses in the current model. This would enable investigation of turning in addition to linear crawling, the two key behaviours involved in larval navigation [20]. Of particular interest is whether positive feedback of strain rate can produce both behaviours.

The model and control schemes presented in this paper may serve as the basis for an efficient crawling robot able to exploit its passive dynamics in order to reduce the energetic and computational burden of control.

References

1. McGeer, T.: Passive Dynamic Walking. The International Journal of Robotics Research **62**(82), 62–82 (1990)
2. Hauser, H., Ijspeert, A.J., Füchslin, R.M., Pfeifer, R., Maass, W.: Towards a theoretical foundation for morphological computation with compliant bodies. Biological Cybernetics **105**, 355–370 (2011)
3. Shepherd, R.: Multigait soft robot. Proceedings of the National Academy of Sciences of the United States of America **108**(51), 20400–20403 (2011)
4. Tytell, E.D., Holmes, P., Cohen, A.H.: Spikes alone do not behaviour make: Why neuroscience needs biomechanics. Current Opinion in Neurobiology **21**(5), 816–822 (2011)
5. Kier, W.M.: The diversity of hydrostatic skeletons. The Journal of Experimental Biology **215**(8), 1247–1257 (2012)
6. Simon, M.A., Woods, W.A., Serebrenik, Y.V., Simon, S.M., van Griethuijsen, L.I., Socha, J.J., Lee, W.K., Trimmer, B.A.: Visceral-locomotory pistoning in crawling caterpillars. Current Biology **20**(16), 1458–1463 (2010)

7. Heckscher, E.S., Lockery, S.R., Doe, C.Q.: Characterization of *Drosophila* larval crawling at the level of organism, segment, and somatic body wall musculature. The Journal of Neuroscience **32**(36), 12460–12471 (2012)
8. Berrigan, D., Pepin, D.J.: How Maggots Move: Allometry and Kinematics of Crawling in Larval Diptera. J. Insect Physiol. **41**(4), 329–337 (1995)
9. Landgraf, M., Bossing, T., Technau, G.M., Bate, M.: The origin, location, and projections of the embryonic abdominal motorneurons of *Drosophila*. The Journal of Neuroscience **17**(24), 9642–9655 (1997)
10. Fox, L.E., Soll, D.R., Wu, C.: Coordination and modulation of locomotion pattern generators in *Drosophila* larvae: effects of altered biogenic amine levels by the tyramine β hydroxlyase mutation. The Journal of Neuroscience **26**(5), 1486–1498 (2006)
11. Boyle, J.H., Berri, S., Cohen, N.: Gait Modulation in *C. elegans*: An Integrated Neuromechanical Model. Frontiers in Computational Neuroscience **6**, 1–10 (2012)
12. Fung, Y.C.: Biomechanics: Mechanical Properties of Living Tissues. Springer-Verlag, New York (1993)
13. Skierczynski, B.A., Wilson, R.J., Kristan, W.B., Skalak, R.: A model of the hydrostatic skeleton of the leech. Journal of Theoretical Biology **181**(4), 329–342 (1996)
14. Alscher, C.: Simulating the motion of the leech : A biomechanical application of DAEs. Numerical Algorithms **19**, 1–12 (1998)
15. Alexandre, C.: Cuticle preparation of *Drosophila* embryos and larvae. In: Dahmann, C (ed) Methods in Molecular Biology : *Drosophila*: Methods and Protocols, Ch. 11, pp. 197–205. Humana Press Inc. (2008)
16. Apostolopoulou, A.A., Hersperger, F., Mazija, L., Widmann, A., Wüst, A., Thum, A.S.: Composition of agarose substrate affects behavioral output of *Drosophila* larvae. Frontiers in Behavioral Neuroscience **8**, 1–11 (2014)
17. Inestrosa, N.C., Sunkel, C.E., Arriagada, J., Garrido, J., Herrera, R.G.: Abnormal development of the locomotor activity in *yellow* larvae of *Drosophila*: a cuticular defect? Genetica **97**, 205–210 (1996)
18. Hughes, C.L., Thomas, J.B.: A Sensory Feedback Circuit Coordinates Muscle Activity in *Drosophila*. Mol. Cell. Neurosci. **35**(2), 383–396 (2007)
19. Inada, K., Kohsaka, H., Takasu, E., Nose, A.: Optical dissection of neural circuits responsible for *Drosophila* larval locomotion with halorhodopsin. PLOS ONE **6**(12), 1–10 (2011)
20. Lahiri, S., Shen, K., Klein, M., Tang, A., Kane, E., Gershow, M., Garrity, P., Samuel, A.: Two alternating motor programs drive navigation in *Drosophila* larva. PLOS ONE **6**(8), 1–12 (2011)
21. Ross, D., Lagogiannis, K., Webb, B.: Online supplementary material. http://maggot.eu/documents/2015/04/SuppMechBodymodel.pdf

Dynamic Walking with a Soft Limb Robot

Yasmin Ansari[1(✉)], Ali Leylavi Shoushtari[1,2], Vito Cacucciolo[1],
Matteo Cianchetti[1], and Cecilia Laschi[1]

[1] The BioRobotics Institute, Scuola Superiore Sant'Anna,
Polo Sant'Anna Valdera, Via Rinaldo Piaggio 34, 56025, Pontedera, Pisa, Italy
{y.ansari,a.leylavishoushtari,v.cacucciolo,
m.cianchetti,c.laschi}@sssup.it
[2] Laboratory of Rehabilitation Bioengineering, Auxilium Vitae Rehabilitation Center,
Volterra, Pisa, Italy

Abstract. We present a novel soft limb quadruped robot "FASTT," with a simple and cheap design of its legs for dynamic locomotion aimed to expand the applications of soft robotics in mobile robots. The pneumatically actuated soft legs are self-stabilizing, adaptive to ground, and have variable stiffness, all of which are essential properties of locomotion that are also found in biological systems. We tested the soft legs for the pace, trot, and gallop gait and found them to move with a forward velocity for each gait with robustness. The legs were able to produce a flight and stance phase as a result of the body-environment interaction and also support the weight of the body while two legs were in flight phase and two in stance phase. The soft robot also exhibited two different postures i.e. sprawl and semi-erect which can also be found in some biological species as the crocodile. Moreover, the robot is safe to interact with. The results highlight the effectiveness of the soft limbs to produce dynamic locomotion which provides potential for application in uncertain environments.

Keywords: Soft robotics · Dynamic · Quadrupedal locomotion · Bio-hybrid

1 Introduction

Terrestrial legged locomotion of biological systems has evolved over millions of years to adapt to complex, unstructured and dynamically changing environments. These systems can traverse difficult terrain with minimum foot contact and energy consumption while keeping a smooth trajectory of the upper body [1] and has been studied by roboticists for a long time. In this research, we restrict our focus to quadrupedal locomotion.

It is fundamentally categorized by the limb posture into sprawling, semi-erect and erect [2]. All species exhibit a variety of gaits with great agility and maneuverability [3]. A gait is identified by the footfall pattern of the legs where each leg goes through a flight phase and stance phase within one limb cycle [4]. If only one leg goes through a stance and flight phase within a limb cycle while the other three legs remain on the support plane, it will result in a slow-paced gait given that it is statically stable. McGhee [5] defined static stability to be achieved when the projection of the center of mass (CoM) through the support plane remains within the boundaries of a virtual polygon formed by connecting the feet in the support plane during locomotion. These

© Springer International Publishing Switzerland 2015
S.P. Wilson et al. (Eds.): Living Machines 2015, LNAI 9222, pp. 13–25, 2015.
DOI: 10.1007/978-3-319-22979-9_2

gaits are further categorized into a walk, crawl or amble, depending on the order in which the legs move as well as the duration of their flight and stance phase [4]. On the other hand, if more than one foot is in flight phase in a limb cycle, it results in a medium to fast paced gait given that it meets the criteria for dynamically stability. According to Kang [6], this will be possible if there exists a point on the support plane where the resultant terrain reaction and moment forces on the CoM disappears. This definition is, however, restricted to flat terrains [7], These gaits, like the previous ones, can also be further categorized into a trot, pace, canter, gallop. This study focuses more on dynamic gaits.

The survey of the history of dynamic locomotion in legged robots can be classified into two approaches. The first vastly explored approach makes use of mechanically rigid components. A complete survey is beyond the scope of the paper, however, we present a general idea of the popular solutions and their limitations. One method uses an increased number of degrees of freedom (DoF) in a leg popularly found in [8] [9]. Tekken II [9] used mammal-like leg design with four actively actuated DOFs with one compliant joint in the ankle to produce dynamic walking. However, this has a high price, weight, and design-control complexity. On the other hand, [10 – 12] make use of under-actuation [13] to reduce design complexity by keeping the knee joint in each leg passive while only actively controlling the hip joint. These robots are able to execute various dynamic gaits at quite impressive speeds [3] [14]. However, most of them are still quite heavy as stated in a comparison in [12]. Moreover, when taken out of the laboratory environment, they introduce a significant risk in interaction with humans due to the overall hard structure.

The second relatively new approach makes use of highly compliant materials (i.e. elastomeric polymers) inspired from biological tissues resulting in a "soft robot" belonging to a rapidly growing field of research called Soft Robotics [15-16]. The bodies of these soft robots have inherent non-linear dynamics that produce rich behavior with a simple design and actuation mechanism [17-18]. It possesses locomotion capabilities as it can naturally adapt to shape of the ground in real-time, has variable stiffness that allows for self-stabilization [19]. Moreover, it is cost effective, and safer to interact with as compared to mechanically rigid components. Such capabilities present a great potential in the design of quadruped robots. [20] used continuum limbs where each leg has three pneumatic chambers to achieve omnidirectional bending for a crawling gait. [21] developed an untethered soft body quadruped that is able to generate static gaits through pneumatic actuation of five chambers. It can successfully traverse uneven terrains and withstand harsh environments. However, soft materials and compliant joints present much higher damping than rigid joints, limiting soft robots' speed of movement and energy efficiency.

In this paper we present Fast Adaptive Soft-hybrid TeTrapod (FASTT), (Fig 1) a proof of concept for the design and control of a soft limbed robot capable of producing dynamic quadrupedal locomotion. The robot consists of four soft limbs that are pneumatically actuated legs and connected to each other through a central stiff spine. Each leg can change morphology in the pitch and yaw directions (Fig 1 b, c) through manually adjustable joints before being actuated. This robot is not intended for a specific posture, gait, or terrain. Rather, the authors are interested in exploring the effect of morphological changes required for quadrupedal locomotion.

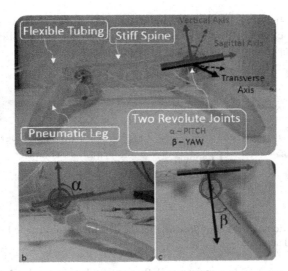

Fig. 1. a. FASTT Robot. It has four soft legs actuated pneumatically via flexible tubing. They are connected to a stiff spine through two revolute joints. b. Angle α rotates ±90° about the vertical axis to change the pitch of the leg. c. Angle β rotates ±90° about the transverse axis to change the yaw of the leg.

We present results of the trot, pace, and gallop gaits achieved with soft limbs for the first time according to our knowledge. We also present the capability of the robot to exhibit two different postures. Section 2 discusses the design and fabrication of the robot followed by the control in Section 3. Section 4 discusses in detail the results and a short conclusion with further research goals provided in Section 5.

2   Design and Fabrication

We have chosen flexible fluidic actuators (FFA) to function as the soft limbs. It consists of an elastic body with an internal hollow chamber. Pressurizing the internal chamber with a compressible gas from one end allows the potential energy of the gas to act along the inner surface of the chamber. Due to the compliance of the body, this is converted into a mechanical motion through elastic deformation. These actuators have high power density, efficiency and are also safe to interact with [22]. For locomotion, we require a non-linear bending movement which is achieved by anisotropic stiffness in the elastic body. In an unpressurized state, the length of the regions of less stiffness (L_{tr}) is equal the length of the regions of higher stiffness (L_{br}) i.e. $L_{tr} = L_{br}$. Pressurizing the internal chamber will then inflate the actuator more in the region of less stiffness and less in region of high stiffness ($L_{ta} > L_{ba}$) as shown in Fig 2a. This uneven elastic deformation results in the bending of the actuator in the direction of high stiffness. In this work, we develop an actuator inspired by [23] which uses an embedded layer of fabric within the elastic body underneath the internal chamber to produce anisotropic stiffness. The fiber is inextensible along the length of the actuator and results in the bending. To avoid radial expansion of the internal chamber when it is pressurized, an external fiber is wrapped around the elastic body. Fig. 2b shows the developed actuator in unpressurized and pressurized state.

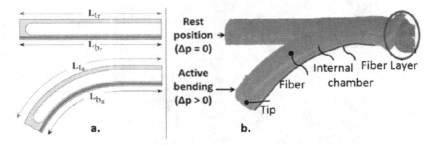

Fig. 2. Bending motion as a result of pressurizing the developed anisotropic FFA. (a) schematic view (b) developed actuator.

The fabrication materials consist of: i) Poly(methyl methacrylate) of 4mm thinckness for the stiff backbone ii) a rigid mold for the pneumatic legs iii) Smooth-On Inc.; Dragon skin 10 Medium (ultimate tensile stress = 3.28 MPa, elastic modulus = 0.15 MPa @ 100 % strain) as the elastic body iv) a fiber featuring high tensile stiffness and strength (ultimate strength σ_R = 4.9 GPa), but negligible bending stiffness (diameter d = 0.18 mm), to constrain the radial expansion v) a fabric with similar properties to (iv) to act as an inextensible layer.

The fabrication process involved (Fig.3): (a) casting the bodies of the four legs in a single mold; (b) wrapping fiber around each of them; (c, d) casting a layer of silicone to include the fabric. To connect the legs to the backbone, we designed a clamp to act as a soft-hard interface. (Fig. 2). Lastly, we connected each of these components to the central backbone through the revolute joints. We used adjustable screws to act as joints, allowing us to change the angles α and β between the experiments while keeping them still during each trial (Fig. 3). We produced both the mold and the clamps components using 3D printing technology, while the backbone and the intermediate links with laser cutting. The fully assembled robot with all the dimensions is shown in Fig 4.

Fig. 3. The fabrication process. (a) silicone poured in the 3D-printed mold to produce the bodies of the four legs. (b) four cured legs with the fiber wrapped around them. (c) the second mold designed to add the fabric layer. (d) the bottom layer is cured and legs are ready.

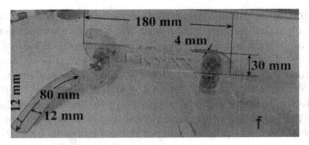

Fig. 4. FASTT fully assembled

3   Control

In order to investigate and exploit the intrinsic dynamic characteristics of the soft limbs, we implement a simple open-loop controller with no feedback. The pneumatic legs are externally actuated via flexible tubing connected to proportional solenoid valves, which is in turn are powered by an analog signal using a micro-controller. The amplitude of the analog signal is proportional to the pressure of the pneumatic flow in this work by a factor of 1/3 i.e. pressure = amplitude/3. The analog signal is used to generate symmetrical trot and pace gaits and asymmetrical gallop gait according to Hildebrand's [4] definition. A gait is generated by periodically alternating the pressure to a limb between high (p_h) and low (p_l) values. The essential requirement for all three gaits is to ensure the duration for both p_h and p_l equals half of the time period of one limb cycle. For symmetrical gaits, diagonally opposite pair of legs are actuated in synchrony for a trot while a pair of front-back legs are actuated in synchrony for a pace. In the asymmetrical gallop gait, a pair of right-left legs are actuated in synchrony and in opposition to the pair of back legs.

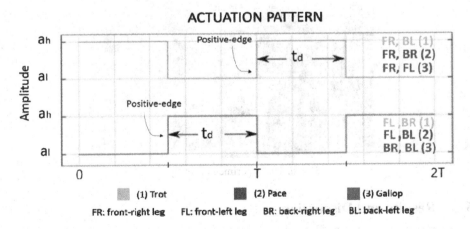

Fig. 5. Actuation Pattern using Arduino for Trot, Pace, and Gallop according to Hildebrand [4]

The Arduino microcontroller generates four analog signals actuating each leg separately. Each signal is a periodic square wave where the time period (T) corresponds to one limb cycle and the width of a square pulse (t_d) is half of the time period (T) i.e. $t_d = 0.5* T$. The square wave alternates between a lower amplitude (a_l) and higher amplitude (a_h) which is proportional to applying a p_l then p_h to the soft leg, respectively. We have summarized the analog signals generated for the three gaits in Fig 5. To achieve a certain gait, the positive-edge of the square pulse is required to start at either the beginning of the time period (0s) or at mid-way of the time period (0.5*T). This is done via potentiometers, at the beginning of each experiment, to avoid pre-programming each gait explicitly into the controller.

4   Experimental Set-up

The experimental set-up (Fig. 3) has: (1) a laptop; (2) two power sources (GW-Instek 24 V, Dual DC Power Supply 15 V); (3) one stand-alone air compressor; (4) four proportional pressure-controlled electronic valves (K8P Series EVP Systems, Input: 0 - 10 V, Output: 0 -3 bar); (5) an Arduino UNO Rev microcontroller; (6) an external electronic circuit; (7) flat ground; (8) a measurement scale; (9) the robot. A complete set-up is shown in Fig 6.

In order to test for straight locomotion, the pitch and yaw angles of a pair of left-right legs were kept the symmetrical which we will refer to as front (α_f, β_f) and back (α_b, β_b) morphology. After pre-setting the robot for a morphology (discussed in the next section) the robot was then placed on one edge of the flat plane and actuated for 60 seconds. The behavior of the robot was captured by two digital cameras (Samsung S5, 16MP, HDR with Selective Focus) mounted for a top and side view. After the completion of an experiment, the distance covered by the robot was measured and used to compute the average speed for that particular configuration.

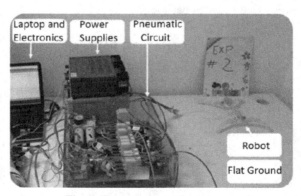

Fig. 6. The experimental set-up

5   Results and Discussion

We tested FASTT having control over seven parameters, T, p_h, p_l, α_f, β_f, α_b, β_b. In order to start investigations, the authors decided to constrain the actuation parameters i.e. keep it the same for each gait at T= 0.5s, p_h = 1.5 bar, and p_l = 0 bar. The pressure values were heuristically selected from the bending characterization of a single actuator (Fig 7) such that the differential pressure allowed for a bending angle of more than 90°. Similarly, T was chosen intuitively such that the bending movement of the leg was not to fast or too slow. This allowed the authors to focus more on the influence of morphological changes on locomotion However, future work will take actuation variations into account. The morphological changes were made sequentially such that for each gait, at α angle at 0°, we started by varying β_f sequentially from 0° to +90° in increments of 10° while β_b = -β_f for symmetry. α angle was not varied for all experiments. However, when used, it was tested both in a symmetric manner as well as an asymmetric manner. We conducted a total of 60 experiments and even though we tested only a small portion of the entire parameter space, we were able to achieve locomotion as well develop a preliminary understanding of the effects of morphology in locomotion. First, we discuss our findings about the pace, trot, and gallop gaits. Then, we present the capability of the robot to achieve stance and flight phases. Finally we make a comparison of the postures to biological systems.

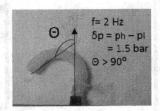

Fig. 7. Bending Characterization of a soft limb

5.1   Trotting

A robust trotting gait was achieved with the following parameters: T = 0.5s, p_h = 1.5 bar, p_l = 0 bar, α_f = 0°, β_f = +60°, α_b = 0°, β_b = -60°. Fig 8(Left) illustrates this gait through an actuation graph and time-series snapshots. The graph shows the pressure profile of a pair of legs in one limb cycle. Below it there are four snapshots that are accompanied by and a black and white footfall pattern and the associated time. The timing of the snapshots has been sampled in from different limb cycles. This was done to highlight the movement of the robot across the flat surface. The black circle in the footfall pattern indicates actuation of the leg at p_h and the white circle at p_l. The first snapshot corresponds to the pressure profile from 0 to 0.5T of a limb cycle and the second snapshot corresponds to the pressure profile from 0.5T to T of a limb cycle. The sequence can then be repeated for the remaining two snapshots. The forward velocity was recorded to be 1.6 cm/s, the slowest of all the gaits achieved. We also tracked the movement of the spine. It is depicted in the diagram from a side view and illustrates that the spine moved in a small pendulum like manner with repetitive motion. According to the studies of [24], this gait with the CoM movement would be categorized as a walking trot. This kind of movement is used mostly by insects, lizards or reptiles where the femur is placed perpendicularly to the body and the angle. This also explains the reason why we did not require to vary the alpha angle to achieve this locomotion.

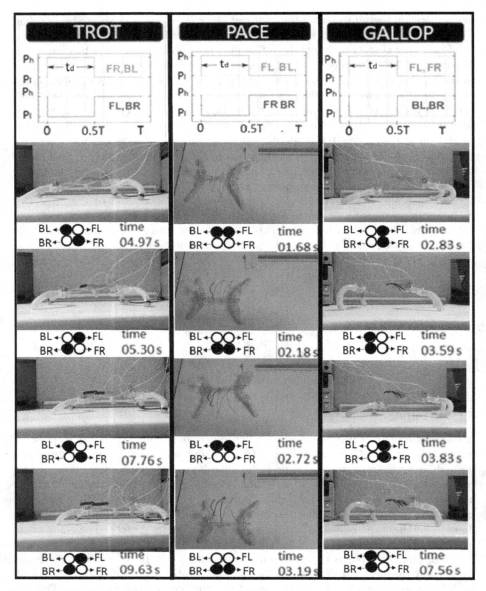

Fig. 8. (Left) Trotting Gait (Center) Pacing Gait (Right) Galloping Gait. Each gait is shown with the pressure profile of a pair of legs in one limb cycle followed by the snapshots of the movement throughout one experiment. The movement of the backbone has been tracked in all gaits using a motion tracking software indicated by the blue line.

5.2   Pacing

Fig 8(Center) illustrates the pressure profile, snapshots, and footfall patterns of this gait for the parameters: T = 0.5s, p_h = 1.5 bar, p_l = 0 bar, α_f = 0°, β_f = +30°, α_b = 0°, β_b = -30°. The snapshots for this gait are shown from the top view to highlight the movement of the spine which illustrates an undulating motion. However, the movement can be seen to be unstable as the body of the robot moves laterally with high oscillations. This is consistent with our results that show that the robot indeed moves straight with a forward velocity of 3 cm/s but only for a short while and eventually turns. By picking the robot up and placing it back down, it repeats the same behavior. We then varied α_f and α_b, both symmetrically and asymmetrically, for the given beta angles in the experiment. We found that changing the pitch angles asymmetrically produced the most effective results in this case i.e. for α_f = 0°, α_b = 20°, an efficient and robust forward velocity of 3.25 cm/s. This indicates that adding height to the hind limbs was the criteria for stability for this gait with the given beta parameters. This is found consistent with the studies of [25] where it is stated the pace gait is used by animals such as camels, giraffes, etc and relates the stability factor of the gait to the height of limbs of the animal.

5.3   Galloping

A gallop gait was achieved for the following parameters: T= 0.5s, ph = 1.5 bar, pl = 0 bar, αf = 0°, βf = +60°, αb = 0°, βb = -60°. Fig 8(Right) illustrates the pressure profile, snapshots, and footfall patterns of this gait from the horizontal view. The forward velocity of this gait was found to have a peak speed of 3.75 cm/s but could not be observed more than a few limbs cycles as it was restricted by the tether length. The movement of the spine shows a very high level of oscillations. The gait did not exhibit a fully aerial phase, however this gait will be further investigated by the use of a flexible spine which is a characteristic of biological systems that exhibit a fully aerial phase [26].

Table 1. A summary of results from sections 5.2 to 5.3 shown in Fig 7

Gait	f (Hz)	P_l (bar)	P_h (bar)	α_f (deg)	β_f (deg)	α_b (deg)	β_b (deg)	V (cm/s)	Posture	Motion of Spine
Trot	2	0	1.5	0	60	0	-60	1.6	Sprawl	Inverted-Pendulum Motion (Stable)
Pace	2	0	1.5	0	30	0	-30	3	Semi-Erect	Oscillatory (Unstable)
	2	0	1.5	0	30	-20	-30	3.25	Semi-Erect	Oscillatory (Stable)
Gallop	2	0	1.5	0	60	0	-60	4	Semi-erect	Large Inverted-Pendulum Motion (Needs more exp)

5.4   Stance and Flight Phase

The stance and flight phase in FASTT cannot be observed in isolation as this was a result of the interaction of the body with the environment as well as highly dependent upon the morphology of the legs. Fig. 9 illustrates the front angle of the trot gait such that all four legs are visible. To achieve this, the video has been taken at a slight angle instead of a complete front view. We used an offline motion tracking system (Kinovea 0.8.15) to track the tip of the front left limb and back right limb from 0.5T to T i.e. total duration of 0.25s, of the trot gait actuation sequence given in the figure. This portion of the limb cycle corresponds to the actuation of the two limbs at p_h (1.5 bar) while the other two limbs are actuated at p_l (0 bar). Fig 9a, 9b, and 9c corresponds to 0.5T, 0.75T, and T respectively. The height of the resulting flight phase was quite small so the tips of limbs have been encircled in each photograph to highlight the area of interest. The motion tracking system shows the trajectory of the movement on the tips of the legs in red which indicates the rise and fall in the height of the limb. There is a difference in the trajectory shapes of the legs which is due to the angle at which the movie was recorded, however, this does not affect the overall result. There are two important observations made from this diagram. One is that the legs are capable of a stance and flight phase. The second is that the diagonally opposite pair of legs exhibit flight phase in synchrony. This implies that the robot is able to bear its weight with two legs in the air. This is due to the compliance of the material. A further investigation will be conducted to develop a relation between the ground reactive forces and gait/posture/morphology of the robot.

5.5   Similarity to Biological Systems

Posture plays an important role in determining the gaits adopted by biological species. Insects and reptiles use the sprawling posture and are mostly known to use the walking trot gait for locomotion. Mammals and some reptiles, on the other hand, use the semi-erect or erect posture and can vary their gaits from a walk, hop, to even a gallop. However, it has been studied that crocodiles [27] make use of both sprawling and semi-erect posture resulting in a large variety of gaits (walk, trot, gallop) with variable speed. This allows the reptile to adapt to various terrain such as land, water, mud, etc. In order to distinguish between a sprawling and semi-erect posture, first we define the height of the body at the erect position to be equal to the height of the limb and clamp (110 mm). The height of the clamp to ground is calculated for each pose. If it is less than 0.5 the limb length, it can be categorized as a sprawling posture otherwise a semi-erect posture [27]. We do not use the kinematic definition of the femur and tibia leg segments to classify postures as our robot does not have them. In the gallop gait, the height of the body is 60 mm which can be classified into an erect pose. In the trot gait, the height of the body is 40mm and can be classified as a sprawling pose. Fig 10 shows the capability of the robot to exhibit two postures. Further investigation will be conducted to test the change in posture to locomote over different terrains similar to that of crocodiles.

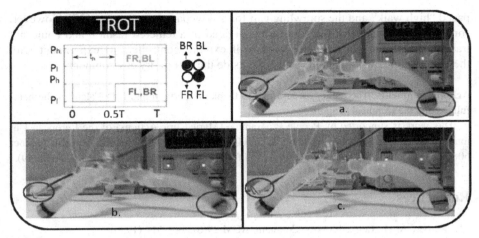

Fig. 9. The snapshots of the stance and flight phase of the front left limb and back right limb in half a limb cycle of trot from 0.5T to T. The path followed by the limbs were tracked using the Kinovea motion tracking system and highlighted in red in the back ground. The shape of the track indicates an increase in the height. **a.** 0.5T **b.** 0.75T **c.** T.

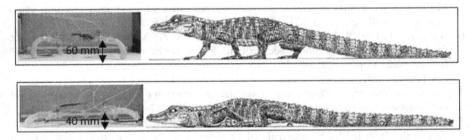

Fig. 10. (Top) Semi-erect Posture: Crocodile on right and FASTT galloping on the left (Bottom) Sprawling Posture: Crocodile on right and FASTT trotting on left. The crocodile postures are used from [27].

6   Conclusion and Future Work

In this work, we presented a new approach to the design and control of a dynamic walker with the use of pneumatically actuated soft limbs. We tested the legs for three dynamic gaits (trot, pace, and gallop) and showed that they were capable to achieve a flight and stance phase by morphological changes and interaction with the environment which allowed for a forward velocity in each gait. The legs were also able to maintain balance of the body while two of the legs were in flight phase and two were in stance phase. The interaction of the legs with the environment for different actuation sequences resulted in two different postures i.e. sprawling and semi-erect. This kind of behavior is observed in crocodiles which use the semi-erect posture for a fast paced "high walk" and the sprawling gait for a slow paced movements [27]. However, some limitations are that the robot was operated in a tethered fashion and could not transition between gaits. Future work will explore controllers to experiment with these capabilities with the aim to take it outside the laboratory environment.

Acknowledgments. The authors would like to thank G. Gerboni and I. DeFalco for the helpfulness and the technical support.

This work is supported by People Programme (Marie Curie Actions) of the European Union's Seventh Framework Programme FP7/2007-2013/ under REA grant agreement number 608022 and RoboSoft - A Coordination Action for Soft Robotics (FP7-ICT-2013-C # 619319).

References

1. Raibert, M.H.: Legged robots. Commun. ACM **29**(6), 499–514 (1986)
2. Charig, A.J.: The evolution of the archosaur pelvis and hind-limb: an explanation in functional terms. In: Joysey, K.A., Kemp, T.S. (eds.) Studies in Vertebrate Evolution, pp. 121–125. Oliver & Boyd, Edinburgh (1972)
3. Kuo, A.D.: Choosing Your Steps Carefully. Robotics & Automation Magazine, IEEE **14**(2), 18–29 (2007)
4. Hildebrand, M.: The quadrupedal gaits of vertebrates. BioScience **39**(11), 766–775 (1989)
5. McGhee, R.B., Frank, A.A.: On the stability properties of quadruped creeping gaits. Mathematical Biosciences **3**, 331–351 (1968)
6. Kang, D.-O., Lee, Y.-J., Lee, S.-H., Hong, Y.S., Bien, Z.: A study on an adaptive gait for a quadruped walking robot under external forces. In: Proceedings of the 1997 IEEE International Conference on Robotics and Automation, vol. 4, pp. 2777–2782, April 20-25, 1997
7. Yoneda, K., Hirose, S.: Three-dimensional stability criterion of integrated locomotion and manipulation. Journal of Robotics and Mechatronics **9**(4), 267–274 (1997)
8. Arikawa, K., Hirose, S.: Development of quadruped walking robot titan-viii. In: Proceedings of the 1996 IEEE/RSJ International Conference on Intelligent Robots and Systems 1996, IROS 1996, vol. 1, pp. 208– 214, November 4-8, 1996
9. Kimura, H., Fukuoka, Y., Biologically inspired adaptive dynamic walking in outdoor environment using a self-contained quadruped robot: 'tekken2'. In: Proceedings of the 2004 IEEE/RSJ International Conference on Intelligent Robots and Systems, (IROS 2004), vol. 1, pp. 986–991, September 28-October 2, 2004
10. Raibert, M., Blankespoor, K., Nelson, G., Playter, R.: BigDog, the rough–terrain quadruped robot. In: Proceedings of the 17th IFAC World Congress. COEX, South Korea, pp. 10823–10825 (2008)
11. Iida, F., Gomez, G., Pfeifer, R.: Exploiting body dynamics for controlling a running quadruped robot. In: Proceedings of the 12th Int. Conf. on Advanced Robotics (ICAR05), pp. 229–235 (2005)
12. Sprowitz, A.,Tuleu, A., Vespignani, M., Ajallooeian, M., Badri, E., Ijspeert, A.:Towards Dynamic Trot Gait Locomotion—Design, Control, and Experiments with Cheetah-cub, a Compliant Quadruped Robot. International Journal of Robotics Research (IJRR) (2013)
13. Zambrano, D., Cianchetti, M., Laschi, C.: The morphological computation principles as a new paradigm for robotic design. In: Hauser, H., Füchslin, R.M., Pfeifer, R. (eds.) Opinions and Outlooks on Morphological Computation, pp. 214-225 (2014)

14. Sitti, M., Menciassi, A., Ijspeert, A.J., Low, K.H., Kim, S.: Survey and Introduction to the Focused Section on Bio-Inspired Mechatronics. IEEE/ASME Transactions on Mechatronics **18**(2), 409–418 (2013)
15. Pfeifer, R., Marques, H.G., Iida, F.: Soft Robotics: The Next Generation of Intelligent Machines. IJCAI (2013)
16. Kim, S., Laschi, C., Trimmer, B.: Soft robotics: a bioinspired evolution in robotics. Trends in Biotechnology **31**(5), 287–294 (2013)
17. Cianchetti, M., Ranzani, T., Gerboni, G., De Falco, I., Laschi, C., Menciassi, A.: Stiff-flop surgical manipulator: mechanical design and experimental characterization of the single module. In: 2013 IEEE/RSJ on Intelligent Robots and Systems (IROS), pp. 3576–3581. IEEE (2013)
18. Cianchetti, M., Calisti, M., Margheri, L., Kuba, M., Laschi, C.: Bioinspired locomotion and grasping in water: the soft eight-arm OCTOPUS robot. Bioinspiration & Biomimetics (Special Issue on Octopus-inspired robotics), accepted for publication
19. Moritz, C.T., Farley, C.T.: Passive dynamics change leg mechanics for an unexpected surface during human hopping. Journal of Applied Physiology **97**(4), 1313–1322 (2004)
20. Godage, I.S., Nanayakkara, T., Caldwell, D.G.: Locomotion with continuum limbs. In: Proceedings of the IEEE/RSJ International Conference on Intelligent Robot Systems (IROS 2012), pp. 293–298, Vilamoura, Portugal (2012)
21. Tolley, M.T., Shepherd, R.F., Mosadegh, B., Galloway, K.C., Wehner, M., Karpelson, M., Wood, R.J., Whitesides, G.M.: A resilient, untethered soft robot. Soft. Robotics **1**(3), 213–223 (2014)
22. De Greef, A., Lambert, P., Delchambre, A.: Towards flexible medical instruments: Review of flexible fluidic actuators. Precision Engineering **33**(4), 311–321 (2009)
23. Deimel, R., Brock, O.: A compliant hand based on a novel pneumatic actuator. In: 2013 IEEE International Conference on Robotics and Automation (ICRA), pp. 2047–2053. IEEE (2013)
24. Biknevicius, A.R., Reilly, S.M.: Correlation of symmetrical gaits and whole body mechanics: debunking myths in locomotor biodynamics. J. Exp. Zool. **305A**, 923–934 (2006)
25. Kar, D.C., Issac, K.K., Jayarajan, K.: Gaits and energetics in terrestrial legged locomotion. Mechanism and Machine Theory **38**(4), 355–366 (2003)
26. Khoramshahi, M., Spröwitz, A., Tuleu, A., Ahmadabadi, M.N., Ijspeert, A.J.: Benefits of an active spine supported bounding locomotion with a small compliant quadruped robot. In: ICRA, pp. 3329-3334 (2013)
27. Reilly, S.M., Elias, J.A.: Locomotion in alligator mississippiensis: kinematic effects of speed and posture and their relevance to the sprawling-to-erect paradigm. J. exp. Biol. **201**, 2559–2574 (1998)

Worm-Like Robotic Locomotion
with a Compliant Modular Mesh

Andrew D. Horchler[1](✉), Akhil Kandhari[1], Kathryn A. Daltorio[1], Kenneth C. Moses[1],
Kayla B. Andersen[1], Hillary Bunnelle[1], Joseph Kershaw[1], William H. Tavel[1],
Richard J. Bachmann[1], Hillel J. Chiel[2], and Roger D. Quinn[1]

[1] Department of Mechanical and Aerospace Engineering, Case Western Reserve University,
Cleveland, OH 44106-7222, USA
{horchler,rdq}@case.edu
[2] Departments of Biology, Neurosciences and Biomedical Engineering,
Case Western Reserve University, Cleveland, OH 44106-7080, USA
hjc@case.edu

Abstract. In order to mimic and better understand the way an earthworm uses
its many segments to navigate diverse terrain, this paper describes the design,
performance, and sensing capabilities of a new modular soft robotic worm. The
robot, Compliant Modular Mesh Worm (CMMWorm), utilizes a compliant
mesh actuated at modular segments to create waveforms along its body. These
waveforms can generate peristaltic motion of the body similar to that of an
earthworm. The modular mesh is constructed from 3-D printed and commercially available parts allowing for the testing of a variety of components that
can be easily interchanged. In addition to having independently controlled segments and interchangeable mesh properties, CMMWorm also has greater range
of contraction (52% of maximum diameter) than our previous robot Softworm
(73% of maximum diameter). The six-segment robot can traverse flat ground
and pipes. We show that a segment is able to detect the wall of a pipe and return
to its initial position using actuator-based load-sensing. A simple kinematic
model predicts the outer diameter of the worm robot's mesh as a function of encoder position.

Keywords: Robot · Worm · Earthworm · Soft · Compliant · Cable · Actuation ·
Hyper-redundant · Mesh

1   Introduction

Utilizing simple changes in body shape to generate locomotion can be a valuable
robotic strategy. Snakes and worms can navigate environments that are difficult for
humans and wheeled or legged robots, such as rubble piles, steep cluttered conduits,
or underground burrows. Robots with limbless body plans can be useful in exploring
tunnels [2], [27], [32] or in medical applications [14], [34]. Soft biologically-inspired
robots can use changes in body shape to swim [31], to crawl like caterpillars [33] and
snails [9], to grasp like elephants [16] and cephalopods [10], and to move with snake-
like gaits [28]. In particular, earthworm locomotion is promising for robotics because

© Springer International Publishing Switzerland 2015
S.P. Wilson et al. (Eds.): Living Machines 2015, LNAI 9222, pp. 26–37, 2015.
DOI: 10.1007/978-3-319-22979-9_3

worms have symmetric, nearly-uniform soft bodies and a small, highly distributed neural system. Body compliance in a terrestrial robot is valuable for safety in fragile environments or around humans and for passive adaptation for grasping or generating traction [20].

Fig. 1. Compliant Modular Mesh Worm (CMMWorm) can crawl on flat ground or in a pipe that is smaller than it's nominal maximum diameter. A worm-like peristaltic wave is generated with six actuators along the body. Pipe shown has 15.2 cm inner diameter.

In earthworms, peristaltic locomotion occurs when waves of muscle contraction travel along the body. Each segment has a constant hydrostatic volume, so when circumferential muscles *contract* the segment diameter, the segment length increases. Similarly when longitudinal muscles decrease the segment length, the diameter *expands*. Because of this hydrostatic coupling between length and diameter [11], the longer contracted segments can be lifted off the ground while the fully radially expanded segments rest on the ground to anchor forward motion. To move forward, waves of segment contraction, expansion, and anchoring travel backward down the body. Our group investigates this strategy in earthworms in a companion paper [19].

Although there are many ways to actuate a robotic worm [26], [37], [29], [22], [36], a mesh body is particularly advantageous [3,4], [7], [30]. Our previous prototype, SoftWorm [3,4], [7], used a mesh of helically-wrapped tubes, pinned at the intersections to form rhombuses with a fixed side length, but a changing aspect ratio. The body of MeshWorm [30] is a woven mesh that consists of rhombuses. The changing aspect ratios of the mesh rhombuses in these robots cause their body-length and diameter to change inversely, similar to the hydrostatic length-diameter coupling in worms [5], [13]. As a result, reduced actuation can be used to generate fast fixed wavelength peristaltic locomotion [3], [7]. The compliance of these mesh designs permits the bodies of these robots to bend and adapt to the environment and actuator forces. This also allows the body diameter to vary along the length of the body, in contrast to the rigid linkages in, for example, a pantograph mechanism in which all the parallelograms extend and fold together. The simplicity of a compliant mesh body design means that the body of the robot can be very durable and can, for example, continue to operate even after being crushed, as demonstrated by [30].

To our knowledge, previous mesh worm body designs have not been able to respond to sensed load or body shape. We have shown in simulation that the ability for individual segments to respond to ground loading can permit more efficient navigation, for example in a narrowing pipe [13]. Our previous single actuator prototype was not able to navigate in pipes or other constrained spaces because the diameter of the rigid cam mechanism was nearly equal to the maximum diameter of the robot [7]. Our goal here is to develop a compliant modular mesh robot in which the segments are responsive to local environmental loads in order to better emulate the high degree-of-freedom behavior seen in worms [12], [15], [19].

In this work, inspired by the kinematics of earthworms, we have designed, built, and tested a new multi-actuator compliant mesh robot (Fig. 1) with a unique modular and reconfigurable design. Segments of the body are controlled with individual actuators that expand the diameter to nearly twice the contracted diameter. When such segments are connected, a continuous compliant mesh is formed that can be used to locomote on flat ground or in constrained environments such as pipes. To adapt to constrained environments, the magnitude of the expansion and contraction can be controlled through the tension in the actuating cables. We expect this robot to be a useful platform for investigating soft-bodied robot control inspired by invertebrate nervous systems.

2   Design

2.1   Bi-directional Actuation

In order to investigate terrain-adaptive peristaltic locomotion, we built a new worm robot that improves upon our previous prototype worm with actuation and sensing at many points along its length. For this purpose we use Robotis Dynamixel MX-64T actuators at each of the segments of CMMWorm. These "smart actuators" have position, speed, and load sensing capabilities. Additionally, maximum load values can be specified at the level of the servo control ensuring faster and more reliable load response, as described in Section 2.3.

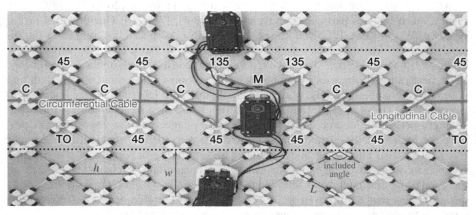

Fig. 2. Mesh of the CMMWorm robot laid flat on a surface during assembly. The longitudinal and circumferential cable placement for one segment is highlighted in red and blue, respectively. Three-ply Spectra® cable is used. Nylon tube (translucent) and polycarbonate rod (highlighted in orange) links connect the vertex pieces. C: circumferential vertex, **45**: longitudinal 45 vertex, **135**: longitudinal 135 vertex, **TO**: tie-off vertex, **M**: actuator mount.

To sense limiting loads at both maximum and minimum diameters, each Dynamixel actuator drives a bi-directional cable actuation system for each segment. In previous mesh-based robots, segment diameter was been actuated by tensioning circumferential cables [3] or coiled shape memory actuators [30] in conjunction with longitudinal compliance to return segments to their maximum diameter. In our robot, we want to be able to sense when the diameter expands to contact the ground or the inside of a pipe. In a circumferentially-actuated mesh, this would require detection of decreases in tension (i.e., slack). However, for the Dynamixel actuators, detecting such decreases in tension is difficult because of the low signal-to-noise ratio, which makes it harder to detect decreases in tension (i.e., slack) than increases in tension.

Thus, two pairs of Spectra® cables actuate each segment allowing one mesh-mounted Dynamixel actuator to act bi-directionally. Spooling in the circumferential cables simultaneously spools out the longitudinal cables and vice versa. When shortened, the two circumferential cables, like circular muscle layer of an earthworm [15], elongate the segment while decreasing its diameter. Similarly, like longitudinal muscles, the longitudinal cables reduce segment length while increasing segment diameter.

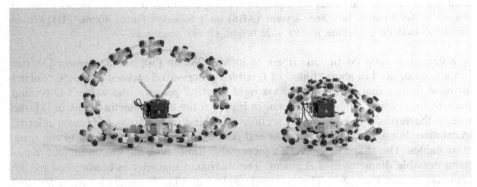

Fig. 3. A single segment of CMMWorm fully expanded (left) and fully contracted (right). The circumferential cable tensioning springs are visible above the black Dynamixel actuator.

The longitudinal and circumferential cable lengths are proportional to the mesh rhombus widths and heights, respectively (Fig. 2). The circumferential cables cross each rhombus height, following the circumference. The longitudinal cables zig-zag to cross each rhombus width by following along the links. Because of the nonlinear (Pythagorean) relationship between rhombus width and height, linear tension springs attached between the actuators and the circumferential cables keep the cable taut over the expansion-contraction cycle (Fig. 3).

2.2 Mesh Structure

Mesh stiffness is critical. The mesh must be sufficiently compliant to permit large differences between adjacent segment diameters, but stiff enough that each segment (including the actuators) can be raised off the ground by neighboring segments with larger diameters. Our previous experience in mesh design [3] required trial and error of material properties, and many manual calibrations of springs and cables, which often required complete disassembly of the robot to implement. Thus a design goal was segment modularity to facilitate the tuning and adjustment of mesh stiffness.

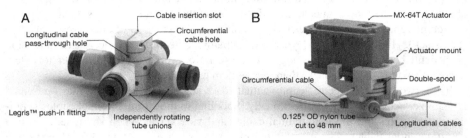

Fig. 4. The mesh rhombuses are joined at hinge joint vertex pieces, such as the circumferential vertex (A), which have slots to insert the actuating cables and Legris™ fittings to securely hold lengths of tube or rod. The actuator mount vertex (B) also houses a spool on which the longitudinal and circumferential cables are wound in opposite directions. The vertex pieces were 3-D printed in Acrylonitrile butadiene styrene (ABS) on a Stratasys Fortus 400mc FDM (fused deposition modeling) machine, (0.010" slice height, ±0.005" tolerance).

Rather than long continuous fibers as in MeshWorm [30] and Softworm [3], our robot is comprised of short "links" of flexible tubing or rod secured by quick connect air hose fittings that are connected via rigid "vertex" pieces – the white 3-D printed parts in Figs. 1 and 3 (shown in detail in Fig. 4). Like the connecting caps in [7], the role of the vertex pieces is to join sections of tubing or rod so as to prevent relative translation, but allow relative rotation and permit attachment and routing of the actuating cables. The shape of the vertex pieces also limits both the minimum and maximum possible diameter of a segment. The minimum diameter is constrained by the actuators within the mesh. The included angle (see Fig. 2) ranges between 50° and 110°. Elastomer feet are affixed to each vertex piece (visible in Figs. 1 and 3) to provide added compliance and ensure even contact with the walls of pipes.

To create an easily modifiable, modular robot, Legris™ push-in fittings were chosen to connect the vertex pieces with the flexible tubing or rod links. The fittings were purchased as equal straight unions (part number 3106 53 00), cut in two, machined flat, and epoxied into the vertex pieces (Fig. 4). The Legris™ fittings allow interchangeable segments to be assembled individually and connected at a later time. This modularity also allows links with different lengths and material properties to be easily tested.

The compliant mesh of the robot presented in this paper is comprised of 0.125" × 0.073" (outer × inner diameter) nylon tubes and 0.125" diameter polycarbonate rods, both cut to a length of 48 ±0.25 mm. The stiffer polycarbonate rods are used for four of the six rhombuses that form each actuated ring (Fig. 2, highlighted in orange). The more flexible nylon tubes are used between the actuated rings to allow adjacent segments to achieve different diameters.

Each segment of the robot is comprised of 18 vertex pieces of five different types: circumferential (Fig. 4a), longitudinal 45, longitudinal 135, tie-off, and actuator mount (Fig. 4b). Each vertex piece has a straight slot on its top, enabling easy insertion and removal of cables, facilitating assembly (Fig. 4a). The longitudinal 45 and 135 have internal radii for the longitudinal actuator cables to pass through them at approximately 45° and 135°, respectively. The circumferential vertex pieces have a hole for a circumferential cable, in addition to a pass-through for a longitudinal cable. The longitudinal cables are each terminated and tensioned at a tie-off vertex.

The circumferential cables are tied to the upper spool affixed to the actuator (Fig. 4b) and pass through the circumferential vertex pieces (Fig. 4a) on each side before being tied-off and tensioned at the top-most circumferential vertex (Fig. 2, see also Fig. 3). The longitudinal cables are tied to the lower spool (Fig. 4b) and emerge together from a single hole on the actuator mount vertex before splitting and "zigzagging" through eight vertex pieces, ending at the two tie-off vertex pieces (Fig. 2).

For longitudinal actuation, the rhombuses closest to the actuator mount deform more than ones farther away due to losses from the links bending and friction at the vertex pieces. A thin Teflon® sheath was added to each longitudinal 45 and 135 vertex piece to mitigate friction losses at cable direction changes. The two longitudinal 135 vertex pieces (Fig. 2), with their larger internal radii, further reduce friction. The stiffer polycarbonate rod links, described above, help distribute forces throughout the actuated ring more uniformly. In addition, tension springs along the sides (visible in Fig. 1) ensure more uniform deformation by increasing the stiffness of the second rhombus from the actuator.

2.3   Electronics and Control

For each of the six segments, Dynamixel MX-64T actuators are connected via a serial bus that supplies power from a 12 V regulated DC power supply (off-board power) and permits communication with a microcontroller. The MX-64T actuators have a 12-bit, 360° absolute encoder and use a PID algorithm to control position, speed, or load. The position sensing capability of the actuators is used to determine the diameter of a segment at any configuration, while the load sensing capability is used to detect the walls of the pipe (see Section 3 below). Using the sensory capabilities of these actuators in this way obviates the need for additional sensors on the mesh itself.

A single Robotis OpenCM9.04 microcontroller (32-bit ARM Cortex-M3, STM32F103CB, 72 MHz) is used for control. This small (66.5 mm × 26.9 mm, 11.1 g) board is mounted to the side of an actuator at one end of the robot (onboard control). The OpenCM9.04 is configured to communicate with the MX-64T actuators at 3 MBps without requiring additional high-speed serial communication circuitry. Programming of the microcontroller and data logging are performed over a USB connection to a PC.

Our open source DynamixelQ library [17] for the OpenCM9.04 microcontroller enables high-speed and robust communication with AX and MX series Dynamixel actuators. The library has syntax to facilitate reading from and writing to multiple actuators simultaneously.

3   Robot Performance

The modular construction of CMMWorm enables rapid modification and mainten-ance. If one segment is damaged, it can be removed and repaired without affecting the rest of the segments' internal connections. The Legris™ fittings are versatile, allow-ing us to test tubes and rods with various bending stiffnesses and yield strengths, re-sulting in the configuration presented here.

Assembled, each CMMWorm segment weighs 317.1 g (including one 126 g actua-tor). Measured from above (see Fig. 6a), the minimum diameter is 52% of the maximum (13.2 cm and 25.6 cm, respectively), including the elastomer feet. In our previous robot, Softworm [7], the shortest mesh rhombuses were about 73% the height of the tallest rhombuses [13]. Our new robot permits greater rotations at the vertex pieces, resulting in a greater range of rhombus heights and allowing segments to contract to roughly half their expanded diameter. This is similar to earthworms, for which the minimum diameter is 50–60% of the maximum diameter [19].

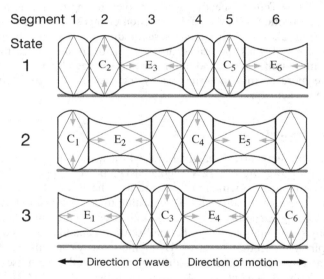

Fig. 5. A cyclic three-state time-based control scheme with fixed period is used to generate waves along the six-segment CMMWorm. In this illustration, two three-segment long waves are propagated from right to left as the robot moves from left to right. State 3 is followed by State 1. Each wave consists of a radially expanding segment, E_i, and a radially contracting segment, C_i, as well as an inactive anchoring segment. At the beginning (shown) of each state, expanding and contracting segments begin moving. State transitions occur at preset durations.

The use of multiple motors allows for different sections of CMMWorm to achieve different diameters independently of one another (Fig. 1). A time-based control scheme (illustrated in Fig. 5) generates waves along the length of the robot to produce locomotion on the ground or in a pipe. In each state, two fully-expanded segments are always inactive, anchoring the position of the robot. For a six-segment body, this results in forward motion of 14 cm/min on a hard tile floor surface. Using the same controller, the six-segment robot can also advance at 7.4 cm/min in a 15.2 cm inner diameter pipe and 7 cm/min on carpet.

4 Sensing

Load sensing can be used to detect when a given segment comes into contact with a surface, such as the inner wall of a pipe, or reaches its minimum or maximum diameter. To sense load reliably, the Dynamixel actuators are set to "torque control" mode (current control) and goal torques are specified. Load measurements are read from each active actuator every 0.1 ms. This data is smoothed to filter noise and compared with a threshold value (based on data collected from the actuators at different goal torques). Exponential smoothing [8] is used with smoothing factor $\alpha = 0.01$. When a segment reaches its maximum or minimum diameter, as limited by the largest and smallest included angles of the vertex pieces (Fig. 2), the measured load increases rapidly and exceeds a threshold. Similarly, an external load, such as that from the interior wall of a pipe, also results in the load threshold being exceeded. This permits the control algorithm to responsively pause or reverse the direction of motion. Independent thresholds are used for contraction (minimum diameter) and expansion (maximum diameter or an external obstruction). The results of this experiment are shown in Fig. 6.

A segment can expand until it achieves a desired level of force contact with the wall and then stop. This ability allows segments to anchor themselves against a pipe wall, which is useful for movement within variable diameter pipes and constrained environments. Alternatively, the maximum diameter can be limited by adjusting this threshold if no pipe is contacted. This sensing is achieved using the MX-64T actuators alone, without the use of additional exterior-mounted sensors. The lack of additional sensors allows for the robot to be free of added sensor weight, wiring, and exposed electronics.

Additionally, though the segments are soft, an estimate of segment diameter can be obtained from the actuators' encoders and a simple kinematic model. Like the hydrostatic skeleton of the worm [21], the rhombuses of the mesh provide a kinematic constraint, or mechanical coupling, between segment length and diameter. Assuming each rhombus of the mesh is in a single plane and that the sides are rigid, the side length, L, width, w, and height, h, of the rhombuses (see Fig. 2) are coupled via the Pythagorean theorem, such that

$$w^2 + h^2 = 4L^2. \tag{1}$$

The diameter of the segment will be proportional to the rhombus height with a proportionality constant C_d.

As the actuator turns the cable spool, the longitudinal cables decrease in length, decreasing the rhombus widths while increasing their heights. As illustrated by the red lines in Fig. 2, both the left and right longitudinal cables pass through three rhombus

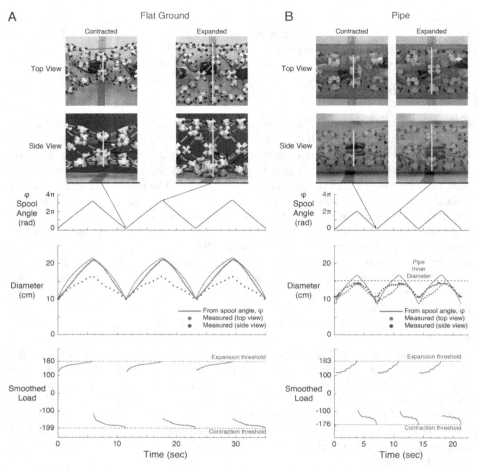

Fig. 6. At each segment, cable tension (actuator load) can be used to consistently limit expansion and spool angle (actuator position) can be used to approximate segment shape. Encoder position and load were logged from a Dynamixel actuator while one segment was controlled cyclically according to the load-sensing scheme described in Section 4. Results are shown for (A) a hard tile surface and (B) a 15.2 cm inner diameter pipe. For (A), the remaining segments were maximally expanded. For (B), the segments on one side of the driven segment were maximally contracted and the remaining ones were maximally expanded to anchor half of the robot in the pipe. Representative stills (cropped) from two video views are shown at top. The yellow overlaid line shows how the diameter was measured in each frame. The spool angle plots show the position encoder tick count logged from the Dynamixel actuator converted to continuous radians. The diameter plots show the segment diameters estimated from spool angle data (blue), and measured from video images (red and purple) with ImageJ (version 1.48, National Institutes of Health). A mesh side length $L = 6.52$ cm, spool radius $r_l = 0.76$ cm, and diameter constant $C_d = 1.72$ were used with the model described in Section 4 to obtain the estimated diameters. The smoothed load plots at bottom show the smoothed load values calculated onboard the microcontroller, which are used in conjunction with expansion and contraction thresholds to detect the limiting diameters or the inner wall of the pipe.

widths along their zigzag path. If all three rhombuses on a side maintain the same width, and the compliance of the connecting links, the offset between the actuator mount and the first connecting link, and out-of-plane deformation of the mesh are neglected, a change in longitudinal cable length due to spool rotation can be related to the maximum rhombus width permitted by the taut inelastic longitudinal cable by

$$3\Delta w \le \Delta l = -r_i \Delta\varphi \qquad (2)$$

where φ is the angular spool position, l is longitudinal cable length, and r_i is the spool radius. Equation (2) is an inequality because cables do not limit in compression, i.e., one cannot push on a cable. Since the circumferential cables have a compliant tensioning system, the cables stay taut during nominal operation and the total spool rotation, $\Delta\varphi$, provided by the actuator can be related to the change in width as $\Delta w = -r_i \Delta\varphi/3$. Thus, by measuring an initial segment diameter, calculating changes to the rhombus based on the spool angle, and multiplying the rhombus height by a constant factor, the segment diameter at any time can be approximated.

To validate the diameter estimates, we compared them with the actual diameter of the robot as measured from video using ImageJ (Fig. 6). Particularly when the segments are fully expanded, we find that there are large differences between the diameters observed from above and from the side (see also Fig. 3), so we used two cameras at different angles. On flat ground, the robot's body deforms due to gravity and the cable tensioning springs. In a pipe, the body becomes more cylindrical as it presses against the pipe walls.

5 Conclusions

The compliant modular mesh of our robot permits both soft body locomotion and soft body sensing, which will be useful both for accomplishing robotic tasks and for learning about soft-bodied animals. In the future, we can use the Legris™ fittings to experimentally test different stiffnesses of the connecting links before making simpler worm robots at different scales. In addition, we can use neurobiologically-inspired oscillators, e.g., [18], to design dynamical controllers that coordinate the many compliant degrees of freedom. Such controllers could take advantage of the wall-sensing ability of the robot to responsively navigate more irregular environments, e.g., varying diameter pipes [13], and ensure anchoring while climbing vertically. Using the interior surface of the mesh, this work may permit experiments with peristaltic grasping and swallowing, as has been studied in slugs as they transport soft, irregular objects [23]. The addition of actuated cables spanning multiple segments would provide full-body turning capability, as we previously demonstrated for fixed bend radii in our earlier robot, Softworm [12]. Incorporating mesh deformation and load measurements may increase the accuracy of our diameter prediction model (Section 4, Fig. 6). The large range of diameter expansion in the CMMWorm design may allow us to better understand and model soft animals [35].

Acknowledgements. This work was supported by NSF research Grant No. IIS-1065489. The authors would like to thank the staff of rp+m and Case Western Reserve University's think[box] for assistance with 3-D printing. We would also like to thank David Cannon, Ian McCurdy, and James Ryan for their help during the course of this project.

References

1. Alart, P., Curnier, A.: A mixed formulation for frictional contact problems prone to Newton like solution methods. Comput. Methods Appl. Mech. Eng. **92**(3), 353–375 (1991)
2. Bertetto, A., Ruggiu, M.: In-pipe inch-worm pneumatic flexible robot. In: Proc. Int. Conf. Adv. Intell. Mechatronics, pp. 1226–1231 (2001)
3. Boxerbaum, A.S., Chiel, H.J., Quinn, R.D.: A new theory and methods for creating peristaltic motion in a robotic platform. In: Proc. IEEE Int. Conf. Robot. Autom, pp. 1221–1227 (2010)
4. Boxerbaum, A.S., Horchler, A.D., Shaw, K.M., Chiel, H.J., Quinn, R.D.: A controller for continuous wave peristaltic locomotion. In: Proc. IEEE Int. Conf. Intell. Robot. Syst., pp. 197–202 (2011)
5. Boxerbaum, A.S., Daltorio, K.A., Chiel, H.J., Quinn, R.D.: A Soft-Body Controller with Ubiquitous Sensor Feedback. Proc. Living Machines **7375**, 38–49 (2012)
6. Boxerbaum, A.S., Horchler, A.D., Shaw, K.M., Chiel, H.J., Quinn, R.D.: Worms, waves and robots. In: Proc. IEEE Int. Conf. Robot. Autom., pp. 3537–3538 (2012)
7. Boxerbaum, A.S., Shaw, K.M., Chiel, H.J., Quinn, R.D.: Continuous wave peristaltic motion in a robot. Int. J. Rob. Res. **31**(3), 302–318 (2012)
8. Brown, R.G.: Exponential Smoothing for Predicting Demand, pp. 1–15. Arthur D. Little Inc., Cambridge (1956)
9. Chan, B., Ji, S., Koveal, C., Hosoi, A.E.: Mechanical Devices for Snail-like Locomotion. J. Intel. Mat. Syst. Str. **18**(2), 111–116 (2007)
10. Cianchetti, M., Licofonte, A., Follador, M., Rogai, F., Laschi, C.: Bioinspired Soft Actuation System Using Shape Memory Alloys. Actuators **3**, 226–244 (2014)
11. Chiel, H.J., Crago, P., Mansour, J.M., Hathi, K.: Biomechanics of a muscular hydrostat: a model of lapping by a reptilian tongue. Biol. Cybern. **67**(5), 403–415 (1992)
12. Collier, H.O.J.: Central Nervous Activity in the Eathworm I. Responses to Tension and to tactile Stimulation. J. Exp. Biol. **16**(3), 286–299 (1939)
13. Daltorio, K.A., Boxerbaum, A.S., Horchler, A.D., Shaw, K.M., Chiel, H.J., Quinn, R.D.: Efficient worm-like locomotion: slip and control of soft-bodied peristaltic robots. Bioinspir. Biomim. **8**(3), 035003 (2013)
14. Dario, P., Ciarletta, P., Menciassi, A., Kim, B.: Modeling and Experimental Validation of the Locomotion of Endoscopic Robots in the Colon. Int. J. Rob. Res. **23**(4), 549–556 (2004)
15. Gray, J., Lissmann, H.W.: Studies in Animal Locomotion VII. Locomotory Reflexes in the Earthworm. J. Exp. Biol. **15**, 506–517 (1938)
16. Hannan, M.W., Walker, I.D.: Kinematics and the Implementation of an Elephant's Trunk Manipulator and Other Continuum Style Robots. J. Robot. Syst. **20**(2), 45–63 (2003)
17. Horchler, A.D.: DynamixelQ Library, Version 1.1 (Retrieved on March 23, 2015). https://github.com/horchler/DynamixelQ
18. Horchler, A.D., Daltorio, K.A., Chiel, H.J., Quinn, R.D.: Designing responsive pattern generators: stable heteroclinic channel cycles for modeling and control. Bioinpir. Biomim. **10**(2), 026001 (2015)

19. Kanu, E.N., Daltorio, K.A., Quinn, R.D., Chiel, H.J.: Correlating kinetics and kinematics of earthworm peristaltic locomotion. In: Proc. Living Machines (July 28–31, 2015)
20. Kim, S., Laschi, C., Trimmer, B.: Soft robotics: a bioinspired evolution in robotics. Trends Biotechnol. 31(5), 287–294 (2013)
21. Kurth, J.A., Kier, W.M.: Scaling of the hydrostatic skeleton in the earthworm Lumbricus terrestris. J. Exp. Biol. 217, 1860–1867 (2014)
22. Mangan, E.V., Kingsley, D.A., Quinn, R.D., Chiel, H.J.: Development of a peristaltic endoscope. In: Proc. IEEE Int. Conf. Robot. Autom., pp. 347–352 (2002)
23. Mangan, E.V., Kingsley, D.A., Quinn, R.D., Sutton, G.P., Mansour, J.M., Chiel, H.J.: A biologically inspired gripping device. Ind. Robot An Int. J. 32(1), 49–54 (2005)
24. Menciassi, A., Gorini, S., Pernorio, G., Dario, P.: A SMA actuated artificial earthworm. Proc. IEEE Int. Conf. Robot. Autom. 4, 3282–3287 (2004)
25. Mizushina, A., Omori, H., Kitamoto, H., Nakamura, T., Osumi, H., Kubota, T.: A discharging mechanism for a lunar subsurface explorer with the peristaltic crawling mechanism. In: Proc. IEEE Int. Conf. Recent Adv. Sp. Technol., pp. 955–960 (2013)
26. Omori, H., Nakamura, T., Iwanaga, T., Hayakawa, T.: Development of mobile robots based on peristaltic crawling of an earthworm. In: Robotics 2010: Current and Future Challenges, pp. 299–319. InTech, Shanghai (2010)
27. Omori, H., Nakamura, T., Yada, T.: An underground explorer robot based on peristaltic crawling of earthworms. Ind. Robot An Int. J. 36(4), 358–364 (2009)
28. Onal, C.D., Rus, D.: A modular approach to soft robots. In: Proc. IEEE RAS EMBS Int. Conf. Biomed Robot. Biomechatron., pp. 1038–1045 (2012)
29. Onal, C.D., Wood, R.J., Rus, D.: An origami-inspired approach to worm robots. IEEE/ASME Trans. Mechatron. 18(2), 430–438 (2013)
30. Seok, S., Onal, C.D., Cho, K.-J., Wood, R.J., Rus, D., Kim, S.: Meshworm: A Peristaltic Soft Robot With Antagonistic Nickel Titanium Coil Actuators. IEEE/ASME Trans. Mechatronics 18(5), 1485–1497 (2013)
31. Suzumori, K., Endo, S., Kanda, T., Kato, N., Suzuki, H.: A bending pneumatic rubber actuator realizing soft-bodied manta swimming robot. In: Proc. IEEE Int. Conf. Robot. Autom., pp. 4975–4980 (2007)
32. Tanaka, T., Harigaya, K., Nakamura, T.: Development of a peristaltic crawling robot for long-distance inspection of sewer pipes. In: IEEE/ASME Int. Conf. Adv. Intell. Mechatronics, pp. 1552–1557 (2014)
33. Umedachi, T., Trimmer, B.A.: Design of a 3D-printed soft robot with posture and steering control. In: Proc. IEEE Int. Conf. Robot. Autom., pp. 2874–2879 (2014)
34. Wang, K., Yan, G.: Micro robot prototype for colonoscopy and in vitro experiments. J. Med. Eng. Technol. 31(1), 24–28 (2007)
35. Webb, B.: What does robotics offer animal behaviour? Anim. Behav. 60(5), 545–558 (2000)
36. Vaidyanathan, R., Chiel, H.J., Quinn, R.D.: A hydrostatic robot for marine applications. Robot. Auton. Syst. 30, 103–113 (2000)
37. Zarrouk, D., Shoham, M.: Analysis and design of one degree of freedom worm robots for locomotion on rigid and compliant terrain. J. Mech. Des. 134(2), 021010 (2012)
38. Zarrouk, D., Shoham, M.: Energy requirements of inchworm crawling on a flexible surface and comparison to earthworm crawling. In: Proc. IEEE Int. Conf. Robot. Autom., pp. 3342–3347 (2013)

WormTIP: An Invertebrate Inspired Active Tactile Imaging Pneumostat

Andrew D. Hinitt[1,3]([✉]), Jonathan Rossiter[2,3], and Andrew T. Conn[1,3]

[1] Mechanical Engineering, University of Bristol, Bristol, UK
ah2087@bristol.ac.uk
[2] Engineering Mathematics, University of Bristol, Bristol, UK
[3] Bristol Robotics Laborartory, Bristol, UK

Abstract. WormTIP is a novel lightweight self-actuating exploratory sensor, using a pneumostatic vessel and a dielectric elastomeric actuator (DEA) to create an active sensory tip capable of object shape determination as part of a flexible soft robot. Utilising the coupling of a static fluid vessel, the DEA is paired with a sensory membrane with internal papillae mimicking the internal morphology found in the fingertip. The sensory membrane is extended onto an object, conforming to its surface. Experimental results are presented which show the detection of shapes using particle velocimetry and papillae density analysis. These are preliminary results which show the potential of the WormTIP, which is the focus of ongoing work. The device is aimed for use as a self-contained palpating sensor, or as an attachment to a bio-inspired robotic worm forming a self-contained exploratory vehicle with the device acting as the sensory appendage or proboscis.

Keywords: DEA · WormTIP · EAP · Proboscis · Exploratory · Tactile sensing · TACTIP

1 Introduction

This paper describes the use of WormTIP, a novel active tactile sensor comprising a flexible skin-like silicone membrane with internal papillae tracked by camera combined with a dielectric elastomer actuator, in a pneumostatic configuration (Figure 1b) . The pneumostatic concept takes inspriation from Nature in the form of similar vessels in mammals and invertebrates (e.g. earthworm, elephant trunk & octopus tentacle), where muscular hydrostats are used to perform manipulation and sensory feedback tasks [1,2] (Figure 1a). Using a sealed vessel filled with pressurised air uses this concept as inspiration, whilst creating a lightweight actuation system with low structural weight. As a sensor the WormTIP benefits from the accuracy of optical tracking of surface deformation, which enables tactile mapping of 3-dimensional objects. This method of sensing follows on from previous works on end-effectors in which the human fingers sensory system inspires the design [3,4]. Mimicking the use of papillae in a finger,

© Springer International Publishing Switzerland 2015
S.P. Wilson et al. (Eds.): Living Machines 2015, LNAI 9222, pp. 38–49, 2015.
DOI: 10.1007/978-3-319-22979-9_4

paired with a specialised singular sensing device (camera), allow the detection of surface shapes and textures. The device provides a method for interacting with various objects which could be fragile - and break easily - or sharp and hard - potentially able to damage more rigid sensors. The skin conforms to objects and allows shape analysis whilst preventing damage to or by the object. The device is aimed for use in soft, remotely controlled robots where precise end effector control cannot be guaranteed. Damage prevention in soft robots is of high importance especially with the use of pressurised surfaces for actuation. DEA actuators also need to be protected from potential damage due to their fragility. With WormTIP sensing can be performed whilst protecting the delicate actuators. Previous work on pressurised vessels has shown that multiple pneumostatic vessels can use the coupling of DEA membranes to produce locomotion [5]. However, this type of robot was designed for applications involving exploratory tasks in remote inaccessible locations. In these applications, locomotion may not be sufficiently useful for the envisaged tasks. Even cameras will not give an indication of the physical properties of objects in remotely viewed environments, thus WormTIP was developed. The WormTIP device has some similarities with grippers, such as the fluid filled Festo FlexShapeGripper [6] and coffee ground filled Granular Gripper [7], which encapsulate objects and paired with a rigid arm can manipulate them. The devices use a fluid or granular filled vessel, but the actuation and uses are very different. These grippers are able to manipulate an object without information on its shape. Some sensors also have similarities such as the iCub Finger [8], which uses a soft structure paired with capacitive sensing to grip objects. These devices aim to develop sensing or manipulation on conventional rigid robots using soft materials. This work is aimed at being a soft sensor for use with soft robots, for use in soft or dynamic environments for which rigid robots are not ideally designed, such as medical applications where a sensing robot may be swallowed for internal examination of the gastrointestinal tract, where internal palpation could identify medical issues an internal camera such as PillCam may not [9]. Other applications could include delicate exploration of structurally unsound environments.

   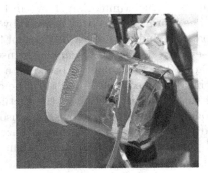

(a) Invertebrate Inspiration   (b) Devised Sensor

Fig. 1. Inspiration and Application

2   Active Touch

In the animal kingdom there are many examples of soft actuation coupling muscles and fluids. These are muscular hydrostats, where an incompressible fluidic vessel or musculatory structure is used to couple actuation and enable direction control of an extended proboscis or tentacle. This actuation matched with a sensory surface enables complex interactions of an animal with its environment. The elephants trunk enables it to pick up and manipulate objects, using sensory feedback from the skin to provide delicate control. The feedback is essential for the gripping tasks it performs [2]. In the case of the octopus, tentacles are used to encapsulate fish, but also to delicately traverse surfaces such as coral, which could cause it damage. The octopus uses its muscular hydrostatic compliance to grip, and sensory feedback to reduce high surface stresses on its body [1]. Another analogous system, is the sensory exploration of roots of plants, where the tip cells provide sensory input of obstacles and allow the redirection of the root growth to circumnavigate obstacles. [10]. Some works have looked to explore the development of a tentacle, which can grip an object, but do not explore the essential sensory feedback [11]. The WormTIP mimics this interaction in a simplified form, enabling sensory information to be processed using actuation in a single direction of actuation, to explore a 3-dimensional surface using a 2-dimensional conformable skin. The environmental sensory feedback found in nature - specifically highly specialised manipulators such as human fingers - is highly sought after in robotics due to its range, density, and resolution of sensors. The sensory information imparted through touch is a highly significant proportion of interaction between the majority of organisms and the environment they inhabit. The human digit is extremely difficult to emulate due to the enormous number of sensory receptors found in skin, the self-healing properties and the high density of connectivity of these sensors. Some solutions have been developed, which do offer some of the properties of skin, but usually entail a trade-off between accuracy, sensor density, & sensory area dimensions [12–14]. Skin on human fingers uses epidermal papillae deflection to stimulate the corpuscles and tactile cells, which are specialised for contrasting stimulation speeds, enabling complex tactile stimulation shown in Figure 2.

The WormTIP has been inspired by soft robotic research and biological counterparts, including the TACTIP sensor [4, 5, 15]. The TACTIP is a passive robotic end-effector developed to enable the use of a robotic arm to track movement on surfaces. The technology is based on a camera used to track pins or papillae on the inside of a spherical digit tip. The camera records the movement of the pins and this visual projection of the tactile interactions is used to analyse objects it touches . The device allows for sensory feedback, enabling applications of for rigid robot damage prevention in hazardous environments with sharp objects, and potentially delicate manipulation.

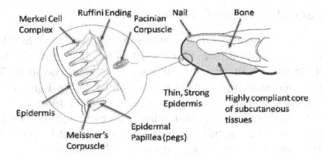

Fig. 2. The Human Finger [16]

2.1 Electro-Active Polymers

Electroactive polymers are a class of materials, which allow actuation through the application of voltage to alter the shape of the material. DEAs - Dielectric Elastomeric Actuators - are a subset of this class and actuate through the application of high voltage across a soft dielectric membrane. The attractive forces across the dielectric compress it, and cause planar areal expansion (Figure 3).

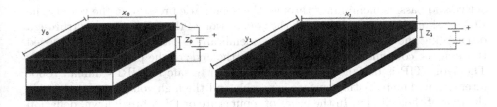

Fig. 3. Dielectric Elastomer Actuation States

In Figure 3, z_0 decreases to z_1, the area expansion for an assumed incompressible membrane is defined by :

$$\lambda_{l_{1,0}} = \frac{l_1}{l_0} \quad (1) \qquad \lambda_x . \lambda_y . \lambda_z = \lambda_p^2 . \lambda_z = 1, \quad (2)$$

where l_1 is final length, l_0 initial length, and λ_L is stretch ratio of length.

This relaxation actuation can be used in a number of configurations, including pre-strained single layers, sprung coils, and hydrostatic/pneumostatic vessel [15, 17]. In WormTIP a single layer, single side active pneumostat configuration is used.

2.2   Pressurised Antagonistic DEAs

Hydrostatic and pneumostatic DEA systems have been used for fluid pumps
[18], locomotion [5], and tactile stimulation [19]. They offer a method of creating
an antagonistic actuator pair without the need for a rigid coupling. This means
the devices are inherently soft and adaptive, whilst being lightweight, with a
homogeneous density. The actuators are constructed in most cases with two
DEA membranes on opposing ends of a cylinder, which is pressurised by a fluid.
The relaxation of one DEA membrane causing a significant displacement of the
opposing membranes. The use of pneumostatic DEA actuators for locomotion
has been developed to show the feasibility of locomotion with DEAs [5].

2.3   WormTIP

The TACTIP sensor and pneumostatic actuator technologies described above
are combined in WormTIP. The WormTIP consists of two principal parts, the
sensor and the actuator. The pneumostatic vessel uses an acrylic tube to form
an inextensible cylindrical body as shown in Figure 4c,d. Each end has a soft
membrane attached. The sensor membrane in Figure 4c,d is comprised of a
silicone skin, cast with papillae forms a passive membrane used with the camera
senor seals one end. Sealing the opposing end of the tube a prestrained DEA
actuator is attached. The pressurisation of the vessel enables the coupling of the
active and passive membranes through the pneumatic pressure in the vessel. The
silicone skin is protracted through the contraction of the DEA actuator (through
discharging the actuator). The silicone membrane is passive and is only moved
through the coupling of the membranes on the ends of the pneumostatic vessel.
The WormTIP is designed so that it is inherently safe for HRI - human robot
interaction. The palpating surface is passive and the high voltage isolated inside
the rear of the cylinder. In the event of a puncture or DEA breakdown, deflation
of the pneumostat will occur retracting the membrane from surface interaction.

The camera is used to track the individual papilla, and map the surface of
the membrane. Actuation of the coupled DEA enables the membrane to envelop
and probe a surface. Figures 4c & d, show the WormTIP unit in actuated and at
rest states, probing an object. The field of view (FOV) of the camera is shown
in the diagrammatic representation and captures images of the domed silicone
membrane. If considered from the camera's perspective, the pin head density
can be seen to increase around the object, decrease at the object edge, and
stay approximately constant at the centre, as portrayed in Figure 4a & b, for
the WormTIP states actuated (retracted) and relaxed (protracted) respectively.
This is useful for detection purposes through the measurement of pin density.

The DEA actuator and silicone skin form domed ends to the pressurised
WormTIP. With the pressurisation the membranes form a coupled pair. In order
for the device to palpate a surface the DEA actuator must be first in an actuated
state (relaxing the DEA membrane) prior to the sensor approaching the surface.
With the DEA in its actuated state the skin is retracted. The skin is protracted
by discharging the DEA actuator.

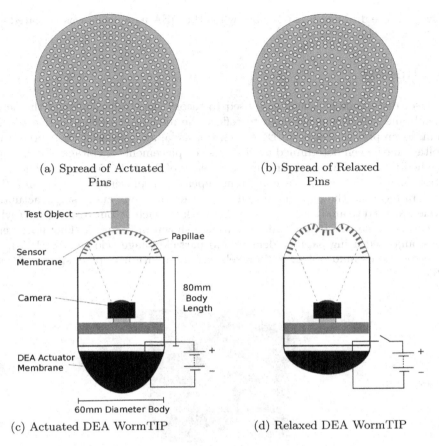

(a) Spread of Actuated
Pins

(b) Spread of Relaxed
Pins

(c) Actuated DEA WormTIP

(d) Relaxed DEA WormTIP

Fig. 4. Diagrammatic Representation of WormTIP Actuation

3   Methods and Experimental Setup

3.1   Fabrication

A silicone (Xiameter 3483) membrane is cast in a mould formed by laser etching
an array of pin holes in an acrylic sheet. The membrane thickness is adjusted
by removing excess uncured silicone with a blade and shim. Once cured the
membrane is highly flexible with uniform distribution of pins each 5mm in length.
The pin tips are painted white to increase image contrast. The membrane is
held in place at one end of the unit using an adjustable clip. A camera with a
resolution of 720x480 pixels is mounted in the cylindrical body of the WormTIP
unit. An acrylic DEA (VHB 4905) membrane is applied to the opposing end and a
nozzle inserted in the body for the pressurisation of the now sealed vessel. A pre-
strain of 300% was applied to the VHB membrane. Once sealed, the pneumostatic
vessel was pressurised to 80 mbar and placed in front of a set of sample shapes.

The pressure dropped to 25-35mbar when the DEA membrane was actuated and the sensing skin retracted.

4   Results

A laser displacement sensor was used to record the displacement of the active membrane, which was assumed to reflect the movement of the passive silicone skin. From previous work on DEA inch worms [5] an optimised DEA pre-strain, voltage have been determined to maximise displacement. A voltage of 3.2kV was applied to the DEA membrane at a frequency of 0.5Hz for 5 seconds. The actuation of the DEA caused a consistent repeated displacement of 8-10mm. The camera recorded the papillae movement during this time and a single actuation cycle was used to analyse the sensory feedback for each of four test objects (cylindrical block, square block, and flat surface) shown in Figure 5. The movements were analysed using papillae density and particle image velocimetry (PIV). An interrogation window of 128x128 pixels was used with a 16 pixel spacing for the PIV.

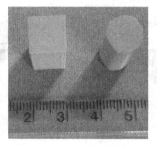

Fig. 5. Test Blocks

4.1   Image Processing

The pin overlay figures for each test case shows the retracted (red pins) and protracted (green pins) states of the membrane. For each of the test cases PIV and density image processing techniques are used. In each case the PIV images show the shape of the object (in red), the vectors showing the travel of the pins (arrows in yellow), which are both depicted on a background image of the membrane conforming to the object. The density mapping is used to show the change in density of the papillae relative to the camera in the FOV. The density images show the processed papillae density on the skin in retracted and protracted states.

4.2   Square Object

Figure 6a shows the relative pin positions. It can be seen there is a square block of pins where there is only minimal movement. Surrounding pins can be seen to spread and conform to the sides of the object. The corners of the square cause a rounding of the pin spread, as the membrane conforms to the sharp intersection. In this test there is negligible lateral movement reflected in the very low vector magnitude inside the square shown in Figure 6b. The density changes are clearly defined (Figures 6c,d), enabling a determination of the shape and its edges.

This test does show that a higher density of pins may be required for more complex surfaces to be accurately sensed. The pin density and length will have to be optimised to improve the sensor accuracy.

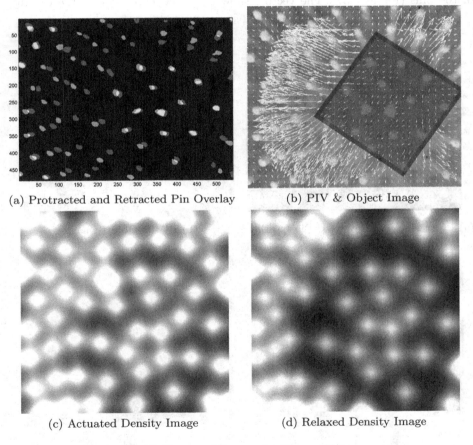

(a) Protracted and Retracted Pin Overlay   (b) PIV & Object Image

(c) Actuated Density Image   (d) Relaxed Density Image

Fig. 6. Processed Images for Square Object

4.3   Circular Object

Figure 7 show post-processed data highlighting the differences in the protracted and retracted states. Figure 7a shows the circular spreading of the pins in the protracted state in the centre of the image, but circular contraction of the pins on the edge of the image. The central spread is due to the domed membrane flattening as it conforms to the shape, whilst around the edge of the cylinder, the direction of movement flips where the membrane is not in contact with the object and the membrane returns to its domed shape. It should also be noted there may be some lateral shift in the membrane to the right of the image. Without perfect alignment of the sensor and the object this will always occur, and shows the type of mechanical shift that will have to be accounted for in future work. The analysis of the pin overlay is backed up by the PIV tracking of the pins (Figure 7b) . The vectors show a circular expansion of the central

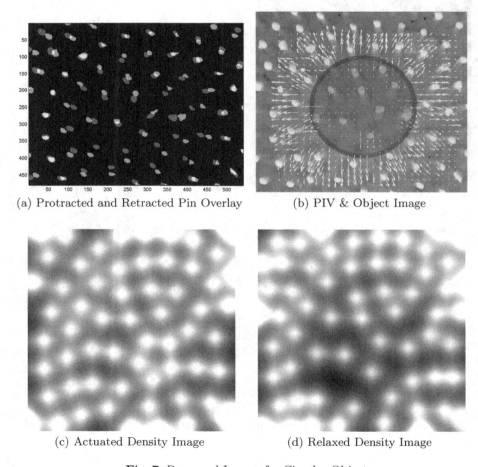

(a) Protracted and Retracted Pin Overlay   (b) PIV & Object Image

(c) Actuated Density Image   (d) Relaxed Density Image

Fig. 7. Processed Images for Circular Object

region of the membrane and the corresponding pin movement. However, the pin density analysis does not show as much detail at the centre, due to the relative magnitude of pin shift. The vectors do highlight the limitations of the pin density used, depicting a quantised multi-sided shape rather than a circle. The pin density images (Figures 7c,d) also show a pin spread at the centre, and increased density just beyond the circumference of the cylinder.

The experiment shows that there is an inherent lower limit of object size, from which shape information can be retrieved for a given papillae density. The membrane will however, still detect the object, but not all of its detail.

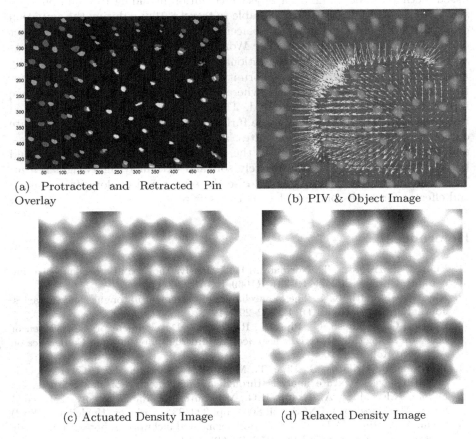

(a) Protracted and Retracted Pin Overlay

(b) PIV & Object Image

(c) Actuated Density Image

(d) Relaxed Density Image

Fig. 8. Processed Images for Flat Surface

4.4   Flat Surface

In the case of the flat surface, the protraction of the membrane allows more pins in the FOV. The pins at the centre do not change position in the FOV, whereas surrounding pins move uniformly inwards increasing the density (as shown in the comparison of Figures 8c,d). The PIV image (Figure 8b) shows how the pins have moved together under actuation.

5   Conclusion

In this work we have presented a preliminary study of the WormTIP, a soft, flexible self-actuating pneumostatic system that can be used to discern shapes. It exploits a lightweight pressurised vessel with an actuating membrane to palpate objects. Image analysis of the trial results show that a combination of image based methods could be used for object recognition including PIV and pin density analysis. The DEA actuator was able to travel 8-10mm during the actuation cycle, providing a large enough displacement for the sensory skin to interact effectively with the test objects. The WormTIP device shows great promise as an active tactile sensor for unconventional exploration, such as conventionally inaccessible spaces, where robot conformity is required, and in medical investigation such as gastro-intestinal tracts, where the potential for the robot to damage its environment should be minimised. These applications will be investigated in a further study. The design can be further optimised through more accurate optical sensing with higher frame rates, adjustment of pin length & spacings, and membrane thickness. Currently the body of the worm is stiff, but could be readily adapted to make an entirely soft and extremely lightweight sensor. The on-going work will fully characterise the sensor and explore its use as an end-effector for a self-contained locomoting soft robot.

References

1. Finn, J.K., Tregenza, T., Norman, M.D.: Defensive tool use in a coconut-carrying octopus. Current Biology **19**(23), R1069–R1070 (2009)
2. Kier, W.M., Smith, K.K.: The biomechanics of movement in tongues and tentacles. Journal of Biomechanics **16**(4), 292–293 (1983)
3. Chorley, C., Melhuish, C., Pipe, T., Rossiter, J.: Development of a tactile sensor based on biologically inspired edge encoding. In: 2009 International Conference on Advanced Robotics (2009)
4. Winstone, B., Griffiths, G., Pipe, T., Melhuish, C., Rossiter, J.: TACTIP - tactile fingertip device, texture analysis through optical tracking of skin features. In: Lepora, N.F., Mura, A., Krapp, H.G., Verschure, P.F.M.J., Prescott, T.J. (eds.) Living Machines 2013. LNCS, vol. 8064, pp. 323–334. Springer, Heidelberg (2013)
5. Conn, A.T., Hinitt, A.D., Wang, P.: Soft segmented inchworm robot with dielectric elastomer muscles. Proc. SPIE **9056**, 90562L (2014)
6. Stoll, W.: FlexShapeGripper Gripping modelled on a chameleon's tongue
7. Brown, E., Rodenberg, N., Amend, J., Mozeika, A., Steltz, E., Zakin, M.R., Lipson, H., Jaeger, H.M.: From the Cover: Universal robotic gripper based on the jamming of granular material (2010)
8. Ascia, A., Biso, M., Natale, L., Ricci, D., Metta, G., Sandini, G.: Comparison between two implementations of icub's fingertip. Procedia Engineering **47**, 1231–1234 (2012)

9. Iddan, G., Meron, G., Glukhovsky, A., Swain, P.: Wireless capsule endoscopy. Nature **405**, 417 (2000)
10. Gilroy, S., Masson, P.H.: Plant Tropisms. Blackwell Publishing Ltd (2008)
11. Giannaccini, M.E., Georgilas, I., Horsfield, I., Peiris, B.H.P.M., Lenz, A., Pipe, A.G., Dogramadzi, S.: A variable compliance, soft gripper. Autonomous Robots **36**(1–2), 93–107 (2014)
12. Burton, A.R., Minegishi, K., Kurata, M., Lynch, J.P.: Free-standing carbon nanotube composite sensing skin for distributed strain sensing in structures. SPIE **9061**, 906123 (2014)
13. Hands, P.J.W., Laughlin, P.J., Bloor, D.: Metal-polymer composite sensors for volatile organic compounds: Part 1. Flow-through chemi-resistors. Sensors and Actuators, B: Chemical **162**(1), 400–408 (2012)
14. Vuong, N.H.L., Kwon, H.Y., Chuc, N.H., Kim, D., An, K., Phuc, V.H., Moon, H., Koo, J., Lee, Y., Nam, J.-D., Choi, H.R.: Active skin as new haptic interface. Proceedings of SPIE - The International Society for Optical Engineering **7642**, 1–9 (2010)
15. Carpi, F., Frediani, G., De Rossi, D.: Hydrostatically coupled dielectric elastomer actuators. IEEE/ASME Transactions on Mechatronics **15**(2), 308–315 (2010)
16. Winstone, B., Melhuish, C., Dogramadzi, S., Pipe, T., Callaway, M.: A novel bio-inspired tactile tumour detection concept for capsule endoscopy. In: Duff, A., Lepora, N.F., Mura, A., Prescott, T.J., Verschure, P.F.M.J. (eds.) Living Machines 2014. LNCS, vol. 8608, pp. 442–445. Springer, Heidelberg (2014)
17. Carpi, F., De Rossi, D., Kornbluh, R., Pelrine, R., Sommer-Larsen, P. (eds.) Dielectric Elastomers as Electromechanical Transducers. Elsevier Ltd (2007)
18. Bowers, A., Walters, P., Rossiter, J., Ieropoulos, I.: Dielectric elastomer pump for artificial organisms. SPIE **7976**, 1–7 (2011)
19. Carpi, F., Frediani, G., De Rossi, D.: Opportunities of hydrostatically coupled dielectric elastomer actuators for haptic interfaces. Spie **7976**, 797618 (2011)

Copying Nature - A Design of Hyper-Redundant Robot Joint/Support Based on Hydrostatic Skeleton

Matthew Olatunde Afolayan[(✉)]

Mechanical Engineering Department, Ahmadu Bello University, Zaria, Nigeria
tunde_afolayan@yahoo.com

Abstract. Mimicking biological system successfully requires that the materials used in building such a system are qualitatively similar to that offered by the biological systems. One of such material is carbon filled natural rubber. Furthermore, biological systems implements various forms of support structures of which the ones imitated in this work is referred to as muscular hydrostatic support as opposed to fluid filled hydrostatic support. A muscular hydrostatic model proposed could be adapted to 3D motion but a planar joint/support was implemented as a proof of concept based on *teleost* fish - 394.01 mm long Mackerel. Static test indicate a well mimicked tail motion even with just three actuators. Turning test of the robotic fish inside tight box was successful as it was able to turn after several attempts. Also the robot was able to swim in a shallow pool of water where it attained 0.985m/s linear speed.

Keywords: Hyper-redundant robot · Hydrostatic skeleton · Muscular hydrostat

1 Introduction

Robot designs can be classified based on their joints. The conventional robots can best be described as discrete manipulators [1], where the designs are based on a small number of actuatable joints that are serially connected by discrete rigid link. Another category are the hyper-redundant robots that have much larger number of joints while the third group are referred to as continuum robots with theoretically no joints or the joints are not distinct. According to Trimmer *et al.* [2], most researchers build their biologically inspired hyper-redundant robots from concatenated rigid modules with multi-axis joints [3], universal joint [4,5] or revolute joint [6], parallel mechanism [7] and some are hybrid [8]. Examples of hyper-redundant robot joint implementations are revolute joint as used by NASA snakebot [6], universal joint [9], parallel mechanism [7], angular swivel joint with universal joint [10], angular swivel joint with bevel gear train [11]. Universal joint is the most popular joint adopted for hyper-redundant robot designs.

© Springer International Publishing Switzerland 2015
S.P. Wilson et al. (Eds.): Living Machines 2015, LNAI 9222, pp. 50–63, 2015.
DOI: 10.1007/978-3-319-22979-9_5

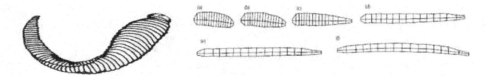

Fig. 1. Leech (*Hirudo medicinalis*) as an example segementally isochoric body [15]. The body grows thinner when extending to keep volume constant. Also each section maintains constant volume (isochoric) while in action.

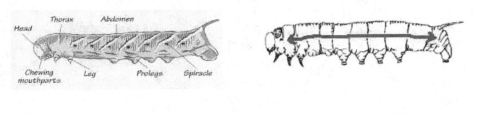

A   B

Fig. 2. (A) *Manduca sexta* Caterpillar. (B) The internally connected chambers that allows exchange of fluid. Sources: (A) The University of Arizona - The Manduca Project (B) Adapted from http://entnemdept.ufl.edu/creatures/field/hornworm.htm

There are three types of joints/supports found in biological models; bony joints (as in mammals and reptiles), fluid filled hydrostatic joints/support (as in invertebrate) and muscular hydrostatic joint/support (found in both vertebrates and invertebrates animals).

Hydrostatic support found in invertebrate organisms uses fluid-filled balloon like elastic structure for support [12,13]. The organism themselves have very simple body structure, mostly tubular. The fluid includes blood, intracellular fluid, seawater etc depending on the animal *taxa*. The incompressibility of these water based fluid and a flexible restraints/container act as the support referred to as hydrostatic skeleton.

Fluid filled hydrostatic supports are of two forms in terms of constancy of volume; sectional isochoric and whole body isochoric. In the sectional isochoric implementation (figure 1), the hydrostatic system is chambered and fluid exchange is not permitted whereas fluid exchange is permitted in the other form (figure 2).

Muscular hydrostat [14] is a form of hydrostatic skeleton found in some vertebrate like elephant trunks and mammalian tongues. Also some invertebrate such as octopus and squid tentacles are muscular hydrostat by design. In muscular hydrostatic system, circular, oblique, longitudinal and dorso-ventral muscles act in unison to orientate it. Figure 3 shows an octopus arm as an example of a muscular hydrostat.

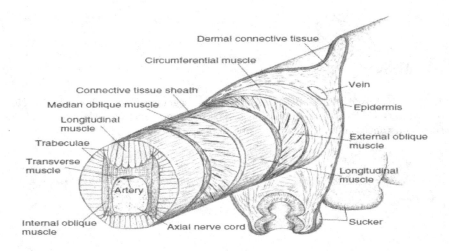

Fig. 3. Muscular hydrostat cross-section [16]

In hyper-redundant robot joint design, it is a common knowledge that the challenge still exist on how to control the multiple degree-of-freedom joints to produce usable motion. A hyper-redundant body can take a very large number of possible shapes without constraints [7,8,14,15,17,18,19]. Another well known challenge is selecting or incorporating a compact actuator that will have enough strength and tenacity to carry the weight of other links (or part) and still be fast enough while not generating too much heat [1,14]. Furthermore, researcher's wishes to simplify the complex control strategy often needed to manipulate a hyper-redundant body.

Most researchers have been extrapolating convectional joints – hinge, universal, even ball and sockets in an attempt to build hyper-redundant robot. These approaches have made many of those robots unsuccessful in their imitation of nature. A work around is to copy nature as closely as possible in terms of design and material used for the basis of the design and then adapt the simplified control strategy nature may offer. According to Srinivasan [20], nature posses shortcuts to mathematically complex issues of life. It is well know that house fly (*Musca domestica*) is perfection at flight control for their small size. The desert ants (*Cataglyphis*), despite their small brains and body size, they can make foraging excursions that take them up to 200m away from their nest. On finding a suitable prey, they return home unfailingly and in a straight line [21,22]. *Cataglyphis* do not use pheromones to retrace its trail in order to return back to its nest [22].

The motivating scientific discovering for this work is the work of Dorfmann *et al.* [23], that shows that the mechanical property of natural rubber (unfilled and carbon filled) and biological tissues are qualitatively similar. However, there is a major difference between biological tissue and rubber in that the biological tissues are composed of isotropic matrix embedding multiple oriented families of protein fiber (or muscles) hence they have anisotropic behavior, but rubber is generally isotropic. Another motivator is the work on magneto rheological rubbers for actuators; examples of such are found in [24,27]. This will one day lead to the ability to build an artificial muscle in conjunction with the muscular hydrostatic support - thus forming a compact system with very wide range of applications.

This work is about a hyper-redundant robot joint and support based on muscular hydrostat. An elastomer is used as the biological equivalent material for the joints within the joint/support structure. The novelty of this approach is that it is a proof of concept in preparation for embedded magneto rheological rubbers as its actuators.

2   The Design of the Artificial Muscular Hydrostatic Joint/Support

The design presented in this work is based on muscular hydrostatic joint (figure 4a). The elastomer is round and the minimal number of parallel actuators needed is three in numbers. A simplified version is shown in figure 4b and is designed as a form of "revolute" joint. It can roll about its longitudinal axis. It can also bend (or yaw) freely about the vertical axis, but pitching about the lateral axis is highly reduced. Pitching and rolling can be greatly reduced if the elastomer is made shorter as in figure 4d. The model with longer elastomer can be used for hyper-redundant robot that requires 3D body motion (figure 4f) since the shorter model will be too rigid to twist. On rigidity, the shorter model will excel since it is more compact and has less elastomer exposure. The model with longer elastomer (figure 4c) will require less number of joint to form a complete circle (figure 4e) while the shorter model will require more number of links to form a circle. Larger number of links will translate to more number of actuators to manage each link and more complex control schemes will ultimately be needed.

2.1   Strength and Weakness of the Evolved Artificial Muscular Hydrostatic Joint/Support

The followings are the advantages expected of the evolved artificial hydrostatic joint

1. It has potential for miniaturization – which may be easier to achieve than using rigid metal joints.
2. Cheaper material – elastomer like natural rubber and silicone and any rigid support – wood, plastic etc can be used.
3. Mass production will be easier.
4. High speed bending rate is possible as long as the damage mechanics (like Payne effect, that is, frequency softening effect) of the elastomer is put into considerations.

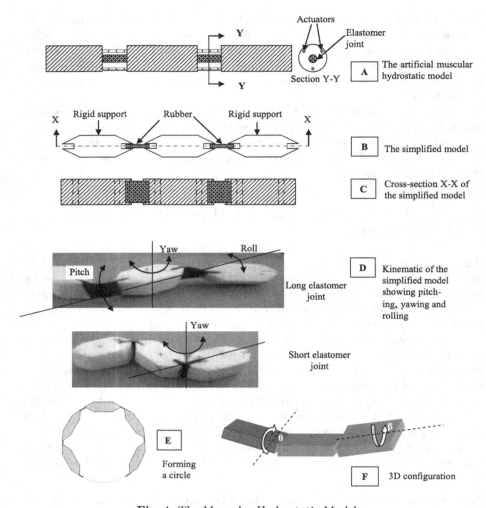

Fig. 4. The Muscular Hydrostatic Model

5. It can be made with medical grade elastomer and used for medical purposes without causing hazard to the organs. In event of collusion with an organ, damage will be minimal or inexistence compared to joints made up of metallic materials.

6. It can be used for precision positioning purposes if used with micro stepper motors or piezo motors.

7. Appropriate experimentation on the damage mechanics of the elastomer used and its implementation in the design will greatly reduce maintenance when deployed.

8. Noiseless operations will be possible with the design especially if the actuation is done with magneto sensitive materials.

There are obvious/inherent weakness of the design which are carry over from biological models, they are:

1. The design cannot be used where there is an extremely high temperature or very low temperature. The elastomer used and the support material will determine the useful temperature range.
2. Since this is not an inflatable design, it cannot be used to lift heavy load
3. The design will not work in a radioactive environment, the elastomer may dissociate.

2.2 Some Hyper-redundant Bodies that can Make Use of this Design

Typical scenarios where the design can be used are shown in figure 5 and describe as follows:

1. Military – application as tree climber or observation post. The artificial hydrostatic joint is flexible enough to be folded and carried about for deployment wherever needed and a rubber cover will enhance gripping of the structure being climbed. The artificial hydrostatic joint controller will have to support 3D motion to achieve this.
2. Under water robots - Rubber does not soak water and does not react with it; it can withstand saline environments also. Under water robots like a robotic fish can use the artificial hydrostatic joint for its flexible tail with minimal maintenance.
3. Snake robot - The artificial muscular hydrostatic joint can be used to build a 3D device that has a rod shape and programmed to behave and move as a snake or serpentine robot. Furthermore, covering the structure with rubber will give a firm grasp of terrain.
4. Minimally invasive surgery -The artificial muscular hydrostatic joint presented in this work can be miniaturized for this purpose. Moreover, a medical grade rubber or silicone will have less impact (bruising and laceration) on the internal organs than a relatively rigid model proposed by [28] and [29]. Advances in artificial muscle research and magneto sensitive rubber can also lead to self propelling device.
5. Space exploration - Following the foot print of the NASA snakebot, the artificial muscular hydrostatic joint can be used to make large numbers of crawling robots that can be deployed over large areas. Beside this, it can access crevices and terrains larger robots will not be able to access.
6. Stealth devices - A stealth device can be made to look like a harmless rod that can move about on its own like a snake robot. It can also be deployed inside water and remain there while giving feedback on the information captured.
7. Manufacturing arm over short distance - If one of its end is fixed, it can be made to follow a convoluted path to reach a component for repair or adjustment or for inspection.

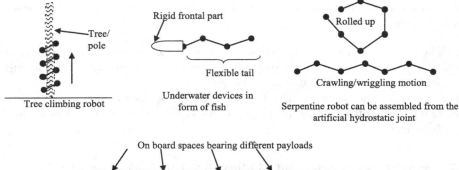

A rod-shaped endoscope with sections bearing different payload. It can be self propelling if actuators are attached.

Fig. 5. Application areas of the developed artificial muscular hydrostat

3   An Adaptation of the Design to a Planar Biological Model

The biological model selected is a 394.01 mm long Mackerel (figure 6a). The model (figure 6b) has a frontal part that is assumed to be rigid and the rear part is completely flexible. It is a six segment model with the joints (between the wooden support structures) being made up of carbon filled natural rubbers that are 5 mm wide and 1.5mm thick. The wooden support is water proofed 3/4 inch (19.05mm) thick seasoned plywood. Four minutes setting Epoxy glue was used to join the parts together.

To actuate the segments, three RC servomotors (Futaba 3003) were connected to the segments 1, 3 and 5 using nylon cables – figure 6c and figure 7. The unconnected joints act as a restoring spring. The servomotors are controlled by Microchip PIC18F4520 microcontroller running at 32MHz to generate three concurrent (or rigidly coupled) Pulse Width Modulated (PWM) signals that is out of phase by 60° for basic swimming operation alone. The flow chart for the three concurrently coupled PWM signal is presented in figure 8. Each port is connected to a servomotor. The three PWM signals have continuously varying duty cycle with each channel having independent duty cycles at any point in time. Furthermore, the PWM signals have built in dead band to allow the RC servomotors to catch up with it.

4   Tests, Results and Discussion

4.1   Biomimicry of the Joints

With the RC-servomotors set to run at 60° out of phase, it was observed (figure 9) that the motion and curvature mimics the live fish perfectly. The 60° phase

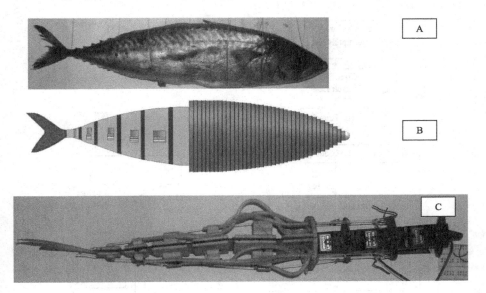

Fig. 6. (A)The biological model selected is a *teleost* fish – Mackerel. (B) The CAD model assumes flexible half part. (C) The robotic implementation is a six segment/five natural rubber joints - imitating muscular hydrostatic skeleton.

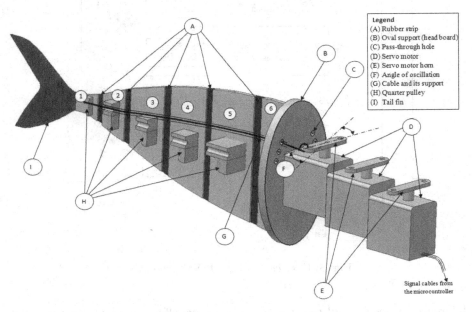

Fig. 7. CAD model of the hydrostatic joints showing cables connected to the first segment only

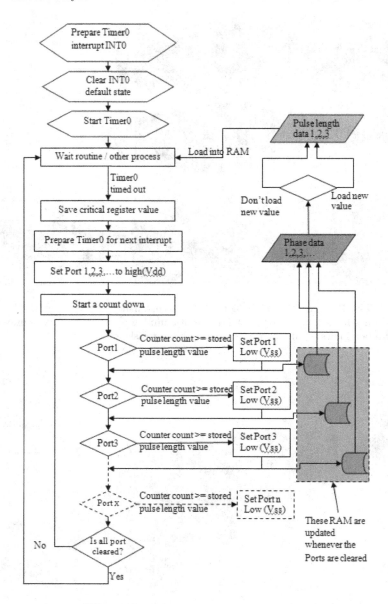

Fig. 8. Flow chart of the PWM generator

difference causes the tail to create traveling wave effect which makes the robot to push water behind it, thus creating a forward motion.

Fig. 9. Test on a practical implementation shows a perfect biomimicry of the bending. This is a still picture from the video of a test on the robot.

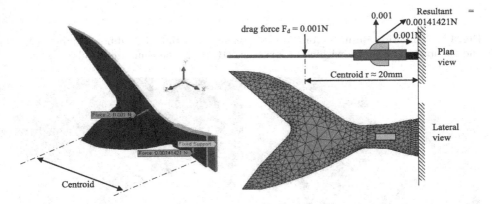

Fig. 10. The simulation inputs: 0.001N on the fin, 0.00141421N (0.001N −z axis, 0.001 N - x axis). The support is made up of seasoned plywood.

4.2 Stress Within the Most Critical Part

For *telesot* species of fishes, the most stressed part, which is also the thinnest portion is the peduncle - the portion just before the tail fin. Also, it is the portion that bend most and accelerates most (second to the tail fin). An ANSYS multi-physics simulation was performed on the model using simulated loads. The rubber used for the joints are hyper-elastic material thus its loading characteristics must be solved using constitutive equations. The tool selected for this procedure is ANSYS Multiphysis version 10. The CAD environment was Autodesk Inventor 7. The rubber uniaxial and biaxial data was entered into the ANSYS environments and the model constrained appropriately (figure 10). The constitutive equation used was determined by using curve fiting tools provided in the ANSYS 10. The tool indicates that Mooney–Rivlin 5 parameter constititutve equation is adequate to predict the behaviour of the rubber material used. The result of the simulation is found in figure 11.

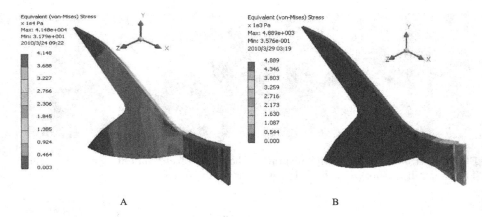

Fig. 11. Simulation Result – von Mises stress acting within the rubber peduncle (A) and the load distribution within the structure (B) using the simulated loads

Fig. 12. The model was tested in a tight box (A) and in a shallow pool of water (B)

4.3   Field Test

Two non qualitative field tests were performed, just to observe the workability of the biological model. One was in a tight box (60.96 cm x 121.92 cm x 60.96 cm) (figure 12A) to see if the model can swim within the micro environment, the other was in a shallow pool of water (figure 12B) with a depth range of 25cm to 50cm (equivalent pressure head of 2.4kPa – 4.9kPa), though the model itself was used at a maximum depth of 10cm.

In the tight box, the model could still turn (figure 13) after several attempt. The test in the shallow pool of water indicates that the model was able to achieve and sustain an average maximum linear speed of 0.985 m/s at maximum tail beat frequency of 1.7Hz inside water. This is approximately 1/3 of that of

Fig. 13. The robotic fish making a turn in a tight box filled with water

live mackerel which is 3.06m/s (Source: http://www.nmri.go.jp/eng/khirata/fish/general/speed/speede.htm).

5 Conclusions and Recommendation

A way of imitating a biological system is to consider the material being used for the artificial model. A muscular hydrostatic joint/support was imitated in this work by creating a simple joint/support model. The biologically equivalent material used was carbon filled rubber. The design was used to build a planar hyper-redundant robotic joint in the form of a robotic fish imitating mackerel. The supporting structure was plywood.

It is hereby recommended that a 3D implementation of this simple biologically equivalent implementation be explored. Other qualitatively similar materials may give better outcome especially at miniature level. Furthermore, magneto-sensitive rubbers should be tried, even in its crudest form, it may offer a very useful result when used with muscular hydrostatic design.

References

1. Robinson, G., Davies, J.B.C.: Continuum robots - a state of the art. In: IEEE Conference on Robotics and Automation, pp. 2849–2854 (1999)
2. Trimmer, B.A., Takesian, A.E., Sweet, B.M., Rogers, C.B., Hake, D.C., Rogers, D.J.: Caterpillar locomotion and burrowing robots. In: 7th International Symposium on Technology and Mine Problem, Monterey, CA, pp 1–10 (2006)
3. Shammas, E., Wolf, A., Brown, H.B., Choset, H.: New joint design for three-dimensional hyper redundant robots. In: Proceedings of the 2005 IEEE International Conference on Robotics and Automation, Las Vegas, Nevada, pp 1–10, October 2003
4. Yamada, H., Hirose, S.: Development of Practical 3-Dimensional Active Cord Mechanism ACM-R4. Journal of Robotics and Mechatronics **18**(3), 305–311 (2006)
5. Mori, M., Hirose, S.: Locomotion of 3D Snake-Like Robots; Shifting and Rolling Control of Active Cord Mechanism ACM-R3. Journal of Robotics and Mechatronics **18**(5), 521–528 (2006)
6. Kevin, B.: How snakebots work (2003). http://electronics.howstuffworks.com/snakebot.htm (accessed August 11, 2007)
7. Masayuki, A., Yoshinori, T., Hirose, S.: Development of "Souryu-VI" and "Souryu-V": Serially Connected Crawler Vehicles for in-Rubble Searching Operations. Journal of Field Robotics **25**, 31–65 (2008). doi:10.1002/rob.20229
8. Choset, H., Lee, J.Y.: Sensor-Based Construction of a Retract-Like Structure for a Planar Rod Robot. IEEE Transactions on Robotics, 435–449 (2001)

9. William, C.: Snake-Like Robot Can Crawl on Land or Swim, http://www. gorobotics.net/The-News/Latest-News/Snake%11Like-Snake-Like-Robot-Can-Cr-awl-on-Land-or-Swim/ (2006)
10. Elie, S., Wolf, A., Brown, Jr., H.B., Choset, H.: New joint design for three-dimensional hyper redundant robots. In: Proceedings of the 2005 IEEE International Conference on Robotics and Automation, Las Vegas, Nevada, pp. 3594–3599, October 2003
11. Wolf, A., Brown, H.B., Casciola, R., Costa, A., Schwerin, M., Shamas, E., Choset, H.: Proceedings of the 2003 IEEE/RSJ International Conference on Intelligent Robots and Systems, Las Vegas, Nevada, pp 1–10, October 2003
12. Alscher, C., Beyn, W.J.: Simulating the motion of the Leech. A biomechanical application of DAEs, Numerical Algorithms 19, 1–12 (1998)
13. Kelly, D.A.: Penises as Variable-Volume Hydrostatic Skeletons. Ann. N.Y Academic. Science 1101, 453–463 (2007)
14. Kier, W.M., Smith, K.K.: Tongues, tentacle and trunks: the biomechanics of movement in muscular-hydrostat. Zoological Journal of the Linnean Society 83, 307–324 (1985)
15. Skierczynski, B.A., Wilson, R.J.A., Kristian Jr., W.B., Skalak, R.: A model of the Hydrostatic Skeleton of the Leech. Journal of Theoretical Biology 181, 329–342 (1996)
16. Yekutieli, Y., Flash, T., Hochner, B.: Biomechanics: Hydroskeletal. In: Squire, L.R. (ed.) Encyclopedia of Neuroscience, vol. 2, pp. 189–200. Academic Press, Oxford (2009)
17. Ma, S., Mitsuru, W.: Time-optimal control of kinematically redundant manipulators with limit heat characteristics of actuators. Advanced Robotics 16(8), 735–749 (2002)
18. Wilbur, C., Vorus, W., Cao, Y., Currie, S.: A lamprey-based undulatory vehicle. In: Ayers, J., Davis, J., Rudolph, A. (eds). Neurotechnology for Biomimetic Robots, pp. 285–296. MIT Press (2002)
19. Crespi, A., Badertscher, A., Guignard, A., Ijspeert, A.J.: Amphibot I: an amphibious snake-like robot. Robots and Autonomous System 50, 163–175 (2005)
20. Srinivasan, M.V.: Distance Perception in Insects, pp. 1–10. Centre for Visual Sciences, Research School of Biological Sciences, Australian National University, Australia. Published by Cambridge University Press (1992)
21. Wehner, R.: Spatial organization of foraging behavior in individually searching desert ants, cataglyphis (sahara desert) and ocymyrmex (namib desert). In: Pasteels, J.M., Deneubourg, J.L. (eds.) From Individual to Collecctive Behavior in Social Insects, pp. 15–42. Birkhauser, Basel (1987)
22. Dimitrios, L., Ralf, M., Thomas, L., Rolf, P., Rüdiger, W.: A mobile robot employing insect strategies for navigation. Robotics and Autonomous Systems 30, 39–64 (2000)
23. Dorfmann, A., Trimmer, B.A., Woods Jr., W.A.: A constitutive Model for Muscle properties in Soft-bodied Arthropod. Journal of Royal Society International 4, 257–269 (2007)
24. Ogden, R.W.: Mechanics of rubberlike solids. In: Xxi ICTAM, Warsaw, Poland, pp 1–2, August 15-21, 2004
25. Ogden, R.W., Dorfmann, A.: Magnetomechanical interactions in magneto-sensitive elastmers. In: Constitutive Models for Rubber IV: Proceedings of the fourth European Conference on Constitutive Models for Rubber, ECCMR 2005, Stockholm, Sweden, pp 531–544, June 27–29, 2005

26. Fauze, A.A., Marcos, R.G., Gilsinei, M.C., Emerson, M.G., Adley, F.R., Edvani, C.M.: Electrochemical and mechanical properties of hydrogels based on conductive poly(3,4-ethylene dioxythiophene)/ poly(styrenesulfonate) and PAAm. Polymer Testing **25**, 158–165 (2006)
27. Norihiro, K., Masaki, Y., Takahiro, K., Kinji, A., Zhi-Wei, L.: Doping effects on robotic systems with ionic polymer-metal composite actuators. Advanced Robotics **21**(1–2), 65–85 (2007)
28. Choset, H., Zenati, M., Ota, T., Degani, A., Schwartzman, D., Zubiate, B., Wright, C.: Enabling medical robotics for the next generation of minimally invasive procedures: minimally invasive cardiac surgery with single port access. In: Rosen, J., et al. (eds.) Surgical Robotics: Systems Applications and Vision, pp 257–270. Springer Science Business Media. doi:10.1007/978-1-4419-1126-1_12
29. Xiaona, W., Max, Q.-H.M.: Robotics for Natural Orifice Transluminal Endoscopic Surgery: A Review. Journal of Robotics 2012, Article ID 512616, p. 9 (2012). doi:10.1155/2012/512616

An Under-Actuated and Adaptable
Soft Robotic Gripper

Mariangela Manti[(⊠)], Taimoor Hassan, Giovanni Passetti, Nicolò d'Elia,
Matteo Cianchetti, and Cecilia Laschi

The BioRobotics Institute, Scuola Superiore Sant'Anna, 56025 Pontedera, PI, Italy
m.manti@sssup.it
http://sssa.bioroboticsinstitute.it/

Abstract. Development of soft robotic devices with grasping capabilities is an active research area. The inherent property of soft materials, to distribute contact forces, results in a more effective robot/environment interaction with simpler control. In this paper, a three-finger under-actuated adaptable soft gripper is proposed, highlighting the design and manufacturing process. A novel design and actuation principle have been implemented to obtain the desired grasping abilities, from mechanical properties of materials and structures. Soft materials have been used to make each finger, for a high adaptability of the gripper to different shapes. We implemented an under-actuated mechanism through a wire loop actuation system, that helps achieving passive adaptation during grasping. Passive adaptability allows to drive the device with a reduced number of control parameters. The soft gripper has been lodged into an experimental setup endowed with one actuation unit for the synchronous flexion of its fingers. Grasping and holding capabilities have been tested by evaluating the grasp stability with target objects varying in shape, size and material. Adaptability makes this soft device a good application of morphological computation principles in bio-inspired robots design, where proper design of mechanical features simplifies control.

Keywords: Soft robotics · Soft gripper · Under-actuation · Cable-driven

1   Introduction

Grasping is an intensely investigated issue in robotics worldwide. In recent years, starting from the main advantages of soft robots in simplified embodied control and in safe interactions, roboticists are taking an interest in the development of soft grippers able to perform grasping actions. Grasping capabilities are directly related to the design of the device and to the actuation technologies. For this reason, specific attention has been paid to the mechanical design with the aim of realizing soft devices with greater capabilities to perform challenging tasks. As shown in [10], there are two possible approaches to perform soft interaction:

© Springer International Publishing Switzerland 2015
S.P. Wilson et al. (Eds.): Living Machines 2015, LNAI 9222, pp. 64–74, 2015.
DOI: 10.1007/978-3-319-22979-9_6

– Controlling the joints' stiffness of hard robots [1].
– Tuning the intrinsical softness, acting on the passive characteristics of the robot bodyware [21].

This distinction appears clear comparing traditional (hard) robotic hands and soft robotic ones and here we report some representative examples (without the ambition of covering all the state of the art).

Several examples of hard robotic grippers have been developed. Salisbury and Craig discuss the kinematics and control issues for articulated, multi-finger, tendon-driven mechanical hand in [16]. [8] describes the design of the UTAH/MIT dexterous hand, which is a four fingers anthropomorphic hand with wire actuation. A comparison between the Oxford and the Manus intelligent hand prosthesis is provided in [9]. A five-finger, underactuated anthropomorphic hand is proposed in [5]; similarly a three-finger, underactuated hand is proposed in [24]. Both systems use motors and wire actuation to obtain a multi-degrees-of-freedom (DoF) system, whereas an underactuated finger for robotic hands, with only one active DoF, is presented in [23]. These grippers are made of hard materials and rigid links. They use mechanical joints and active compliance to exploit under-actuation and achieve adaptability, which results in complex control. Furthermore, the use of metallic materials and mechanical mechanisms makes them bulky. A more comprehensive review can be found in [3].

Regarding soft robotic grippers, few examples can be found in literature. The use of soft materials improves the capability to undergo high deformations during interaction. This last aspect in combination with design features helps the soft gripper to exploit a passive deformation of the body adapting its shape to the object surface. For example, the Amedius [15] is a hydraulic gripper endowed with flexible actuator fingers, each made of three parallel cylindrical bellows. The finger bends when the internal fluid pressure in the bellows is varied. Other pneumatically actuated grippers have been developed using the same principle [17,18,22]. These systems are inherently compliant and adaptable, but hydraulic and pneumatic systems require pumps/compressors and valves, which increase the overall size of the system. Furthermore, the parallel bellow type actuators require complex control. A pneumatically actuated soft gripper with the ability to change its fingers stiffness is reported in [11]. The gripper uses fiber-reinforced silicone rubber tubes. As a consequence, the control tasks that in the traditional robotic scheme are supervised by control units, in the soft robotics paradigm can be partially taken on by the mechanical properties of the physical body itself [10,13]. The universal gripper [4] developed by University of Chicago and Cornell University, is a commercially available end-effector that is able to handle a wide variety of different objects. The gripper uses the physical phenomena called jamming transition; when the gripper is soft it can be pressed against the object passively taking the objects shape. By applying vacuum, the granular material contracts and hardens quickly, as consequence grasping and holding capabilities can be performed. Between the hard and soft robotic grippers, it is possible to identify an intermediate class. It refers to a new design that arises from the combination of both soft and rigid materials. These systems usually have a hard skeleton, fully or partially covered by soft materials with sensors embedded [14]. [19] proposes a multi-finger robotic gripper with the ability to change the softness of its rubber skin. A compliant and under-actuated three-finger hand is presented in [12]. The system is constructed by combining soft and hard polymers through shape deposition manufacturing (SDM) with the sensors embedded inside the fingers. The Shadow hand [2] achieves compliance by using Mc-Kibben type actuators that are inherently compliant.

Starting from the three main possibilities (uni-directional or omni-directional compliance and semi-soft paradigm) to make a soft device, and taking bio-inspiration into account, we propose a 3-finger soft, adaptable and under-actuated gripper with low cost, compact and novel design.

In this paper, we present the design and implementation of a soft robotic gripper, combining basic principles of soft robotics with one of morphological computation [10]. The design has been focused on setting the mechanical properties of the fingers so as to simplify the control of grasping. Section 2 discusses the mechanical design, the materials choice and the fabrication process. Results and discussions about the grasping capabilities are shown in section 3. Conclusions highlight the strengths and shortcomings of the gripper, adding some considerations for future versions, in order to overcome current limits.

2   Materials and Methods

2.1   Design Concept of the Mechanical Structure

Finger design has been defined taking into account the biological model of the human one. It means that we mimic the natural finger by considering its shape with the three phalanges and dimensions as shown in figure 1a. Despite that, the proposed model, entirely soft, does not contain any hard skeleton. Moreover, there are no physical mechanical joints but their action is carried out by the elastic energy stored into the soft material. In fact, the intrinsic passive properties of the silicone empower adaptability and enveloping capabilities to the soft device.

Each element is made of three phalanges that are separated by a triangular cut with an angle of 45°. As consequence, each phalanx is able to produce an angular displacement of 90° with respect to the previous one. The lengths of the phalanges are equal and were selected so that the tip of the finger touches the palm of the gripper when the finger is fully bent, in order to achieve a stable enveloping grasp. Mechanical springs were embedded longitudinally in the back part of the fingers in order to increase their lateral stiffness.

In order to simplify the design only two of the three phalanges are actuated while, the proximal phalanx acts as a part of the palm. This simplification reduces the manipulation capability and range of motion of the individual finger but it still enough for producing a stable and enveloping grasp. Furthermore, the gripper consists of three identical fingers in total, as shown in figure 1b, with two fingers aligned, attached to one side of the base, and the third finger on the opposite side, acting as a thumb bending over the plane between the other two.

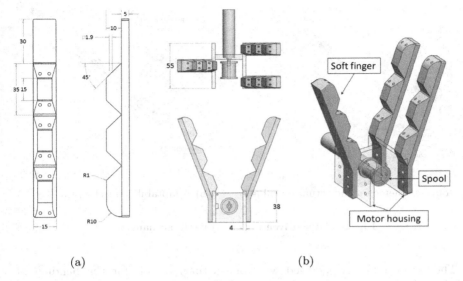

(a)   (b)

Fig. 1. Finger dimensions (a) and CAD model of the soft gripper (b)

2.2   Cable-Driven Under-Actuated Mechanism

The actuation of the finger is achieved by using a single motor and actuation cable. The actuation cable is a single wire loop which passes through tubes that are embedded inside the soft finger and U-tubes placed inside the distal phalanx and between fingers as shown in figure 2a. The use of the tubes reduces the friction between the wire and the soft material and avoid cutting the soft silicone when tension is applied. Both ends of the wire are wound over a spool at the base of the gripper. As the motor is actuated, both ends of the wire are wound on the spool, hence reducing the overall length of the wire and producing bending in the fingers. When the motor is actuated in the opposite direction, the tension in the wire is reduced and extension of the fingers is achieved due to elastic force stored in the silicone material and the embedded metallic springs. The actuation cable and the U-tubes constitute a type of differential mechanism [7], which distributes the tension in the actuation cable to all the fingers. Moreover, as it consists of no hard pulleys, the overall size of the base of the gripper is minimized. The base of the gripper also consists of tubes for routing the wire; they are not shown in the CAD figure. Furthermore, as the wire can move freely inside the tubes, the amount of bending of the finger is determined by the tension in the cable loop and its interaction with the environment, as shown in figure 2b (third finger is not shown). If an external force is applied on one of the finger and the motor is actuated, the other finger starts bending because of the free movement of the cable loop inside the tubes of the fingers hence enabling an adaptive grasp. Due to this feature the fingers of the gripper can easily adapt to irregular-shaped objects.

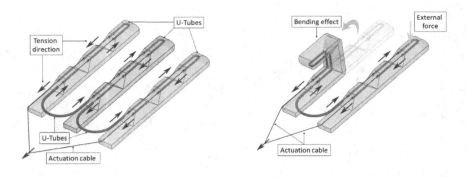

(a) Routing scheme for the actuation cable   (b) Adaptability mechanism

Fig. 2. Cable-driven under-actuated mechanism

The soft gripper is endowed with one actuation unit for the bending of the three fingers. It consists of a rotary DC motor (1331T006SR, Faulhaber) equipped with a magnetic incremental encoder (400 lpr, IE2-400, Faulhaber) and coupled with a 246:1 planetary gearhead (Series 14/1, Faulhaber).

2.3   Material Choice and Finger Fabrication Process

The use of soft and deformable materials helps in the distribution of forces over a large contact area eliminating stress concentration that is responsible for target object damage and forming stable grasping. With this in mind, the choice of the appropriate elastomer derives from a trade-off between two different requirements: too soft material produces a finger structure unable to perform a stable grasping due to deformations produced and the gravity effects; too stiff material is unable to adapt its contact surface to the object shape, causing an inaccurate grasp with a strong object/finger interaction. As consequence, after tensile tests on several materials using an Instron® 4464 machine, we decided to use Dragon Skin 30 (Smooth-On, Inc.), with an elastic modulus of 1 MPa, to develop each finger of the gripper. This silicone is bi-component (Part A, Part B) and can be processed easily, it has a low cost and a curing time relatively short. The fabrication has been carried out in two phases: the first consisting of the casting procedure and the second one regarding the assembly work of the fingers. The process starts from the finger mold, with silicone tubes inserted in order to create channels inside the polymerized silicone (figure 3a). The second step is pouring silicone into the mold and embedding the elastic reinforcement springs in the upper portion (figure 3b). This first fabrication phase ends with curing at room temperature. After that, silicone tubes are removed, leaving a channel for the under-actuation cable. In order to reduce friction between the actuation cable and the silicone, pieces of silicone tubes have been inserted into the channels embedded in the phalanges (figure 3c). Thanks to the design choices, the U-tube

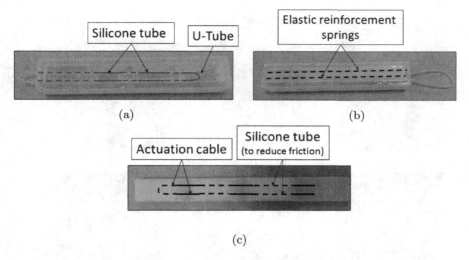

Fig. 3. The figure shows the three main fabrication phases: (a) the mold with the tubes inserted, (b) the mold with the silicone poured in and the elastic reinforcement springs added, (c) the cured silicone finger with the actuation cable

for the actuation cable has been directly embedded into the distal phalanx, reducing the encumbrance and improving the aesthetics of the prototype.

3   Results and Discussions

The soft gripper performances have been evaluated by measuring stable grasping and manipulation capabilities. The novel, under-actuated device has been tested with different targets in order to demonstrate how the soft fingers adapt themselves to the shape of the object, producing a stable enveloping grasp.

In order to assess the dexterous grasping capabilities of the soft gripper, a set of objects different in size and shape have been grasped as shown in figure 4. Moreover, we selected a number of everyday objects in order to test challenging grasping tasks. Among possible grasp postures, the power grasp has been tested and evaluated. The softness and adaptability of the gripper allow stable grasping with delicate objects as a raw egg and a fresh tomato without damaging them. Another parameter that is important to verify with soft devices is their capability to lift objects of reasonable weight maintaining a stable envelope grasp. A cylindrical metal object was grasped as shown in figure 5a and more weights were suspended from the object using strings. It was evidenced that the grasp remained stable for more than 3 kg of weight, while the weight of the soft gripper is 156 grams. At the same time, the performances of the soft gripper have been tested with different directional disturbance forces on two objects: a cylinder and a box. In figure 5b and 5c the measurements for the box case are shown. The same configuration has been used with the cylinder.

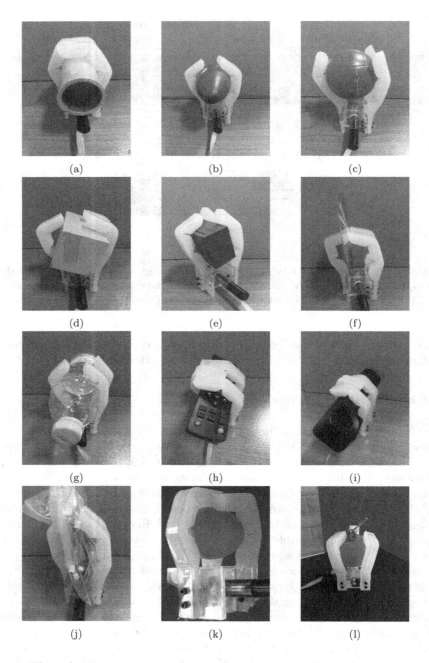

Fig. 4. The soft gripper grasping objects of various materials, with different shapes. From (a) to (l): a metal cylinder, an ellipsoid, a ball, two boxes, a compact disc, a plastic bottle, a remote control, a handset, a bag with various objects inside, an egg, a fresh tomato.

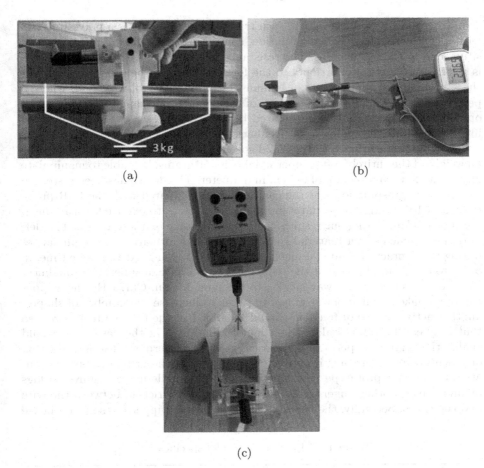

(a)

(b)

(c)

Fig. 5. (a) shows the experimental setup for maximum load, where te white lines represent the suspended weight; (b) and (c) show the experimental setup to respectively evaluate horizontal and vertical resistive forces

The applied forces are different in the both cases, as relate in table 1. In the first case the object is perfectly enveloped by the device. This produces threshold forces around 28 N and 20.65 N beyond which, the object will slide vertically or horizontally respectively. On the other side, the cylindrical object is able to endure forces around 8 N. It is worth to underline that tests were carried out controlling a single input parameter: the number of revolutions for the motor which is directly related to the opening of the hand. The user set this parameter in a LabView Interface and a PID controller (implemented by Maxon Motor EPOS Studio®, control library) managed the actuation cable pulling length.

In order to quantify the interaction between the phalanges and the object, a force sensor has been used as a probe. The sensor used to evaluate grip strength is an Interlink FlexiForce®A201. Such transducer essentially is a polyester strip with a conductive film embedded inside, connecting a round 9.53 mm wide sensitive area to the output pins. The resistance measured between the output pins is roughly inversely proportional to the force exerted on the sensitive area. For these preliminary experiments, a human in the loop control was used, where a person using vision adjusted the power of the actuated motor. In a further development, the process can be automated by using tactile sensors [20] embedded into the phalanx, providing force/feedback to close the loop.

In Table 1, the gripper features and specifications are summarized. The overall weight of the entirely soft gripper weights is 156 grams. It is able to manipulate objects with a cross-section of 6-7 cm in diameter. The device has been tested up to 3 kg. The grasping force at the contact point, evaluated with the FSR probe is around 1-6 N. Moreover, data about reaction forces to external disturbances are shown. Considering our results, there are still some issues to tackle. The definition of a much stiffer material could reduce the deformation of the phalanges during the contact obtaining a much more stable grasping. At the same time, in order to avoid the slippage at the object/finger interface, a superficial roughness could be reproduced in a way similar to the human skin. Currently the gripper can only make a stable power grasp. In order to increase the number of shapes, another active degree of freedom should be added to the two parallel fingers so that they can be rotated along their longitudinal axes at the base. This would enable the gripper to perform a spherical grasp. Furthermore, the introduction of pinch grasp will greatly improve the usefulness of the gripper. During actuation of the first prototype it was noted that the middle finger (figure 2a) lags behind the two other fingers. Probably, this is due to friction between the wire and the tubes. Secondly, the tubes at the base of the gripper are soft and hence

Table 1. Gripper features and specifications

FEATURES	Weight (equipped with the motor)	156 g
	Dimensions	5.5x4.5x13.5 cm
	Rest position (tip to tip)	4.5 cm
PERFORMANCES	Weight able to lift	3 kg
	Object's diameter	10-15 cm
	Grasping force	1-6 N
RESISTIVE FORCE (Box)	Vertical	28 N
	Horizontal	20.65 N
RESISTIVE FORCE (Cylinder)	Vertical	25 N
	Horizontal	8.05 N

produce deflection during actuation. This can be improved by either using hard pulleys at the base or exploiting two motors, i.e. one for the two parallel fingers and one for the thumb. This will complicate the controls and will increase the overall size and weight of the gripper base but will improve the manipulability of the gripper. Moreover, it was noted that when the finger is actuated, the middle and the distal phalanx of the fingers start to bend together. Generally this is not required, the middle phalanx should bend first, after reaching its maximum bending or when it interacts with an obstacle; only then the proximal phalanx should start to bend, insuring contact of each phalanx with the target object to be grasped. This issue can be solved providing different stiffness for the two joints. i.e the distal joint should be stiffer than the proximal one [6].

4   Conclusions

This paper describes the development of a three-finger, low-cost, under-actuated and adaptable soft gripper with a novel cable-driven actuation system. Some of the actuation components are embedded into the silicone fingers, reducing the overall system size. The arrangement of the fingers is specific for power grasp. Different objects, varying in size, shape and material are successfully grasped by the gripper; moreover the gripper showed stable grasping of delicate objects such as a raw egg and a fresh tomato without damaging them. Despite some shortcomings, this soft device is able to apply forces between 1-6 N and lift out up to 3 kg maintaining a stable grasp.

In conclusion, the soft gripper that has been developed represents a good example of how the application of morphological computation principles in bio-inspired design can give rise to a new generation of soft robots adaptable to different scenarios and with a reduced number of control parameters.

Acknowledgments. The authors would like to acknowledge the support by the European Commission through the RoboSoft CA ("A Coordination Action for Soft Robotics") in FP7 ICT FET-Open project #619319 and the People Programme (Marie Curie Actions) of the European Union's Seventh Framework Programme FP7/2007-2013/ under REA grant agreement number #608022.

References

1. Albu-Schaffer, A., et al.: Soft robotics. IEEE Robotics & Automation Magazine **15**(3), 20–30 (2008)
2. Kochan, A.: Shadow delivers first hand. Industrial Robot: An International Journal **32**(1), 15–16 (2005). ISSN: 0143–991X
3. Bicchi, A.: Hands for dexterous manipulation and robust grasping: A difficult road toward simplicity. IEEE Transactions on Robotics and Automation **16**(6), 652–662 (2000)
4. Brown, E., et al.: Universal robotic gripper based on the jamming of granular material. Proceedings of the National Academy of Sciences **107**(44), 18809–18814 (2010). ISSN: 0027–8424, 1091–6490

5. Controzzi, M., Cipriani, C., Carrozza, M.C.: Mechatronic design of a transradial cybernetic hand. In: IEEE/RSJ International Conference on Intelligent Robots and Systems, IROS 2008, pp. 576–581, September 2008
6. Dollar, A.M., Howe, R.D.: A robust compliant grasper via shape deposition manufacturing. IEEE/ASME Transactions on Mechatronics 11(2), 154–161 (2006)
7. In, H., et al.: Exo-Glove: A Wearable Robot for the Hand with a Soft Tendon Routing System. IEEE Robotics & Automation Magazine 22(1), 97–105 (2015). ISSN: 1070–9932
8. Jacobsen, S.C., et al.: Design of the Utah/MIT dextrous hand. In: Proceedings 1986 IEEE International Conference on Robotics and Automation, vol. 3, pp. 1520–1532. IEEE (1986)
9. Kyberd, P.J., Pons, J.L.: A comparison of the oxford and manus intelligent hand prostheses. In: Proceedings ICRA 2003. IEEE International Conference on Robotics and Automation, vol. 3, pp. 3231–3236. IEEE (2003)
10. Laschi, C., Cianchetti, M.: Soft robotics: new perspectives for robot bodyware and control. Frontiers in Bioengineering and Biotechnology 2 (2014)
11. Nagase, J., et al.: Design of a variable-stiffness robotic hand using pneumatic soft rubber actuators. Smart Materials and Structures 20(10), 105015 (2011)
12. Odhner, L.U., et al.: A compliant, underactuated hand for robust manipulation. The International Journal of Robotics Research 33(5), 736–752 (2014)
13. Pfeifer, R., Bongard, J.: How the body shapes the way we think: a new view of intelligence. MIT press (2006)
14. Quigley, M., et al.: Mechatronic design of an integrated robotic hand. The International Journal of Robotics Research 33(5), 706–720 (2014)
15. Robinson, G., Davies, J.B.C.: The Amadeus project: an overview. Industrial Robot: An International Journal 24(4), 290–296 (1997)
16. Salisbury, J.K., Craig, J.J.: Articulated hands force control and kinematic issues. The International journal of Robotics research 1(1), 4–17 (1982)
17. Shepherd, R.F., et al.: Soft machines that are resistant to puncture and that self seal. Advanced Materials 25(46), 6709–6713 (2013)
18. Suzumori, K., Iikura, S., Tanaka, H.: Development of exible microactuator and its applications to robotic mechanisms. In: Proceedings 1991 IEEE International Conference on Robotics and Automation, pp. 1622–1627. IEEE (1991)
19. Takeuchi, H., Watanabe, T.: Development of a multi-fingered robot hand with softness-changeable skin mechanism. In: Robotics (ISR), 2010 41st International Symposium on and 2010 6th German Conference on Robotics (ROBOTIK), pp. 1–7. VDE (2010)
20. Tenzer, Y., Jentoft, L.P., Howe, R.D.: Inexpensive and easily customized tactile array sensors using MEMS barometers chips. Under Review, p. 2013. (accessed: October 21, 2012)
21. Trivedi, D., et al.: Soft robotics: Biological inspiration, state of the art, and future research. Applied Bionics and Biomechanics 5(3), 99–117 (2008)
22. Wakimoto, S., et al.: Miniature soft hand with curling rubber pneumatic actuators. In: IEEE International Conference on Robotics and Automation, ICRA 2009, pp. 556–561. IEEE (2009)
23. LiCheng, W., Carbone, G., Ceccarelli, M.: Designing an underactuated mechanism for a 1 active DOF finger operation. Mechanism and Machine Theory 44(2), 336–348 (2009)
24. Zollo, L., et al.: Biomechatronic design and control of an anthropomorphic artificial hand for prosthetic and robotic applications. IEEE/ASME Transactions on Mechatronics 12(4), 418–429 (2007)

Measuring the Local Viscosity and Velocity of Fluids Using a Biomimetic Tactile Whisker

Tom Rooney$^{(\boxtimes)}$, Martin J. Pearson, and Tony Pipe

Bristol Robotics Laboratory, University of Bristol
and University of the West of England, Bristol, UK
tom.rooney@brl.ac.uk
http://www.brl.ac.uk/bnr

Abstract. A novel technique for determining the relative visco-density of fluids using an actuated flexible beam inspired by the tactile whiskers of marine mammals is presented. This was developed for the in-situ calibration of a tactile whisker based system for measuring flow velocity around autonomous robots working in complex underwater environments.

Keywords: Whiskers · Viscosity · Re-afferent · Flow sensing

1 Introduction and Motivation

Current autonomous inspection of underwater structures and the sea floor is constrained by the quality of available sensory information. This is particularly problematic in near shore marine environments where wave action and the abundance of plankton, algae, and vegetation degrade both visual and acoustic sensor performance. For example, the gas filled lacunae of near shore sea grasses have been reported to generate significant acoustic backscatter [1]. Backscatter is further compounded by the salinity turbulence experienced near estuaries [2] and to sea floor clutter such as rocks, holes and buried objects [3]. For these reasons we have been investigating the utility of tactile sensing for underwater search and inspection which would be impervious to the visual and acoustic impediments highlighted. However, precise motion planning in dynamic marine environments, such as active harbours or in the surf zone, is a challenging problem that must be overcome to enable precise tactile inspection. To mitigate this problem we take inspiration from the tactile whiskers of marine mammals; the compliant shaft introducing a degree of flexibility in the required precision of positioning during tactile interaction. Previous work using land based artificial tactile whisker sensors has demonstrated that surface textures and object morphology can be reliably classified [4], [5]. Other experiments with whisker arrays mounted onto mobile platforms have been used to build tactile maps of the environment [6]. Both of these themes of work would be extremely useful if demonstrated in the marine environment, however, preliminary experiments have revealed the significant influence of drag, or viscous friction, between the

© Springer International Publishing Switzerland 2015
S.P. Wilson et al. (Eds.): Living Machines 2015, LNAI 9222, pp. 75–85, 2015.
DOI: 10.1007/978-3-319-22979-9_7

water and the whisker during motion [7]. The relative motion between whisker and water could be caused by the flow of the water itself or by the self-motion of the whisker through the water, influences that we will refer to as ex-afferent and re-afferent sensory stimuli respectively. Put into context, we consider the sensory response caused by deflections of a whisker as it is purposely moved through the water as re-afferent and of little interest for tactile inspection tasks. If the same whisker was deflected by tidal currents or turbulence whilst held stationary we would consider this as a potentially interesting ex-afferent sensory influence, however, again this would not be a source of information that would be relevant to tactile inspection tasks. Finally, if the whisker is deflected as a result of making a contact with a solid object then this source of ex-afferent sensory stimuli is of interest. Therefore, there is a need to de-correlate these 3 sources of sensory stimuli to enable further investigation of tactile whisker based underwater inspection.

The paper is composed as follows: The next section provides some background context to the whiskers of marine mammals, fluid dynamics and the typical operating conditions of near shore and harbour environments. The following methods section details the proposed approach to measuring both fluid velocity and visco-density using an artificial whisker. The whisker sensor itself, the experimental set-up and apparatus used for subsequent experiments is also described. The results section summaries the key findings and analysis, which are then discussed along with a future plan of work in the final section.

2   Background

The marine mammals such as seal (Fig. 1), sea-lion and walrus use their prominent facial whiskers to find their food underwater. The seal uses its whiskers to detect turbulence patterns in the background flow-scape of the water caused

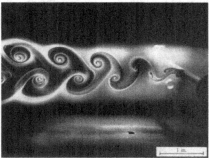

Fig. 1. Left) The turbulence pattern of a typical Von Karman sheet highlighting the vortex shedding pattern from a cylindrical beam visualised using smoke [8]. **Right)** The prominent facial whiskers of the harbour seal have evolved to minimise the oscillations induced in their whiskers caused by the periodic Von Karman sheet vortex shedding.

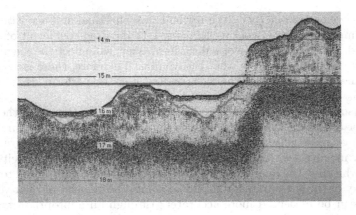

Fig. 2. An example image of the seabed derived from measurements taken by a dual frequency echo sounder of Portbury harbour in Bristol, UK. The dark band at the bottom indicates the nautical bottom of the harbour with the lighter bands above indicating fluid mud or dense suspended particulate.

by swimming fish [9]. The walrus pushes its whiskered snout into the silt at the bottom of the sea and feels for shell-fish, identifying and feeding on the soft parts of the shell-fish buried in the inhomogeneous sea bed substrate [10] [11]. Accordingly, the form of the whiskers in each of these species is very different; walruses have short stiff, almost tusk-like whiskers, whilst seals have evolved a whisker form that minimises the influence of what we refer to as re-afferent sensory stimuli [12]. When a fluid flows around a cylindrical beam a turbulence pattern known as a Von Karman vortex sheet is generated (see fig. 1 for reference). This consists of a regular shedding of vortices generated from alternate sides of the beam exerting an oscillatory force in the plane normal to the direction of relative flow. The undulated morphology of the harbour seal whisker disrupts the symmetry of the pressure field caused by the primary vortex separation resulting in a larger gap between the whisker and the formation of the first vortices. This greatly attenuates the oscillatory force exerted on the whisker by the vortex shedding providing the seal with whiskers that are non-sensitive to laminar flow [13]. This has inspired recent research into artificial models of such whisker morphologies for future sensory applications [14] and potentially could be useful for tactile inspection tasks. However, in this study we use a non-undulated cylindrical whisker model to explore techniques for measuring the relative velocity of the surrounding fluid.

In the near shore environment the viscosity of the sea water is highly variable, with the depth of the sea itself often being described in terms of viscosity. For example, the Belgian harbour of Zeebrugge defines the critical limit (or nautical bottom of the harbour) as having a dynamic viscosity of $0.06 Pa.s$, which is close to the dynamic viscosity of olive oil [15]. Fig. 2 shows a typical sea bed echo scan of a harbour in the Severn estuary in the UK. The high tidal range of this harbour generates the broad range of fluid viscosities in the region between the critical limit and the clear channel, i.e., the region in which visual and acoustic sensing is greatly impaired. The viscosity effectively sets the drag coefficient of a object in that fluid, which, in the case of a flexible whisker, will affect the magnitude of the sensory response caused by the relative flow. Therefore, there is also a need to determine the viscosity of the surrounding fluid to perform useful whisker based tactile inspections.

To measure the viscosity we have taken further inspiration from the biological whisker analogue, this time from the land based whiskered mammals that actively move their whiskers during exploration, a behaviour called whisking. By actively moving a whisker through the fluid with a controlled velocity profile similar to whisking, the re-afferent sensory response can be interpreted to classify the viscosity of the surrounding fluid. In so doing, the velocity of relative flow of that fluid can be measured more accurately through an appropriate transform of the observed deflection patterns.

Two experimental configurations were used to investigate the feasibility of this proposal; Firstly, a flow tank was used to control the relative velocity of flow of a fluid with a fixed viscosity across a stationary whisker; And secondly, a whisk-tank was constructed in which the viscosity of the fluid could be controlled as the whisker is moved (whisked) through a known velocity profile.

3   Method

3.1   Experiment 1: Flow Tank

A small tubular flow tank was constructed such that the flow speed could be controlled and measured using a standard rotary propeller drive and flow sensor. A simple collimator was also fitted to reduce turbulence caused by the apparatus itself [16]. An artificial whisker based on the Hall-effect sensor model described in [4] and shown in figure 4, was mounted into the flow tank and connected to a PC for data sampling (650Hz) via a simple micro-controller interface. This whisker sensor generates 2 voltages that are proportional to the degree of deflection of the whisker shaft along 2 orthogonal axes. We shall refer to these axes as x which was set parallel to the direction of flow, and y which was normal to the flow direction. The voltage to whisker deflection ratio is pre-programmed into the sensor during assembly, i.e., the sensitivity of the whisker is configurable. After calibrating to an appropriate deflection range, the whisker sensor assembly was then calibrated using a load cell and micrometer displacement tool to relate the voltage/deflection measurement to force. This could then be compared to the magnitude of whisker deflection observed in the flow tank at different flow speeds and the calculated force caused by drag used to determine flow speed. In addition, as the whisker sensor is sensitive to shaft deflections in 2-dimensions we were also interested to see if we could detect the vortex shedding oscillations caused by the Von Karman sheet in the axis normal to flow direction (y) giving us a second measure of flow velocity from the same whisker.

The force per unit length, F_d, exerted on the whisker shaft of diameter d by a fluid flow with velocity u and density ρ can be found using the following

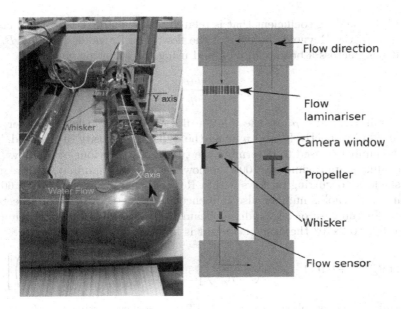

Fig. 3. The flow tank measures 1600mm length, by 500mm width. A single whisker is positioned perpendicular to the oncoming flow and adjacent to a small viewing window. (Example videos of experimental set-up and data sampling are available at: goo.gl/fsCk3d)

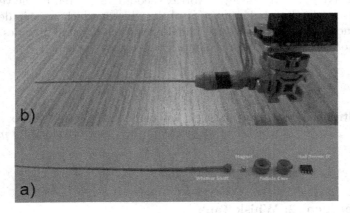

Fig. 4. The artificial whisker sensor used in this study. **a)** Details the main components of the sensor assembly **b)** Assembled sensor mounted onto servo motor for whisk tank experiments

relationship:

$$F_d = \frac{1}{2}\rho C_d u^2 d \tag{1}$$

Where C_d is the drag coefficient that is related to the shape of the object cross-section and to the Reynolds number of the flow. The Reynolds number, R_e, of a fluid flowing across a beam can be found using the following equation:

$$R_e = \frac{ud\rho}{\eta} \tag{2}$$

Where, again u, d and ρ are the relative flow velocity, whisker diameter and density of fluid respectively, and η being the dynamic viscosity of the fluid. The Reynolds number is used to determine the turbulent behaviour of the flow [17], below a value of approximately 1000 the flow is considered laminar, with regular vortex shedding occurring in flows with a Reynolds number greater than 60. In addition the Reynolds number also influences the drag coefficient of objects in that flow. In the case of a cylindrical beam in a flow with Reynolds numbers between 0.02 to 2×10^5 the drag coefficient is found using the following:

$$C_d = 11R_e^{-0.75} + 0.9 \left[1 - exp\left(-\frac{1000}{R_e} \right) \right] + 1.2 \left[1 - exp\left(-\left(\frac{R_e}{4500} \right)^{0.7} \right) \right] \tag{3}$$

To determine the flow velocity we looked at 2 approaches; firstly, the deflection of the whisker in-line to the direction of flow (referred to as the x-axis above) caused by the whisker drag. And secondly; the vortex shedding of the Von Karman street may cause a measurable oscillation of the whisker in the plan normal to flow direction (y-axis). The frequency of vortex shedding is related to the velocity of fluid flow using the Strouhal number S_t of the whisker which is again dependent on the Reynolds number of the flow. The Rayleigh relationship provides a robust transform for the cylindrical form of the whisker cross-section:

$$S_t = 0.195 \left(1 - \frac{20.1}{R_e} \right) \tag{4}$$

With the Strouhal number we can now calculate the expected vortex shedding frequency, f_s Hz, for a whisker of diameter dm in a flow of velocity v ms^{-1} :

$$f_s = \frac{S_t v}{d} \tag{5}$$

3.2   Experiment 2: Whisk Tank

To investigate the feasibility of the whisker sensor for measuring the viscosity of stationary fluids, the whisker sensor was mounted onto the spindle of a water-proof servo motor such that it can be rotated through an arc with a controlled angular velocity. This apparatus was positioned in a small test bath filled with water (see photograph in Fig. 5) to which measured quantities of catering glycerol were added to adjust the dynamic viscosity of the fluid based on data available from previous studies [18]. The same sampling rate and data capture hardware were used as in experiment 1 with 10 second bouts of whisking performed in

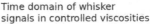

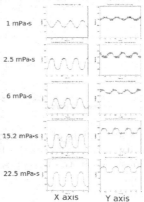

Fig. 5. The whisk tank experimental configuration consists of a small tray containing water glycerol mixture in which the artificial whisker can be whisked through a controlled velocity profile. The whisker is actuated using a small servo motor. The time series plots to the right of the photograph show how the whisker deflection is affected by the viscosity of the fluid. (Example videos of experimental set-up and data sampling are available at: goo.gl/fsCk3d)

each test of the different glycerol water solutions. The whisk rate in each bout was set at $1.4Hz$ which relates to a peak velocity at the tip of the whisker of $0.2ms^{-1}$. As is clear from the sample time series data plots shown in Fig. 5, the magnitude of the re-afferent sensory response of the whisker increases in both measured axes of the sensor in fluid with higher viscosities. In keeping with the flow tank experiments, two approaches were studied to determine fluid viscosity, the magnitude of deflection in the plane parallel to whisker motion (relative velocity of fluid), and the frequency of oscillation in the plane normal to whisker movement.

4   Results

For the flow tank experiments the flow velocity was gradually increased from stationary through to $0.5ms^{-1}$. Both axes of the whisker sensor were sampled at 650Hz in 5 second batches for each flow velocity. From this the time series x-axis data set (in-line with flow direction) was examined and the average deflection magnitude extracted. These voltage measurements were then converted into the equivalent force applied at the calibration point along the whisker shaft. Figure 6a shows the observed force measurements against the predicted forces caused by the flow velocities using equation 1.

The second approach was to look for oscillations in the normal plane that may be caused by vortex shedding and therefore indicative of the flow velocity.

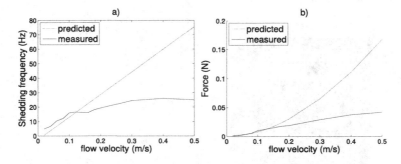

Fig. 6. a) Comparative plots of predicted and measured forces in x-axis of whisker against flow velocity in the flow tank. **b)** Predicted vortex shedding frequency against the oscillations measured in the normal whisker plane (y-axis) to flow in response to different flow speeds.

For this the y-axis time series data set was divided into 1 second windows and transformed into the frequency domain. The centre of the dominant peak in the spectrum was extracted from each sample set and an average taken for each flow velocity. Again this data was compared to the predicted vortex shedding frequency calculated using equation 5 as shown in figure 6b.

For the whisk tank experiments the 2 whisker axes were again sampled at 650Hz in 10 second batches. The peak to peak deflection magnitude measured in the x-axis from each batch was extracted and plotted against fluid viscosity (figure 7a). Whisker y-axis data taken from the whisk tank experiments was transformed to frequency and the same windowing and dominant peak extraction algorithm applied as in the flow tank (figure 7b).

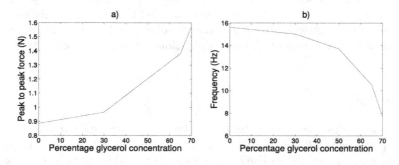

Fig. 7. a) Plot of measured forces in x-axis of whisking in stationary fluid containing different concentrations of glycerol. **b)** Vortex shedding frequency oscillations measured in the normal whisker plane (y-axis) to whisking in response to different viscosities.

5   Analysis

From these results it is clear that the whisker sensor is indeed sensitive enough to detect changes in flow velocity and viscosity, however, the relationship between these measurements and the physical characteristics of the fluid has not been successfully modelled. To begin, the deflection force in the x-axis caused by drag deviates significantly from the predicted force as the velocity increases. This is particularly the case at flow velocities greater than $0.15ms^{-1}$ and coincides with the deviation from predicted response to the vortex shedding frequency measured in the y-axis. Interestingly this flow velocity also generates a shedding frequency very close to the natural frequency of the whisker sensor itself when immersed in water (15Hz measured experimentally). The whisker would appear to be experiencing the 'lock-in' phenomenon whereby its resonant frequency dominates the spectrum through a wide range of Reynolds numbers [19]. If this is the case then the first mode of natural frequency of the whisker in fluid will dictate its usefulness in measuring flow velocity, i.e., a higher natural frequency being preferential. We see evidence for this in the behaviour of seals as they use their whiskers in the pursuit of their prey. Their tendency to fully protract their whiskers will increase the stiffness at the mounting of the whisker at the base, which is effectively a pinned rather than clamped boundary condition. Increasing the stiffness of this boundary condition will increase the natural frequency of the whisker thereby broadening the range of sensitivity to fluid flow velocity before the lock-in frequency dominates the response.

For the viscosity measurement, however, this phenomenon can be exploited by purposely exciting this natural frequency as a way of robustly classifying the viscosity of the fluid as supported by the trend in shedding frequency response shown in figure 7b.

6   Conclusion

A sensor based on the mammalian whisker has been evaluated as simple fluid flow velocity and viscosity sensor. Information from the two instrumented orthogonal axes of the sensor have been used to investigate the response to fluids with known viscosity and flow velocities. This included fluid flowing over a static whisker and one that is moved in a manner analogous to the the whisking behaviour of many land based mammals. The results reveal that the whisker is sensitive enough to detect changes in the velocity of fluids but that this relationship is non-linear when exposed to high flow velocities. Similarly, the whisker is sensitive to the expected vortex shedding frequency at low flow speeds but subject to a frequency lock-in phenomenon as this frequency approaches the natural frequency of the whisker. The natural frequency of the whisker changes when immersed in a viscous fluid, which has been used in the past as the basis to measure viscosity [20]. The whisker presented here also appears to be capable of robustly classifying the viscosity of fluids by exciting this frequency through a controlled whisking like self motion pattern.

Acknowledgments. This work is funded by DSTL as part of their national Ph.D. programme. We would like to also acknowledge the Bristol port company of Avonmouth and Portbury Docks, and the crew of the survey vessel Isambard Brunel for their kind assistance.

References

1. Mulhearn, P.: Mapping seabed vegetation with sidescan sonar (2001)
2. Goodman, L., Sastre-Cordova, M.M.: On observing acoustic backscattering from salinity turbulence. The Journal of the Acoustical Society of America **130**(2), 707–715 (2011)
3. Martin, A., Cexus, J., Le Chenadec, G., Cantéro, E., Landeau, T., Dupas, Y., Courtis, R.: Autonomous underwater vehicle sensors fusion by the theory of belief functions for rapid environment assessment. In: Proceedings of ECUA (2010)
4. Sullivan, J., Mitchinson, B., Pearson, M.J., Evans, M., Lepora, N.F., Fox, C.W., Melhuish, C., Prescott, T.J.: Tactile discrimination using active whisker sensors. IEEE Sensors Journal **12**(2), 350–362 (2012)
5. Pearson, M.J., Mitchinson, B., Sullivan, J.C., Pipe, A.G., Prescott, T.J.: Biomimetic vibrissal sensing for robots. Philosophical Transactions of the Royal Society B: Biological Sciences **366**(1581), 3085–3096 (2011)
6. Pearson, M.J., Fox, C., Sullivan, J.C., Prescott, T.J., Pipe, T., Mitchinson, B.: Simultaneous localisation and mapping on a multi-degree of freedom biomimetic whiskered robot. In: 2013 IEEE International Conference on Robotics and Automation (ICRA), pp. 586–592. IEEE (2013)
7. Rooney, T., Pearson, M., Welsby, J., Horsfield, I., Sewell, R., Dogramadzi, S.: Object localisation using active whiskers on underwater autonomous walking robots. In: 14th International Conference on Climbing and Walking Robots (CLAWAR), pp. 190–195 (2011)
8. Koopmann, G.: The vortex wakes of vibrating cylinders at low reynolds numbers. Journal of Fluid Mechanics **28**(03), 501–512 (1967)
9. Dehnhardt, G., Mauck, B., Hanke, W., Bleckmann, H.: Hydrodynamic trail-following in harbor seals (phoca vitulina). Science **293**(5527), 102–104 (2001)
10. Kastelein, R., Stevens, S., Mosterd, P.: The tactile sensitivity of the mystacial vibrissae of a pacific walrus (odobenus rosmarus divergem). part 2: Masking. Aquatic Mammals **16**, 78–87 (1990)
11. Born, E., Rysgaard, S., Ehlmé, G., Sejr, M.K., Acquarone, M., Levermann, N.: Underwater observations of foraging free-living atlantic walruses (odobenus rosmarus rosmarus) and estimates of their food consumption. Polar Biology **26**(5), 348–357 (2003)
12. Miersch, L., Hanke, W., Wieskotten, S., Hanke, F., Oeffner, J., Leder, A., Brede, M., Witte, M., Dehnhardt, G.: Flow sensing by pinniped whiskers. Philosophical Transactions of the Royal Society B: Biological Sciences **366**(1581), 3077–3084 (2011)
13. Hanke, W., Witte, M., Miersch, L., Brede, M., Oeffner, J., Michael, M., Hanke, F., Leder, A., Dehnhardt, G.: Harbor seal vibrissa morphology suppresses vortex-induced vibrations. The Journal of experimental biology **213**(Pt 15), 2665–2672 (2010)

14. Alvarado, P.V., Subramaniam, V., Triantafyllou, M.: Performance analysis and characterization of bio-inspired whisker sensors for underwater applications. In: 2013 IEEE/RSJ International Conference on Intelligent Robots and Systems (IROS), pp. 5956–5961. IEEE (2013)
15. Delefortrie, G., Vantorre, M., Laforce, E., et al.: Revision of the nautical bottom concept in zeebrugge based on the manoeuvrability of deep-drafted container ships (2005)
16. Vogel, S., LaBarbera, M.: Simple flow tanks for research and teaching. Bioscience, 638–643 (1978)
17. Reynolds, O.: On the experimental investigation of the circumstances which determine whether the motion of water in parallel channels shall be direct of sinuous and of the law of resistance in parallel channels. Philosophical Transactions of the Royal society of London, 174 (1883)
18. Segur, J.B., Oberstar, H.E.: Viscosity of glycerol and its aqueous solutions. Industrial & Engineering Chemistry 43(9), 2117–2120 (1951)
19. Bourguet, R., Karniadakis, G., Triantafyllou, M.S.: Lock-in of the vortex-induced vibrations of a long tensioned beam in shear flow. Journal of Fluids and Structures 27, 838–847 (2011)
20. Badiane, D., Gasser, A., Blond, E.: Vibrating beam in viscous fluid for viscosity sensing: application to an industrial vibrating viscometer. In: Proceedings of the 12th Pan American Congress of Applied Mecahnics (2012)

Biomimicry of the *Manduca Sexta* Forewing Using SRT Protein Complex for FWMAV Development

Simone C. Michaels[1(✉)], Kenneth C. Moses[1], Richard J. Bachmann[1],
Reginald Hamilton[2], Abdon Pena-Francesch[2], Asheesh Lanba[2],
Melik C. Demirel[2], and Roger D. Quinn[1]

[1] Biologically-Inspired Robotics Lab, Case Western Reserve University, Cleveland, OH, USA
{scm61,kcm7,rjb3,rdq}@case.edu
[2] Pennsylvania State University, State College, PA, USA
MDemirel@engr.psu.edu

Abstract. A new thermoplastic protein complex, Squid Ring Teeth (SRT), has been adapted for use in the artificial reconstruction of a *Manduca sexta* wing. The SRT protein complex exhibits consistent material properties over a wide range of temperatures (25 °C to 196 °C) and retains it mechanical integrity across a large frequency spectrum (0.1 Hz to 150 Hz). Insect-inspired wings comprised of SRT can therefore be reliable and robust, which are essential characteristics for flapping wing MAVs (FWMAV). The preliminary results in this paper suggest that a thorough analysis of an SRT-based wing be conducted using load cell, optical digitization, and PIV techniques. With these results, we believe it will be possible to accurately mimic the *M. sexta* wing in order to pave the way for next generation FWMAV development.

Keywords: Protein · Wing · FWMAV · Biomaterial · *Manduca sexta* · Hawkmoth · Biomimicry

1   Introduction

Research to precisely mimic flapping-winged animals is being conducted in order to take advantage of the evolutionary benefits found in nature. Advancements in electronics, manufacturing, materials, and aerodynamics allow for more accurate biomimicry of winged animals. Flapping flight is of particular interest because it allows for a wider range of agility, robustness, and flight capabilities. Fixed-wing UAVs have inferior flight performance to flapping-wing micro air vehicles (FWMAVs) for missions requiring slow speed flight or maneuvering in close quarters, largely due to their power requirements in these conditions [1]. Furthermore, efficient fixed-wing systems tend to be relatively large, exacerbating the possibly destructive repercussions of ground effect along walls [1]. Small FWMAVs can surmount these issues by providing a system that is relatively adaptable and maneuverable.

Although birds have inspired numerous FWMAV designs, these models tend to be better suited for outdoor purposes. These systems are fast, heavy, and restricted in motion and are not well suited for congested environments [1]. The size and agility of

© Springer International Publishing Switzerland 2015
S.P. Wilson et al. (Eds.): Living Machines 2015, LNAI 9222, pp. 86–91, 2015.
DOI: 10.1007/978-3-319-22979-9_8

insects, on the other hand, make them an extraordinary biological inspiration for high-ly maneuverable flight. Insects are lighter and smaller than their class Aves counter-parts, partially because the majority of actuation occurs at the wing root attached to the thorax instead of requiring the endoskeleton and muscles that comprise the wing of a bird [1]. Furthermore, the flight capabilities of insects allow them to hover and make sharp turns. These features make FWMAVs inspired by winged insects excel-lent candidates for flight in urban and indoor environments.

Much of the benefit of insect flight comes from the material and structural properties of the wings themselves. In general, insect wings are flexible, passive structures com-posed of a venation structure embedded within a protein-based membrane [2]. The chi-tin-based veins provide support to the flexible protein matrix. These veins are thicker and more concentrated at the root and along the leading edge of the wing, alleviating bending stresses closer to the thorax [2]. This reduction in mass along the length of the wing also causes the amount of applied bending torque to be much lower at the tip than the base, significantly decreasing the power necessary to accelerate the wing in flight [3].

Several attempts have been made to mimic insect wings in order to take full advantage of their unique structural characteristics. Wings constructed through MEMS-based chemical etching and 3-D printing were too heavy and unable to gener-ate sufficient lift [4]. The most successful approach at replicating the structure of an insect wing has been a polymer-based membrane embedded within, or on top of, a carbon fiber venation structure [5]. Despite the ability to duplicate certain features of an insect wing, this method offers an uncertain degree of reproducibility. Further-more, the approach of creating a composite part increases the amount of fabrication time and difficulty. Instead, utilizing a unique protein complex to more accurately imitate insect wing characteristics opens the door for next-generation biomimicry.

A common subject used for research and biomimicry is the hawkmoth *Manduca sexta* (*M. sexta*) because it is easily reared and readily available. The large size of this insect makes it more feasible to replicate its morphology and gather consistent obser-vations during tests. Furthermore, mimicking the *M. sexta* is simplified given that this insect does not necessarily need its hind wings for flight. Reliable data can therefore be gathered with just the forewing [6].

2   Recombinant Squid Ring Teeth (SRT) Protein Complex

The Squid Ring Teeth (SRT) protein complex exhibits novel elastomeric behavior and is based on protein extracted from jumbo squid. The SRT protein has unique physio-chemical properties capable of providing multifunctional performance and is a ther-moplastic elastomer. The inner surface of squid tentacles are covered with suction cups, or suckers, that are critical for prey capture. Each sucker includes a ring of hard, sharp teeth that allow the squid to maintain a positive mechanical grip on the prey. Unlike most hard biological structures, the ring teeth on the squid are composed en-tirely of protein, making it a good candidate for insect wing biomimicry.

The SRT polymer can be synthesized with a modulus of elasticity ranging from 10 MPa (similar to latex rubber) up to 1 GPa (similar to Nylon 6). When the SRT is kept in a vacuum overnight, resulting in less than 5% sample humidity ("dry" condi-tions), the complex has a dynamic elastic modulus of 1.93 ± 0.1 GPa. These values are very similar to the average elastic modulus of the *M. sexta* wing membrane (2.446 ± 1.37 GPa) [7]. The SRT dry storage modulus is stable over a large range of temperatures (25 °C to 196 °C) and drops sharply above the denaturing temperature at 220 °C. This characteristic is essential in the reconstruction of the *M. sexta* membrane, giving SRT an advantage over many other biomaterials whose material properties vary significantly with temperature. Both the storage and loss moduli of SRT are constant as a function of frequency between 0.1 and 150 Hz [8][9]. This range includes the average *M. sexta* flapping frequency of 25.65 Hz [6].

The SRT protein also demonstrates 100% cohesion, i.e. subsequent deposits of the material bond completely to the existing protein structure. At about 180 °C, the elastic modulus of the material starts to drop off precipitously, allowing reshaping of the structure with minimal applied force. As the material cools, it recovers all of its original mechanical properties that it had before heat application. This behavior adds to the robustness and ease of fabrication of a protein-based insect wing.

3   Wing Design and Fabrication

A 2:1 scale replica mold was fabricated that carefully replicates the venation structure and membrane of a *M. sexta* wing. Using an optically scanned image of a *M. sexta* wing with a known pixel density, a program was written to generate the tool paths necessary to construct a mold for solution casting [7]. A Hurco VM1 CNC Mill was

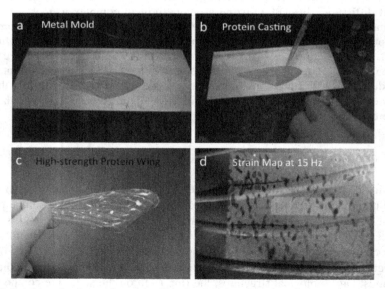

Fig. 1. Solution casting is employed for biomimetic wing manufacturing in an aluminum mold (a). The protein powder is dissolved in a polar-protic solvent (HFIP) and casted (b). A high strength film of 200 micron is obtained (c). Digital image correlation map (d) at 15 Hz (natural frequency of the moth wing at 2:1 ratio) shows a uniform distribution of strain.

used to machine the mold out of 6061 aluminum. The SRT protein was dissolved in hexafloroisopropanol (HFIP) and casted to form a 200 micron thick film. The recombinant protein wing in Figure 1 is lightweight (0.4 grams) and strong (modulus of 1 GPa). Vibrations were applied to the wing by means of a vibrator (PASCO Mechanical Vibrator- SF) at low amplitude (roughly 1–3 mm) between 1–100 Hz. The motion was recorded with a digital image correlation (Figure 1). The strain map taken did not show any strain localization in the tested frequency range, indicating that the elastic modulus remains constant regardless of vibration frequency.

4   Summary and Future Work

The goal of this project is to mimic the *M. sexta* wing as accurately as possible in order to increase the chance of success when constructing an efficient FWMAV. The initial attempt at fabrication shows that the SRT retains its material properties in the form of a *M. sexta* wing. Despite the complex shape and thin structure of the wing, the vibration tests showed that the strain was constant between 1 and 100 Hz, encompassing the mean *M. sexta* wing beat frequency (24.8 to 26.5 Hz). The mass of the scaled artificial wing is 0.4 grams. Dimensional analysis tells us that at a 1:1 scale, its mass would be 0.05 grams. This indicates that the average mass of the *M. sexta* forewing (0.035 grams) should be achievable using SRT. These results are promising and encourage further analysis to be performed.

The next step, already in progress, is to construct a mechanism to recreate the flapping motion of the moth in order to observe in-flight performance characteristics of the wing. This test stand will recreate the distinctive 3D flapping motion that gives the moth its maneuverability. The test stand will be based off of a crank rocker design and consist of a single 6 V, 3000 RPM DC motor that drives a sequence of linkages that are attached to two artificial moth wings. As the wings are flapped, data acquisition software will log the produced lift using a load cell. Concurrently, a stereoscopic system will be used to capture high speed video which will then be optically digitized to study the flapping behavior. Particle image velocimetry (PIV) will be implemented to observe the fluid-structure interaction as the wing is driven through air. This will ensure that the aerodynamics of the SRT wing are consistent with that of the hawkmoth. Successive design iterations will be performed using the aforementioned results. Additional testing will include using the artificial wing as a prosthetic on a live moth with the intention of proving successful construction of a *M. sexta* wing.

Insects often have the capability of changing the shape of their wings during flight and while resting or dormant. This capability provides additional control over flight trajectory and affords them the ability to protect their wings when not in use. The SRT protein complex allows for advanced concepts to be implemented in an attempt to morph an artificial wing. One of these concepts involves embedding a shape memory alloy (SMA) within the protein as it is casted into a wing. Preliminary work has already been conducted on this concept. Figure 2 shows a nickel titanium (Ni-Ti or nitinol) SMA being heated from 40 °C to 80 °C and the resulting wing deformation. The full deformation takes approximately 6 seconds to occur. In its current state, this reversible process could be used for ground transitions. However, a reduction in response time is necessary for this technique to be applicable to in-flight performance modifications. Numerous other advanced concepts will be investigated for use with the protein-based artificial wing and performance metrics determined to compare the efficacy of each.

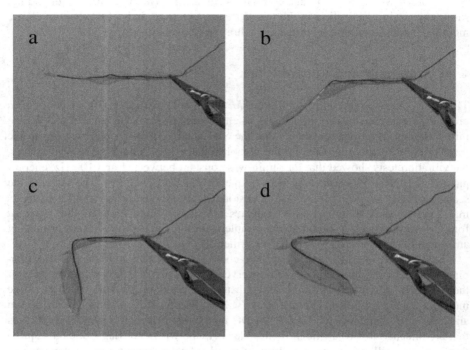

Fig. 2. Ni-Ti SMA embedded wing morphing progression

This work will further our understanding of insect flight and provide a method to precisely mimic the well-known flight behavior of the *M. sexta*. With this newfound ability to produce wings that exhibit the advantageous characteristics of the hawkmoth, new biologically-inspired FWMAVs can be designed to overcome current limitations.

Acknowledgements. This research was supported by the Office of Naval Research (N000141310595) and the Pennsylvania State University.

References

1. Ansari, S.A., Zbikowski, R., Knowles, K.: Aerodynamic modelling of insect-like flapping flight for micro air vehicles. Prog. Aerosp. Sci. **42**(2), 129–172 (2006)
2. Combes, S.: Materials, structure, and dynamics of insect wings as bioinspiration for MAVs. Encycl. Aerosp. Eng., 1–10 (2010)
3. Ennos, A.R.: Inertial and aerodynamic torques on the wings of diptera in flight. J. Exp. Biol. **142**(1), 87–95 (1989)
4. O'Hara, R.P.: The Material Property and Structural Dynamic Characterization of the Manduca Sexta Forewing for Application to Flapping Micro Air Vehicle Design. Air Force Institute of Technology (2011)
5. Chirarattananon, P., Ma, K.Y., Wood, R.J.: Adaptive control of a millimeter-scale flapping-wing robot. Bioinspir. Biomim. **9**(2), 025004 (2014)
6. DeLeon, N.E.: Manufacturing and Evaluation of a Biologically Inspired Engineered MAV Wing Compared to the Manduca Sexta Wing Under Simulated Flapping Conditions. Air Force Institute of Technology (2011)
7. O'Hara, R.P., Palazotto, A.N.: Species for the Application of Biomimetic Flapping Wing Micro Air Vehicles. Bioinspir. Biomim. **7**(4), 046011 (2012)
8. Pena Francesch, A., Florez, S., Jung, H., Sebastian, A., Albert, I., Curtis, W., Demirel, M.C.: Materials Fabrication from a Recombinant Thermoplastic Squid Protein. Advanced Functional Materials **24**(47), 7401–7409 (2014)
9. Pena Francesch, A., Akgun, B., Miserez, A., Zhu, W., Gao, H., Demirel, M.C.: Pressure Sensitive Adhesion of an Elastomeric Protein Complex Extracted from Squid Ring Teeth. Advanced Functional Materials **24**(39), 6227–6233 (2014)

Correlating Kinetics and Kinematics of Earthworm Peristaltic Locomotion

Elishama N. Kanu[1(✉)], Kathryn A. Daltorio[2], Roger D. Quinn[2], and Hillel J. Chiel[1]

[1] Department of Biology, Case Western Reserve University,
10900 Euclid Avenue, Cleveland, OH 44106, USA
enk18@case.edu
[2] Department of Mechanical and Aerospace Engineering, Case Western Reserve University,
10900 Euclid Avenue, Cleveland, OH 44106, USA

Abstract. The study of biological organisms may aid with designing more dynamic, adaptable robots. In this paper, we quantitatively studied the coupling of kinematics and kinetics in the common earthworm, *Lumbricus terrestris*. Our data correlates changes in worm segment shape to variable, non-uniform load distribution of worm weight. This presumably leads to variable friction forces. Understanding the way the worm exerts these forces may help us implement peristalsis in robots in diverse real-world environments. In our preliminary data, at the front of the worm, the segments with the widest diameter bear the most weight and anchor the worm to the ground during motion, as we hypothesized. The rear segments also exhibit variation in ground reaction forces. However, for rear segments, the peak kinetic waves are phase-shifted from the kinematic waves. Future work will explore this phenomenon.

1 Introduction

Biological systems display profound mastery in balancing both compliance and strength. This mastery, exhibited even in simple organisms, remains a challenge for robotic design. The soft-bodied common earthworm, *Lumbricus terrestris*, for example, can bend and contort its body to navigate its terrain and squeeze into narrow, constrained spaces. At the same time, the earthworm can exert forces radially and laterally against its environment to break up compacted soil, create and enlarge burrows, and resist extraction from its burrow by predators. Studying these abilities may prove useful for designing dynamic, flexible yet resilient robots (see companion paper [3]) that can help with tasks where high levels of adaptability are needed. Such applications range from endoscopy to navigating through debris (like searching through a collapsed building, for example).

Earthworms move by a mechanism known as peristalsis, a wave of contractions that move posteriorly down the worm body. The contracted segments increase in diameter and serve to "anchor" the body to the ground, while the segments anterior to them protrude forward by increasing lengthwise and decreasing radially [4]. If the force each segment exerts on the ground upon contact is equal to the weight of that segment, then the speed and efficiency of forward motion are constrained by the worm's ability to decrease its friction coefficient [1],[4]. However, hydrostatic

© Springer International Publishing Switzerland 2015
S.P. Wilson et al. (Eds.): Living Machines 2015, LNAI 9222, pp. 92–96, 2015.
DOI: 10.1007/978-3-319-22979-9_9

constraints dictate that as each segment contracts in length, it must increase in diameter. Thus, if the body is stiff enough, this change in shape can result in a non-uniform weight distribution, with the larger diameter (anchoring) segments supporting more of the weight. Ideally, if the anchoring segments carry all of the weight, then the moving segments will not be acted upon or slowed by any Coulomb friction. This therefore results in more efficient movement, as the moving worm will not lose any energy to friction (similar to a rolling wheel). Simulating such kinematics has been shown to allow a flexible robot to move on flat ground as well as in constrained environments, such as in a narrowing pipe [2].

Peristalsis thus consists of kinematic and kinetic waves whose timing and synchrony will determine the effectiveness of the motion. Quillin [5, 6] separately studied the effects of worm size on kinematic wave speed and kinetic burrowing force. Here, we further explore how the kinematics of wave contractions corresponds with changes in segment shape (diameter which is coupled with length) to produce non-uniform variability in load distribution of weight, and friction forces relative to the ground. We expect that, in order to minimize frictional drag losses, the segments with the widest diameter, being the "anchoring" segments, will carry the most of the load and consequently exert the maximum observed force normal to the ground.

2   Methods

Lumbricus terrestris (earthworms) were ordered from Carolina Biological suppliers, and stored in a moist, dark bin filled with soil and with proper ventilation and drainage of excess fluids. The worms were fed regularly with organic matter (apples, carrots, and leaves). Before an experiment, the earthworm was washed with distilled water, carefully blotted dry with a lab wiper, and weighed.

To capture peristaltic kinematics, an earthworm was set on a smooth, uniform, Plexiglas surface. An HD camera was mounted directly above the setup such that the entire length of the worm fit inside the field of view. A light source was provided to reduce shadows. The earthworm was gently prodded with a cotton applicator tip to initiate motion and was then allowed to move horizontally across the substrate freely and without further prodding.

To record kinetics, the experiment was modified to include a force transducer (Grass Technologies, Model FT03). The setup consisted of two Plexiglas sheets with an aluminum bridge in between for the worm to cross over, making sure the sheets and the bridge were level with each other. The bridge was hung directly beneath the force transducer, with taut suture strings, so the force transducer measured the weight of the bridge and the vertical forces exerted by the worm. Two aluminum pillars were placed on each side of the bridge, to keep it from swaying laterally. Care was taken not to have either the Plexiglas or pillars directly pressing against the bridge, so as not to contribute additional forces. An HD camera was mounted directly above the setup, and with a light source to reduce shadows. At the end of each trial, a weight of 0.70 -g was dropped onto the bridge to align the force and video data. Trials in which the worm brushed against the strings supporting the bridge were discarded.

The videos were analyzed with Matlab to determine the worm diameter, shape and position for each video frame. First, we thresholded each color of the image to separate the worm from the background. After filling in "holes" in the body due to reflections, we determined the largest inscribed circle and, from the center of that circle, the largest chain of adjacent circles along the body of the worm (Figure 1). The diameters of these circles provide the overhead width of the worm at a string of points along the length of the body. If the cross section of the worm can be approximated as a circle, the overhead width will be equal to the segment diameter. The force data and visual data were synchronized by aligning the video-recorded sound of the weight falling with the impact force of the weight falling on the bridge. Since the force transducer records at 5,000 Hz, and the video camera at 30 Hz, the mean force over each video frame was used.

Fig. 1. At every video frame the largest circles inscribed by the worm's body were determined and plotted. Simultaneously the force transducer recorded the portion of the weight distribution supported bythe aluminum bridge. Here at 83.5 seconds, which corresponds to the first peak force in Fig. 3, worm diameter is larger on the bridge than at the adjacent Plexiglas plates. The red diamond is the center of the largest circle (63 pixel radius, colored orange). We hypothesize that by supporting more weight at large diameter segments, small diameter segments should be freer to move.

3   Results

In a trial in which the motion of the worm was largely linear and horizontal across the video frame, the horizontal position of each of the inscribed circles (corresponding to points on the worm body) is shown at each frame, color coded by diameter (Figure 2). As reported in other studies [5, 6], waves travel posteriorly down the body. The anterior segments tend to be larger than the posterior segments, both when contracted and relaxed. However, this pattern is interrupted by the clitellum, a reproductive organ located about one third of the way from the head. This is the widest part of the worm, during motion and while stationary, so it can be used as a reference point on the body, since we did not identify individual segments. From this plot, we can see that for the tip, tail and clitellum the greatest motion occurs when the diameter is least. For both anterior and posterior sections of the worm, the minimum diameter is 50–60% of the maximum diameter. Transition between the two extrema happens relatively rapidly. The clitellum, for example, is stationary and has a large diameter for 60% of the frames. In 30% of the frames, the diameter is small and moving horizontally. Transitioning between the two states occurs for the remaining 10% of frames.

In a representative kinetic trial in which the worm crossed the bridge in one conti-nuous motion, without pausing or traveling backwards, we found that maximum diameter and maximum force were only synchronized at the first registered peak force. After, the peak force and diameter were found to be 180 degrees out of phase with each other (Figure 3). In two succeeding analyzed videos, the observed trends remained the same.

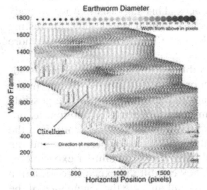

Fig. 2. An earthworm crawling linearly from right to left on a Plexiglas surface. A regular spatiotemporal pattern of the propagation of the peristaltic wave can be seen. The anterior segments increase to a larger diameter than posterior segments. The clitellum, containing the largest segments, can be seen about one third of the length from the front as a larger di-ameter band from adjacent areas. A new wave is propagated when the existing wave travels three-fourths down the body.

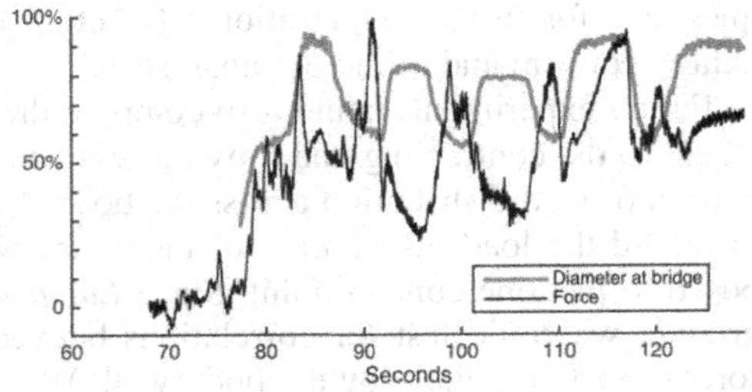

Fig. 3. Forces (black) registered as the worm crossed the force transducer were plotted with the diameter of the segment on the force transducer. The maxima for the force and diameter were in phase at the first peak (shown in Fig. 1), but fol-lowing after became 180 degrees out of phase with each other until the fifth peak. At the end of the trial the worm remained stationary on the bridge.

4   Discussion

We hypothesized that peristaltic wave contractions would result in changes in segment shape (such as increases in diameter and decreases in length), which would consequently lead to a variable weight distribution. The contracting "anchor" seg-ments, with wider diameter, we proposed, would carry most of the load distribution and result in maximum friction forces. The increased frictional force would in turn aid in gaining adequate traction to properly stabilize and anchor the body to the ground. Simultaneously, this would result in decreased friction in the moving segments with smaller diameter, to aid in easier and more efficient movement, without losing energy to ground resistance forces.

Our data suggests that both segment shape and weight distribution are variable and nonuniform and coupled in some way. However, the worm did not take advantage of these kinematics to reduce friction for many moving segments. The smallest diameter segments often bear more weight than larger segments at the rear, the opposite of our predictions. Possible reasons could be that since *Lumbricus terrestris* are adjusted to living in vertical burrows, the anterior worm segments may not be sufficiently stiff or biologically designed for horizontal walking. It may also be that the front of the worm is actually the stronger part, which would aid in creating new burrows and exploring obstacles. There may be other reasons for utilizing a gait that is not energetically efficient in horizontal movement on flat, smooth surfaces. This suggests that worm body design and control may have many unknown factors, which raises new, interesting questions for further exploration. In future trials, these methods will provide more data to confirm and refine our analysis.

Future experiments could also compare the weight distribution in the segments adjacent to the contracting and moving segments, in order to gain a better picture of the pattern of load distribution across the body. Perhaps in the future it would be possible to record the load distribution of the entire worm body while it is in motion, as opposed to just one contact point. Since *Lumbricus terrestris* burrow vertically into the ground, we could test for correlations between the kinematics of peristalsis and the forces exerted radially by the body wall. We could also carry out similar experiments on worms that make horizontal burrows, to see if they exhibit the expected correlation between kinetics and kinematics. Additionally, we could analyze the kinematics of the worm turning or climbing.

From this we can design more dynamic robots that are able to uniquely vary load distribution and traction across the body. This will result in robots able to successfully navigate a variety of heterogeneous environments by adjusting their kinematics and kinetics through peristalsis.

References

1. Alexander, R.McN: Animal Mechanics, 2nd edn. Blackwell Scientific Publications, Oxford (1983)
2. Daltorio, K.A., Boxerbaum, S.A., Horchler, D.A., Shaw, K.M., Chiel, H.J., Quinn, R.D.: Efficient worm-like locomotion: slip and control of soft-bodied peristaltic robots. Bioinspiration & Biomimetics **8**(3), 035003 (2013)
3. Horchler, A.D., Kandhari, A., Daltorio, K.A., Moses, K.C., Andersen, K.B., Bunnelle, H., Kershaw, J., Tavel, W.H., Bachmann, R.J., Chiel, H.J., Quinn, R.D.: Worm-like robotic locomotion with a compliant modular mesh. In: Proc. Living Machines, July 28-31, 2015
4. Miller, G.S.P.: The motion dynamics of snakes and worms. Computer Graphics **22**(4), 169–178 (1988)
5. Quillin, K.J.: Kinematic scaling of locomotion by hydrostatic animals: Ontogeny of peristaltic crawling by the earthworm Lumbricus terrestris. Journal of Experimental Biology **202**, 661–674 (1999)
6. Quillin, K.J.: Ontogenetic scaling of burrowing forces in the earthworm Lumbricus terrestris. Journal of Experimental Biology **203**, 2757–2770 (2000)

Visualizing Wakes in Swimming Locomotion of Xenopus-Noid by Using PIV

Ryo Sakai[1]([✉]), Masahiro Shimizu[2], Hitoshi Aonuma[3], and Koh Hosoda[2]

[1] Graduate School of Information Science and Technology,
Osaka University, 2-1 Yamada-oka, Suita 565-0871, Japan
ryo.sakai@arl.sys.es.osaka-u.ac.jp
[2] Graduate School of Engineering Science, Osaka University,
1-3, Machikaneyama, Toyonaka 560-8531, Japan
{shimizu,hosoda}@arl.sys.es.osaka-u.ac.jp
[3] Research Institute for Electronic Science, Hokkaido University,
Kita 8 Nishi 5, Sapporo 060-0808, Japan
aon@es.hokudai.ac.jp

Abstract. Frogs can swim adaptively in the water dexterously utilizing interactions between body biomechanics and fluid dynamics.
We have been developing an aquatic frog robot, Xenopus-noid, which has similar musculoskeletal structure as its biological counterpart, *Xenopus laevis*. This robot allows us to study the interaction between the biomechanical structure of the frog and the fluid dynamics during swimming locomotion in a natural context. In this report, particle image velocimetry (PIV) is used for visualizing wakes generated by the Xenopus-noid. Experimental results demonstrate that the robot can produce appropriate wakes for swimming if we implement a rigid beam that mimics the function of the Semimembranosus (SM) muscle in *Xenopus laevis*. The function is utilized for proper posture, that is to say, this muscle prevents a hyperextension of the knee.

Keywords: Xenopus-noid · Visualized wakes · PIV · Swimming locomotion, *Xenopus laevis* · Musculoskeletal structure · Fluid dynamics

1 Introduction

Frogs achieve locomotion on land and in water by utilizing its body dynamics. The swimming locomotion of frogs is governed by both body biomechanics and fluid dynamics. In previous studies, Richards et al. developed a bio-robotic platform that controls rotation of the robotic foot imitating movement of an aquatic frog, *Xenopus laevis*, and revealed that the shape of the frog foot contributes to the generation of propulsion forces[1]. Richardson et al. revealed that the musculoskeletal structure of a hind limb of a frog stabilizes movement of a leg[2]. Frogs can swim adaptively by utilizing not only the shape of the foot but also the whole musculoskeletal structure of a hind limb.

© Springer International Publishing Switzerland 2015
S.P. Wilson et al. (Eds.): Living Machines 2015, LNAI 9222, pp. 97–100, 2015.
DOI: 10.1007/978-3-319-22979-9_10

Therefore, we have been developing Xenopus-noid to clarify the importance of the biomechanical structure for swimming locomotion of *Xenopus laevis*[3]. It is a swimming robot whose hind limb is imitating the structure of *Xenopus laevis*, and is driven by dissected living muscle of this frog. By changing the musculoskeletal structure of the robot, we can investigate relationship between biomechanics of *Xenopus laevis* and fluid dynamics.

In this paper, we measure the wakes generated by a Xenopus-noid equipped with the SM muscle by PIV[4]. Then, we removed the SM muscle of the robot and measured the wakes again. We could observe that Xenopus-noid with the SM muscle produced the wakes. We understand that the SM muscle has the function, which avoids a hyperextension of the knee.

2   Method

Adult South African clawed frogs, *Xenopus laevis* were obtained from a commercial supplier (Hamamatsu Seibutsu Kyozai Ltd., Shizuoka, Japan). Animals were doubly pithed and decapitated. The Plantaris longus (PL) muscle with tendons and the ischiadic nerve was dissected out in Ringer solution at 20°C and pH 7.0. Xenopus-noid has crinkles on its fumer and its foot for clamping the Achilles tendon and a part of the tibiofibula. We fixed the PL muscle to Xenopus-noid[3] by clamping the Achilles tendon and the part. In this study, we focus on the SM muscle, which originates from the body trunk and inserts at the tibiofibular in *Xenopus laevis*. We hypothesize that the SM muscle works as a rigid link when the PL muscle contracts, which enables *Xenopus laevis* to kick backward efficiently. The PL muscle mainly generates propulsion force in swimming locomotion[5].

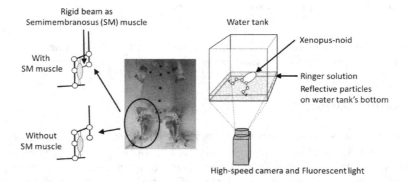

Fig. 1. The left side: Xenopus-noid with/without the rigid beam replacing SM muscle function. We recorded swimming locomotion of the robot with/without the rigid beam to check a difference of wakes generated during kicking. The right side: experimental setup of our PIV system. We can visualize the instantaneous flow changes by monitoring the degree of reflection of particles immersed in the water when wakes are generated.

The left side of Fig. 1 shows the robot with/without the rigid beam mimicking the function of the SM muscle. We recorded swimming movement of the robot with/without this beam to observe a difference of wakes generated during kicking. We utilized PIV to visualize these wakes. The right side of Fig. 1 shows the experimental setup of our PIV system. The water tank which is made of glass was filled with Ringer solution, containing reflective particles, DIAION Sepabeads HP20 (Mitsubishi Chemical Co., Tokyo, Japan). There is no flows in the water tank before swimming locomotion of the robot. The swimming locomotion of the robot was recorded through the bottom of the water tank at 200 frames per second with a high speed video camera HAS-L1 (DITECT Co.Ltd., Tokyo, Japan). The bottom of the water tank was illuminated by a fluorescent light. We used the visualizing software Flownizer2D, which can measure the robot's wakes (DITECT Co.Ltd., Tokyo, Japan). The interrogation area was 32 × 32 pixels and the search area was 52 × 52 pixels. This resulted in 7931 vectors per image. In this paper, we visualized the magnitudes of the current velocity vectors of the wakes generated by swimming locomotion of Xenopus-noid.

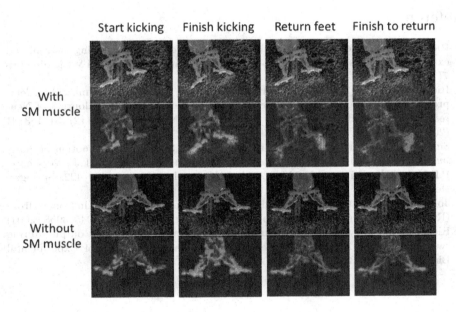

Fig. 2. The magnitudes of the current velocity vectors of the robot's wake. There is not the magunitude of the current velocity vectors in blue area. Red area shows the magunitude of the current velocity vectors. "Start kicking" phase is that the robot begin to move its feet. "Finish kicking" phase is that the PL muscle stops the contraction and the robot finish to row by the feet. "Return feet" phase correspond to the phase when the feet start returning to the ignition position due to extension of the PL muscle. The "Finish to return" phase captures the moment when the feet return completely to the initial position. The robot with the SM muscle could swim and could generated the wake from the "Return feet" phase to the "Finish to return" phase, however the robot without the SM muscle could not.

3 Result

Figure 2 shows the magnitudes of the current velocity vectors of the wakes generated by swimming locomotion of the robot with/without the SM muscle. The robot with the SM muscle swam forward and generated the wake. From "Finish kicking" phase to "Finish to return" phase, the wake arose near the feet. The robot without the SM muscle couldn't swim and couldn't generate the wake. We confirmed that the coordination between the motions of the ankle and that of the knee transforms kicking motions of the robot and effects wake generation. Also, we observed that the SM muscle provides mechanical stabilization and supports the efficiency of swimming locomotion.

Acknowledgments. This work was supported partially by Grant-in-Aid for Scientific Research on 24680023, 23220004, 25540117 from the Ministry of Education, Culture, Sports, Science and Technology of Japan.

References

1. Richards, C.T., Clemente, C.J.: A bio-robotic platform for integrating internal and external mechanics during muscle-powered swimming. Bioinspiration & Biomimetics **7**(1), March 2012
2. Richardson, A.G., Slotine, J.J.E., Bizzi, E., Tresch, M.C.: Intrinsic musculoskeletal properties stabilize wiping movements in the spinalized frog. The Journal of Neuroscience : The Official Journal of the Society for Neuroscience **25**(12), 3181–3191 (2005)
3. Sakai, R., Shimizu, M., Aonuma, H., Hosoda, K.: Swimming locomotion of *xenopus laevis* robot. In: Duff, A., Lepora, N.F., Mura, A., Prescott, T.J., Verschure, P.F.M.J. (eds.) Living Machines 2014. LNCS, vol. 8608, pp. 420–422. Springer, Heidelberg (2014)
4. Johansson, L.C., Lauder, G.V.: Hydrodynamics of surface swimming in leopard frogs (Rana pipiens). The Journal of Experimental Biology **207**(Pt 22), 3945–3958 (2004)
5. Richards, C., Biewener, A.: Modulation of in vivo muscle power output during swimming in the African clawed frog (Xenopus laevis). Journal of Experimental Biology **210**(18), 3147–3159 (2007)

Biomimetic Approach for the Creation of Deployable Canopies Based on the Unfolding of a Beetle Wing and the Blooming of a Flower

Giulia Evelina Fenci$^{(\boxtimes)}$ and Neil Currie

Directorate of Civil Engineering, University of Salford, Salford, UK
g.e.fenci@edu.salford.ac.uk, n.g.r.currie@salford.ac.uk

Abstract. Modern architectural designs create dynamic and flexible spaces, able to adapt to the ever-changing environment by virtue of temporary and convertible structures. Biomimetics is the applied science that, through the imitation of Nature, finds the solution to a human problem. The unfolding of a beetle wing and the blooming of a swirl flower were recognised as having outstanding features to be mimicked for the creation of deployable canopies. This paper focuses on the analysis methodology of the two biomimetic, deployable structures with multiple degrees of freedom. The general validity of a pseudo-static analysis was proved based on time-stepping the geometry at set deployment stages with optimisation of multiple, potential deployment sequences.

Keywords: Deployable · Morphology · Biomimetic · Beetle wing · Flower blooming

Introduction

Since the beginning of time, humans have used Nature as an inspiration by observing how it behaves and operates in both its simple or more complex mechanisms. For example, da Vinci was interested in making flying machines and, when he was creating his designs, they were the result of the direct observation of Nature, especially the flight of birds of which he produced his own illustrations representing their wing bone conformation [1]. More recently, within the discipline of robotics, solutions are often inspired by the functioning of man's articulations [2]. This conscious process of drawing inspiration from Nature is known as *biomimicry* [3] or *biomimetics*, defined by Santulli [4] as the approach to finding the solution to a problem through the observation of Nature, transferring sustainable solutions from Nature to technology [5].

Biomimicry in architecture is increasingly becoming more popular, opening the way to new, functional, and sustainable possibilities [6]. Nature is typically in a state of continuous motion, sometimes imperceptibly, but always evolving in response to its environment. Thus, in the sphere of structural engineering, where structures are commonly designed to be rigid (monuments or functional buildings) biomimetics has the potential to introduce the notion of movement. An example is offered by articulated facades that respond to the position and intensity of the sun in the same way as a

© Springer International Publishing Switzerland 2015
S.P. Wilson et al. (Eds.): Living Machines 2015, LNAI 9222, pp. 101–112, 2015.
DOI: 10.1007/978-3-319-22979-9_11

leaf would. Hence, the need for *deployability* to confer the power to vary the form and function of a system by a series of manipulations [7] occurring in an autonomous and safe manner [8]. Under an architectural point of view, folding and retracting may be employed as a means to a morphological solution creating a habitable structure that accommodates pragmatic and aesthetic evolution.

Literature Review

In this study, two examples offered by Nature from both the plant and animal world have been taken into consideration: the unfolding of an insect's wing and the blooming of a swirl flower. Extraordinary features have been identified in the hind wings of beetles. They are aerodynamic force generators and can be folded and tucked under the elytra, a protective forewing, multiple times without any structural degradation [9]. The unfolded wing can be up to ten times the area of its packed configuration through the use of a complex folding pattern [10]. Conversely, plants are constantly seeking space and light in order to live and a process of growth drives their deployment, but this motion occurs using different principles and at a different rate compared to the beetle's wings. Expansion and cantilevering are necessary to withstand gravity, wind, and other external factors. In most plants, the flowers or leaves are folded in order to fit in the small buds, for example hornbeams leaves possess a regular folding pattern similar to a Miura Ori origami pattern [11] that controls the motion of the surface by creating a single degree of freedom deployable surface.

Wing Folding

Insects' wings are extremely flat limbs, developed from the expansion of the meso- and metathorax. Research on beetle wings carried out by Wootton [12] explains how wings are living organs in which the haemolymph runs through special ducts called veins. The veins running along the wings are cuticular tubes that are sclerotized (hardened and darkened) on one or both sides of the wing surface. Veins provide hemolymph circulation and also protect parts of the insects' body, such as the trachea or the nerves from potential damage [13]. Direct, indirect, and accessory muscles mainly govern wing movement. The hemolymph running within the veins and the fact that such circulation may aid the wings' folding and unfolding movements recalls the concept of pneumatic structures. The common feature consists in the structure deploying along the dominant veins or creases due to the pressure added or subtracted to the internal space, being the increase/decrease in pressure a consequence of the varying amount of air or liquid flowing in the ducts.

The veins confer structure to the alar limbs, providing the correct rigidity to flexibility ratio. The costa vein, which runs down the anterior edge, often forms an enlargement, called *stigma* that has a mechanical function associated with the strength distribution during flight [14].

Fig. 1. Pacific Beetle-Cockroach Wing [15]

Fedorenko [16] explains how all longitudinal veins are connected by means of cross-veins that enhance the stiffness of the wing-supporting carcass forming a kind of grillage structure. Some veins also act as braces at the location of the pivots or joints by restraining the displacement of the main axial elements relatively to each other.

The reasons for focusing on the wings of cockroaches, in particular those of the species known as *Diploptera Punctata*, are the following:

- The simplicity of the geometry, which corresponds to a straightforward deployment sequence.
- The wing performs a significant area reduction when passing from the un-folded to folded state. In fact, about half of the wing folds back on itself along the median flexion line which experiences a joint at mid span.
- The areas in which the wing is divided do not bend, but remain stiff throughout the deployment. The material constituting the spanning sections does not have to be flexible and the veins could be idealized as rigid joints around which stiff plates are able to rotate.

Flower Blooming

Flower blooming occurs mainly to expose the colourful petals and attract pollinators, allowing for reproduction. In angiosperms (flowering plants) different reproductive systems can be found, each related to different blooming mechanisms [17]. In many flowers the opening and closing mechanism is governed by differential cell growth. Species that bloom in the morning undergo a rapid growth in length in the inner part of the petals due to the temperature rise, while the outer part is not greatly affected. However, the opposite will occur later in the day once the temperature cools down, forcing the flower to close. Such a mechanism can be observed in tulips, as described by Pfeffer [18]. In other flowers the mechanism is governed by movements of the midrib, as occurs in Morning Glory flowers [19], shown in Fig. 2.

Different kinds of flower exist, each one with their own distinguishing characteristics. For example Chrysanthemums, posses a large number of small petals, which make for a relatively large bud. Tulips, on the other hand, have fewer individual petals and blooming is mainly governed by a growth process rather than by unfolding of tightly packed petals. There are flowers equipped with only two petals, symmetric about the axis along which they are joined, like a Venus Flytrap.

Fig. 2. a) Front View [20]; b) Side View

After an in depth investigation of flower morphology, the advantages of mimicking a swirl flower for the design of a deployable awning are:

- The lesser number of petals allows for the creation of simple and effective geometry, bypassing surplus of members, thus, increasing the ability for the structure to pack tighter.
- The continuity of the petals, forming a continuous membrane connecting the petal ribs, implies dependency between the petal ribs themselves. One rib will affect the other only by tension (membranes have no compressive resistance).
- The additional twisting characteristic of the deployment enables compact packing, leading towards the design of the structure being relatively small when folded.

Consequently, this research has focused on a member of the Ipomoea family: the Morning Glory Flower (*Ipomoea Violacea*), whose petals are arranged in a conical manner. With its green oval leaves, its distinguishing bell shape flowers create a dynamic sense of movement with the petals, typically being blue or purple [21].

The bud of a Morning Glory flower resembles the shape of a spindle and is characterized by a spirally striated pattern. When blossoming, the spiral pattern loosens and the petals are allowed to unfold and stretch. Each flower has 5 petal units, which consist of a petal rib (a relatively stiff central part providing support, see Fig. 2b) and a thin membrane conferring the petal its shape. The portion of the flower that connects to the stem extends through a series of compatible petal ribs to form an elongated hollow tube. Research by Watanabe et al. [22] regarding Morning Glory flowers has brought to the conclusions that the petal rib is approximately three times stronger and more rigid than the petal itself due to the different cell types. The ribs have slender and reinforced cells, while the petals are composed of round and isotropic cells. An increase in the number of petal ribs results in an increased thickness of the flower bud due to the overlapping of petals. Nature's packaging logistics is also evident in maple leaves, where the slight angle discrepancy avoids overlapping of the main veins when in the fully folded state [23]. However, the energy required for unfolding is inversely proportional to the number of ribs, hence, a compromise between geometrical and mechanical factors is found within Nature.

A Morning Glory flower is a sympetalous flower as the petals are not separate individual elements, but are interconnected and continuous over the stiff ribs, like a duck's webbed feet have membrane spanning between their toes. Investigating the blooming and stretching of these kinds of flowers has concluded that a constant Gaussian curvature is imposed on the tissue by growth. Hence, the bell shape of these flowers can be expressed in mathematical form allowing to find that the optimum number of veins (petal ribs) to minimise elastic energy is four, five, or six [24].

Methodology

For this project, Nature was imitated with the purpose of creating deployable canopies, able to modify habitable space based on weather conditions or intended use, creating a flexible and responsive environment as a result. Physical and geometrical models were developed to express the changing geometry through the deployment process. The focus was not on the faithful imitation of every aspect of the mechanisms selected but, rather, on singling out advantageous characteristics, which could be replicated in the design of deployable awnings. Thus, some of the principles are similar, however, the application is different from the purpose of flight and that of pollination.

The deployment sequences were time-stepped: the starting, tightly packed position was classed as zero percent deployed (folded) and, through a set of carefully controlled motions, the form evolved in incremental controlled steps until the structure reached its fully deployed state (100% deployed). This time-stepping approach was primarily a geometrical exercise using translational and rotational manipulations to determine the new position of the structure. The larger the number of degrees of freedom, the greater the number of potential deployment sequences. Numerous deployment sequences were established geometrically (for example, over 140 unique deployment paths were analysed for the flower). At each deployment stage the structure was equated to a static model in AutoDesk ROBOT 2014, with the dynamic aspects of the moving masses discounted through slow, steady, controlled motions during deployment. This resulted in a large number of pseudo-static models that were elastically analysed with the geometrical input controlled via an Excel spreadsheet to iterate the motion through the application of translational and rotational matrices.

The Beetle Wing

In order to mimic the mechanism of the unfolding of the wing of a *Diploptera Punctata*, a model composed of four triangular plates was created, based on the Miura Ori fold that is present in many beetle wings. According to Pellegrino [8], opposite angles of 96° and 84° offer the most efficient arrangement between packaging efficiency and deployment coupling. The wing was discretised into 4 triangles and 8 boundary struts, as shown in Fig. 3.

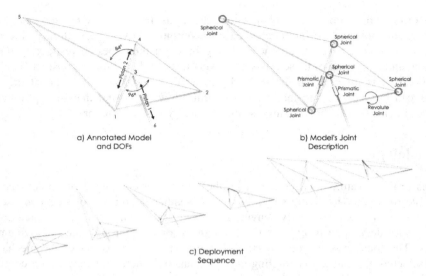

a) Annotated Model
and DOFs

b) Model's Joint
Description

c) Deployment
Sequence

Fig. 3. Deployable Model Inspired by Beetle Wing

Fabric material covers each triangle, providing protection. The triangles behave as rigid-foldable origami [25] creating a purely geometrical mechanism that does not depend on the material's elasticity. The strut spanning between joints 1 and 2 never changes location, but only rotates around its local x axis while the others all take active part in the deployment sequence. Numerous deployment sequences are possible, one of which is shown in Fig. 3. The structure unfolds to an area up to three times greater than the retracted configuration.

Each node was idealised as a pin or spherical joint so that the struts, spanning between them, were able to rotate, acting as revolute joints themselves. Pin joints are beneficial because they do not develop bending moments embracing the ethos of lightweight structures being formed by only tension and compression elements.

Two pistons provided the means of deployment and introduced two controlled degrees of freedom. Pistons are planar joints each with a single degree of freedom; they move in a single longitudinal direction and do not develop bending. Piston 1 is fixed between node 1 and 6 (point 6 does not belong to the beetle wing and acts as an anchor point for the piston), while piston 2 spans between node 1 and 4. The pistons' location was chosen in order to maintain the triangular relationship within the model. Piston 1 governs the orientation of the whole structure while piston 2 controls the main deployment movement. The pistons' ends cannot be at the same level otherwise the structure would not be able to unfold fully. The slight remaining slope to the surfaces also presents a benefit of creating a free draining surface. If a flatter structure was required, nodes would need to be offset.

The struts were modelled as Circular Hollow Sections (CHS) as the bending resistance of a CHS is uniform throughout regardless of local axis unlike a flanged section, preventing torsion problems developing as the sections rotate during deployment. Finally, a circular section is aesthetically more pleasing compared to other section types and such a factor becomes significant if the elements are going to be visible, as in this case.

The Flower

A physical and digital model of a structure mimicking the form and motion of the blooming of a Morning Glory flower were created after the observation of fast motion videos. From the literature, four struts are efficient and will enable a small closed configuration, ensuring equilibrium, as too few struts would compromise the structural integrity of the canopy when fully deployed. Fabric spans between the struts and accommodates the deployment while unfolding with the structure. The structure has three degrees of mechanical freedom: the rotation of the lower plate θ, the vertical translation of the plate D, and the change in length of the pistons $R2$.

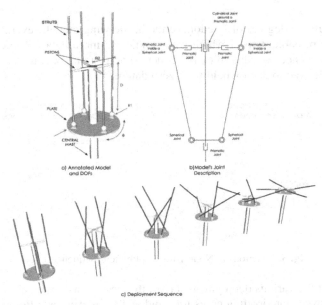

Fig. 4. Deployable Model Inspired by Flower Blooming

Fig. 4 shows the degrees of freedom present within the structure, the kind of joints necessary for the deployment to occur, and one of the numerous potential deployment sequences. The complexity of the joints belonging to the flower model compared to those of the beetle wing highlights how an additional degree of freedom can significantly change the characteristics of a structure. The struts, representing the flower's midribs, were modeled as CHS's for the same reason as for the beetle wing. The rotation of the sections around their longitudinal axis must be kept into consideration when finding a way to attach the lightweight fabric to the struts. Some freedom must be allowed between the connections of the fabric to the struts to avoid wrapping and tearing of the tensile material that will provide protection from the environment once the structure is fully deployed.

As for the beetle wing model, the analysis was purely geometrical and some deployment stages considered would not be achievable in practice due to the structural elements' volume and their interaction. An example is the 100% lift stage where, in reality, the lower plate would always be offset from the piston's location and the struts would never be able to reach a fully horizontal position. Such scenarios are limit cases, which would be avoided in reality, but first require the establishment of steel section sizes and detailed design of the connection to determine the true working range of the structure. The structure in this model is theoretically capable of deploying from 0% to 100%, but through limitations brought about by the steel sections having a real thickness, this range may reduce to 5% to 95%. Whilst this limitation is important, it shows that the analysis approach is valid for all steelwork section sizes that can be considered.

Results

Due to the multiple degrees of freedom of the beetle wing model, several deployment sequences are possible, two of which were studied in more detail. The choice of the deployment sequences was based on the structure performing one movement at a time and was not allowed to deploy below the zero datum.

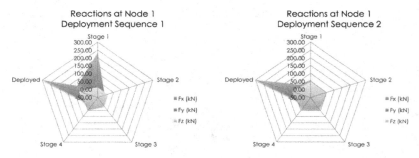

Fig. 5. Reactions at Node 1 for Deployment Sequence 1 & 2

Throughout the various deployment stages, the struts resist their self-weight and in the fully deployed state loading cases for wind and snow were additionally taken into account. Each element develops different axial, shear, and bending moments that were represented by means of radar plots. The example in Figure 5 shows the graph for the reactions occurring at node 1. It is to be noted that the first stage is different for each sequence due to having discounted the fully folded configuration for geometrical reasons: overlapping of the steel sections.

On the other hand, the flower model output was best presented by means of contour plots. Due to the complexity of the analysis, 2D plots resulted in being inappropriate as limited to one R1/R2 ratio and one degree of freedom.

The horizontal axis represents the rotation of the plate and the vertical axis shows the elevation of the plate along the central mast. Each contour plot allows viewing the change in the reactions at all deployment configurations for one R1/R2 ratio. Stacking the graphs for each ratio enables a further parametric comparison of the structure's behaviour, making the analysis three-dimensional. By studying the differences between the contour plots for every reaction, force, and displacement that the structure undergoes, the most efficient deployment routine can be found for the realisation of a deployable structure with maximum efficiency.

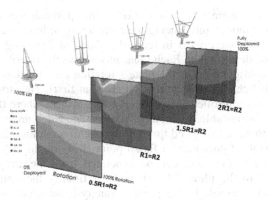

Fig. 6. Vertical Reactions on Plate

Discussion

In this research project two mechanisms occurring in Nature were identified as being advantageous for the creation of deployable canopies. Nature offers numerous examples of mechanisms, which have survived over others due to being the most efficient and their imitation has proven beneficial. The significance of architecture changing its configuration in space is that it enables the use of the same volume for different purposes and adapts to environmental conditions. Deployable structures tend to be lightweight so that the motion can occur with limited energy consumption.

The resistance opposed by air as a structure unfolds is proportional to the speed of motion. Hence, the velocity of deployment must be slow and gradual avoiding the formation of high pressures on the cladding material. Similarly, the inertial effects generated by the movement and the self-weight of the structural elements can be considered negligible, negating juddering, and, thus, a pseudo-static analysis method becomes appropriate for investigating the performance of the structure.

The analysis should consider in service loads as well as the deployment forces. The method adopted for the analysis is referred to as pseudo-static as the deployment sequences were divided into stages, each one of which was statically analysed. The analysis of each deployment stage aids the identification of the governing forces and allows the structure to be designed for the worse case scenarios. The aim of such an analysis is to determine the most efficient deployment sequence allowing for the optimisation of the structure based on parameters ranging between overall weight of the structure and total energy necessary for deployment.

With regards to the beetle wing, however, first the deployment sequences were chosen and then the static stages identified accordingly. The design of the structure is, thus, limited to the predetermined deployment sequences and the only form of optimisation that can occur is in the decision of choosing one sequence over the other. Even if only two deployment sequences were considered for the wing model, the process resulted in being time consuming. In fact, the geometry of the wing, although comprising of four triangles, proved to be complex and each stage had to be specifically identified.

Once the limitations were recognised, a different approach was employed for the analysis of the flower model, introducing the complexity of having a further degree of freedom. Contour plots were used instead of radar plots, starting with considering only 2 degrees of freedom (DoFs). Each radar plot shows the distribution of the forces for every possible configuration of the plate's rotation and translation along the central mast (the pistons' change in length was ignored at first). The structure's geometry at different stages was controlled by means of a spreadsheet where the location of the nodes is dependent on variable deployment parameters. Such an approach allows all deployment stages to be investigated through the use of a parametric approach. There is a starting point (fully folded) and a finish point (fully deployed) and the potential path to go from one to the other can be generated from the array of intermediate stages provided within the plots. The chances of finding the optimum solution is greatly increased with respect to the previous analysis adopted for the beetle wing model.

The inclusion of the third degree of freedom –the pistons' change in length– was obtained by means of nesting the graphs for different ratios of the radius of the lower plate to the pistons' length (R1/R2). By stacking the graphs, creating a three dimensional volume, the optimised deployment path could be distinguished by drawing a line progressing from the folded to the fully deployed state and avoiding the hotspots, allowing for optimisation of the structure.

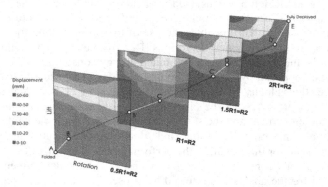

Fig. 7. Potential Deployment Routine Optimization Process

For example, in Figure 7, the stacked plots show the displacements of the struts under different conditions. The aim is to reach point E, which corresponds to the complete deployment of the awning, whilst minimising deflections. By enabling the pistons to expand, varying the R1/R2 ratio, the displacements are kept in the green areas of the graph, which correspond to lower displacement values and consequently a more efficient design.

Reduced stresses in the elements and minimisation of the deployment energy will result as a consequence of avoiding high displacement areas (red and yellow). The same process is to be carried out for all the forces and stresses that the structure develops during the motion and, then, a comparison between the deployment sequences will lead to the finding of the optimum unfolding routine.

Although, the outlined optimisation procedure has been justified as being effective for the design of deployable structures, it resulted in being long-winded and reliant on the number of deployment stages taken into consideration. The greater the number of intermediate stages considered, the more accurate the determination of the boundaries where deflections and associated stresses started to change dramatically.

Conclusion

This research project has considered the analysis of multiple degrees of freedom systems based on mechanisms that have been inspired by Nature. The two examples considered were the unfolding of a beetle wing and the blooming of a swirl flower. After having digitally replicated the morphology of the natural organisms, the optimum deployment sequence, starting from a fully folded position to being fully deployed, was determined.

Just as in natural selection and evolution, the analysis process was gradually improved for it to be able to consider more deployment stages and to accommodate the addition of a further degree of freedom. Starting from the beetle wing, radar plots resulted in limiting the consideration of set unfolding routines and preventing the possibility of combining efficient deployment stages. Subsequently, contour plots were developed for the flower model. The graphs enabled the visualisation of multiple deployment stages and potential deployment sequences that would avoid critical situations for the structure.

The deployment sequence generating the minimum deployment force, maximum coverage and minimum deflections (or a compromise between these criteria) was determined through plotting a deployment path connecting the minima within the contour plots to create a path between a fully folded status to fully deployed. The methodology established is transferable to any deployable structure with 3 or fewer degrees of freedom. The number of iterations should be increased for highly sensitive structures in order to avoid interpolation errors. Automation of the process with regards the plotting of the minima for the optimisation criteria would improve the ease of analysis.

Ultimately buildings and structures will need to be adaptable and reconfigurable to meet the demands of modern design. It is hoped that this paper identifies both a source of inspiration for the selection of potential solutions and also a reliable method for understanding their structural behaviour during the deployment process.

References

1. Cremante, S., Leonardo, D.V.: Leonardo da Vinci : the complete works. David & Charles, Newton Abbot (2006)
2. Vepa, R.: Chapter 4 - Biomimetic Robotics. In: Lakhtakia, A., Martìn-Palma, R.J. (eds.) Engineered Biomimicry, pp. 81–105. Elsevier, Boston (2013)
3. Benyus, J.M.: Biomimicry : innovation inspired by nature. Morrow, New York (1997)
4. Santulli, C.: Biomimetica: la lezione della Natura. CIESSE Edizioni, Padova (2012)

5. Bar-Cohen, Y.: Biomimetics : Nature Based Innovation. CRC Press, Boca Ratona (2011)
6. Gruber, P.: Biomimetics in architecture: architecture of life and buildings, 1st edn. Springer, Vienna (2010)
7. Chiesa, G.: Biomimetica, tecnologia e innovazione per l'architettura. Celid, Torino (2010)
8. Pellegrino, S.: Deployable structures. Springer, Wien (2001)
9. Truong, Q.-T., Argyoganendro, B.W., Park, H.C.: Design and Demonstration of Insect Mimicking Foldable Artificial Wing Using Four-Bar Linkage Systems. Journal of Bionic Engineering 11(3), 449–458 (2014). doi:10.1016/S1672-6529(14)60057-3
10. Haas, F., Gorb, S., Wootton, R.J.: Elastic joints in dermapteran hind wings: materials and wing folding. Arthropod Structure & Development 29(2), 137–146 (2000). doi:10.1016/s1467-8039(00)00025-6
11. Kobayashi, H., Kresling, B., Vincent, J.F.V.: The geometry of unfolding tree leaves. P. Roy. Soc. B-Biol. Sci. 265(1391), 147–154 (1998)
12. Wootton, R.: How do Insects Fold and Unfold their Wings. The Amateur Entomologists' Society, London (2012)
13. Chapman, R.F., Simpson, S.J., Douglas, A.E.: The Insects: Structure and Function. Cambridge University Press (2013)
14. Gullan, P.J., Cranston, P.S.: The insects : an outline of entomology, 2nd edn. Blackwell Science, Oxford (2000)
15. McCormack, G.: Pacific Beetle-Cockroach. In: Cook Islands Biodiversity Database, vol. 69KB. Cook Islands Natural Heritage Trust, Rarotonga (2007)
16. Fedorenko, D.N.: Evolution of the Beetle Hind Wing, with Special Refrence to Folding (Insecta, Coleoptera), 1st edn. Pensoft Publisher, Sofia-Moscow (2009)
17. Niklas, K.J.: Plant biomechanics : an engineering approach to plant form and function. University of Chicago Press (1992)
18. Pfeffer, W.: Physiologische Untersuchungen. W. Engelmann, Leipzig (1873)
19. Doorn, W.G.V., Meeteren, U.V.: Flower opening and closure: a review. Journal of Experimental Botany 54(389), 1801–1812 (2003)
20. Lorello: Ipomea. In: _DSC2906.JPG (ed.) Flickr (2012)
21. Stern, K.R., Jansky, S., Bidlack, J.E.: Introductory plant biology, 9th edn. McGraw-Hill, New York (2003)
22. Watanabe, K., Ziegler, F., Kobayashi, H., Daimaruya, M., Fujita, H.: Unfolding of morning glory flower as a deployable structure. In: IUTAM Symposium on Dynamics of Advanced Materials and Smart Structures. Solid Mechanics and Its Applications, vol. 106, pp. 207–216. Springer Netherlands (2003)
23. Kobayashi, H., Daimaruya, M., Vincent, J.F.V.: Folding/Unfolding Manner of Tree Leaves as a Deployable Structure. Solid. Mech. Appl. 80, 211–220 (2000)
24. Amar, M.B., Müller, M.M., Trejo, M.: Petal shapes of sympetalous flowers: the interplay between growth, geometry and elasticity. New Journal of Physics 14(8) (2012)
25. Tachi, T.: Generalization of rigid foldable quadrilateral mesh origami 50(3), 7 (2009)

Biomimetic Tactile Sensing Capsule

Benjamin Winstone[1]([⊠]), Tony Pipe[1], Chris Melhuish[1], Sanja Dogramadzi[1], and Mark Callaway[2]

[1] Bristol Robotics Laboratory, University of the West of England, Bristol, UK
Benjamin.Winstone@brl.ac.uk
[2] Department of Radiology, Bristol Royal Infirmary, Bristol, UK
Mark.Callaway@UHBristol.nhs.uk

Abstract. Here we present a tactile sensing capsule endoscopy system. Whilst current capsule endoscopy utilises cameras to diagnose lesions on the surface of the gastrointestinal tract lumen, this proposal uses remote palpation to stimulate a bio-inspired tactile sensing surface that deforms under the impression of hard raised objects. This provides the capability to characterise tissue density and lesions more deeply than the lumen surface.

1 Introduction

One of a surgeon's most important skills is their highly enhanced sense of touch. However modern advances in medical technology are separating the surgeon from direct contact from the patient. Whilst systems like remote operated laparoscopic surgery introduce minimally invasive procedures that minimise patient discomfort and improve recovery, they are making much less use of this very human perception of touch. Capsule endoscopy describes a swallowable capsule with self contained microsystem, similar to a traditional drug delivery capsule. This capsule provides a platform to achieve similar procedures to more traditional methods of endoscopy but without the restrictions of a tethered system. This presents a capability to explore the entirety of the the gastrointestinal (GI) tract which is not currently possible using any other system. In its current form capsule endoscopy is limited to only visual inspection of the lumen surface. Tissue health deeper than the lumen surface cannot be detected by visual means, however using a tactile palpation technique deep tissue deformities could be sensed.

The idea of swallowing a small capsule with an internal sensing system that can communicate to the outside of the body goes as far back as 1957 when R. Stuart Mackay presents the idea of RF transmission of the temperature and pressure within the human body[1]. Later in 1998 NASA registered a patent with John Hopkins University that describes a swallowable pill that transmits the internal temperature of the body for use in monitoring astronauts during space expeditions. An obvious ambition of a swallowable capsule is to perform and transmit imaging from inside the body. Modern capsule endoscopy presents an ability to view the small bowel beyond the duodenum and proximal to the colon. There are a number of pathologies that can occur in the small bowel, such as tumours, strictures and ulceration as part of an inflammatory bowel condition

© Springer International Publishing Switzerland 2015
S.P. Wilson et al. (Eds.): Living Machines 2015, LNAI 9222, pp. 113–122, 2015.
DOI: 10.1007/978-3-319-22979-9_12

such as crohns disease. Shi et al. [2] discusses the importance of the GI tract motility and how failings in motility can be an indication of more serious GI diseases. Shi presents a sensing capsule to measure pressures encountered during the journey through the GI tract as an indication of GI tract motility. Direct contact sensing has long been a common practice for diagnosis in the vast majority of medical fields. Cox et al. [3] presents the reason that we palpate the skin that not only does it present reassurance to the patient but more importantly it is an underestimated examination of modalility that identifies tenderness, consistency, induration, depth and fixation. Palpation of tissue is used to identify strained muscles, skeletal breaks and deformed growths amongst other signs of ill health. Tactile sensing and palpation are trained and developed expert skills that separate medical practitioners from the untrained.

In 2011 Kume et al. [4] investigated the development of an articulated robotic endoscopic tool. Although adding articulation showed evidence that the challenge of insertion becomes easier, and learning to operate quicker, they clearly state that tactile feedback in an endoscopy tool would aid in limiting unintended application of force. Further evidence of the lack of, but need for, tactile sensing in remote surgical systems is discussed by Chaudhary et al. [5] where they review medical robotics as a current technology. In 2006 Schostek et al. [6] presented work on adding tactile sensing capabilities with a visual feedback display to laparoscopic tooling in order to aid in local tumour surgery. More recently, in 2013 Roke et al. [7] took this concept further with the addition of a tactile feedback display using a mechanical matrix of linear actuators to remotely stimulate the operator's fingers based on sensor deformation in contact with artificial tissue and tumours. This later work orientates itself towards remote operated surgical systems such as the DaVinci® robot. Although examples exist of technology capable of tactile sensing outside the body or internally through a limited reach tethered system, no work has yet focused on remote palpation using an untethered capsule like robot capable of minimally invasive tactile examination of the internal organs of the body.

In this paper we propose the further development of a biomimetic tactile fingertip device, the Tactip, to create a biologically inspired system capable of providing minimally invasive untethered remote operated palpation that will increase localisation and characterisation of surface shape beyond what is typically identifiable through vision alone. Using a tactile sensing system that is based on the human finger provides the most suitable data for a remote haptic feedback system. This concept could provide a new diagnostic technique not yet explored and as such, could present new approaches to medical diagnosis and even minimally invasive surgery. Whilst this paper focuses on acquisition of haptic sensing data, it is assumed that it would later be coupled with a haptic feedback technology to complete a remote sensing system. The advantage of using the Tactip sensing device over a rigid sensor is that its compliancy works in favour of the GI tract environment. A rigid sensor will more likely cause blockage whilst a soft sensor can deform to its surroundings whilst still sensing. This could avoid obstruction of the peristaltic flow of the GI tract whilst moving with or against the flow of the gut.

2   Biologically-Inspired Tactile Sensing Device

The Tactip is a biologically-inspired sensing device, based upon the deformation
of the epidermal layers of the human skin. Deformation from device-object interaction is measured optically by tracking the movement of internal papillae pins
on the inside of the device skin. These papillae pins are representative of the
intermediate epidermal ridges of the skin, whose static and dynamic displacement are normally detected through the skin's mechanoreceptors, see Fig.1. In
this paper we have realised a previous proof of concept cylindrical Tactip [8],
developing a pill like capsule based upon the biological model of tactile sensing
that is capable of lump detection.

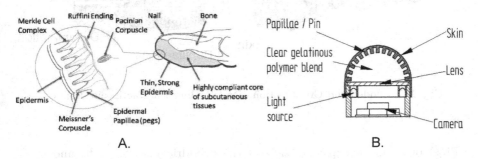

Fig. 1. Tactip concept, A. biological model and B. engineered model.

Fig.1 shows the design of a traditional Tactip which comprises an artificial
cast silicone skin, optically clear flesh like gel, camera and internal illumination.
Whilst in the biological model it is the papillae pin movement that stimulates the
mechanoreceptors in the finger, the Tactip design replaces the mechanoreceptor
for a camera system capable of tracking pin interaction visually. Translation of
this design into a cylindrical tactile skin system offers two main challenges. The
first is how to effectively build such a structure which has until now been reliant
on a complex silicone casting process. A cylindrical Tactip offers an even more
complex mould design. Through the use of new 3d printing technology it has
been possible to avoid casting a skin, and instead use an Objet Connex 260
printer capable of simultaneously printing in hard and soft materials. This new
skin has been designed to fit and seal to an internal clear acrylic tube which acts
as both a main body structure and also a barrier to contain the flesh like gel.
The quality and robustness of such printed material is not yet as good as a cast
rubber silicone, however for the purpose of experimentation it greatly speeds up
the whole development process.

Whilst the traditional design of the Tactip uses a standard camera system to
observe papillae pin movement horizontally and vertically central to the field of
view, the second challenge is in designing an optical system capable of observing
activity 360° around the centre of the field of view, as presented by a cylindrical

skin. The chosen solution to this problem is to use a panoramic catadioptric mirror system that is capable of a observing 360° field of view presented as a distorted two dimensions image. This distorted image can be simply unwrapped for ease of image process, which is described further on in this paper.

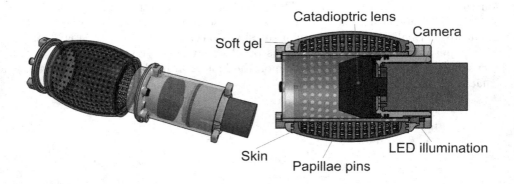

Fig. 2. Exploded and Cross section view of cylindrical Tactip pill design

Fig.2 presents an exploded view of the cylindrical Tactip, left, and cross section of the cylindrical Tactip design, right. As with a traditional Tactip design, it comprises an artificial 3d printed skin, optically clear flesh like gel, camera and internal illumination, but in addition uses a catadioptric mirrored lens. The camera used is a Microsoft LifeCam Cinema HD.

3   Experiment Design

The purpose of this experiment is to identify the capability of a cylindrical Tactip to identify lumps representative of the kind of suspicious tissue that could reside within the GI tract through active exploration. The experiment environment has three components, a cylindrical Tactip for sensing, a six axis ABB IRB120 robot arm for capsule locomotion, and a large 74mm internal dia. acrylic tube with six artificial lumps placed randomly along a length of 250mm, see Fig.3. The robot arm is used during this experiment as a mode of locomotion for the capsule. The future vision is that the capsule would be capable of untethered locomotion though the GI tract with the addition of an appropriate locomotion system, however this paper focuses purely on the tactile sensing capabilities. The cylindrical Tactip is pushed along the length of the tube 1mm at a time. At each position an image is captured from the camera to be processed later. Using the algorithm described below, the sensor data is then used to reconstruct the tube environment in three dimensions, highlighting regions of skin deformation caused by each of the lumps. There are two sizes of lumps, 10mm and 8mm in height. The cylindrical Tactip at its widest diameter is 63mm. The maximum a large lump can depress the Tactip is 4.5mm and the small lump is 2.5mm. These lumps are larger than would typically occur in the GI tract, however in order to match the scale of the prototype capsule they size has been scaled up. Future work would focus on both miniaturisation of the capsule and more accurate tumour representations. There are three types of lump material, hard plastic, hard rubber and soft rubber. By using different size lumps with different density materials it will give an indication of whether the Tactip can distinguish between different types of deformation. In a medical application this could help determine the progress of a tumour.

Fig. 3. Experiment setup, ABB IR120 pushing tactile sensing through rigid tube containing raised lumps

3.1 Sensing Algorithm

The algorithm used to interpret tactile interaction with the capsule focuses on papillae pin separation replicating the role of the Meissner's corpuscle mechanoreceptor in the human finger. When objects contact the capsule skin and deform the surface shape, pins within the region of contact separate from each other whilst moving closer towards adjacent neighbouring pins. This can be seen in Fig.4 where the first image shows the Tactip in a relaxed state with no objects in contact with the skin, whilst the second image shows the Tactip in contact with a lump highlighted by the red circle where the papillae pins have separated in the region of contact.

The first stage of image processing is unwrapping the distorted image acquired through the catadioptric lens. The following algorithm is an effective method of unwrapping a 360° degree image. The algorithm takes a square source image containing the 360° degree field of view, operating the function to return a translated two dimensional unwrapped image similar to that of a panoramic photograph. Equation 1 shows how to calculate the corresponding pixel in the source image to the destination image such that if one were to iterate through the destination image it would become populated with corresponding pixels from the source image.

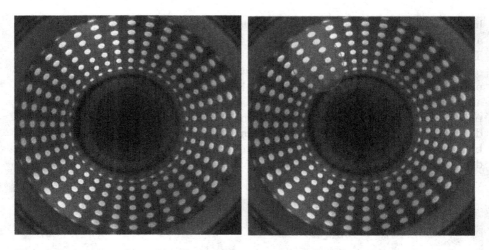

Fig. 4. Raw camera image showing calibration state on the left, and contact with lump state on the right highlighting the location of the contact

$$d(x, y) = s((cos(x * (2\pi/W)) * y) + R, (sin(x * (2\pi/W)) * y) + R) \quad (1)$$

s = 360 source square image array
R = Radius of 360 image
W = Width of destination image
d = Destination flat image W * R
x = Horizontal pixel co-ordinate
y = Vertical pixel co-ordinate

Fig. 5. Unwrapped 360° panoramic image capture with identified pin positions

Fig.5 shows the result of equation 1 and also subjected to the next stage of the image processing, blob tracking. Blob tracking is achieved using the OpenCV feature detection library and provides accurate identification of the papillae pin positions within the camera view. In order to identify the location of deformation or lump contact, pin locations are compared against the relaxed calibrated state. In particular it is the increase in pin separation that highlights the location of skin deformation, so pixel distance between pins is used as the comparison value.

Gaps between pins that exceed the calibrated state provide a representative value of deformation at that location. Fig.6 shows a localised group of pins subjected to skin contact where the pins within the contact region separate and

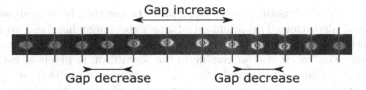

Fig. 6. Localised group of pins subjected to skin contact showing the pins within the contact region separate

the adjacent neighbouring pins move towards each other. It is that pin separation value that can be used as a relative skin depression from the nominal radius of the Tactip to derive a reconstruction of the environment. In relation to the ABB IRB120 position co-ordinates localisation of lumps and an accumulated profile of the tube can be achieved through active exploration. By linking the collected points through the tube scan, mesh vertices can be defined and a complete visual representation of the tube based on the Tactip sensor data alone.

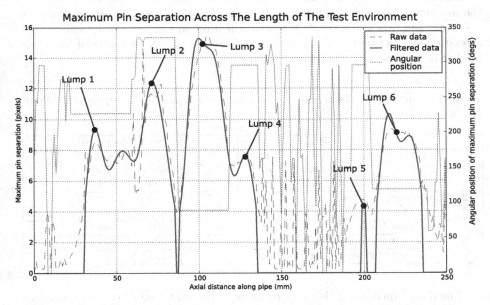

Fig. 7. Location of maximum pin separation and angular position along length of test pipe

4   Results

Using an ABB IRB120 robot arm the cylindrical Tactip has been moved through an 74mm diameter acrylic tube, containing six artificial tumours or lumps. Data in the form of an image from the internal camera has been captured at 1mm increments along the length of this tube. This data has then been analysed using the discussed algorithm. The intention of this experiment has been to localise and potentially characterise lumps within a simplified artificial GI tract-like environment. Fig.7 presents the maximum pin separation in pixels at each stage along the pipe. The data received after the image processing is represented by the green raw data line. This data shows some level of noise, which is most likely due to jitter in the blob detection algorithm. The second plot shows that data after being passed through a Butterworth low pass filter, and thresholding the data so that only significant changes in pin separation are acknowledged. This graph shows six clear defined peaks in the data that coincide accurately with the locations of the the lumps in the test environment. The final plot on the graph shows the angular position of the maximum pin separation. Again this is at times noisy data, however the points where no noise exists coincide directly with the detection lump positions. This shows that a combination of stable data from angular position and a held peak in pin separation only occur at points when the Tactip is subject to contact from an artificial lump.

Table 1 summarises the results of axial and angular lump detection. The table shows the exact location of the lumps with their size and material, along side the measured location of the lumps. The results show a worst case accuracy of +/- 9mm and +/- 4 degrees. When also reflecting on Fig.7 the results also show a relationship between pin separation and lump size, whilst material hardness further influences greater pin separation.

Table 1. Summary results of real measured lump positions and Tactip detected lump positions

Lump No.	Size	Type	Real Axial Position	Measured Axial Position	Difference (mm)	Real Angular Position	Measured Angular Position	Difference (degs)
1	Large	Soft Rubber	47	38	-9	233	234	1
2	Large	Hard Rubber	64	71	7	346	348	4
3	Large	Hard Plastic	102	101	-1	92	90	-2
4	Large	Soft Rubber	121	128	7	307	308	1
5	Small	Soft Rubber	195	200	5	306	306	0
6	Large	Soft Rubber	220	221	1	125	121	-4

The data captured by the Tactip can be collected together to create vertices and a mesh surface to present a three dimension reconstruction of the test environment. Fig.8 shows a three dimensional representation of the collected data.

5   Conclusion

Whilst the original Tactip device can be considered as bio-mimetic due to its design directly mimicking the human sensing model, creating a cylindrical Tactip has advanced to inspire a new more capable system that can provide data that

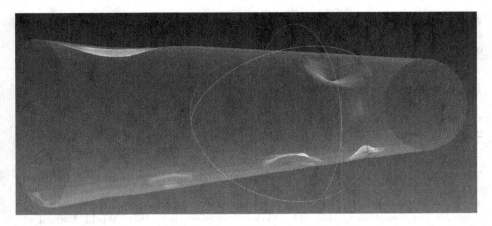

Fig. 8. 3D reconstruction of test environment built from data captured by the cylindrical Tactip sensor

a human finger could not. This paper has presented a new application for a biologically inspired tactile sensing device. Whilst first thoughts of applications lead toward replication of typical fingertip type activities, we have proposed a new method of medical diagnostics. The human finger is the most suitable device for a medical practitioner to use in order to inspect a patient for suspicious tissue, however inside the body this is not possible without open surgery. A sensor capable of mimicking the capabilities of the human finger that can access the inside of the body without drastic surgery is the first step in creating a minimally invasive remote palpation system. This is a new capability which could lead to a new approach in diagnostics. We have proposed an experiment which provides evidence that the cylindrical Tactip is capable of sensing surface deformation. In successfully doing this we have also presented the sensor data as quite accurate three dimensional reconstruction of the test environment. This is a capability that would exceed that of the human fingertip.

6   Future Work

The work presented here has proven the capability of the Tactip to sense a rigid cylindrical environment, however the human GI tract is soft and flexible. The next natural progression of the work is to create a soft flexible environment to present to the Tactip and investigate how the data acquired changes with tissue density. Sensing is one part of a larger complete system. In order for this to be truly realised the capsule needs a system to move through the GI tract untethered and the data returned needs to be presented to the surgeon in a more tangible way, such as a haptic feedback display system. Examples of both haptic feedback in medical robotics [7], and suitable approaches to locomotion [9] exist, the next step is to integrate these ideas and test in a suitable environment.

Acknowledgments. Bristol Robotics Laboratory gratefully acknowledge that this work was supported by the Above & Beyond charity, http://www.aboveandbeyond. org.uk/.

References

1. Mackay, R.S., Jacobson, B.: Endoradiosonde. Nature **179**, 1239–1240 (1957)
2. Shi, Q., Wang, J., Chen, D., Chen, J., Li, J., Bao, K.: In Vitro and In Vivo characterization of wireless and passive micro system enabling gastrointestinal pressure monitoring. Biomedical Microdevices, August 2014
3. Cox, N.H., Soc, J.R.: Palpation of the skinan important issue. Journal of the Royal Society of Medicine **99**, 598–601 (2006)
4. Kume, K., Kuroki, T., Sugihara, T., Shinngai, M.: Development of a novel endoscopic manipulation system: The Endoscopic operation robot. World Journal of Gastrointestinal Endoscopy **3**, 145–150 (2011)
5. Chaudhary, A., Atal, D., Kumar, S.: Robotic Surgical Systems A Review. International Journal of Applied Engineering Research (IJAER) **9**(11), 1289–1294 (2014)
6. Schostek, S., Ho, C.-N., Kalanovic, D., Schurr, M.O.: Artificial tactile sensing in minimally invasive surgery - a new technical approach. Minimally Invasive Therapy & Allied Technologies : MITAT : Official Journal of the Society for Minimally Invasive Therapy **15**, 296–304 (2006)
7. Roke, C., Spiers, A., Pipe, T., Melhuish, C.: The effects of laterotactile information on lump localization through a teletaction system. In: 2013 World Haptics Conference (WHC), pp. 365–370, April 2013
8. Winstone, B., Melhuish, C., Dogramadzi, S.: A Novel Bio-inspired Tactile Tumour Detection Concept for Capsule Endoscopy. Biomimetic and Biohybrid Systems, 2–4 (2014)
9. Boxerbaum, A., Horchler, A.: Worms, waves and robots. In: 2012 IEEE International Conference on Robotics and Automation (ICRA), pp. 3537–3538 (2012)

Wings of a Feather Stick Together: Morphing Wings with Barbule-Inspired Latching

Aimy Wissa$^{(\boxtimes)}$, Amy Kyungwon Han, and Mark R. Cutkosky

Stanford University, Stanford, CA 94305, USA
awissa@stanford.edu
http://bdml.stanford.edu

Abstract. Birds' feathers are equipped with hook-like structures called friction barbules, which prevent separation and rubbing between feathers under nominal flow conditions. This paper presents a segmented wing prototype that uses controllable dry adhesives to mimic the function of friction barbules. The adhesives latch wing segments together during moderate flight conditions and allow them to separate in extreme conditions. We present the characteristics of adhesive patches and their performance as they are incorporated into a flexible wing prototype. The attachment force is a function of the applied shear stress. We then present results of a wind tunnel test to evaluate the aerodynamic effect of gaps formed as wing segments unlatch and separate. The separation of wing segments delays stall and reduces overall drag, which could improve the ability of an unmanned air vehicle to fly in gusty conditions.

1 Introduction

Birds are able to fly in gusty winds and around buildings and other structures in conditions that are far beyond the capabilities of small unmanned air vehicles. They excel in these conditions, in part, because they have wings that respond both passively and actively to broaden the range of conditions for stable flight and mitigate the effects of large transient disturbances. The bird's wing responds to the airflow at various levels. The first level involves the passive, geometric properties of the feathers and the musculoskeletal system in the wing. Higher levels involve coordination and integration of visual and vestibular information with information obtained from mechanoreceptors in the wings [1].

During the past few decades, several authors and organizations investigated and realized the benefits of wing morphing to improve the overall aerodynamic performance of manned and unmanned air vehicles [2]. Barbarino et al. present a detailed review of the state of-the-art on morphing aircraft [3]. However, only a few works (e.g. [4]) have addressed the function and structure of birds' feathers to improve the performance of unmanned air vehicles and none, to the authors' knowledge, has addressed the function of the feathers' friction barbules to improve flight performance.

This paper introduces a mechanism whereby a controllable, fibrillar adhesive is applied to segments of a wing so that they can repeatedly attach, or separate,

© Springer International Publishing Switzerland 2015
S.P. Wilson et al. (Eds.): Living Machines 2015, LNAI 9222, pp. 123–134, 2015.
DOI: 10.1007/978-3-319-22979-9_13

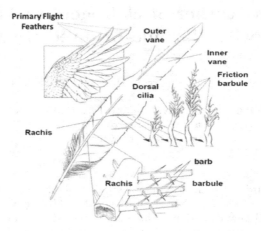

Fig. 1. Schematic of an avian feather showing friction barbules responsible for reducing inter-feather slippage and separation during flight (adapted from [5]).

like the feathers of a bird, in response to dynamic loads. During moderate flow conditions, controllable adhesives mimic the function of the friction barbules found in birds' feathers. Barbules are hook-like structures that exist both within a feather and between feathers. Figure 1 shows the structure of a bird feather with barbules for interlocking branches of a feather to each other, and additional friction barbules that enable interlocking between feathers [5]. Friction barbules are found on the inner vane of a primary flight feather, and they rub against the barbs of the outer vane of an overlying feather. They are used by birds to reduce slippage and separation between feathers during flight [6].

In the mechanism introduced in this paper, wing segments can be separated both passively and actively. Separation occurs passively when pressure on the wing segments exceeds the adhesive latching force, which would occur in extreme conditions such as gusts. Figure 2 shows a schematic of the aforementioned wing structure both in the continuous and segmented configurations. Upon separation, gaps are introduced and air is allowed to flow between the feathers, delaying stall.

The same gaps can be introduced actively by controlling the applied in-plane shear stress because the latching force is a function of the shear stress. Actively introducing the gaps can be useful during soaring and flight at slow speeds. The function of the gaps during the aforementioned flight regime is similar to the function of slots found between the primary feathers of various raptors and flyers with low wing aspect ratio (figure 3). The function is also similar to that of winglets on commercial aircraft, i.e., to reduce the induced drag resulting from span-wise flow [7]. Reducing the induced drag allows birds to soar efficiently and prevents stall at low flight speeds [8].

In the following sections, we present a bio-inspired wing prototype that uses a controllable dry adhesive to change the aerodynamic performance of the wing. We also present the results of tests conducted on the wing prototype to evaluate the performance of the adhesives in the interlocking configuration. We then

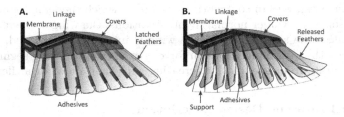

Fig. 2. A schematic of a barbule-inspired segmented wing section. A. In the latched state the feathers mimic the function of a continuous wing. B. In the unlatched state, the sections separate to reduce drag and delay stall.

present results from a wind tunnel test performed to determine the aerodynamic effects of latching and unlatching the wing segments at various angles of attack and flight speeds. A discussion of extensions to improve the design and integrate it into a small air vehicle follows, along with proposed future work.

2 Technical Approach

This section provides a brief overview of the controllable adhesives used to latch the wing segments and introduces the wing prototype. More details about the adhesives can be found in [9,10].

Fig. 3. Primary feather slots near the wing tips reduce induced drag and delay stall during slow flight.[1]

2.1 Controllable Dry Adhesives

The $8 \times 22\,$mm adhesive patches included in the wing prototype are comprised of rows of polydimethylsiloxane (PDMS) micro-wedges directly cast onto a 25μm thick polyimide film. Each row of microwedges has a base width of 60μm and is approximately 100μm high. As seen in figure 4, the application of a shear force causes the microwedges to bend over, increasing the real area of contact and the adhesion, which is due to van der Waals forces. Hence, one can control adhesion by controlling the applied shear force. In previous work, the adhesives were used to adhere to smooth flat or curved surfaces for climbing robots and similar applications [10,11]. In the present case, we adhere small patches of adhesive, face to face, on adjacent wing segments to make them less sensitive to nonuniform pressure distributions (e.g. alignment errors) on the thin, compliant wing sections. This is because on the face to face patches some adhesion is present even in the absence of shear load. In this regard, they can function similarly to friction barbules, latching or releasing the wing segments. By controlling the in-plane shear stress

---
[1] adapted from http://animalia-life.com/eagles.html

applied to the wing, we can control the pressure at which the wing segments will detach and separate. Under normal flight conditions, the wing segments remain attached, unless the shear stress in the wing is intentionally relaxed. In gusty weather, when the maximum adhesion force is exceeded, the wing segments will passively separate from one another, creating gaps that allow for airflow.

2.2   Wing Prototype Design Description

The wing prototype consists of three major components: a four-bar mechanism, the wing segments or "feathers," and the controllable adhesive patches (figures 5, 7A). The 12 × 3 cm feather-like segments are made from two 0.38 mm thick fiberglass reinforced sheets, bent around a cylindrical form and glued together to impart a curvature. The curvature and elasticity of the sheets provide a gentle elastic force that keeps the edges of the segments in contact with each other when aligned.

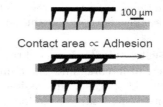

The four-bar mechanism is built from 3.2 mm wide laser cut acrylic links. Actuating the mechanism rotates the wing segments and applies a shear stress between the facing adhesive patches on adjacent segments. Figure 5 shows plan and side views of the prototype,

Fig. 4. A. When microwedges initially contact a surface, there is little contact area and negligible adhesion. B. Applying a shear load causes the microwedges to bend over, increasing adhesion. C. Releasing the shear load allows the wedges to straighten, eliminating the adhesion.

along with the directions of the applied shear forces. When the segments are aligned vertically (A, C) the shear force keeps the patches latched; when rotated (B, D) the shear force is relaxed and the segments separate. Therefore, by controlling the magnitude and direction of the applied force on the four-bar mechanism, we can actively adjust the level of adhesion and separation between the feathers. The four bar linkage was inspired by the bone structure found in birds' wings. The two horizontal bars in the four bar linkage resemble the arrangement of the Radius and Ulna bones, shown in figure 6. Unlike the prototype, birds feathers are not directly attached to the bones but are implanted into the the skin. The base of a feather is then connected to a complex network of muscles and tendons that enables their motion [12].

3   Experimental Setup

This section describes three experimental setups. The first two were designed to evaluate the performance of the adhesives in the interlocking configuration. The third experiment took place in a low speed wind tunnel and was designed to determine the aerodynamic performance of the wing prototype when the wing segments are latched and unlatched.

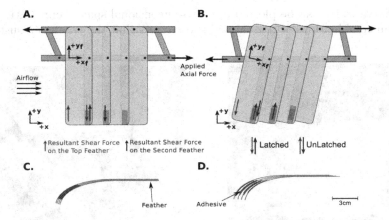

Fig. 5. Wing prototype showing the top view of the feathers in the A. latched and B. unlatched states. C. and D. show side views of the latched and unlatched states, respectively. In the latched state, adhesive patches are in contact and the segments will not separate, except during large gusts. In the unlatched state, gaps are present allowing air to flow between the segments.

3.1 Interlocking Adhesives Limit Curve

The goal of this experiment was to characterize the adhesion properties of the patches when they are applied to each other. First, a pair of interlocking adhesives were tested to find their normal and shear stress limits. With these stresses, a limit curve which maps shear stress to a corresponding normal adhesive stress can be generated. The interlocking adhesive patches were attached to flat, rigid tiles and brought into contact, face to face. A tendon was attached to each of the tiles. A weight was attached to one of the tendons applying a known shear force. The second tendon was pulled, using a digital force gauge (Mark-10 M4-50), perpendicular to the tiles to apply a normal force. The force at which separation occurred was recorded. Applying a series of normal and shear forces results in

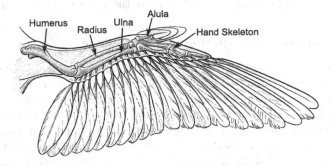

Fig. 6. Top view of the typical skeletal bone structure of a bird's wing. The feathers are not directly connected to the bones but implanted in the skin and connected to a complex network muscles and tendons (adapted from [13]).

a limit curve which can be plotted in a two-dimensional space, representing the maximum combinations of normal and shear stress that the patch can sustain.

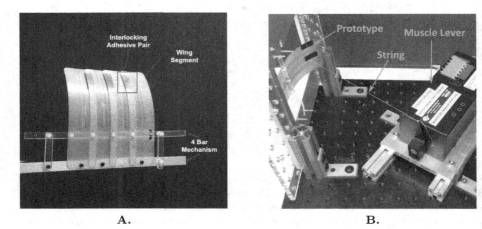

Fig. 7. A. Wing prototype with segments and adhesive patches. B. Latching tests: various weights were attached to the mechanism to provide varying shear forces and a muscle level applied normal forces perpendicular to the wing segments.

A second experiment was conducted to evaluate the adhesive patches *in situ* on the wing prototype. Figure 7A shows an image of the wing prototype with the interlocking adhesive patches. Figure 7B shows the prototype with a tendon pulling on the lower wing segment and attached to a muscle lever (Aurora Scientific 305B). A weight is attached to the four-bar mechanism to adjust its orientation and the corresponding in-plane shear stress. At every applied weight, the bottom segment is pulled at constant velocity, and the force at which it separates is measured using the muscle lever and recorded. The experiment was conducted for a pair of interlocking adhesive patches at the edge of the wing segment and for a single adhesive patch against the surface of the wing segment.

3.2   Wind Tunnel Test Experimental Procedure

After characterizing the adhesion properties of the interlocking adhesive patches, we conducted a wind tunnel test to determine the effect of latching and separating the wing segments on aerodynamic performance. We positioned the wing prototype in a low speed open section wind tunnel with a test section of 76×76 cm in cross section and 300 cm in length. The wing prototype was mounted to the tip of a cylindrical beam that extended from an ATI®Gamma 6-axis load cell. During the test, the wind speed and angle of attack were varied from 1.5 m/s to 13 m/s and from -18° to 36°, respectively. The forces and moments at each speed and angle of attack were measured for both the latched and unlatched wing states.

4   Results and Discussion

4.1   Interlocking Adhesive Limit Curve Results

Figure 8A shows the normal versus shear stress limit curve of the adhesives mounted to rigid tiles in the interlocking face to face configuration. A second set of points shows the corresponding limit curve for a tile of adhesive sticking to a glass surface, as is the usual case for climbing robots, etc. [10].

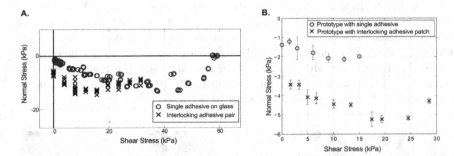

Fig. 8. A. The limit curve of an interlocking adhesive pair and of a single patch on glass. The interlocking adhesives achieve comparable maximum adhesion and have a non-zero normal stress under zero shear stress. B. The limit curves for an interlocking adhesive pair and a single adhesive patch when integrated in the curved feather prototype. The adhesive in the interlocking configuration provides more adhesion when compared to the single patch.

As the data reveal, the maximum adhesive pressure for both configurations is approximately the same. However, the interlocking adhesives reach their peak adhesion at much lower levels of shear stress. This result indicates that they are less sensitive to the ratio of applied shear to normal force, a desirable result given the flexibility of the wing prototype and the difficulty in maintaining precise alignment between wing segments.

A second difference in the curves is that the maximum shear stress for the interlocking adhesives is smaller than for adhesives on glass. Once the interlocking adhesives start to slide, adhesion is lost. In contrast, the adhesive can slide slowly against a flat glass surface without losing adhesion.

A third difference is that the interlocking adhesives have a nonzero normal stress at zero applied shear stress, whereas adhesion for a single adhesive on glass is almost zero. This difference makes the interlocking adhesives more suitable for the barbule-inspired application because it guarantees some level of adhesion even in the absence of an applied shear load.

Figure 8B depicts the limit curves for one and two patches of adhesives as mounted to the wing prototype. In this case, using two interlocking patches is clearly advantageous because it provides more adhesion. Moreover, placing the patches face to face in the prototype enables adhesion without the need to tailor the underlying surface orientation, roughness, or stiffness because the patches are adhering to each other and not the surface. However, the maximum adhesive pressure is only about one half the value obtained for patches on rigid, aligned tiles. Nonetheless, as seen in the section 4.2, the pressure is more than sufficient to prevent the wing segments from separating under normal wind speeds and pressures.

In order to better understand the interlocking adhesion properties and interpret the results in figure 8, we observed a pair of adhesives in face to face contact using a microscope video system. The video reveals three stages of adhesion, as shown in figure 9. In stage A, at zero shear stress, there is a small amount of adhesion because there is immediately some contact area between the tips of the micro-wedges, however not all the wedges are in contact. As the shear stress is increased (stage B), the area quickly increases, resulting in an increase in adhesion. Applying more shear does not further increase the contact area, so adhesion does not continue to increase. Eventually, at stage C, the tips of the micro-wedges start to slide past each other and adhesion is lost.

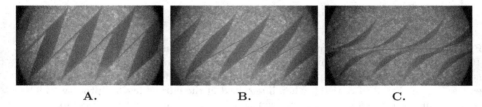

A.   **B.**   **C.**

Fig. 9. The Interlocking adhesion can be divided into three stages. There is a small amount of initial contact (A) and adhesion as the surfaces are brought together. With an increased applied shear stress (B), the contact area and adhesion increase and then level off. With excessive shear stress (C), the wedges slide past each other. The dashed line indicates the boundaries between the microwedges.

4.2   Aerodynamic Performance of the Latched and Unlatched Wing Prototype

The wingtip feather slots in birds enable them to reduce induced drag while soaring, as well as delay stall during high angle of attack maneuvers, such as perching. This section discusses the aerodynamic effects of the synthetic gaps introduced by unlatching wing segments at various angles of attack and wind speeds.

Figure 10A shows the lift coefficient, C_L, of the latched and unlatched wing states over various angles of attack, α. The lift coefficient is defined as:

$$C_L = \frac{2L}{\rho V^2 S} \tag{1}$$

where L is the lift force, V is the wind speed, S is the wing surface area and ρ is the air density. Figure 10A shows that unlatching the wing segments delays stall. The latched wing reaches a maximum lift coefficient, C_{Lmax}, at $\alpha=30°$ and starts to stall, while C_L of the unlatched wing keeps increasing past $36°$. Delaying stall enables high angle of attack maneuvers such as perching. In addition, the reduction in C_L observed between $\alpha=0°$ and $\alpha=24°$ is useful during gusty conditions.

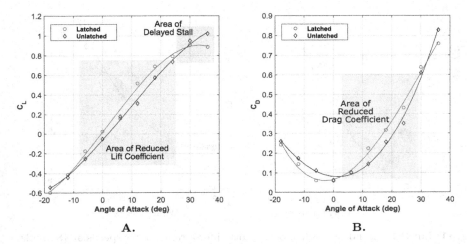

Fig. 10. A. The wing in the latched stalls at a lower angle of attack than in the unlatched state. B. The reduction in the coefficient of lift, C_L causes a reduction in the the coefficient of drag, C_D. The solid lines are generated using a third order polynomial fit to the data.

Figure 10B shows the drag coefficient, C_D, of the latched and unlatched wing states. Unlatching the wing segments reduces the overall drag produced by the wing section, especially for $\alpha > 0°$. The component of C_D, responsible for this reduction is the coefficient of induced drag, C_{Di}. C_{Di} is reduced for the unlatched wing state because it is proportional to C_L^2:

$$C_{Di} = \frac{C_L^2}{\pi e AR} \qquad (2)$$

where e is the Oswald efficiency number and AR is the wing aspect ratio.

C_L and C_D are non-dimensional quantities and are not a function of the wind speed. Lift and drag, on the other hand, depend on wind speed. Figure 11 shows the lift and drag produced by the wing prototype over various flight speeds at an angle of attack of $18°$, as an example. The figure confirms the aerodynamic effects of the gaps. Unlatching the wing segments, reduces the amount of lift produced at a given speed, which is beneficial for an unmanned air vehicle (UAV) flying in gusty weather conditions. Moreover, the wing in the unlatched state experiences less drag. Drag reduction is beneficial because it delays stall and increases range.The maximum differences in lift and drag values between the latched an unlatched cases are 17% and 32%, receptively.

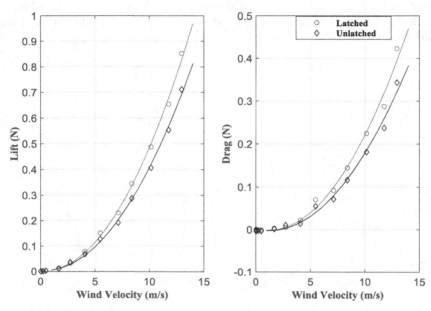

Fig. 11. Lift (left) and drag (right) forces plotted for a range of air speeds at 18° angle of attack. Unlatching the wings reduces the overall lift and drag, which is beneficial for flying in gusty conditions and soaring, respectively. The solid lines are generated using a third order polynomial fit to the data.

5   Conclusions and Future Work

This paper presented a novel approach in which controllable gecko-inspired dry adhesives are adapted to a wing prototype intended for a small, unmanned air vehicle. The adhesives allow the segments of the wing to latch, or separate, mimicking a function of the friction barbules found in avian feathers.

When mounted in a segmented wing prototype, the adhesives kept the wing segments attached to each other over a broad range of wing speeds and angles of attack. In the unlatched state, gaps were present between the segments. These gaps are similar to the primary feather slots found near the wing tips of various raptors and used to delay stall at high angles of attack and reduce induced drag during slow flight. To characterize the effect of gaps on the aerodynamic performance of the wing prototype, we performed wind tunnel tests on latched and unlatched configurations. Results confirmed that, in the latched state, the adhesives remain in contact and the segmented prototype behaves like a continuous wing. Conversely, when the wing segments are unlatched, the prototype stalls at a higher angle of attack and produces less induced drag when compared with the latched state. These changes in performance are favorable for a UAV flying in gusty weather conditions, soaring at slow speeds, or maneuvering at high angles of attack. Although in this paper we demonstrated active unlatching of the wing, passive unlatching will occur if the normal load on a given wing segment exceeds the maximum normal adhesion force. Passive unlatching is expected to occur during gusts when an abrupt change in wind speed may cause an increase in the angle of attack. At higher angles of attack, the pressure gradient between the upper and lower wing surfaces increases. If the pressure gradient exceeds the maximum adhesion stress then the adhesives will fail and the wing segments will separate passively.

The barbule-inspired prototype is a first step towards a multi-functional wing that is able to respond to external disturbances both actively and passively during various flight missions and conditions. To improve the overall design of the prototype, future work must include various milestones. First, the adhesive micro-wedges should be redesigned explicitly for interlocking adhesion. Improving the overall design of the adhesives will promote uniform load sharing and expand the limit curve to include a broader range of shear stresses. Second, we need to integrate a refined version of the prototype into a flying UAV. Various considerations such as the wing segment thickness and camber, adhesives placement and size, and appropriate actuators for latching and unlatching the wing segments on demand, must be addressed. Finally an aeroelastic model is needed in order to understand and predict the effects of design changes on flight performance for a desired flight envelope.

Acknowledgments. Aimy Wissa is supported by a postdoctoral fellowship from the University of Illinois at Urbana-Champaign Mechanical Science and Engineering Department. Amy Kyungwon Han is supported by a master's fellowship from Kwanjeong Educational Foundation. The authors would like to acknowledge resources and cooperation of the members of the Biomimetics and Dexterous Manipulation Laboratory, especially A. Suresh, J. Henrie, and E. Eason, for their help with fabricating the adhesive patches, machining parts for the wind tunnel test, and capturing the microscope video of the interlocking adhesives, respectively. We also gratefully acknowledge D. Lentink at Stanford and D. Altshuler at UBC Vancouver for their invaluable insights and discussions into bird flight, which lead to the concept of a latching wing prototype.

References

1. Hedrick, T.L.: Damping in flapping flight and its implications for manoeuvring, scaling and evolution (2011)
2. Gomez, J.C., Garcia, E.: Morphing unmanned aerial vehicles (2011)
3. Barbarino, S., Bilgen, O., Ajaj, R.M., Friswell, M.I., Inman, D.J.: A review of morphing aircraft (2011)
4. RoboSwift: Bio-inspired morphing-wing micro aerial vehicle. RoboSwift. N.p., n.d. Web. March 16, 2015. http://www.roboswift.nl/
5. Proctor, N., Lynch, P.: Manual of ornithology : Avian structure and function. American Scientist **82**, 288–289 (1994)
6. Gill, F.B.: Ornithology, 3rd edn. Freeman, W. H. & Company (2006)

7. Tucker, V.: Gliding birds: reduction of induced drag by wing tip slots between the primary feathers. Journal of Experimental Biology **180**(1), 285–310 (1993)

8. McCormick, B.: Aerodynamics, aeronautics, and flight mechanics, 2nd edn. Wiley (1994)

9. Day, P., Eason, E.V., Esparza, N., Christensen, D., Cutkosky, M.: Microwedge machining for the manufacture of directional dry adhesives. Journal of Micro and Nano-Manufacturing **1**(1), 011001 (2013)

10. Hawkes, E.W., Eason, E.V., Asbeck, A.T., Cutkosky, M.R.: The gecko's toe: Scaling directional adhesives for climbing applications. IEEE/ASME Transactions on Mechatronics 18(2) (2013)

11. Hawkes, E.W., Christensen, D.L., Han, A.K., Jiang, H., Cutkosky, M.R.: Grasping without squeezing : Shear force gripper with thin film dry adhesives. In: IEEE International Conference on Robotics and Automation (in press) (2015)

12. Videler, J.J.: Avian Flight, 1st edn. Oxford University Press (2006)

13. Herzog, K.: Flapping wing flight in nature and science. Mechanikus, 44–57 (1963-1964)

Obstacle-Avoidance Navigation by an Autonomous Vehicle Inspired by a Bat Biosonar Strategy

Yasufumi Yamada[1], Kentaro Ito[2], Arie Oka[1], Shinichi Tateiwa[1],
Tetsuo Ohta[1], Ryo Kobayashi[2], Shizuko Hiryu[1,3(✉)], and Yoshiaki Watanabe[1]

[1] Faculty of Life and Medical Sciences, Doshisha University, Kyotanabe 610-0321, Japan
shiryu@mail.doshisha.ac.jp
[2] Department of Mathematical and Life Sciences, Hiroshima University,
Higashi Hiroshima, Japan
[3] JST, PRESTO, 4-1-8 Honcho, Kawaguchi, Saitama 332-0012, Japan

Abstract. An autonomous vehicle controlled using real-time obstacle-avoidance algorithms and ultrasound was constructed to understand the active sensing system of the bat. The vehicle was designed to mimic bat behavior in which 1) the outgoing pulse was emitted toward the obstacle detected by the previous echo (obstacle aiming) and 2) the interpulse interval was adjusted using the distance to the detected object. As a result, the obstacle-aiming system facilitated obstacle avoidance by keeping the obstacle in the center of the beam sight of the vehicle.

Behavioral experiments involving a bat avoiding obstacles demonstrated that the bat responds to multiple echoes and then decides the direction of the next outgoing pulse. Based on this behavior, a multi-object-detection navigation system was proposed to enable a vehicle to move in more complicated space that it failed to navigate previously. Our findings suggest that the bat behavioral strategies provide new perspectives for engineering involving simple sensing.

Keywords: Echolocation · Active sensing · Pulse direction · Navigation · Time sharing processing

1 Introduction

The spatial perception behavior of visually guided animals, including humans, has been studied by measuring eye movements [1-3]. These studies found that animals adjusted their gaze to precede their direction of movement. By contrast, bats guide their movement (flight) acoustically using ultrasound; they emit pulses and analyze the returning echoes for spatial perception, a process known as echolocation [4, 5]. The directions of the echolocation pulses emitted by the bats can be considered to be the equivalent of the eye gaze of visually guided animals. Similarly, by measuring the pulse direction and emission timing of the bats, we can analyze the decision-making process involved in the spatial perception of bats using active sensing, which is thought to differ from that of visually guided (passive sensing) animals. This means that analysis of the bat echolocation strategy should provide new perspectives on the advantages of acoustical sensing for effective spatial perception.

© Springer International Publishing Switzerland 2015
S.P. Wilson et al. (Eds.): Living Machines 2015, LNAI 9222, pp. 135–144, 2015.
DOI: 10.1007/978-3-319-22979-9_14

Previously, we demonstrated that echolocating bats have several unique sensing strategies, compared to conventional human sonar systems [6]. However, few studies have investigated bat echolocation from an engineering perspective. In this study, we constructed a simple autonomous vehicle with embedded real-time obstacle-avoidance algorithms that used ultrasound. By comparing it with bat behavior, a field demonstration using the autonomous vehicle helped us to understand the essential design concept of ultrasound active sensing. The final objective of this study was to determine the behavioral algorithms of bat active sensing for engineering applications, such as sonar navigation systems of autonomous robots, including unmanned aircraft and automated guided vehicles.

Echolocating bats emit ultrasound from one emitter (the mouth or nose) and receive the echoes at two receivers (the right and left ears). In this study, we constructed an autonomous vehicle with an ultrasound sensing system that consisted of one transmitter and two receivers. The vehicle was programmed with embedded obstacle-avoidance algorithms using three different strategies—"obstacle aiming" and "interpulse interval (IPI) control"—inspired by bat behavior. Then, we demonstrated that these simple algorithms based on bat behavior enabled the vehicle to avoid obstacles without collisions. We embedded the proposed multiple-object detection and navigation system (Multi-ODNS) in the vehicle, so that it could respond to and avoid multiple objects.

2   Obstacle-Aiming System for Autonomous Vehicles

2.1   Vehicle Design

The proposed autonomous vehicle (30 × 15 × 25 cm H × W × L) consisted of three ultrasonic sensor units (one transmitter and two receivers), motor units, and a central processing unit (CPU) (Fig. 1A). The beam pattern of the transmitter corresponded to that of echolocating bats; $i.e.$, $Pipistrellus$ $abramus$, −6 dB (half-amplitude) off-axis angles from the direction of the emitted pulse on the pulse directivity pattern is ±40–50° (Yamada et al., in preparation). In addition, one servomotor connected to the sensor unit was used to control the direction of ultrasound emitted from the transmitter (pulse).

Figure 1B shows an example of a pulse and the returning echoes recorded by the right and left receivers. The received signals were half-wave rectified by an electric circuit for subsequent processing. For sensing, a 2-ms pulse with a center frequency of 40 kHz was used. When the vehicle detected an obstacle, the object position relative to vehicle was calculated from the echo arrival time, which was defined as the interval of the instant timings exceeding the voltage threshold between the pulse and echo. The distance from the vehicle to the object and the direction toward to the object were calculated using equations (1) and (2), respectively.

$$r_{obs} = \frac{c\Delta t_{pe}}{2} \tag{1}$$

$$\theta_{obs} = \sin^{-1} \frac{c\Delta t_{rl}}{d} \tag{2}$$

where d indicates the distance between the right and left receivers (Fig. 1C). In our vehicle, $d = 8$ cm. The distance from the vehicle to the object, r_{obs}, is calculated from the echo arrival time, Δt_{pe}. The direction toward the object, θ_{obs}, is calculated from the difference in the times that the echo arrived at the right and left receivers, Δt_{rl}. In this case, the distance and direction were calculated only for the fastest returning echo (the vehicle calculated the echo arrival time from the beginning of the echo stream, which means that the vehicle could detect only the nearest object).

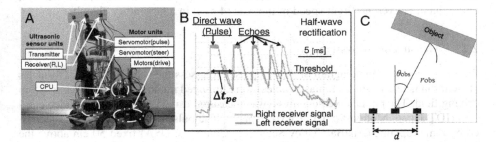

Fig. 1. (A) Photograph of the autonomous vehicle. (B) Representative observed signals (one pulse and four main returning echoes) recorded by the right and left receivers. (C) Schematic figure showing the obstacle direction and distance from the vehicle sensor unit (top view).

Previously, we demonstrated that bats change the pulse direction dynamically during obstacle-avoidance flight. Bats often directed their pulse toward certain objects, which is considered to be important for flight-path planning, while controlling the interpulse interval (IPI) to perceive the environment [7]. Therefore, we defined this pulse direction control behavior as an obstacle-aiming system, and embedded it in the vehicle using an IPI control mimicking the bat behavior. Here, we designed the vehicle so that it 1) emitted an outgoing pulse toward the nearest obstacle when obstacles are detected (Fig. 2B) and 2) then changed its travel direction to avoid the obstacle if the distance between the vehicle and obstacle was less than 1 m (Fig. 2A and B). When the detected obstacle is out of range of the beam sight or the angle between the aimed pulse and traveling direction reaches 90°, the sonar stage returns to its original traveling position. For comparison, we constructed a vehicle with a normal sensing system (without an obstacle-aiming system) in which the pulse direction was

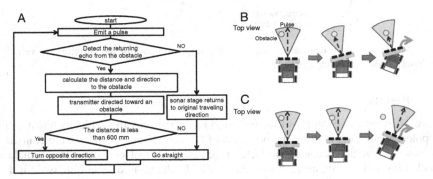

Fig. 2. (A) Flowchart of real-time obstacle avoidance by the vehicle using an obstacle-aiming system. Schematic illustrations of obstacle avoidance by the vehicles (B) with and (C) without the obstacle-aiming system.

fixed in the direction of movement (Fig. 2C). We compared the real-time obstacle-avoidance performance between vehicles with and without the obstacle-aiming system. Note that the vehicle controls the IPI in response to the distance between the vehicle and object in 0.1-s steps from 0.3 s to 0.8 s. The IPI of the vehicle was tenfold longer than that of bats because the vehicle driving speed was tenfold slower (38 cm/s) than the average flight speed of the bats (4 m/s) in a laboratory chamber.

2.2   Measurement System

Figure 3 is a schematic diagram of the setup used to assess the vehicle driving. Obstacles (12-cm-diameter, 1.5-m-high plastic poles) were arranged randomly within the 6 × 4.5-m driving field. The path the vehicle drove was measured using two high-speed video cameras (IDT Japan, MotionPro X3, Tokyo, Japan).   The pulse emission timing was recorded by a microphone (AnaBat; Titley Scientific, Columbia, USA) fixed 50 cm above the floor, located outside the driving field. The distance and direction of the obstacles, and pulse direction, were detected instantly by the vehicle and transmitted using wireless serial communication from the vehicle CPU to the personal computer (PC) in real-time. To quantify the obstacle-avoidance performance of the vehicle, the time driven without colliding with a pole was compared between vehicles with and without the obstacle-aiming system using the average driving time of three trials for each condition.

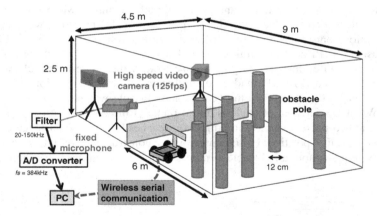

Fig. 3. The system used to measure the pulse direction and path of the autonomous vehicle in the chamber while avoiding obstacles

2.3   Results

Figure 4A shows an example of the path (red line) driven by the vehicle while avoiding the poles. Black dots indicate the positions of the vehicle when pulses were emitted. In this case, the vehicle avoided obstacles repeatedly without collision. The vehicle responded instantly to the nearest obstacle by simply changing direction to avoid it. During the drive, the vehicle aimed pulses toward each pole (Fig. 4B) and controlled the IPI according to the distance to the obstacle (Fig. 4C). The average driving time of the vehicle with the obstacle-aiming system (42.75 (range 28–55) s) was longer than that of the vehicle without the obstacle-aiming system (20 (range 13–35) s). This suggests that the obstacle-aiming system facilitated navigation using acoustic sensing, because keeping the object in the center of the beam facilitated tracking of the object by the vehicle.

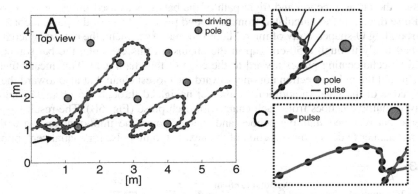

Fig. 4. Top view of the driving path (A) and pulse direction (B) during obstacle-avoidance driving of the vehicle using obstacle-aiming and IPI control. The vehicle decreased the IPI as it approached an obstacle (C).

Figure 5 shows a representative failure in which the vehicle with the obstacle-aiming system collided with a pole. In this case, the vehicle avoided pole 1 and the chamber wall. However, after making a U-turn to avoid the wall, the vehicle struck pole 2. This happened because the vehicle paid attention only to the closest obstacle, not to the next obstacle (pole 2). Therefore, there are limitations to considering multiple echoes as a single echo; *i.e.*, the vehicle responds only to the nearest object. It is unclear how bats respond to multiple echoes returned from complex surroundings.

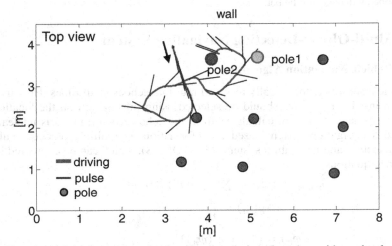

Fig. 5. Top view of driving path and pulse direction during obstacle avoidance by the vehicle with the obstacle-aiming system. In this case, the vehicle collided with pole 2 after making a U-turn to avoid the wall.

Figure 6A shows a representative flight path and pulse direction of a bat (*Pipistrellus abramus*) during obstacle-avoidance flight in the same chamber. Obstacles (plastic chains) were suspended vertically from the chamber ceiling, and we investigated how the bats controlled their flight and pulse directions to avoid obstacles. The system used to measure the pulse direction and flight path of the bats was as used in our previous study [7]. Pulse direction was calculated from the sound pressure level of the pulse recorded by a surrounding O-shaped array of microphones. In the experiment, the bat (wing length 15 cm) had to fly through a 50-cm gap in the chains. In this example, the bat sometimes aimed its echolocating pulses toward to the edge of the chain gate. This means that the bat adjusted the pulse direction not only toward the closest chain, but also toward the one at the edge of the gate. In addition, as the bat navigated through the obstacle chains, it received multiple echoes from each chain with each pulse (Fig. 6B). The result suggests that the bat received multiple echoes and then selected the important one in a time-sharing manner to decide the direction of the next outgoing pulse for spatial perception.

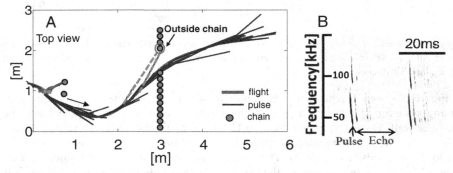

Fig. 6. (A) Top view of the flight trajectory and pulse direction during obstacle-avoidance flight by a bat. (B) Spectrogram of the pulse and echoes during this flight recorded by a telemetry microphone on the back of the bat.

3   Multi-Object-Detecting Navigation System

3.1   Vehicle Navigation Algorithm

The vehicle sensors automatically received multiple echoes at all times. How the next direction in which to move should be selected using information on the locations of multiple obstacles that were calculated from the echo arrival timings is unclear. We designed a navigation system based on information on multiple objects (multiple-object detection and navigation system; Multi-ODNS), which can be expressed by the following equations:

$$\theta_{move} = \arg(\alpha - \sum_{n=0}^{s}(g(r_n)h(\varphi_n)e^{i\varphi_n})) \tag{3}$$

$$g(r_n) = \frac{1}{r_n} \tag{4}$$

$$h(\varphi_n) = (1 + \cos(\varphi_n))^2 \tag{5}$$

where s indicates number of objects detected and n indicates the number of each object. Briefly, the angle to the current movement direction of the vehicle, θ_{move}, can be calculated by adding the repulsive force vectors from each echo and the force vector α of the present direction of movement of the vehicle. The length of each repulsive force vector can be obtained from both the distance from the vehicle to the obstacle, $g(r_n)$, and the obstacle direction relative to the driving direction, $h(\varphi_n)$. In this experiment, the pulse direction was fixed in the driving direction to simplify the analysis.

In conventional sensing, the vehicle responds only to the closest obstacle and to avoid it moves in the opposite direction. This requires the vehicle to change direction suddenly because the position of the closest obstacle changes in a discontinuous manner during a drive in an obstacle-filled environment. In comparison, the Multi-ODNS model continuously provides the movement angle of the vehicle by using information from all detected obstacles. The continuous repulsive forces can be used as long as the obstacles are located within the beam sight of the vehicle, which helps the vehicle to generate a smooth steering path during obstacle avoidance. Using these algorithms, the vehicle could drive safely based on integrated information on all of the detected obstacles located within the beam sight. Moreover, the obstacle-avoidance path is adjusted only by one parameter, the force vector α of the current movement direction. The Multi-ODNS model appears to be useful and results in a smooth path by using a simple sensing design.

In this study, the conventional navigation system, which detects only the closest obstacle, was called Mono-ODNS. We compared the avoidance performance of the vehicle using Multi-ODNS and Mono-ODNS.

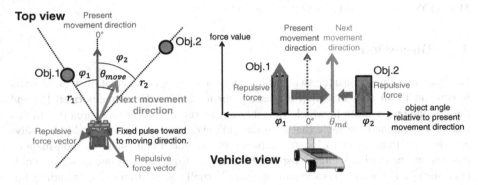

Fig. 7. Schematics of top (A) and side (B) views of the vehicle and obstacles. The next movement direction of the vehicle could be defined by the repulsive forces of each object.

3.2   Results

Figure 8A shows an example of the path of the vehicle using Multi-ODNS. Gray circles indicate the locations of the obstacles and the open symbols indicate the obstacle locations detected by the vehicle. The estimated obstacle positions that were detected by each pulse emitted by the vehicle were labeled as the nearest, second, third and fourth in ascending order of distance from the vehicle. This means that the vehicle decided its movement angle based on not only the nearest echo but also the second, third, and fourth individual echoes. In this case, the vehicle successfully avoided the obstacles by considering multiple obstacles detected by each pulse. Figure 8B compares the paths of Multi-ODNS (red line) and Mono-ODNS (green line). This figure shows that Mono-ODNS collided with the wall because the vehicle paid attention only to the pole and it detected the wall behind the pole too late to take avoiding action. In comparison, Multi-ODNS avoided both the poles and the wall because the vehicle had already detected the wall behind the poles and decided to pass through the gap between the pole and the wall. These findings suggest that Multi-ODNS facilitates obstacle avoidance by ultrasound, suggesting that this simple mathematical model is a robust avoidance algorithm in a complex environment.

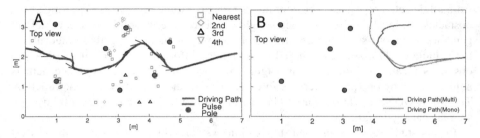

Fig. 8. (A) Top view of the path driven to avoid obstacles by the vehicle using the multiple-object-detection navigation system. The open symbols indicate the obstacle locations detected by the vehicle. (B) Top view of the paths driven during obstacle avoidance by vehicles using Multi-ODNS (red line) and Mono-ODNS (green line).

4   Discussion

Navigation studies consider two basic conditions: mapped navigation that restricts movement to an area in which the arrangement of obstacles is memorized [8] and non-mapped navigation in which the vehicle is driven in an unknown area [9]. In this study, we designed a vehicle that used obstacle-avoidance algorithms and two receivers and one transmitter to locate obstacles in the horizontal plane. In non-mapped navigation, the aiming strategy inspired by bats was found to facilitate obstacle avoidance using such simple sensor-unit sensing. Visually guided animals, including humans [10], zebra finches [3], and falcons [11], show saccadic eye movement in which the visual fovea is kept directed toward certain key objects. The saccadic eye movement keeps the key object in the sweet spot of the sensor unit (the fovea). This seems to be common to acoustic sensing by bats, in which the pulses are directed toward a key object. Therefore, the object-aiming strategy appears to be important for both visually and acoustically guided animals during movement. A comparative study of acoustical and visual gaze control by animals might provide unique insights into simple-design sensing in engineering.

Few studies have devised navigation algorithms for obstacle avoidance using bat behaviors from a mathematical perspective. For example, [12] stated that bat sensorimotor (steering and speed) control was modeled mathematically based on the Relatively Parallel Adapted Frame (RPAF) [13-16] during pursuit flight, in which a bat was tasked to capture a tethered target (meal worm) in an environment with obstacles. However, the mathematical model proposed in that study assumed that the pulse direction was fixed in the flight direction of the bat. Recently, both the pulse and flight directions were considered for bat navigation during flight in an enclosed mapped chamber (the bat already knew the chamber size before the flight) using nonlinear dynamics in a mathematical model [17]. In this model, the bat is assumed to pay attention only to the closest object. These studies investigated the bat behavior using numerical simulations, and few field studies have used a vehicle based on bat echolocation [18]. In SONAR sensing, the sensing performance depends on the situation. A comprehensive assessment of behavioral, mathematical, and field demonstrations is necessary to understand bat biosonar for practical engineering applications.

In this study, we embedded one of two rules in the vehicle: an obstacle-aiming system and a multi-object-detecting navigation system. We found that the bat behavioral strategies facilitated the avoidance performance of the vehicle using a simple sensing style. Recent practical use of sensor systems designed for autonomous robots has been based on sensor array scanning systems, which requires high-performance CPUs and graphics processing units (GPUs) [9]. The sensorimotor behaviors of animals might lead to development of new sensing systems with simple designs, which may be superior for signal processing and design materials.

Bats change their pulse direction dynamically toward key objects in succession. This suggests that the bats choose an aiming object from among multiple objects using time-sharing. We plan to improve our vehicle system to include a time-sharing obstacle-aiming system, and to process multiple echoes based on the pulse direction selectivity rules of bats.

Acknowledgments. We thank Nobutaka Urano for his assistance in capturing bats in the wild. This work was partly supported by a Grant-in-Aid for Young Scientists (A) (Grant No. 70449510) from the Japan Society for the Promotion of Science (JSPS), JST PRESTO program.

References

[1] Land, M.F., Collett, T.S.: Chasing Behaviour of Houseflies (Fannia canicularis). J. Comp. Physiol. **89**, 331–357 (1974)
[2] Land, M.F., Tatler, B.W.: Steering with the head: The visual strategy of a racing driver. Current Biology **11**(15), 1215–1220 (2001)
[3] Eckmeier, D., Geurten, B.R., Kress, D., Mertes, M., Kern, R., Egelhaaf, M., Bischof, H.J.: Gaze strategy in the free flying zebra finch (Taeniopygia guttata). PLoS One **3**(12), e3956 (2008)
[4] Simmons, J.A.: Perception of Echo Phase Information in Bat Sonar. Science **204**(4399), 1336–1338 (1979)
[5] Moss, C.F., Surlykke, A.: Auditory scene analysis by echolocation in bats. The Journal of the Acoustical Society of America **110**(4), 2207 (2001)
[6] Hiryu, S., Hagino, T., Riquimaroux, H., Watanabe, Y.: Echo-intensity compensation in echolocating bats (Pipistrellus abramus) during flight measured by a telemetry microphone. The Journal of the Acoustical Society of America **121**(3), 1749 (2007)

[7] Yamada, Y., Oka, A., Hiryu, S., Ohta, T., Riquimaroux, H., Watanabe, Y.: Investigation of acoustic gaze strategy by Pipistrellus abramus and Rhinolophus ferrumequinum Nippon during obstacle avoidance flight, p. 010009

[8] Cox, I.J.: Blanche-an experiment in guidance and navigation of an autonomous robot vehicle. IEEE Transactions on Robotics and Automation 7(2), 193–204 (1991)

[9] Baralli, F., Couillard, M., Ortiz, J., Caldwell, D.G.: GPU-based real-time synthetic aperture sonar processing on-board autonomous underwater vehicles, pp. 1–8

[10] Land, M.F.: Motion and vision: why animals move their eyes. Journal of Comparative Physiology A 185(4), 341–352 (1999)

[11] Kane, S.A., Zamani, M.: Falcons pursue prey using visual motion cues: new perspectives from animal-borne cameras. The Journal of Experimental Biology 217(2), 225–234 (2014)

[12] Freyman, L., Livingston, S.: Obstacle Avoidance and Boundary Following Behavior of the Echolocating Bat. Merit Bien Program, 1-9 (2008)

[13] Bishop, R.L.: There is more than one way to frame a curve. American Mathematical Monthly, 246–251 (1975)

[14] Justh, E., Krishnaprasad, P.: Natural frames and interacting particles in three dimensions, pp. 2841–2846

[15] Justh, E.W., Krishnaprasad, P.: Steering laws for motion camouflage. Proceedings of the Royal Society A: Mathematical, Physical and Engineering Science 462(2076), 3629–3643 (2006)

[16] Reddy, P., Justh, E., Krishnaprasad, P.: Motion camouflage in three dimensions, pp. 3327–3332

[17] Aihara, I., Yamada, Y., Fujioka, E., Hiryu, S.: Nonlinear dynamics in free flight of an echolocating bat. Nonlinear Theory and Its Applications, IEICE 6(2), 313–328 (2015)

[18] Kuc, R.: Biomimetic sonar recognizes objects using binaural information. The Journal of the Acoustical Society of America 102(2), 689–696 (1997)

Development of Piezoelectric Artificial Cochlea Inspired by Human Hearing Organ

Young Jung$^{(\boxtimes)}$, Jun-Hyuk Kwak, Hanmi Kang, Wandoo Kim, and Shin Hur

Korea Institute of Machinery and Materials, Daejeon, South Korea
{yjung,jhkwak,kanghanmi,wdkim,shur}@kimm.re.kr

Abstract. Miniaturized artificial hearing organ with excellent sensitivity and wide dynamic frequency range over human hearing range, while requiring small amount of energy, is important step to develop artificial systems interacting in human living space. This paper presents the development of piezoelectric artificial cochlea (PAC) capable of analyzing incoming vibratory signals over human hearing range without external power source. The design, component and function of PAC were inspired by those of human cochlea. The PAC was made of corona-poled piezoelectric thin film with vibrating membrane part of unique shape. The vibration displacement of membrane was measured using laser Doppler vibrometer and analyzed to show the frequency separation of the developed PAC. The experimental results of mechanical vibratory behavior demonstrated successful separation of incoming signals into 13 different frequency bands depending on their frequency over 300 Hz ~ 6,000 Hz.

Keywords: Hearing organ · Artificial cochlea · Sound analyzer · Piezoelectric · Basilar membrane

1 Introduction

Hearing is one of vital senses that a living thing such as insect, fish, bird, and mammal has. Hearing organs of different types of animals have been evolved over time for their successful survival and vary in their shapes, compositions and abilities in hearing.

Human ear is composed of outer, middle, and inner ear and it is known as a miniaturized acoustic transducer of great sensitivity (20 μPa ~ 60 Pa) and wide dynamic frequency range (20 Hz ~ 20 kHz). Understanding of human hearing is an important step to develop an auditory system that can be implanted or installed on artificial systems operating or interacting in human living space.

In the effort to understand the mechanosensory system in human ear, scientists have found the phenomenon of sound analysis and signal transduce within a cochlea – a snail shell-shaped organ in the inner ear [1]. Unlike man-made acoustic systems that require various energy-consuming components such as microphone, sound amplifier, and sound frequency analyzer, human cochlea receives vibratory signals induced by sound through middle ear bones, analyzes the signals depending on their frequencies, and converts them into electrical signals with a basilar membrane and stereocilia.

© Springer International Publishing Switzerland 2015
S.P. Wilson et al. (Eds.): Living Machines 2015, LNAI 9222, pp. 145–152, 2015.
DOI: 10.1007/978-3-319-22979-9_15

The basilar membrane has a unique structure of trapezoidal shape with varying thickness over its length, which is thick and narrow near the base (close to oval window) and becomes thinner and wider near the apex. This enables the basilar membrane to vibrate or resonate differently depending on the frequency of incoming signals and to function as a mechanical sound analyzer. There are hair cells array along the length of a basilar membrane and as a local area of the basilar membrane vibrates depending on the frequency of incoming signal, the stereocilia on the hair cells nearby deform and generate bioelectrical signal to be transmitted to brain through hearing nerve.

Inspired by human cochlea, several research groups have developed sound analyzers based on the structure of a basilar membrane [2, 3, 4, 5, 6]. However, the focus of their research was to mimic the frequency separation functions of the basilar membrane and did not show signal conversion of mechanical movement of basilar membrane to electrical signal. Recently, the signal conversion of the basilar membrane upon sound signal input was reported over narrow frequency bandwidth [7, 8, 9]. However, the reported frequency separation ranges were 1.4 kHz ~ 4.9 kHz in silicone oil [8] and 2.5 kHz ~ 13.5 kHz in air [9] while the voice frequency band ranges from 300 Hz ~ 3,400 Hz and most first and second formant frequencies of the vowel sounds are below 1 kHz and 2.5 kHz [10].

In this paper, we present the development of a piezoelectric artificial cochlea capable of frequency separation over 300 Hz ~ 6,000 Hz to distinguish human voice sound through the design modification of artificial basilar membrane and inclusion of package components for total implantable artificial cochlea.

2   Design and Fabrication

The developed piezoelectric artificial cochlea (PAC) is composed of a piezoelectric artificial basilar membrane (ABM) part and a package part. The ABM part has a corona-poled PVDF film (25.4 μm thick) as a membrane material and a printed circuit board (PCB) substrate defining the shape of the membrane (Fig. 1).

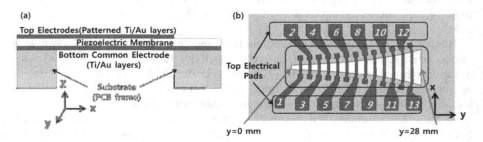

Fig. 1. 2D schematic of artificial basilar membrane part (a) from side view and (b) from top view

Thirteen patterned top electrodes and common bottom electrode on the top and bottom of the film were fabricated by shadow mask deposition of Cr / Au layers (10 nm / 100 nm thick) respectively. The length in y axis of opening in the PCB substrate is 28 mm and the width in x axis varies logarithmically from 0.97 mm (y = 0 mm) to 8.0 mm (y = 28 mm). The electrical connection in the PCB substrate was designed to measure the piezoelectric signals from multiple electrodes on the membrane simultaneously. The package part is composed of a liquid chamber and two ports. The bottom and two long side walls of a liquid chamber was closed with no opening, while the top of the liquid chamber was completely open before assembly with an ABM part and two short side walls have one circular opening on each side. The fabricated ABM part is attached firmly to the package part over the liquid chamber and the membrane makes direct contact with the liquid in the package part. Both ports were capped with thin elastomer film to seal the liquid in the package and to work as an artificial oval window for vibratory signal input similar. The fabricated PAC is shown in Fig. 2.

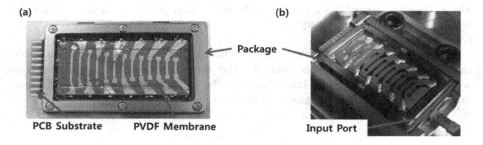

Fig. 2. Fabricated PAC (a) from top view and (b) with input port in contact with micro actuator

3   Experimental Results

The frequency separation capability of the fabricated PAC was characterized by applying vibratory input signal into liquid chamber through input port and measuring the displacement over the membrane. The vibration was applied to the base input port of the package part, where the ABM has narrow width, using a PICMA® Stack Multilayer Piezoactuator (P-883.11, Physik Instrumente (PI) Ceramic, Lederhose, Germany) positioned precisely with a micrometer (Fig. 3). For characterization, the PAC was securely fastened to a package holder and package platform. The package platform then can be bolted to anti-vibration optical table to minimize the effect of environmental vibration on the PAC.

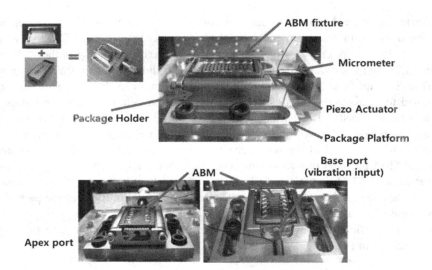

Fig. 3. The PAC assembled in a package holder and package platform with piezoactuator and micrometer for measuring vibratory behavior

The vibratory input signal of the piezoactuator represents the middle ear vibration into oval window of human cochlea. The membrane displacement in z axis was measured with scanning laser Doppler vibrometer system (PSV-I-400 LR and OFV-505, Polytec, Germany) as shown in Fig. 4.

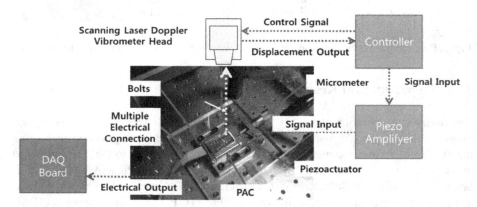

Fig. 4. Experimental setup for charactering the develop PAC; both mechanical vibration and piezoelectric signal measurement setup

In human cochlea, a local area of basilar membrane near base area responds most at high frequency range, while that of basilar membrane near apex responds most at low frequency range. Fig. 5 shows the membrane displacement at various frequencies. The PAC also show the vibratory behavior similar to that of human cochlea; at low frequency, the wider area near apex responded more than the narrower area and the position of maximum displacement moved toward the narrower area near base as the frequency of input signals increased.

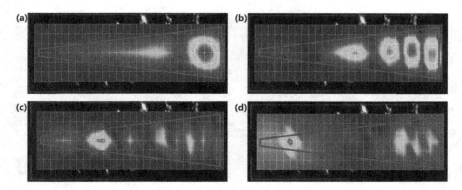

Fig. 5. Measured vibration of ABM upon input signal of (a) 581 Hz, (b) 1413 Hz, (c) 2619 Hz, and (d) 4606 Hz

The measured data were analyzed to find the maximum displacement on the membrane and the position of maximum displacement at each frequency.   White noise input signal of 0 ~ 10 kHz was applied to the PAC using a piezoactuator and the maximum displacement measured was plotted over frequency (Fig. 6).

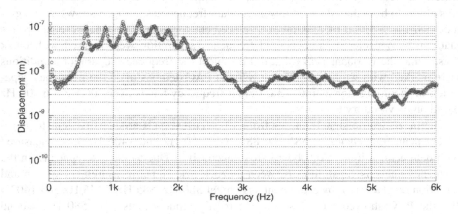

Fig. 6. The maximum displacement of membrane over frequency

The maximum displacements were 100 nm, 97.1 nm, 130 nm, 137 nm, 104 nm, 84.9 nm and 53.9 nm at first seven local resonant frequencies of 581 Hz, 881 Hz, 1150 Hz, 1406 Hz, 1638 Hz, 1856 Hz and 2100 Hz. Over the frequency range of 300 Hz ~ 6,000 Hz, 19 resonant peaks were found in the maximum displacement plot, but while the maximum displacements at local resonant frequencies below 2 kHz were around 100 nm, those above 3 kHz were around 10 nm.

The position of maximum displacement at each frequency is shown in Fig. 7. As expected from the displacement map in Fig. 4, the position of maximum displacement moves from apex area to base area as the frequency of input signals increased.

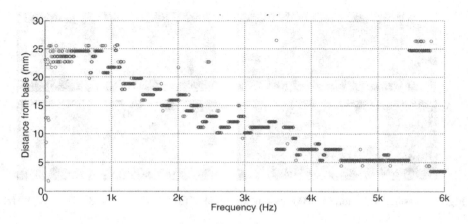

Fig. 7. The position of maximum displacement on the membrane over frequency

Out of 19 resonant peaks over 300 Hz ~ 6,000 Hz, the number of frequency separation bands of PAC was determined by counting the number of resonant frequencies at which the PAC had different positions of maximum displacement and the position was closer to base than those at lower resonant frequencies.   Fig. 8 shows the magnitude and the position of maximum displacement together. The local resonant peaks that can be considered as successful frequency separation were marked with black dots, while the resonant peaks that did not meet the frequency separation conditions were marked with white dots. The developed PAC demonstrated the frequency separation of incoming signals into 13 different frequency bands over 300 Hz ~ 6,000 Hz based on its vibratory behavior.

The piezoelectric signal outputs from 13 electrical pads around first four local resonant peak frequencies of 581 Hz, 881 Hz, 1150 Hz and 1406 Hz were measured with data acquisition (DAQ) board simultaneously.   Due to the discrepancies in the frequency analysis capabilities between piezoelectric signal measurement system and vibration displacement measurement system, at 580 Hz, 885 Hz, 1145 Hz and 1402.5 Hz, the PAC showed local resonant electrical output signals.   At 580 Hz, second electrical pad from apex (#12) generated highest output voltage of 12.7 µV among 13 electrical pads.   At 885 Hz and 1145 Hz, third electrical pad from apex (#11) generated highest voltage of 16.0 µV and 28.7 µV, while at 1402 Hz, fourth electrical pad from apex (#10) generated highest voltage of 16.7 µV.

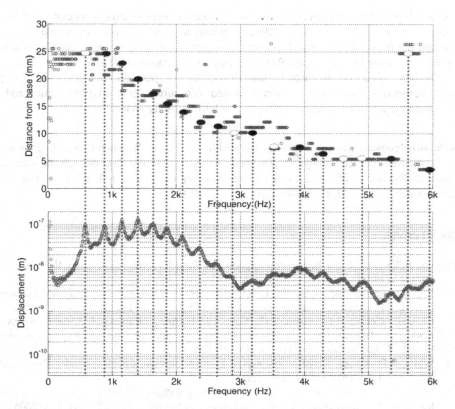

Fig. 8. The number of frequency separation bands of the developed PAC with the local resonant peaks and position of maximum displacement plots

4 Conclusion

In this research, we have developed the piezoelectric artificial cochlea inspired by the shape, component, and function of human cochlea. The developed PAC has a tapered trapezoidal shape membrane with varying width from 0.97 mm to 8 mm defined by PCB substrate opening and a package part with a liquid chamber and input ports. This unique shape gave the PAC the ability to analyze incoming signals based on their frequencies. Through the measurement and analysis of the membrane vibration displacement with scanning laser Doppler vibrometer system, 19 different resonant peak frequencies were found over 300 Hz ~ 6,000 Hz. The developed PAC demonstrated the frequency separation of incoming signals into 13 different frequency bands over 300 Hz ~ 6,000 Hz based on its vibratory behavior by counting the number of the resonant peak frequencies that meets the conditions of frequency separation. The maximum electrical output voltages at first four resonant peak frequencies were measured from the second, third, third and fourth electrical pads from apex (#12, #11, #11, #10) . With the design modification and inclusion of package part to PAC, the developed PAC can cover the human audible frequency range, especially the first and second formant frequencies. Also, by removing power consuming sound processor and microphone, the overall power consumption of from artificial cochlea sensor system including the developed PAC is expected to be around 30 mW.

The developed PAC inspired by human cochlea is expected to provide another step in implementing low power-consuming sound recognition system within human audible range for artificial systems interacting in human living space.

Acknowledgement. This research was supported by the Pioneer Research Center Program and by the Convergence Technology Development Program for Bionic Arm through the National Research Foundation of Korea funded by the Ministry of Science, ICT & Future Planning (2009-0082960, 2010-0018347, NRF-2014M3C1B2048177)

References

1. von Békésy, G.: Some biophysical experiments from fifty years ago. Annu. Rev. Physiol. **36**, 1–16 (1981)
2. Xu, T., Bachman, M., Zeng, F., Li, G.: Polymeric Micro-cantilever Array for Auditory Front-end Processing. Sens. Actuators A Phys. **114**(2–3), 176–182 (2004)
3. Chen, F., Cohen, H.I., Bifano, T.G., Gastle, J., Fortin, J., Kapusta, C., Mountain, D.C., Zosuls, A., Hubbard, A.E.: A Hydromechanical Biomimetic Cochlea: Experiments and Models. J. Acoust. Soc. Am. **119**(1), 394–405 (2006)
4. White, R.D., Grosh, K.: Microengineered Hydromechanical Cochlear Model. Proc. Natl. Acad. Sci. USA **102**(5), 1296–1301 (2005)
5. Wittbrodt, M.J., Steele, C.R., Puria, S.: Developing a Physical Model of the Human Cochlea using Microfabrication Methods. Audiology and Neurotology **11**(2), 104–112 (2005)
6. Jang, J., Kim, S., Sly, D.J., O'leary, S.J., Choi, H.: MEMS Piezoelectric Artificial Basilar Membrane with Passive Frequency Selectivity for Short Pulse Width Signal Modulation. Sens. Actuators A Phys. **203**, 6–10 (2013)
7. Inaoka, T., Shintaku, H., Nakagawa, T., Kawano, S., Ogita, H., Sakamoto, T., Hamanishi, S., Wada, H., Ito, J.: Piezoelectric Materials Mimic the Function of the Cochlear Sensory Epithelium. Proc. Natl. Acad. Sci. USA **108**, 18390–18395 (2011)
8. Shintaku, H., Nakagawa, T., Kitagawa, D., Tanujaya, H., Kawano, S., Ito, J.: Development of Piezoelectric Acoustic Sensor with Frequency Selectivity for Artificial Cochlea. Sens. Actuators A Phys. **158**, 183–192 (2010)
9. Jung, Y., Kwak, J.-H., Lee, Y.H., Kim, W.D., Hur, S.: Development of Multi-channel Piezoelectric Acoustic Sensor based on Artificial Basilar Membrane. Sensors **14**, 117–128 (2014)
10. Catford, J.C.: A Practical Introduction to Phonetics. Clarendon Press, Oxford (1988)

Visual Odometry and Low Optic Flow Measurement by Means of a Vibrating Artificial Compound Eye

Fabien Colonnier, Augustin Manecy, Raphaël Juston, and Stéphane Viollet[✉]

Aix-Marseille Université, CNRS, ISM UMR 7287, 13288 Marseille Cedex 09, France
stephane.viollet@univ-amu.fr
http://www.biorobotics.eu

Abstract. In this study, a tiny artificial compound eye (diameter 15mm) named CurvACE (which stands for Curved Artificial Compound Eye), was endowed with hyperacuity, based on an active visual process inspired by the retinal micro-movements occurring in the fly's compound eye. A periodic (1-D, 50-Hz) micro-scanning movement with a range of a few degrees (5°) enables the active CurvACE to locate contrasting objects with a 40-fold greater accuracy which was restricted by the narrow interommatidial angle of about 4.2°. This local hyperacuity was extended to a large number of adjacent ommatidia in a novel visual processing algorithm, which merges the output signals of the local processing units running in parallel on a tiny, cheap micro-controller requiring very few computational resources. Tests performed in a textured (indoor) or natural (outdoor) environment showed that the active compound eye serves as a contactless angular position sensing device, which is able to assess its angular position relative to the visual environment. As a consequence, the vibrating compound eye is able to measure very low rotational optic flow up to $20°/s$ and perform a short range odometry knowing the altitude, which are two tasks of great interest for robotic applications.

Keywords: Hyperacuity · Robotics · Eye micro-movements · Insect vision

1 Introduction

The curved compound eye depicted in the present study is the first example of an artificial compound eye which is able to locate targets or determine its orientation relative to the visual environment, with a much greater accuracy than the one imposed by its optics (i.e. by the interommatidial and acceptance angles). Findings on the visual processes at work in the fly inspired us to design and construct an active version of the previously described cylindrical compound eye CurvACE, which was based on an array of artificial ommatidia [4] and [12]. This active compound eye features two properties that are usually banned by

© Springer International Publishing Switzerland 2015
S.P. Wilson et al. (Eds.): Living Machines 2015, LNAI 9222, pp. 153–163, 2015.
DOI: 10.1007/978-3-319-22979-9_16

optic sensor designers because they detract from the sharpness of the resulting images: optical blurring and vibration. The first micro-scanning sensor based on the periodic retinal micro-movements observed in the fly (for a review, see [11]) was presented in [9]. This scanning eye was capable of detecting low levels of translational optic flow, such as those encountered by a mobile robot around its heading direction (the focus of expansion).

In the present study, hyperacuity was based on the existence of an overlap between the receptive visual fields of 2 neighbouring ommatidia. In a fly's ommatidium, each photoreceptor has a quasi-Gaussian ASF (Angular Sensitivity Function), which is given by the convolution of the point spread function of the facet lens with the photoreceptors diameter [5] - [10].

An artificial curved compound eye can provide a useful optic flow measurement and motion detection ability (see [4]). Moreover, we established here that the same compound eye subjected to active periodic micro-scanning movements combined with appropriate visual processing algorithms can be endowed with angular position sensing abilities.

Brückner et al. assessed the hyperacuity of an artificial compound eye in terms of its ability to locate a point source or an edge [1]. The robustness of these visual sensors' performances with respect to the contrast and the distance from the object targeted has not been addressed so far, however. In our previous paper [2], we report that the active CurvACE enabled a robot to perform visual odometry and track a target moving over a textured ground. These two properties are of great interest for designing autonomous robotic aerial vehicles. Section 2 describes the active version of the CurvACE eye inspired from the fly's retinal micro-movements. A complete description of the visual processing algorithms resulting in hyperacuity is presented in section 3, and section 4 describes the ability of the active CurvACE to assess its angular orientation and speed based on the novel sensory fusion algorithm developed here. Another application shows that this measurement can also be used in a linear positioning task, relatively to the environment.

2   Eye Movements Inspired by the Fly's Visual Micro-scanning Movements

In this study, visual hyperacuity results from an active process whereby periodic micro-movements are continuously applied to an artificial compound eye. This approach was inspired by the retinal micro-movements observed in the eye of the blowfly *Calliphora*. Unlike the fly's retinal scanning movements, which result from the translation of the photoreceptors in the focal plane of each individual facet lens (for a review on the fly's retinal micro-movements see [11]), the eye tremor applied here to the active CurvACE by means of a micro-stepper motor (figure 1) results from a periodic rotation of the whole artificial compound eye.

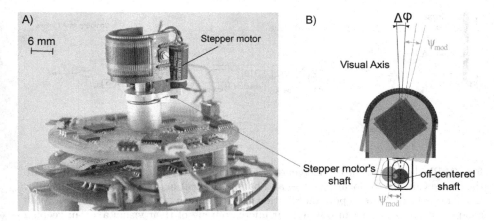

Fig. 1. Artificial vibrating compound eye. A) Active CurVACE sensor with its vibrating mechanism based on the use of a tiny stepper motor. A small periodic movement is applied to the artificial CurVACE compound eye composed of 630 ommatidia, which has a diameter of only 15 mm, giving a panoramic horizontal FOV of 180° and a vertical FOV of 60° (see [4] for details). B) Diagram of the micro-scanning compound eye subjected to active periodic rotational movements generated by a miniature eccentric mechanism. The angular vibration ψ_{mod} is generated by the miniature stepper motor shown here in the form of an orange shaft and a purple off-centered shaft, which translates along the oblong hole. The scanning frequency can be easily adjusted by changing the rotational speed of the motor. The scanning amplitude depends on the offset between the motor shaft and the off-centered shaft.

3   Insights Into the Visual Processing Algorithm

3.1   Edge and Bar Location by an Active Compound Eye

As shown in figure 2, each local processing unit (LPU) is connected to a pair of adjacent ommatidia. For further details about the implementation and characterization of the LPU, see [7] and [6].

One LPU's output signal results from the difference-to-sum ratio of the demodulated photoreceptor signals, as follows:

$$Output_{Pos} = \frac{Ph_{1demod} - Ph_{2demod}}{Ph_{1demod} + Ph_{2demod}} \tag{1}$$

Equation (1) in the original model for the VODKA sensor [7] placed in front of an edge can be greatly simplified as follows:

$$\theta_i(\Psi_c) = K_1 \cdot tanh\left(\frac{4\Delta\varphi log(2)}{\Delta\rho^2}\Psi_c\right) \tag{2}$$

with θ_i the output signal of an LPU (see figure 2). Due to the limited FOV of each LPU, the angular position θ_i of a contrasting object ranges from $-2°$ to $2°$.

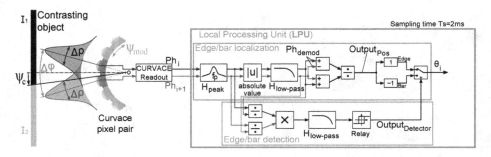

Fig. 2. Block diagram of the elementary local processing unit (LPU) integrated into the active Curvace for locating edges and bars with great accuracy. The stepper motor (see figures 1) generates a periodic rotation $\psi_{mod}(t)$ (green double arrows) of the overall compound eye, resulting in the angular microscanning of their visual axes, in keeping with a sinusoidal law (scanning frequency 50Hz, amplitude about 5° peak to peak with $\Delta\varphi = 4.2°$ and $\Delta\rho = 4.2°$). Two parallel processing pathways (one for edge/bar localization and one for edge/bar detection) were implemented. The edge/bar localization block gives the local angular position θ_i of an edge or bar placed in the visual field of two adjacent ommatidia. The edge/bar detection block detects the presence of a bar or edge and triggers the appropriate sign: +1 for edges and -1 for bars. The principle of this detector is described in [6]. The central frequency f_p of the peak filter is equal to the scanning frequency (50Hz), whereas the cut-off frequency of the second order digital low-pass filter is equal to 10Hz. Adapted from [6].

The photoreceptor output signals (Ph_i and Ph_{i+1}) were first digitized by means of a 10-bit analog-to-digital converter integrated into each column of the CurvACE eye (see [4] for further details). A peak filter acts as both a differentiator and a selective filter centered at the scanning frequency ($f_p = f_{mod}$). The classical amplitude demodulation procedure used here was based on an absolute value function cascaded with a low-pass filter to smooth out the photoreceptor's output signals. Arithmetic operations were performed to obtain the sensor's output signal, as described in [7] - [6]. An innovative edge/bar detector was also included and used to select the sign of the $Output_{Pos}$ (see [6] for further details about the Edge/bar detection process).

3.2   Hyperacute Localization of Contrasting Bars and Edges

The characteristic static curve of the active CurvACE obtained with a contrasting edge and a black bar 1cm in width subtending an angle of 2.8° is presented in figure 3.

As was to be expected in view of equation (2), the curve in figure 3a has a tangent hyperbolic profile with respect to the angular position of the edge. It can be clearly seen by comparing the two curves plotted in figure 3 that the slopes of the characteristic static curves obtained with a bar and an edge are inverted. A theoretical explanation for this inversion of the slopes is given in

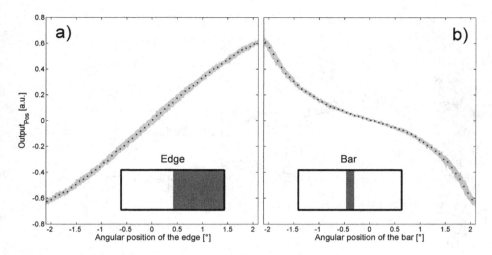

Fig. 3. Characteristic static curves of one LPU (see figure 2). The $Output_{Pos}$ signal is plotted versus the angular position of a) an edge or b) a bar placed 20 cm in front of active CurvACE rotating in $0.075°$ steps, each lasting 0.5s. It can be clearly seen from this figure that the slope of the characteristic static curve obtained with a bar is inverted in comparison with that obtained with an edge. Bars therefore have to be distinguished from edges in order to select the appropriate sign of the gain K (see figure 2). Adapted from [2].

[6]. This inversion fully justifies the use of an edge/bar detector (see figure 2) to compensate for the gain inversion.

3.3   Merging the Output of Local Pairs of Processing Units

To endow a robot with the capability to sense its angular speed and position, a novel sensory fusion algorithm was developed using several LPU in parallel. An example of implementation shows in figure 4 a 2D region of interest (ROI) composed of 8×5 artificial ommatidia. The algorithm used here implements the connection between the 8×5 photosensors' output signals to an array of 7×5 LPUs in order to provide local measurements of edge and bar angular positions. Then, a selection is performed by computing the local sum S of two demodulated signals Ph_{demod} obtained from two adjacent photosensors:

$$S_{n,n+1} = Ph_{(n)_{demod}} + Ph_{(n+1)_{demod}} \tag{3}$$

Indeed, as a signal $Output_{Pos}$ is pure noise when no feature is in the field of view, an indicator of the presence of a contrast was required. The sum of the demodulated signals was used here to give this feedback, because we observed that the contrast is positively correlated with the sum and the Signal-to-Noise Ratio. Therefore, at each sampling step, each local sum S is thresholded in order to select the best LPU's outputs to use. All the sums above the threshold value

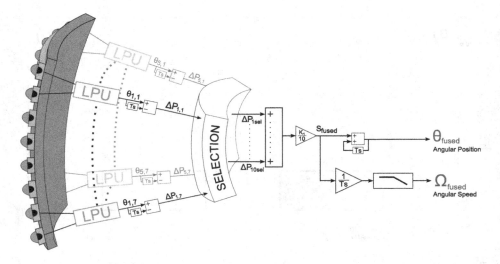

Fig. 4. Description of the sensory fusion algorithm to assess the angular speed Ω_{fused} as well as the position θ_{fused} resulting here from a translation of the textured panel with respect to the arbitrary reference position (i.e. the initial position if not resetting during the experiment). The 35 (7×5) LPU output signals corresponding to the ROI (8×5 photosensors) of the active CurvACE were processed by the 35 Local Processing Units (LPUs) presented in figure 2. Here, the signal obtained before reaching the discrete integrator, denoted S_{fused}, was used to compute the angular speed. This procedure involved normalizing the time $\frac{1}{T_s}$, with T_s equal to the sample time of the system. A first order low-pass filter with a cut-off frequency of $1.6 Hz$ limited the noise. The sensor's position θ_{fused} was scaled in degrees by means of the gain K_1. Adapted from [2].

are kept and the others are rejected. The threshold is then increased or decreased by a certain amount until 10 local sums have been selected. The threshold therefore evolves dynamically at each sampling time step. Lastly, the index i of each selected sum S gives the index of the pixel pair to process. Thus, the computational burden is dramatically reduced. Moreover, this selection helps to reduce the data processing because only the data provided by the 10 selected LPUs are actually processed by the micro-controller.

In a nutshell, the sensory fusion algorithm presented here selects the 10 highest contrasts available in the FOV. As a result, the active CurvACE is able to assess its relative linear position regarding its initial one and its speed with respect to the visual environment.

It is worth noting that the selection process acts like a strong non-linearity. The output signal θ_{fused} is therefore not directly equal to the sum of all the local angular positions θ_i. The parallel differentiation coupled to a single integrator via a non-linear selecting function merges all the local angular positions θ_i, giving a reliable measurement of the angular orientation of the visual sensor within an infinite range. The active CurvACE can therefore be said to serve as a visual compass once it has been subjected to a rotational movement (see section 5). Mathematically, the position is given through the 3 equations as follows:

$$\begin{cases} \Delta P_{i_{sel}} = \theta_{i_{sel}}(t) - \theta_{i_{sel}}(t-1) \\ \theta_{fused}(t) = \theta_{fused}(t-1) + \frac{K_1}{10} \sum_{i=1}^{10} \Delta P_{i_{sel}} \\ \Omega_{fused}(t) = \frac{1}{Ts}(\theta_{fused}(t) - \theta_{fused}(t-1)) \end{cases} \tag{4}$$

As shown in figure 4, the eye's speed is determined by applying a low-pass filter to the fused output signal S_{fused} (which is the normalized sum of the local displacement error ΔP_{isel}), whereas the eye's position is determined in the same way as θ_{fused}, with the gain K_1 (equal to 3.814).

To sum up, the algorithm developed here sums the local variation of contrast angular positions in the sensor field of view to be able to give the angular displacement.

4   Experimental Results

4.1   Measurement of Very Low Optic Flow

In optic flow based navigation, the measurement of very low angular speed can be crucial, for example, to compensate for very low drift or to measure the translational optic flow values near the focus of expansion, which is by essence very low. We characterized here our vibrating compound eye outdoors by making the eye rotate at variable speed ranging from 0 to $20°/s$. The ROI was a single line composed of 42 ommatidia leading to a visual field of $180°$. Figure 5 shows that the angular position and speed are well measured by the sensor. Even if the speed is relatively noisy because of the derivative process, it still can be used as a feedback in a control loop, as demonstrated in [2]. It is worth noting that the range of rotational optic flow that can be measured by active CurvACE is very complementary to the one measured by bio-inspired optic flow sensors. For a similar outdoor environment, the optic flow sensors were able to measure variations from $50°/s$ to $300°/s$ (see [3] for further details). As a consequence, activating the periodic micro-eye movements can enhance largely the eye's motion sensing sensitivity to detect very slow movements such as those occurring near the poles of the optic flow field or resulting from a slow drift.

4.2   Short Range Visual Odometry

In the previous experiment (see figure 5), the FOV was composed of one complete CurvACE line (consisting of 42 ommatidia). Here we focus on a FOV composed of 40 ommatidia, but organized in a ROI of 8 by 5 ommatidia (giving a FOV of about $32°$ by $20°$).

By placing infrared reflective markers on a rotating arm supporting the active CurvACE, it was possible to measure with great accuracy the ground truth

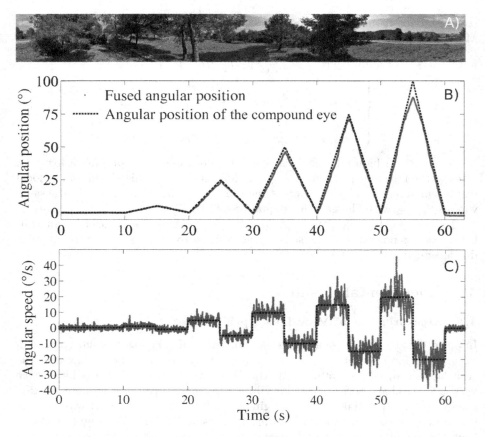

Fig. 5. Measured static characteristics of the active CurvACE placed outdoors in a natural environment (A). B) Fused output signals θ_{fused} superimposed with the angular position of the active CurvACE. The latter can be seen to have served here as a visual compass giving its relative angular position with respect to the environment. C) Eye's rotational speed calculated from B) by applying a discrete derivative the eye's angular position. The eye can provide a good estimation of its angular speed and thus the rotational optic flow even at very low speed ranging from 0 to $20°/s$.

position (curvilinear abscissa) of the eye, which can be compared with the active CurvACE's visual output signal. Once the active CurvACE's output signal has been calibrated, it can be used directly to give the linear position of the eye with respect to the visual environment (a textured panel) placed 39cm below the eye. All the signal processing algorithms have been entirely implemented onboard an electronic board based on a 16-bit microcontroller (dSPic from Microchip) connected to the eye.

The eye's output signal θ_{fused}, which was scaled in millimetres, assuming the height to be known, is superimposed in Figure 6 on the absolute (ground truth) position of the eye measured by the VICON system. The excellent accuracy of the linear position measurement given by the active CURVACE makes it a good candidate for stabilizing an aerial robot when it is flying over an arbitrary terrain (see [2]).

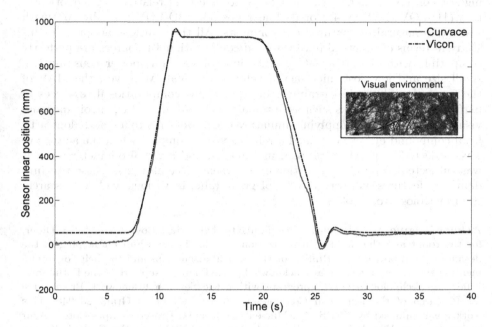

Fig. 6. The active CurvACEs output signal (red) superimposed on the linear position of the eye (dashed black) measured by the VICON system. The good match observed between the two curves shows that the vibrating compound eye can be served as a visual odometer during short range displacements over landscapes such as the textured 2-D pattern composed of branches and leaves tested here.

5   Conclusion

In this paper, we describe the development and the characterization of the active version of a tiny cylindrical curved compound eye named CurvACE. The active process referred to here means that miniature periodic eye movements have been added in order to improve CurvACE's perception of the environment in terms of the localization of the contrasting objects encountered. By subjecting this arti-ficial compound eye to oscillatory movements (micro-scanning movements with a frequency of 50Hz) with an amplitude of a few degrees (5°), it was endowed with hyperacuity, i.e., the ability to *locate* an object with a much greater accu-racy than that which was possible so far because of the restrictions imposed by the interommatidial angle. Hyperacuity was achieved here by several (41 or 35, depending on the experiment) local processing units applying the same local visual processing algorithm across a ROI of active CurvACE sensors, each of which was connected to a pair of adjacent ommatidia. The novel sensory fusion algorithm used for this purpose, which was based on the selection of the 10 highest contrasts, enables the active eye to assess its relative angular orientation (1D-FOV: 180° by 4°) or its linear position (2D-FOV : 32° by 20°) with respect to natural or textured environments. All the solutions adopted in this study in terms of practical hardware and computational resources are perfectly compatible with the stringent specifications of very low power consumption, small size and low cost micro-aerial vehicles (MAVs). Moreover, the MAV of the future (e.g. [8]) will certainly require very few computational resources to perform demanding tasks such as automatic navigation, obstacle avoidance and visual stabilization. By applying miniature eye movements to a stand-alone artificial compound eye, we obtained a reliable visual compass when the active eye was subjected to purely rotational movements, and a visual odometer when it was subjected to purely translational movements (see also [2]). These two fundamental features will certainly be of great value in various fields of research, from metrology to robotics.

Acknowledgments. The authors would like to thank Marc Boyron and Julien Diperi for the robot and the test bench realization; Nicolas Franceschini for helping in the design of the robot, Franck Ruffier for the fruitful discussions and the help for implementing the flying arena. We also acknowledge the financial support of the Future and Emerging Technologies (FET) program within the Seventh Framework Programme for Research of the European Commission, under FET-Open Grant 237940. This work was supported by CNRS, Aix-Marseille University, Provence-Alpes-Cote d'Azur region and the French National Research Agency (ANR) with the EVA, IRIS and Equipex/Robotex projects (EVA project and IRIS project under ANR grants' number ANR608-CORD-007-04 and ANR-12-INSE-0009, respectively).

References

1. Brückner, A., Duparré, J., Bräuer, A., Tünnermann, A.: Artificial compound eye applying hyperacuity. Opt. Express **14**(25), 12076–12084 (2006). http://www.opticsexpress.org/abstract.cfm?URI=oe-14-25-12076
2. Colonnier, F., Manecy, A., Juston, R., Mallot, H., Leitel, R., Floreano, D., Viollet, S.: A small-scale hyperacute compound eye featuring active eye tremor: application to visual stabilization, target tracking, and short-range odometry. Bioinspiration & Biomimetics **10**(2), 026002 (2015). http://stacks.iop.org/1748-3190/10/i=2/a=026002
3. Expert, F., Viollet, S., Ruffier, F.: Outdoor field performances of insectbased visual motion sensors. Journal of Field Robotics **28**, 529–541 (2011)
4. Floreano, D., Pericet-Camara, R., Viollet, S., Ruffier, F., Brckner, A., Leitel, R., Buss, W., Menouni, M., Expert, F., Juston, R., Dobrzynski, M., L'Eplattenier, G., Recktenwald, F., Mallot, H., Franceschini, N.: Miniature curved artificial compound eyes. Proc. Natl. Acad. Sci. USA **110**(23), 9267–9272 (2013). http://dx.doi.org/10.1073/pnas.1219068110

5. Franceschini, N., Chagneux, R.: Repetitive scanning in the fly compound eye. In: Göttingen Neurobiology Report, vol. 2. Thieme (1997)
6. Juston, R., Kerhuel, L., Franceschini, N., Viollet, S.: Hyperacute edge and bar detection in a bioinspired optical position sensing device. IEEE/ASME Transactions on Mechatronics **PP**(99), 1–10 (2013)
7. Kerhuel, L., Viollet, S., Franceschini, N.: The vodka sensor: A bio-inspired hyperacute optical position sensing device. IEEE J. Sensor **12**(2), 315–324 (2012). http://www.sciencemag.org/content/340/6132/603.abstract
8. Ma, K.Y., Chirarattananon, P., Fuller, S.B., Wood, R.J.: Controlled flight of a biologically inspired, insect-scale robot. Science **340**(6132), 603–607 (2013)
9. Mura, F., Franceschini, N.: Obstacle avoidance in a terrestrial mobile robot provided with a scanning retina. In: Proc. IEEE Intelligent Vehicles Symposium, pp. 47–52, September 19–20, 1996
10. Stavenga, D.G.: Angular and spectral sensitivity of fly photoreceptors. i. integrated facet lens and rhabdomere optics. J. Comp. Physiol. A, Neuroethol. Sens. Neural. Behav. Physiol. **189**(1), 1–17 (2003). http://dx.doi.org/10.1007/s00359-002-0370-2
11. Viollet, S.: Vibrating makes for better seeing: from the flys retinal micro-movements to hyperacute bio-inspired visual sensors. Frontiers in Bionics and Biomimetics (submitted 2014)
12. Viollet, S., Godiot, S., Leitel, R., Buss, W., Breugnon, P., Menouni, M., Juston, R., Expert, F., Colonnier, F., L'Eplattenier, G., Brückner, A., Kraze, F., Mallot, H., Franceschini, N., Pericet-Camara, R., Ruffier, F., Floreano, D.: Hardware architecture and cutting-edge assembly process of a tiny curved compound eye. Sensors **14**(11), 21702–21721 (2014). http://www.mdpi.com/1424-8220/14/11/21702

Closed-Loop Control in an Autonomous Bio-hybrid Robot System Based on Binocular Neuronal Input

Jiaqi V. Huang and Holger G. Krapp[✉]

Department of Bioengineering, Imperial College London, London SW7 2AZ, UK
h.g.krapp@imperial.ac.uk

Abstract. In this paper, we describe the implementation of a closed-loop control architecture on a bio-hybrid robotic system. The control loop uses the spiking activity from two motion-sensitive H1-cells recorded in both halves of the blowfly's brain as visual feedback signals that are sent to an ARM processor, programmed to establish a brain machine interface. The resulting output controls the movements of the robot which, in turn, generates optic flow that modifies the activity of the H1-cells. Instead of being inhibited by front-to-back optic flow would the robot move forward in a straight line, the closed-loop system autonomously produces an oscillatory trajectory, alternatingly stimulating both H1-cells with back-to-front optic flow. The spike rate information of each cell is then used to control the speed of each robot wheel, on average driving the robot in the forward direction. Our extracellular recordings from the two cells show similar spike rate oscillation frequencies and amplitude, but opposite phases. From our experiments we derive parameters relevant for the future implementation of collision avoidance capabilities. Finally, we discuss a control algorithm that combines positive and negative feedback to drive the robot.

Keywords: Motion vision · Brain machine interface · Blowfly · Closed-loop control · Autonomous

1 Introduction

The blowfly's ability to perform agile flight manoeuvres [1], using robust reflexes, such as gaze control [2] and collision avoidance [3], make it a suitable model system for opto-motor research.

We use a bio-hybrid system to investigate the H1-cells of the blowfly, *Calliphora vicina*, in a state of locomotion stimulating non-visual sensory systems (e.g. the halteres), in order to explore visual processing and multisensory integration in a closed-loop condition, which can be created by a robot. The research expands on a previous study of visual stabilization of a robot using H1-cell activity in closed-loop [4] and extracellular neural recordings on a mobile platform [5]. A similar implementation of insect sensory control of a wheeled robot was performed using mainly the olfactory system of the silk-moth [6]. Other relevant closed loop control projects are also of importance, such as collision avoidance based on a well characterized cell in locust [7]. Previous work in fruit flies has shown that locomotor state affects the processing of motion vision in individual neurons [8], so we expect to find a difference in response upon movement of the platform. Integration of visual and haltere input has

© Springer International Publishing Switzerland 2015
S.P. Wilson et al. (Eds.): Living Machines 2015, LNAI 9222, pp. 164–174, 2015.
DOI: 10.1007/978-3-319-22979-9_17

also been shown to be non-linear in the neck motor system of the blowfly [9] and a mobile platform for recordings will enable further investigation of such gating mechanisms dependent on locomotion.

Following on from previous research, where a mobile extracellular recording platform was designed [10] and a potential control architecture was proposed [11], here we describe a closed loop control implementation on a bio-hybrid robot system that involves the signals of both H1-cells during forward movement of the robot. In addition we accumulate data to enable collision avoidance in future applications of the system.

The H1-cell is one of the lobula plate tangential cells (LPTCs) located in the third optic lobe. There are about 60 identified directional-selective interneurons in the lobula plate [12]. So-called heterolateral tangential cells convey visual motion information from one lobula plate to its contralateral counterpart. The other LPTCs connect directly or via descending neurons, to the various motor systems supporting gaze control, flight stabilization and collision avoidance [3, 13]. The H1-cell is a heterolateral spiking interneuron sensitive to horizontal back-to-front motion over the ipsilateral eye, the cell's preferred direction (PD), which increases its spike rate. Front-to-back motion in its null direction (ND) inhibits spiking in the H1-cell [14]. If there is no visual motion at all, the H1-cell generates action potentials at a low spontaneous rate of 10-30 spikes per second. Its signals are comparatively easy to measure by means of extracellular recordings, and has been studied for decades [15].

Compared to other LPTCs, the H1-cell has a large receptive field. It's azimuthal sensitivity to visual motion ranges from -15° (contralateral) to 135° (ipsilateral), and from +45° to -45° elevation above and below the eye equator of the animal [14]. Previous research has shown the temporal frequency tuning of the H1-cell under wide field stimulations, e.g. for a constant spatial frequency (black and white vertical stripe pattern) [10] or a combination of different spatial frequencies (lab environment) [11]. Due to the contrast dependence of its responses, H1-cell signals are believed to decrease as the distance to visual objects increases, which has important implications for using the H1-cell as motion vision sensors in a closed-loop robotic system that avoids collision with objects in the surroundings [16].

As mentioned above, spiking in the H1-cell increases upon horizontal visual motion from back-to-front and is inhibited by front-to-back motion. The challenge is to control the forward movement of a robot with a biological sensor that is sensitive to back-to-front motion. If a forward facing blowfly is placed on top of a robot driving forward, both H1-cells will be inhibited. If the blowfly is facing backwards, during motion, both H1-cells will be excited, but the frontal area of the robot would be out of the receptive field of the H1-cells.

It was discussed in previous work [11], one possible way to control the robot by H1-cell activity would be to implement a preprogrammed sinusoidal trajectory. Instead of inhibiting both H1-cells in straight forward motion, the sinusoidal oscillating robot trajectory would stimulate the H1-cells one after the other, thus, producing feedback signals for the control loop. But there are still several unknown parameters that need to be determined before launching such an experiment, including: the optimum turning radius and turning frequency of the preprogrammed sinusoidal robot trajectory.

In this paper, we are introducing a primitive closed-loop control method which generates an autonomous forward oscillation for the bio-hybrid robot system, in which more detailed information is gathered for the development of a closed-loop control system that supports collision avoidance, e.g. turning radius and turning frequency of the robot.

2   Methods

2.1   Bio-hybrid Robot System

The bio-hybrid robot system consists of three subunits: i) the blowfly, ii) the mobile extracellular recording platform, iii) a two-wheeled robot (Pololu© m3pi), as shown in figure 1.

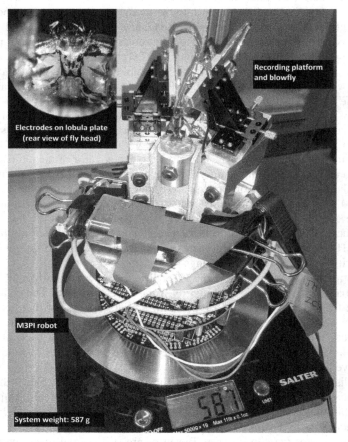

Fig. 1. The assembly of the bio-hybrid robot system. The mobile extracellular recording platform is mounted on top of the m3pi robot. The blowfly is positioned in the center of the recording platform. The whole bio-hybrid robot system weights 587 grams. The inset in the top left corner shows a rear view of the opened blowfly head capsule with two tungsten recording electrodes placed in the left and right lobula plate, respectively.

The mobile extracellular recording platform is similar to the version described previously [7]. It consists of an aluminium chassis holding three parts in position: i) the blowfly, ii) the tungsten recording electrodes and micromanipulators to position them, and iii) custom-designed high-gain amplifiers (Figure 1). A number of features were found to improve experiments. These include: a) re-routing the PCB for better signal-to-noise ratio and to eliminate electrodes cross talk between channels, b) lifting up the electrodes for a better view of blowfly's brain during electrode placement, c) blowflies are mounted from the front, protecting the electrodes from damage during frequent removal, d) improved electrical shielding of the recording platform.

The m3pi robot contains an ARM processor (NXP© LPC1768) which is now programmed with firmware, driving three modules: i) the ADC module, for digitizing bi-lateral H1-cell action potentials, sampled at 5 KS/s each side, ii) the H1-cell model which converts the spike rates into wheel speed control commands, iii) the UART module, for sending a control command to the robot, that generates PWM voltages to control the DC motors, driving the wheels.

2.2   Blowfly Preparation

Female blowflies, *Calliphora vicina*, from 4-11 days old were used in the experiments. The fly's legs and proboscis were cut off and the wounds were sealed with bee wax to reduce any movements which could degrade the quality of the recordings. Wings were immobilized by a small droplet of bee wax on the wing hinges. The head of the fly was placed in between two pins of a custom-made fly holder while adjusting its orientation with reference to the 'pseudopupil methods' [17] before being waxed to the pins. The back of the head capsule was exposed by bending down the thorax and fixing it to the pins. Holes were cut into the back of the head capsule under optical magnification using a stereo microscope (Stemi 2000, Zeiss©) on either side. Fat and muscle tissue were cleared to get access to the lobula plate and physiological Ringer solution (for recipe see Karmeier et al. [18]) was added to the brain frequently to prevent desiccation.

Tungsten electrodes (3 MΩ tungsten electrodes, UEWSHGSE3N1M, FHC Inc., Bowdoin, ME, USA) were used for extracellular recording from H1-cells. The placement of the electrode was adjusted to approach the H1-cell by using a micromanipulator. Recordings of acceptable quality had signal-noise-ratios of > 2:1, i.e. the amplitude of the recorded H1-cell spikes was at least 2 times higher than the largest amplitudes of the background noise. The signals were amplified by a nominal gain of 10,000 with band pass filter from 300 Hz to 3000 Hz. A copy of the signal was sampled by a data acquisition board (NI USB-6215, National Instruments Corporation, Austin, TX, USA) at a rate of 20KS/s.

2.3   Control Algorithm Design

The H1-cell model is derived from the inverse function of the temporal frequency tuning curve obtained in the lab environment [11], which transfers a spike rate (input signal) into an angular velocity (output signal). The curve of the model is simplified to a straight line for the range from 50 Hz to 300 Hz at H1-cell spike rate, and keeps the zone below 50 Hz as dead zone, shown in figure 2. The simplified model is applied for primitive investigation so that the gain can be easily configured to adjust the stability of the system, in addition to reducing processing time in the ARM processor to keep the control loop delay as short as possible.

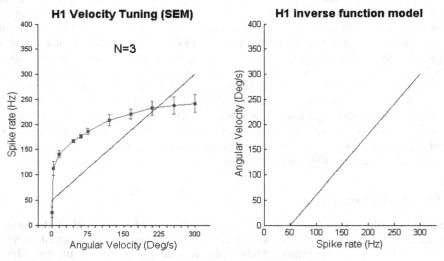

Fig. 2. The H1-cell model. **[Left]** the H1-cell velocity tuning curve in the lab environment [11], the blue curve is the simplified straight line approximation of the tuning curve. **[Right]** the inverse function of the simplified tuning curve on the left plot, which is programmed in an ARM processor on-board of the robot.

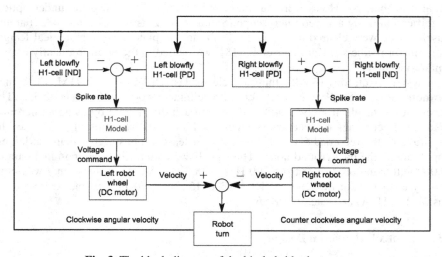

Fig. 3. The block diagram of the bio-hybrid robot system

On top of the monotonic H1-cell inverse function model, the system is capable of eliciting self-oscillation with the control loop designed as figure 3. The control loop of the bio-hybrid robot system can be broken down into two sub-loops based on two

H1-cells, which influence each other. Specifically, the increment of the spike rate on the left H1-cell will increase the speed of the left robot wheel, which will generate a clockwise yaw motion on the robot. This yaw motion inhibits the left H1-cell and excites the right H1-cell, and vice-versa for the right H1-cell. The way in which the control loops are coupled means that an increase in spike rate of a given H1-cell provides negative feedback on its own activity, but positive feedback to the H1-cell in the contralateral part of the brain.

3   Results

3.1   Robot Trajectory

After providing an initial stimulation to one eye of the blowfly, in order to pass the 50 Hz threshold in activity to start movement, the wheel on one side drives the robot away from the movement seen by the fly. The resulting turn produces motion across the other eye, stimulating the contralateral H1-cell, and this sequence repeats as a series of oscillations, gradually driving the robot forwards. The overall motion of the robot was forward, with a little bias to the right wall of the tunnel. (Video link: https://www.youtube.com/watch?v=yNZAB3YrplY)

From the overhead video-footage of the experiments, the orientation and position of the robot relative to the center-line in between the walls of the tunnel was found by analyzing individual frames using a program written in Python OpenCV. 90° is defined as the viewing direction of the fly being aligned with the center-line of the tunnel. The coordinates of the robot center and the robot's rear were recorded from each frame. Those coordinates were connected to reconstruct the trajectory of the robot on a frame-by-frame basis, plotted as green (robot central trajectory) and red (robot posterior trajectory) curves in figure 4, respectively.

Fig. 4. The robot trajectory. The experimental arena was a tunnel with vertical stripes (gratings) attached to the walls (spatial wavelength: 30 mm or 16.2° from initial point). The green curve is the trajectory of robot center; the red curve is the trajectory of the robot's rear.

The orientation of the robot for each frame was derived from the robot's central and rear points via an arctangent function, where the 0° of the polar coordinate is defined on the right of the video frames. The oscillation frequency of the robot was found to be around 5 Hz, with amplitude of 5° to 10° change in orientation with each cycle (Figure 5).

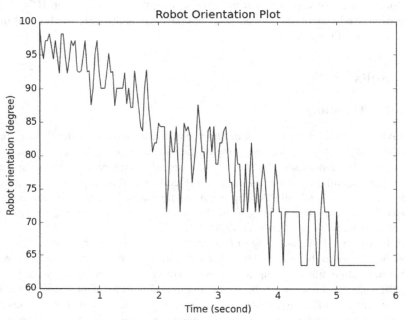

Fig. 5. Robot orientation plot. Data were derived from the frame-by-frame video analysis. 90° in orientation refers to an orientation of the robot that is aligned with the center line of the tunnel.

3.2 H1-cell Responses

H1-cell activities were recorded during the autonomous closed-loop control, shown in figure 6.

The bilateral H1-cell extracellular recording is performed on the fast moving robot. The recording quality was acceptable by means of the following criteria: i) the signal-noise-ratios of each cell are both over 2:1, which means a single threshold was sufficient for successful spike detecting, ii) the inter-spike interval rates of each cell did not exceed 400 Hz at any time, which means the recordings were obtained from single neurons with sensible refractory periods, iii), the peak spike rate, spontaneous spike rate and horizontal motion-induced excitation/inhibition properties indicate the single neuron recordings were obtained from (the left and right) H1-cells.

Double thresholds were used here for spike sorting to accurately detect the occurrence time of the peak of each action potential. The peak is found by searching for the minimum value between two consecutive negative threshold crossings and two consecutive positive threshold crossings. The top threshold is set at 2.5V, where the reference potential is, and the bottom threshold line is set at around 60% of the average action potential peak. Two threshold lines are located at the top and bottom edge of the magenta area in figure 6.

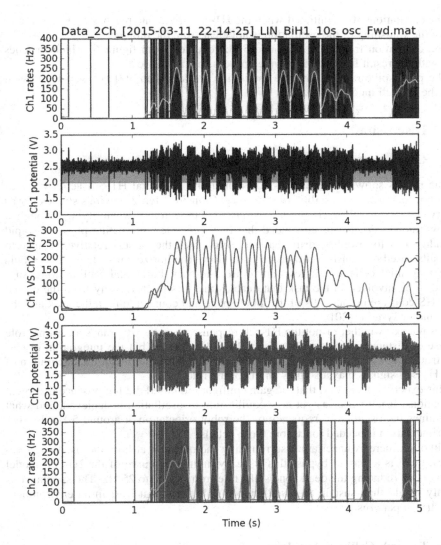

Fig. 6. Recording of bi-lateral H1-cell activities during autonomous closed loop control, plotted from 0 to 5 second. **[Top]** the blue raster is the occurrence of the left H1-cell action potential, the magenta curve is the inter-spike interval (ISI) rate, and the yellow curve is the Gaussian-smoothened spike rate. **[Upper]** the neuron activity of the left H1-cell, the purple area indicates the threshold values for spike detection. **[Middle]** the comparison of two spike rates from each side of H1-cell. **[Lower]** the neuronal activity of the right H1-cell, the purple area indicates the threshold values for spike detection. **[Bottom]** the red raster is the occurrence of the right H1-cell action potential, the magenta curve is the ISI rate, the cyan curve is the Gaussian smoothened spike rate.

The oscillations were initiated when the H1-cell spike rate reaches 50 Hz. The reason of the vanishing of the oscillation is not clear from the plot yet.

The oscillation frequency is around 5 Hz, counted from figure 6, which matches well with the result from previous video processing, as figure 5.

The peak spike rates during the oscillation are less than 300 Hz, which fits well with the H1-cell model.

4   Discussion

4.1   Control Loop Analysis

As the results show, under closed-loop conditions, bi-lateral H1-cell activity drives the robot forward in an oscillatory trajectory. Coincidently, in previous studies it was observed that blowflies do produce sinusoidal trajectories when flying through a narrow tunnel [19]. It was concluded that they generated alternating phases of rapid saccadic turns followed by drift phases during which they assess relative distance to the walls based on translation-induced optic flow. The horizontal systems (HS) lobula plate tangential cells which respond to front-to-back horizontal motion were suggested to be involved in the task that enables the flies to steer away from the closed wall. HS-cells are output cells of the lobula plate and connect to both the neck and the flight motor systems [20].

It is unclear whether or not the H1-cell response characteristic plays a cardinal role in case of distance estimation based on HS-cell activities, but the integration of signals from both eyes certainly does. In case of our robot experiment, the integration of both H1-cell signals may also play an important role.

This experiment showed that the gain configuration used did not produce divergent oscillations. It is unclear as yet how equilibrium of oscillation is achieved, and what determines the oscillation frequency of the robot orientation at around 5 Hz. Further experimentation is needed to explore those details.

Although there is uncertainty surrounding the question at how the oscillation vanishes, there is a way to bypass this issue by tuning the curve of the H1-cell model down, when reducing the dead zone spike rate from 50 Hz to 25 Hz. The spontaneous activity would then sustain the oscillation for all times – at least in an environment with striped patterns.

4.2   Towards Collision Avoidance

Theoretically, the system should achieve collision avoidance by only using the two H1-cells themselves. But from the trajectory, we can see the robot system is not yet capable of doing that. Ultimately it generates a bias in oscillation towards the right hand side of the wall.

Two things need to be considered here. First is the balance (mass distribution) of the robot, which has not been taken care of yet in the robot platform. Certain components could be upgraded. The friction on each wheel is required to be identical to avoid drift. Alternatively, an advanced robot with feedback of wheel speed could overcome this problem.

Second is the neuronal activity from both H1-cells, which needs to be normalized, due to the variance of the neuronal response caused by the neuron's asymmetric structure or firing rate adaptation. Two 360° yaw rotations in opposite direction would have to be performed right after powering up the robot, to measure the peak spike rate of each H1-cell for subsequent normalization of the dynamic output range generated by the left and right cells.

H1-cells are quite sensitive to rotation. As we can see, the bio-hybrid system can achieve 5 Hz oscillation frequency during forward movement and the full swing of the robot orientation in the oscillation is around 5° to 10°. It is possible to calculate the robot tuning radius based on the 5 Hz turning frequency and the 5° to 10° peak to peak angle of robot orientation oscillation. These parameters will guide us when developing and implementing a collision avoidance control algorithm including a preprogrammed trajectory as proposed in the previous paper [11].

5   Conclusion

This work presents the first time an autonomous closed-loop control implemented on a bio-hybrid robot system using simultaneous recordings of bilateral blowfly H1-cells, which can drive the robot forwards based on visual feedback information. The data collected with the system will be used to optimize the platform performance to achieve a collision-free trajectory in arbitrarily structured environments.

Acknowledgments. We'd like to thank Ben Hardcastle and Peter Swart for proof reading and discussion of this paper. And thanks to Gary Jones who provided professional support on fabrication of mechanical components. Also appreciate Yilin Wang, Martina Wicklein and Léandre Varennes-Phillit for all the help and experience sharing on the work presented. This work was partially supported by US AFOSR/EOARD grant FA8655-09-1-3083 to HGK.

References

1. Bomphrey, R.J., Walker, S.M., Taylor, G.K.: The Typical Flight Performance of Blowflies: Measuring the Normal Performance Envelope of Calliphora vicina Using a Novel Corner-Cube Arena. PLoS One **4**, e7852 (2009)
2. Hengstenberg, R.: Gaze control in the blowfly Calliphora: a multisensory, two-stage integration process. Semin. Neurosci. **3**, 19–29 (1991)
3. Lindemann, J.P., Weiss, H., Möller, R., Egelhaaf, M.: Saccadic flight strategy facilitates collision avoidance: closed-loop performance of a cyberfly. Biol. Cybern. **98**, 213–227 (2008)
4. Ejaz, N., Peterson, K.D., Krapp, H.G.: An Experimental Platform to Study the Closed-loop Performance of Brain-machine Interfaces. J. Vis. Exp. (2011)
5. Lewen, G.D., Bialek, W., de Ruyter van Steveninck, R.R.: Neural coding of naturalistic motion stimuli. Netw. Bristol. Engl. **12**, 317–329 (2008)
6. Minegishi, R., Takashima, A., Kurabayashi, D., Kanzaki, R.: Construction of a brain–machine hybrid system to evaluate adaptability of an insect. Robot. Auton. Syst. **60**, 692–699 (2012)

7. Reid, R. Harrison, R.J.K.: Wireless Telemetry of In-Flight Collision Avoidance Neural Signals in Insects 26 (2010)
8. Maimon, G., Straw, A.D., Dickinson, M.H.: Active flight increases the gain of visual motion processing in Drosophila. Nat. Neurosci. **13**, 393–399 (2010)
9. Huston, S.J., Krapp, H.G.: Nonlinear Integration of Visual and Haltere Inputs in Fly Neck Motor Neurons. J. Neurosci. **29**, 13097–13105 (2009)
10. Huang, J.V., Krapp, H.G.: Miniaturized electrophysiology platform for fly-robot interface to study multisensory integration. In: Lepora, N.F., Mura, A., Krapp, H.G., Verschure, P.F., Prescott, T.J. (eds.) Living Machines 2013. LNCS, vol. 8064, pp. 119–130. Springer, Heidelberg (2013)
11. Huang, J.V., Krapp, H.G.: A predictive model for closed-loop collision avoidance in a fly-robotic interface. In: Duff, A., Lepora, N.F., Mura, A., Prescott, T.J., Verschure, P.F. (eds.) Living Machines 2014. LNCS, vol. 8608, pp. 130–141. Springer, Heidelberg (2014)
12. Borst, A., Haag, J.: Neural networks in the cockpit of the fly. J. Comp. Physiol. A **188**, 419–437 (2002)
13. Lindemann, J.P., Kern, R., van Hateren, J.H., Ritter, H., Egelhaaf, M.: On the Computations Analyzing Natural Optic Flow: Quantitative Model Analysis of the Blowfly Motion Vision Pathway. J. Neurosci. **25**, 6435–6448 (2005)
14. Krapp, H.G., Hengstenberg, R., Egelhaaf, M.: Binocular Contributions to Optic Flow Processing in the Fly Visual System. J. Neurophysiol. **85**, 724–734 (2001)
15. Hausen, K.: Functional characterization and anatomical identification of motion sensitive neurons in the lobula plate of the blowfly Calliphora erythrocephala. Z Naturforsch, 629–33 (1976)
16. Egelhaaf, M., Kern, R., Lindemann, J.P.: Motion as a source of environmental information: a fresh view on biological motion computation by insect brains. Front. Neural Circuits. **8**, 127 (2014)
17. Franceschini, N.: Pupil and pseudopupil in the compound eye of drosophila. In: Wehner, R. (ed.) Information Processing in the Visual Systems of Anthropods, pp. 75–82. Springer, Berlin Heidelberg (1972)
18. Karmeier, K., Tabor, R., Egelhaaf, M., Krapp, H.G.: Early visual experience and the receptive-field organization of optic flow processing interneurons in the fly motion pathway. Vis. Neurosci. **18**, 1–8 (2001)
19. Kern, R., Boeddeker, N., Dittmar, L., Egelhaaf, M.: Blowfly flight characteristics are shaped by environmental features and controlled by optic flow information. J. Exp. Biol. **215**, 2501–2514 (2012)
20. Taylor, G.K., Krapp, H.G.: Sensory systems and flight stability: what do insects measure and why?. In: Casas, J., Simpson, S.J. (eds.) Advances in Insect Physiology, pp. 231–316. Academic Press (2007)

MantisBot: A Platform for Investigating Mantis Behavior via Real-Time Neural Control

Nicholas S. Szczecinski$^{(\boxtimes)}$, David M. Chrzanowski, David W. Cofer,
David R. Moore, Andrea S. Terrasi, Joshua P. Martin,
Roy E. Ritzmann, and Roger D. Quinn

Case Western Reserve University, 10900 Euclid Ave., Cleveland, Ohio 44106, USA
nss36@case.edu

Abstract. We present Mantisbot, a 28 degree of freedom robot controlled in real-time by a neural simulation. MantisBot was designed as a 13.3:1 model of a male Tenodera sinensis with the animal's predominant degrees of freedom. The purpose of this robot is to investigate two main topics: 1. the control of targeted motion, such as prey-directed pivots and striking, and 2. the role of descending commands in transitioning between behaviors, such as standing, prey stalking, and walking. In order to more directly use data from the animal, the robot mimics its kinematics and range of motion as closely as possible, uses strain gages on its legs to measure femoral strain like insects, and is controlled by a realistic neural simulation of networks in the thoracic ganglia. This paper summarizes the mechanical, electrical, and software design of the robot, and how its neural control system generates reflexes observed in insects. It also presents preliminary results; the robot is capable of supporting its weight on four or six legs, and using sensory information for adaptive and corrective reflexes.

Keywords: Real-time neural control · Robot · Mantis

1 Introduction

Praying mantises make visually-guided posture adjustments to align themselves with prey [24], [4]. These adjustments require the animal to process visual information in the brain [23] to produce descending commands to low-level systems that control its body and legs, which execute these translations and rotations. We are especially interested in the central complex (CX), a midline neuropil in the arthropod brain, and its role in controlling behavior. The CX receives multimodal sensory information and communicates directly with premotor centers, suggesting that it plays a key role in controlling orientation and locomotion. Work with

N.S. Szczecinski—This work was supported by a NASA Office of the Chief Technologists Space Technology Research Fellowship (Grant Number NNX12AN24H).

© Springer International Publishing Switzerland 2015
S.P. Wilson et al. (Eds.): Living Machines 2015, LNAI 9222, pp. 175–186, 2015.
DOI: 10.1007/978-3-319-22979-9_18

Fig. 1. MantisBot can support its own weight on four legs, with the front legs off the ground for striking. Communication with the robot is performed by the controller on the rear of the robot.

cockroaches suggests that activity in the CX precedes directed motion, specifically linear and angular velocity in the frontal and sagittal planes [9]. Mantises' directed motion facilitates studying the CX; therefore we have begun to investigate these questions in the praying mantis, one of the cockroach's closest living relatives [18], and an animal that exhibits deliberate, targeted motion as a predator.

In our previous work, we have constructed detailed neuromechanical models of insects to investigate questions about motor control [19] [20]. Building realistic models of animal locomotion requires the scientist to confront the details of network parameters and dynamics, an approach that has led to improved understanding of how animals generate rhythms for walking [17] [6] and modify them to change direction [13]. While neuromechanical software simulations are useful investigative tools, robots offer a more physically realistic way to explore animal control strategies [14]. Phenomena like ground contact and body strain are important details of controlling motion, yet they are difficult to model. In addition, the real world is noisy, and while noise can be modeled in simulation, a real-world environment is an excellent test for an experimental controller. Therefore, MantisBot was developed to be controlled by the AnimatLab Robotics Toolkit. This makes the transition from simulation to robot straightforward.

Many other robots have served as models by which to explore animal control systems, and only a few are mentioned here for brevity. ROBOT II, modeled after the stick insect and controlled by a finite state machine, was one of the first to implement leg reflexes based on insect behavior [8]. One of the most complete robotic models of insect behavior is WALKNET, which is a heterarchical artificial neural network that replicates behavioral data from stick insects [15]. WALKNET is used to control HECTOR, perhaps the most sophisticated biologically-inspired hexapod robot [16]. The robot SCORPION explores rhythm generation and reflexes through a neural system [7], including abstracted CPGs. Our goals for MantisBot are parallel to those for these other robots, except that we seek to explore how neural dynamics themselves affect the control of posture, reflexes, and rhythm. As such, we model the animal's control system as one hierarchical network of nonspiking neurons and synapses.

MantisBot is a robotic test platform for neural controllers that model insects' nervous systems, like those in our previous work [19] [20]. It mimics the anatomical proportions and kinematics of a male praying mantis *Tenodera sinensis*, and is controlled by a real-time neural simulation implemented with the AnimatLab Robotics Toolkit. This paper describes the robot, explores design decisions that make it like the animal, and explains the control system. Results are also presented from preliminary experiments, in which MantisBot uses sensory signals to coordinate its joints and exploits CPG dynamics to produce reflexive correction steps.

2 Robot Hardware

2.1 Mechanical and Electrical Design

MantisBot is actuated by Robotis Dynamixel MX-64T and smaller AX-12 smart servo motors (Robotis, South Korea). Each unit can measure position, mechanical load and temperature, and possesses its own microcontroller for communication. Our experiments revealed that MX-64Ts can output sufficient torque at stall, while only weighing 1.24 N.

Motors are controlled by an Arbotix-M (Vanadium Labs LLC, New York), an Arduino-compatible microcontroller based on the ATMEGA644p. Conveniently, chains of Dynamixels can be plugged into the TTL connectors built into the board, providing power and communication. In addition, this board is supported by the AnimatLab Robotics Toolkit, which provides low-level controls for MantisBot. We power both the servos and the Arbotix-M with a 12 VDC 83 A power supply. To avoid running all motors' current through the Arbotix-M, some motors plug directly into the microcontroller while others plug into a power hub, which when interfaced with the Arbotix-M via a TTL cable functions as a signal repeater.

The Arbotix-M also has eight analog inputs, allowing us to use strain gages for continuous load detection on each leg. Each strain gage is mounted on the proximal dorsal surface of the femur, providing the same kind of information as the femoral campaniform sensilla (fCS), which are crucial to timing stance and swing motions [25] [1]. A 5V rail is used to power an LM324 op-amp and Wheatstone quarter-bridge for each strain gage.

MantisBot's structural components are all made from polycarbonate (PC). We chose PC over aluminum because PC is sufficiently strong for a robot of this size and offers a better strength to weight ratio. Using PC also allowed most of the components to be 3D printed, allowing for complex geometries. For example, MantisBot's body segments are each a 3D truss, a very strong and light shape that would be difficult to produce with subtractive manufacturing. PC also is flexible enough that the amplifier gain for the strain gages can be set low (200), producing a very clean signal.

The microcontroller communicates with a desktop computer (i7 2770K 3.5 GHz, 32 GB RAM) at 256 kbps over a virtual serial connection (USB) using a modified version of the Firmata protocol. The Arbotix-M collects inputs from the robot and writes them to a buffer that AnimatLab uses to update the neural control system. MantisBot's inputs are the position of all 28 servos, as well as femoral strain, one gage for each of the six legs. The strain is used as a 10-bit analog signal to provide the network with continuous (i.e. not discrete) load signals. The network then writes new motor position commands to the buffer for the Arbotix-M to read.

2.2   Mantis Kinematics

Mantises are highly flexible insects with many degrees of freedom (DOF). Fig. 2 shows a to-scale schematic of the animal with segments and joints labeled.

The prothorax and mesothorax are connected by a multi-DOF joint which allows the mantis to rear and pivot the prothorax and the attached forelegs and head. Each thoracic segment (prothorax or T1, mesothorax or T2, and metathorax or T3) has a pair of multi-jointed legs.

Each leg has four main segments, moving distally: the coxa, trochanter, femur, and tibia, terminating in a tarsus (foot) for gripping the substrate. The T1 legs are highly mobile, possessing three thorax-coxa (ThC) joints, which together function like a ball-and-socket joint. The trochanter and femur are fused, keeping the coxa-trochanter (CTr) and femur-tibia (FTi) joints parallel. CTr extension lowers the tarsus toward the ground, and FTi flexion pulls the tarsus backward toward the body. The T2 and T3 legs each has the same DOF as those in T1, with the addition of a mobile trochanter-femur (TrF) joint.

Because the cockroach has been studied more thoroughly and is closely related [18], it may be helpful to contrast mantis leg anatomy to that of the cockroach. Unlike the cockroach, the mantis's T2 legs possess inwardly-mobile ThC3 joints, which rotate the leg ventrally about the coxae's long axis. The ThC3 joint is used to maintain the mantis's upright hunting posture. The T3 legs are nearly identical to the T2, except that the ThC3 moves the leg dorsally, and the TrF joint is less mobile. Fig. 2 shows a schematic of each leg and the DOF it possesses.

The T2 and T3 legs also differ from the T1 in that the segments are proportioned differently. The raptorial T1 legs are specialized for grasping, while T2 and T3 provide a wide, stable base for four-legged posture. Table 1 shows the measurements of leg segments from a male *Tenodera* on which MantisBot is

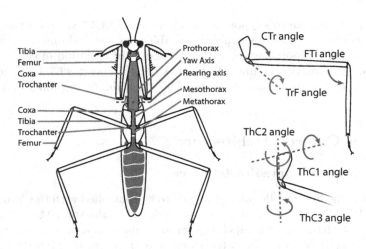

Fig. 2. Scaled schematic of a male *Tenodera sinensis*, with segments and degrees of freedom labeled. Joints with green arrows are not included on MantisBot. To see how the robot captures these proportions and DOF, see Fig. 3.

based. On the front leg, the femur is roughly 150% the length of the coxa, and 250% the length of the tibia. On the middle leg, the coxa is much shorter and the tibia is much longer, making the femur about 350% the length of the coxa, and 130% the length of the tibia.

2.3   Robot Kinematics

MantisBot has two body segments, a prothorax and mesothorax/metathorax, which are connected by a two-DOF joint. This enables the prothorax to rear and yaw with respect to the main body segment. The yawing motion is directly driven by an MX-64T. The rearing is driven by a four-bar mechanism underneath the thorax, with an MX-64T on the rear of the robot. The four-bar mechanism both provides additional mechanical advantage required to lift the prothorax and moves the center of mass of the robot rearward, which is beneficial for quadrupedal posture.

MantisBot's T1 legs include all of the degrees of freedom of the animal. This is important because the front legs are the most mobile and volitional [4], and will be necessary for studying directed motions such as striking at prey, or mobile locomotive tasks such as climbing. MantisBot's T2 and T3 legs possess ThC1, ThC3, CTr and FTi joints, which our previous works suggests are the most crucial for postural tasks [20]. Fig. 3 shows photos of each of the robot's legs overlaid with a scale schematic of the corresponding leg from *Tenodera*. The most noticeable discrepancy is that MantisBot's T2 and T3 coxae are longer than the animal's. This is necessary because placing the motors as close together as possible would establish a 23.1:1 scale, which would make the legs very long and reduce the mechanical advantage of the proximal leg motors so much that the robot would be unable to support itself. However, a 13.3:1 scale for the femora and tibia would mitigate this problem, establishing the scale used for most of the robot. These proportions are within the range of variation of other mantid species (G. Svenson, per. comm.). In total, MantisBot weighs 63.2 N, and when all motors are zeroed, has an envelope of 90 cm wide, 60 cm long, and 50 cm tall.

3   Robot Control Architecture

3.1   AnimatLab-MantisBot Interface

MantisBot is the first mobile robot designed to be controlled with the AnimatLab Robotics Toolkit (ART). MantisBot was designed with AnimatLab 2 by first constructing a virtual model of its body, servos, sensors, and nervous system. The ART lets us assemble a model of the robot in a graphical user interface (GUI),

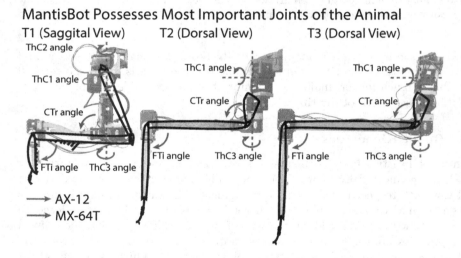

Fig. 3. Photos of each leg of the robot with joints labeled. 13.3:1 leg schematics from Fig. 2 are overlaid to compare the robot's proportions to the animal's. This comparison is quantified in Table 1.

Table 1. Segment lengths for the animal (left) and robot (center) in mm. The ratios between the segments is also shown for comparison (right). Weight is also included in the top row. Compare the lengths to the overlays in Fig. 3.

Animal	0.108 N			Robot	63.2 N			Ratio	585:1		
Segment	T1	T2	T3	Segment	T1	T2	T3	Segment	T1	T2	T3
Coxa	13.3	5.50	5.60	Coxa	135	127	127	Coxa	10.2:1	23.1:1	22.7:1
Femur	18.2	18.6	24.0	Femur	175	249	320	Femur	9.62:1	13.4:1	13.3:1
Tibia	7.3	14.2	18.0	Tibia	100	190	240	Tibia	13.7:1	13.4:1	13.3:1

and fully simulate the rigid body model and servos via the open source Bullet physics engine (bulletphysics.org). In simulation, the nervous system interfaces with the virtual body, producing a neuromechanical simulation used for initial testing of the control system. The ART is designed to allow the same nervous system to control both the simulation and the real robot by swapping out the underlying physics engine for a "robotic engine", which handles communication with the robot.

A link to the robot control hardware is added within the GUI by defining a hardware interface, which contains one or more I/O control modules that interact with a specific type of microcontroller. Part interfaces can be added to an I/O controller to link specific sensors or motors to their counterparts within the simulation.

MantisBot uses the Firmata I/O protocol to communicate with an Arbotix-M. Firmata allows the robotics engine to interface with most servos and sensors without requiring any new programming. The robotics engine runs on the master computer, and configures the I/O of the Arbotix-M slave. For a Dynamixel servo, the membrane voltage from a motor neuron is converted to a position command to control torque output (see Section 3.2). Motor commands that have changed are sent to the Arbotix-M, which updates the servos simultaneously. Servos are read round-robin with the data from one servo being read and sent back during each update. Digital and analog signals are sent back each time they change. The engine converts the incoming sensor values into currents that are injected into sensory neurons, completing the sensory/motor feedback loop. The robotics engine ensures that neural processing is kept in synchrony with the real-time I/O of the hardware, but on a per time step basis it easily simulates MantisBots control system of 775 neurons and 1258 synapses 150 times faster than real time.

3.2 Neural Controller - Single Joint Control

In order to apply animal data as directly as possible, MantisBot's controller is composed of conductance-based nonspiking neuron models. Neural dynamics are simulated as

$$\frac{dV}{dt} = G_{mem}(E_{rest} - V) + \sum g_{syn}(E_{syn} - V) + G_{Na}m_\infty h(E_{Na} - V) \quad (1)$$

$$\frac{dh}{dt} = \frac{(h_\infty - h)}{\tau(V)} \quad (2)$$

in which V is the membrane voltage, G are constant conductance values, g are instantaneous conductance values, and m and h are the sodium channel activation deactivation, respectively. The subscript mem stands for membrane, syn stands for synaptic, Na stands for sodium, and ∞ stands for steady state. The summation is over all incoming synapses to the neuron. Modeling neurons this way is appropriate because nonspiking neurons are known to exist throughout motor control systems in insects [3], and a single nonspiking model approximates the mean activity of a population of coupled spiking neurons. In addition, we make use of persistent sodium channels to build nonspiking central pattern generators (CPGs) like those in [11] [17] [6] [19]. MantisBot's controller is hierarchical and distributed, mimicking that in insects [2]. Each of MantisBot's joints has the same controller topology, shown in Fig. 4, tuned to the range of motion of that joint. For preliminary controller verification, most joints possess a CPG (in red). Future locomotion work will require that every joint has its own CPG. The CPG is based on persistent sodium models in other modeling studies [6] [17], and its parameters are designed to operate close to the oscillatory regime, such that it does not oscillate without descending excitation, but sensory information may cause a single flexion/extension transition. Phase space and phase response analysis have revealed that an inhibitory input to the CPG's interneurons (INs) will cause a single transition when it is applied, but an excitatory input will cause two transitions, one when it is applied and one when it is removed [12]. These

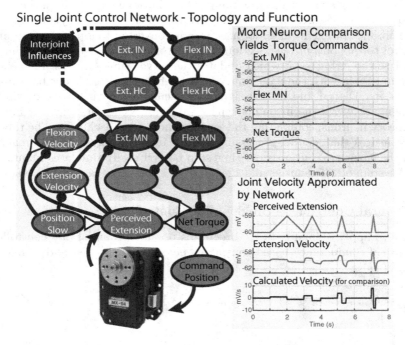

Fig. 4. Diagram showing the joint controller network implemented on MantisBot, as well as plots demonstrating how the structures produce relevant signals. CPGs (red, interneurons "IN" and half-centers "HC") inhibit motor neurons (dark blue, "MN"), the comparison of which yields a torque command (light blue). The first data set (violet shading) shows how this structure functions. The servo returns a perceived position (green), which provides position and velocity feedback to the MNs. The second data set (green shading) shows that this network reproduces a smoothed version of the time derivative of the perceived position. Interjoint influences, shown as a black box, may affect the CPGs or MNs.

transitions can be exploited to implement reflexes seen in insects (described in Section 3.3).

Each joint's torque output is commanded by an antagonistic pair of motor neurons (MNs). As an abstraction of MN activation causing muscle force, the motor neuron voltages are compared and the difference is added (for extension) or subtracted (for flexion) from the current servo position. This signal is then sent as a position command to the servo. But since the MNs' activation is in addition to the current position, this is actually a torque command in the form of the MN activation times the proportional gain of the servo. Data in Fig. 4 shows how the network converts MN activations to a desired torque.

To mimic passive forces that dominate joint control in small animals [10], each joint's MNs receive persistent position and velocity feedback, seeking a constant flexed position and zero velocity. Velocity signals from servos are noisy at low speeds, so our controller approximates velocity with a simple network based on vision filtering in *Drosophila* [22]. The position signal is passed through an interneuron, whose time constant is an order of magnitude larger than the sensory neuron. When the sensory neuron's membrane voltage fluctuates, the slower neuron's response lags behind, and taking the difference between the two yields an approximate velocity calculation. Data in Fig. 4 (green) compares this network's output with the actual, calculated differential of a position signal.

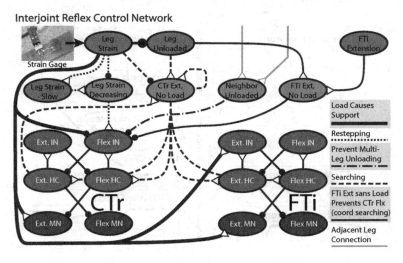

Fig. 5. A network diagram that shows interjoint reflexes based on insect neurobiology. Connections related to a particular reflex are drawn with the same line type. Load causes support [25], decreasing load causes corrective resteps [26], a leg may not unload when the posterior or contralateral leg is unloaded [5], leg extension without ground contact initiates searching, and FTi extension without leg load causes CTr extension [21].

3.3   Neural Controller - Intraleg Control

Posture is generated by coordinating the joints within a leg (Fig. 5). The mechanisms discussed in Section 3.2 provide the resistance reflexes that insects are known to exhibit when standing still [2], but insects are also known to use loading information to control both the timing and magnitude of leg extension [25]. Therefore, the FTi and CTr extensor MNs receive excitatory input directly from the strain gage on each leg. This lets each leg produce support forces proportional to the load acting on it. Loading information also feeds into the FTi CPG, even when it is inactive. It is known that load information from the fCS entrains the motion of the FTi [1], therefore in this control network load information excites the extension IN in the CPG, such that cycling the load causes the joint to actively oscillate. Adding interjoint pathways enables MantisBot to exhibit several intraleg reflexes observed in insects, which can be used to improve postural stability without volitional command of the each leg. For instance, if the load on the leg decreases rapidly while the leg is in support, the CTr will temporarily flex, as observed in crickets [26]. This lifts the leg, unloading it. Because the FTi CPG is excited by load, unloading the leg will cause the FTi to flex. When

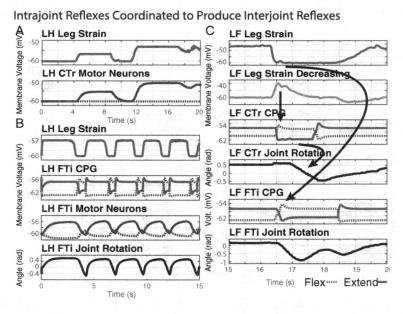

Fig. 6. MantisBot's low level reflexes can be coordinated through the interjoint pathways in Fig. 5 to produce coordinate reflexive posture adjustments. (A) The MNs of support joints receive direct inputs from strain sensors within that leg, to control motor output amplitude. (B) CPGs for the FTi joints can be entrained by strain sensors within that leg, to control motor output timing. (C) These and other joint level reflexes can be coordinated to produce leg level motions, such as a corrective restep when the leg senses it is slipping. The flow of reflexes is described in the text.

the CTr extends and loads the leg again, the FTi is more flexed, moving the foot closer to the body and stable ground. All of these reflexes are illustrated with data from MantisBot in Fig. 6, and additional reflexes are illustrated and explained in Fig. 5.

4 Conclusions

MantisBot is a research robot designed after the praying mantis *Tenodera sinensis* and controlled with the AnimatLab Robotics Toolkit to explore the neural control of motion by descending commands from the central complex. It is capable of emulating reflexes seen in the animal via biological neural networks, including central pattern generators. These basic capabilities provide a basis for future behavioral development and show the feasibility of controlling a robot in this manner.

References

1. Akay, T., Bässler, U., Gerharz, P., Büschges, A.: The role of sensory signals from the insect coxa-trochanteral joint in controlling motor activity of the femur-tibia joint. Journal of Neurophysiology **85**(2), 594 (2001)
2. Büschges, A., Gruhn, M.: Mechanosensory Feedback in Walking: From Joint Control to Locomotor Patterns. Advances In Insect Physiology **34**(07), 193–230 (2007)
3. Büschges, A., Wolf, H.: Nonspiking local interneurons in insect leg motor control. I. Common layout and species-specific response properties of femur-tibia joint control pathways in stick insect and locust. Journal of Neurophysiology **73**(5), 1843–1860 (1995)
4. Cleal, K.S., Prete, F.R.: The Predatory Strike of Free Ranging Praying Mantises, Sphodromantis lineola (Burmeister). II: Strikes in the Horizontal Plane. Brain Behavior and Evolution **48**, 191–204 (1996)
5. Cruse, H.: What mechanisms coordinate leg movement in walking arthropods? Trends in Neurosciences **13**(1990), 15–21 (1990)
6. Daun-Gruhn, S.: A mathematical modeling study of inter-segmental coordination during stick insect walking. Journal of Computational Neuroscience, 255–278, June 2010
7. Dirk, S., Frank, K.: The bio-inspired SCORPION robot: design, control & lessons learned. In: Climbing and Walking Robots, Towards New Applications, pp. 197–218, October 2007
8. Espenschied, K.S., Quinn, R.D., Beer, R., Chiel, H.J.: Biologically based distributed control and local reflexes improve rough terrain locomotion in a hexapod robot. Robotics and Autonomous Systems **18**(1–2), 59–64 (1996)
9. Guo, P., Ritzmann, R.E.: Neural activity in the central complex of the cockroach brain is linked to turning behaviors. The Journal of Experimental Biology **216**(Pt 6), 992–1002 (2013)
10. Hooper, S.L., Guschlbauer, C., Blümel, M., Rosenbaum, P., Gruhn, M., Akay, T., Büschges, A.: Neural control of unloaded leg posture and of leg swing in stick insect, cockroach, and mouse differs from that in larger animals. The Journal of Neuroscience : The Official Journal of the Society for Neuroscience **29**(13), 4109–4119 (2009)

11. Hunt, A., Schmidt, M., Fischer, M., Quinn, R.D.: Neuromechanical simulation of an inter-leg controller for tetrapod coordination. In: Duff, A., Lepora, N.F., Mura, A., Prescott, T.J., Verschure, P.F.M.J. (eds.) Living Machines 2014. LNCS, vol. 8608, pp. 142–153. Springer, Heidelberg (2014)

12. Hunt, A.J., Szczecinski, N.S., Andrada, E., Fischer, M.S., Quinn, R.D.: Using data and neural dynamics to design and control a neuromechanical rat model. In: Living Machines (accepted 2015)

13. Knops, S.A., Tóth, T.I., Guschlbauer, C., Gruhn, M., Daun-Gruhn, S.: A neuro-mechanical model for the neural basis of curve walking in the stick insect. Journal of Neurophysiology, 679–691, November 2012

14. Ritzmann, R.E., Quinn, R.D., Watson, J.T., Zill, S.N.: Insect walking and biorobotics: A relationship with mutual benefits. Bioscience 50(1), 23–33 (2000)

15. Schilling, M., Hoinville, T., Schmitz, J., Cruse, H.: Walknet, a bio-inspired controller for hexapod walking. Biological Cybernetics 107(4), 397–419 (2013)

16. Schneider, A., Paskarbeit, J., Schaeffersmann, M., Schmitz, J.: HECTOR, a new hexapod robot platform with increased mobility - control approach, design and communication. In: Advances in Autonomous Mini Robots, pp. 249–264 (2012)

17. Spardy, L.E., Markin, S.N., Shevtsova, N.A., Prilutsky, B.I., Rybak, I.A., Rubin, J.E.: A dynamical systems analysis of afferent control in a neuromechanical model of locomotion: I. Rhythm generation. Journal of Neural Engineering 8(6), 65003 (2011)

18. Svenson, G.J., Whiting, M.F.: Phylogeny of Mantodea based on molecular data : evolution of a charismatic predator. Systematic Entomology 29, 359–370 (2004)

19. Szczecinski, N.S., Brown, A.E., Bender, J.A., Quinn, R.D., Ritzmann, R.E.: A Neuromechanical Simulation of Insect Walking and Transition to Turning of the Cockroach Blaberus discoidalis. Biological Cybernetics (2013)

20. Szczecinski, N.S., Martin, J.P., Ritzmann, R.E., Quinn, R.D.: Neuromechanical mantis model replicates animal postures via biological neural models. In: Duff, A., Lepora, N.F., Mura, A., Prescott, T.J., Verschure, P.F.M.J. (eds.) Living Machines 2014. LNCS, vol. 8608, pp. 296–307. Springer, Heidelberg (2014)

21. Tryba, A.K., Ritzmann, R.E.: Multi-joint coordination during walking and foothold searching in the Blaberus cockroach. I. Kinematics and Electromyograms, June 2000

22. Tuthill, J.C., Nern, A., Rubin, G.M., Reiser, M.B.: Wide-field feedback neurons dynamically tune early visual processing. Neuron 82(4), 887–895 (2014)

23. Yamawaki, Y., Toh, Y.: Response Properties of Visual Interneurons to Motion Stimuli in the Praying Response Properties of Visual Interneurons to Motion Stimuli in the Praying Mantis, Tenodera aridifolia. Zoological Science 20(7), 819–832 (2003)

24. Yamawaki, Y., Uno, K., Ikeda, R., Toh, Y.: Coordinated movements of the head and body during orienting behaviour in the praying mantis Tenodera aridifolia. Journal of Insect Physiology 57(7), 1010–1016 (2011)

25. Zill, S.N., Schmitz, J., Büschges, A.: Load sensing and control of posture and locomotion. Arthropod Structure & Development 33(3), 273–286 (2004)

26. Zill, S., Frazier, S.: Characteristics of dynamic postural reactions in the locust hindleg. Journal of Comparative Physiology A 170, 761–772 (1992)

The Vertical Optic Flow: An Additional Cue for Stabilizing Beerotor Robot's Flight Without IMU

Fabien Expert and Franck Ruffier[⊠]

Aix Marseille University, CNRS, ISM UMR7287, 163 av. de Luminy CP910,
13288 Marseille, France
franck.ruffier@univ-amu.fr
http://www.biorobotics.eu, http://www.ism.univ-amu.fr

Abstract. Bio-inspired guidance principles involving no reference frame
are presented here and were implemented in a rotorcraft called Beerotor,
which was equipped with a minimalistic panoramic optic flow sensor and
no accelerometer, no inertial measurement unit (IMU) [9], as observed in
flying insects (The halters of Diptera are only sensitive to rotation rates).
In the present paper, the vertical optic flow was used as an additional cue
whereas the previously published Beerotor's visuo-motor systems only
used translational optic flow cues [9]. To test these guidance principles,
we built a tethered tandem rotorcraft called Beerotor (80g), which flies
along a high-roofed tunnel. The aerial robot adjusts its pitch and hence
its speed, hugs the ground and lands safely without any need for an
inertial reference frame. The rotorcraft's altitude and forward speed are
adjusted via several optic flow feedback loops piloting respectively the lift
and the pitch angle on the basis of the common-mode and differential
rotor speeds, respectively as well as an active system of reorientation
of a quasi-panoramic eye which constantly realigns its gaze, keeping it
parallel to the nearest surface followed. Safe automatic terrain following
and landing were obtained with the active eye-reorientation system over
rugged terrain, without any need for an inertial reference frame.

Keywords: Panoramic optic-flow · Optic flow of expansion · No refer-
enced states · Bio-inspired autopilot

1 Introduction

Miniature insect-scale robots [15], just like Micro Aerial Vehicles (MAVs), have
to be able to make their way autonomously through cluttered, partially moving
environments, e.g. foliage moving with the wind, and cope with unpredictable
events, e.g. vehicle or human movements. These challenging tasks may call for
novel sensors and novel control methods that differ from those used in conventional approaches, where all the states of the aerial robot are either measured
or estimated in the inertial reference frame [16,23].

© Springer International Publishing Switzerland 2015
S.P. Wilson et al. (Eds.): Living Machines 2015, LNAI 9222, pp. 187–198, 2015.
DOI: 10.1007/978-3-319-22979-9_19

Ethological findings have shown that complex navigation tasks such as terrain following and speed control are performed by flying insects on the basis of optic flow (OF) cues by means of their tiny compound eyes have a very poor spatial resolution in comparison with modern high resolution cameras. In particular, recent studies on insects have shown that the ventral [1,2,18] and dorsal [19] optic flows (OFs) play an important role in altitude control.

Several authors inspired by studies on honeybee landing [24] recently started to use the optic flow as a means of landing automatically [6–8,11,13,14,17,22,27].

In all robotic studies involving the use of OF, the inputs used by the autopilots of rotary-winged robots were always referred to the inertial frame provided by either an IMU [12,13,26], a barometric altimeter [12] or an external actuator placed on a tether [8,20–22]. In some studies, fixed-wing robots did not have to use any inertial frame of reference [3–5,10,28] because fixed-wing robots are naturally more stable than rotorcraft. The use of a rate gyro in an inner loop is compulsory to stabilize the roll and pitch flight dynamics of most rotary-wings based robots.

In recent studies conducted on Beerotor at our laboratory [9], the rotorcraft's altitude and forward speed are adjusted via several feedback loops based on ventral and drosal translational optic flow piloting:

- the lift and the pitch angle on the basis of the common-mode and differential rotor speeds, respectively
- an active system of reorientation based on a quasi-panoramic eye which constantly realigns its gaze, keeping it parallel to the nearest surface followed.

Safe automatic terrain following and landing were obtained with the active eye-reorientation system over rugged, changing terrain, without any need for an inertial reference frame [9].

In the present paper, we aim at incorporating the use of the vertical optic flow into our Beerotor aerial robot. Indeed, during experiments on Drosophila freely flying in a 3D virtual reality environment, Straw et al. have shown that a ventral expansion avoidance reflex is used by insects to control their altitude. The flies generated an increasing climb rate when flying over an expanding stimulus [25]. We suggest here a embodied model - a real aerial robot - where such reflex does not conflict with a ventral optic flow regulator and both could be used together with the expansion avoidance reflex only triggered by a strong stimulus overriding all others control signals and behaviors. In our case, as we are flying in translation in a tunnel with optic flow sensors looking only downward and upward, we can not measure expansion that would only appear in front of the aircraft near the focus of expansion.

Section 2 shortly describes the mechanical and electronic design of the Beerotor robot and the 12m-long circular experimental set-up in which the flying robot was tested. Section 3 present the Beerotor II's main feedback loops based on the optic flow measurements performed by its quasi-panoramic eye during autonomous flights. Section 4 defines the vertical Optic Flow. Section 5 presents the feedforward control loop based on the vertical OF.

2   Beerotor's Airframe

To test the validity of the eye-reorientation guidance principle, the Beerotor II robot (see Figure 1) was equipped with a quasi-panoramic eye (see Figure 1c) consisting of 4 visual motion sensors, each of which comprised 6 pixels only and covered a solid angle of 23°. This eye was placed 7 cm from the robot's body to prevent the propellers from entering the visual field of the eye. The robot's miniature quasi-panoramic eye and the optic flow processing scheme measured the median optic flow based on 5 local optic flow measurements delivered by neighbouring 2-pixel Local Motion Sensors (LMSs) (see[8] for more details). The robot's eye constantly realigned itself with respect to the slope of the nearest surface (see figure 7 in [9]) by means of the presence of a stepper motor coupled with a gear-reducer giving a resolution of $0.02^o/steps$, which pitches the orientation of the eye up or down with respect to the robot's body. The visual cues (the optic flow) used by the aerial robot therefore always refer to the slope of the nearest surface and not to the absolute vertical. This eye-reorientation guidance principle enabled to perform all the optic flow measurements in the new frame of reference associated with the robot's eye, (E, x_e, y_e, z_e), which is defined by the local slope of the surface followed.

3   Beerotor's Main Visuomotor Control Loops

The Beerotor autopilot relies almost exclusively on its optic flow sensors to control its eye orientation, its forward speed and its altitude by means of three main feedback loops (see figure 2). The first feedback loop (green) controls the orientation of the eye relatively to the body and always keeps the eye parallel to the closest surface by means of an angle estimation done by Least Squares approximation performed on the optic flow measurements. In particular, the reorientation strategy makes the aerial robot fly safely in the case of highly variable environments and very steep obstacles (30° slope). The second feedback loop (red) controls the altitude of the aircraft in order to always keep constant the optic flow generated by the closest surface. By taking the maximal value between the forward ventral and dorsal optic flows, the aircraft safely follows the closest surface (either the groud or the ceiling). To do so, the optic flow is compared to the Maximum OF setpoint $\omega_{setMaxOF}$ and the error is minimized by controller which acts on the vertical lift of the aircraft and therefore its altitude. The distance to the closest surface depends on the chosen setpoint and the forward speed of the robot.

The last feedback loop (blue) controls the forward speed of the robot based on the sum of the ventral and the dorsal optic flows by means of two nested feedback loops. The difference between the sum of the optic flows and a setpoint value $\omega_{setSumOF}$ is used to optionally control the rotorcraft's airspeed measured by means of a custom-made airspeed sensor and to regulate the pitch rate of the aircraft $\dot{\theta}$ by means of the rate gyro measurements. This feedback loop coupled with a ventral or dorsal regulator automatically adapts the forward speed

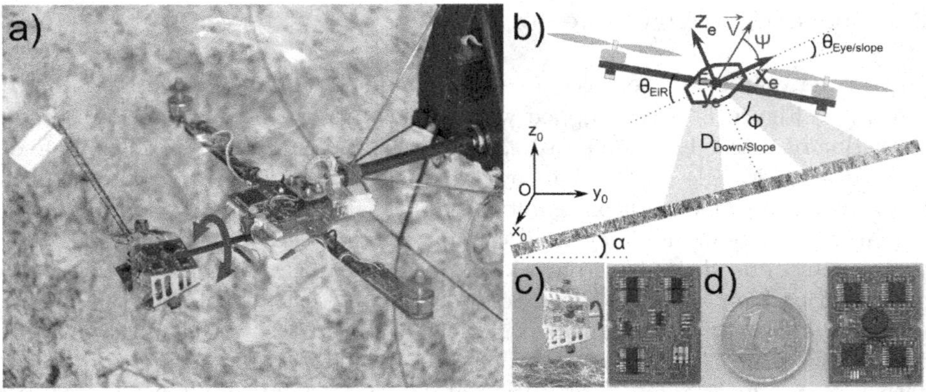

Fig. 1. a) Photograph of the 80g Beerotor II robot. The Beerotor robot is equipped here with a quasi-panoramic eye decoupled from the body, which is composed of 4 visual motion sensors sampling the visual environment with a $4 \times 24°$ FOV. **b)** Drawing of the Beerotor robot flying over a terrain slanting at an angle α. The angle of the eye relative to the body θ_{EiR} is measured via a magnetic sensor and the angle $\theta_{Eye/Slope}$ between the eye's equator and the slope of the nearest surface is estimated on the basis of the optic flow (see [9] for details of the method), and the result is used to align the eye, keeping it parallel with the terrain. The aerial robot is assumed to be flying at a velocity $\overrightarrow{V}$ in the direction defined by the angle Ψ (Ψ is the angle between the direction of the speed vector and the eye's equator). **c)** Photograph of the quasi-panoramic eye mounted on the Beerotor II robot, which constantly realigned itself with respect to the slope of the nearest surface. The orientation of the eye relative to the body can be finely adjusted via a lightweight stepper motor combined with a $\frac{1}{120}$ gear-reducer. **d)** Top and bottom view of the electronic board (size: 33×40mm) of one Visual Motion Sensor with its lens mounted on the LSC photosensor array.

of the aircraft to the size of the tunnel where the rotorcraft is flying by reducing its speed when the tunnel is narrowing and accelerating when the tunnel is getting wider.These last two intertwined feedback loops guarantee that the Beerotor robot will always keep a safe distance from both walls while adapting its forward speed to the size of the tunnel without any measurements of distance or groundspeed. Here, we suggest that the vertical lift of the aircraft can also be controlled by a feedforward controller using the measurement of the Vertical optic flow $\omega_{VerticalOF}^{meas}$ to reduce the altitude oscillations of the aircraft. The idea is to anticipate the changes in the ventral or dorsal optic flow during strong variations of altitude like when overflying the obstacle.

4   Definition of the Vertical Ventral or Dorsal OF

In our case, as we are flying in translation in a tunnel with optic flow sensors looking only downward and upward, we can not measure expansion that only

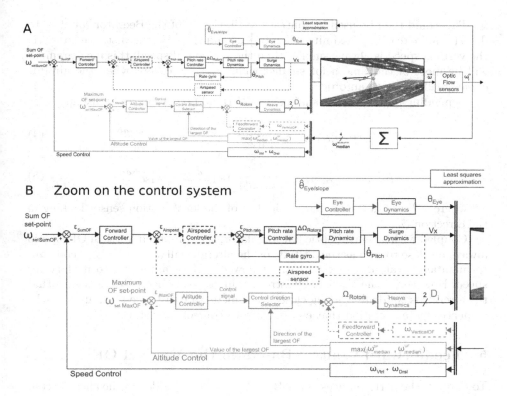

Fig. 2. (A-B) The Beerotor autopilot relies almost exclusively on its optic flow sensors to control its eye orientation, its forward speed and its altitude by means of three main feedback loops

appears in front of the aircraft near the focus of expansion. However, the difference between the two ventral or dorsal optic flows ($\omega(\phi) - \omega(-\phi)$) depends on the ratio between the vertical speed V_z and the distance to the surface h.

$$\omega(\phi) = \frac{\left\|\vec{V}\right\|}{h} cos(\phi).cos(\psi - \phi) \tag{1}$$

where ω is the angular speed, ϕ is a visual direction, $h = D.cos(\phi)$ is the altitude of the aircraft and ψ is the angle between the eye's equator and the direction of the speed vector $\vec{V}$.

$$\omega(\phi) - \omega(-\phi) = \frac{\left\|\vec{V}\right\|.sin(2\phi).sin(\psi)}{h} \tag{2}$$

$$\omega(\phi) - \omega(-\phi) = sin(2\phi)\frac{V_z}{h} \tag{3}$$

This could then be used to control the climb rate of the Beerotor robot or to detect the increasing proximity of an object and increase the distance to it. As the vertical speed is always smaller than the horizontal speed, this measurement will be most of the time a lot smaller than ω_{MaxOF} except when the aircraft comes really close to an object.

Depending on the closest surface, the Vertical OF is computed using:

$$\omega_{Ventral\ VerticalOF}^{meas} = \frac{1}{sin(2\phi)} \cdot \left(\omega_{\phi}^{median} - \omega_{-\phi}^{median}\right) \tag{4}$$

$$\omega_{Dorsal\ VerticalOF}^{meas} = \frac{1}{sin(2\phi)} \cdot \left(\omega_{180^o+\phi}^{median} - \omega_{180^o-\phi}^{median}\right) \tag{5}$$

where ω_{ϕ}^{median} is the median optic flow of the local motion sensor looking in the direction ϕ.

Coupled with a ventral optic flow regulator, the computed Vertical OF should always be close to zero as the altitude of the aircraft will always be kept constant to maintain a safe distance with the followed surface according to its forward speed. However, if the Vertical OF strongly increases or decreases, it means that the aircraft is suddenly getting closer to or increasing distance from an obstacle, respectively, information that can be used to control the robot.

5   Feedforward Control Based on the Vertical OF

To improve the performances of the Beerotor robot, we added in the altitude control loop a second feedback loop based on the measured Vertical OF $\omega_{VerticalOF}^{meas}$ which was used to control the altitude of the aircraft by means of a feedforward controller (see Figure 2). The output of this controller is added to the output signal of the same Altitude controller used previously. This inner loop can be said to act as a ventral or dorsal expansion avoidance reflex. Indeed, when the aircraft is flying away from obstacles, the measured Vertical OF will be low and the altitude of the aircraft will mainly be determined by the previous altitude control loop using the maximum value between the ventral and dorsal optic flows to control the thrust of the robot. On the other hand, a strong increase of the Vertical OF indicating an approaching object will cause through the feedforward controller an increase or a decrease in the vertical lift keeping the aircraft away from danger. In the same way, an important decrease of the Vertical OF will lead to a reduction or an increase of the mean speed of the propellers Ω_{Rotors}.

Figure 3 shows the Beerotor robot's performance while automatically following the terrain for several initial conditions and values of the Maximum OF setpoint $\omega_{setMaxOF}$ during ten consecutive turns with an optic flow setpoint of the speed control loop $\omega_{setSumOF} = 250^o/s$. Regardless of the value of the Max OF setpoint, the autopilot incorporating a feedforward controller based on the Vertical OF avoided the relief. In the bottom part is represented the value of the Vertical OF which was most of the time around 0 /s. In particular, when following the ceiling at a constant altitude, the Vertical OF did not significantly

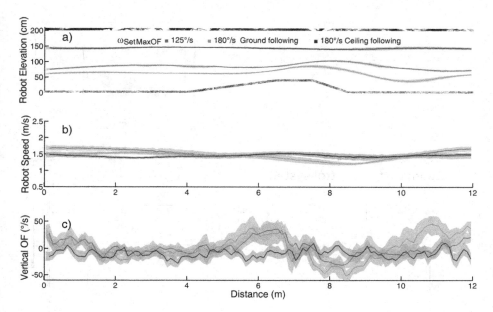

Fig. 3. Automatic ground-hugging under optic flow regulation with a feedforward controller based on the Vertical optic flow and acting on the Vertical lift. Each curve shows the mean altitude, speed and Vertical optic flow of the aircraft with the corresponding standard dispersions during 10 consecutive turns of the aircraft on its environment with an optic flow setpoint of the speed control loop $\omega_{setSumOF} = 250°/s$. **a)** Altitude of the aircraft for three different conditions: following the closest surface with $\omega_{setMaxOF} = 125°/s$ (magenta), following respectively the ground (green) and the ceiling (blue) with $\omega_{setMaxOF} = 180°/s$. As expected, the aircraft perfectly avoided the slanted relief by means of the altitude control loop. When we increased the Maximum optic flow setpoint, the aircraft immediatly came closer to the closest surface in order to reach the setpoint value. By means of the feedforward controller on the Vertical OF, the oscillation after the obstacle was reduced. **b)** Forward speed of the rotorcraft in the three conditions. The steady-state forward speed was the same in the three experiments as the Sum OF setpoint was the same. **c)** Vertical OF of the aircraft. When following the ceiling, $\omega_{VerticalOF}^{meas}$ was always close to 0 /s whereas it increased during the ascending ramp and decreased during the descending ramp of the relief while the aircraft followed the ground.

differ from 0 /s having no influence on the behavior of the aircraft. On the other hand, when following the ground with the slanted ground profile and in particular when the Max OF setpoint increased (green curve), the robot flew closer to the objects leading to variations of the Vertical OF that helped the rotorcraft to avoid the obstacle. In particular, we noticed that the Vertical OF decreased in relation to the changes in the floor profile when the aircraft was flying down to restore its ventral optic flow leading to an increase in the thrust of the propellers by means of the feedforward controller. This is particularly interesting as

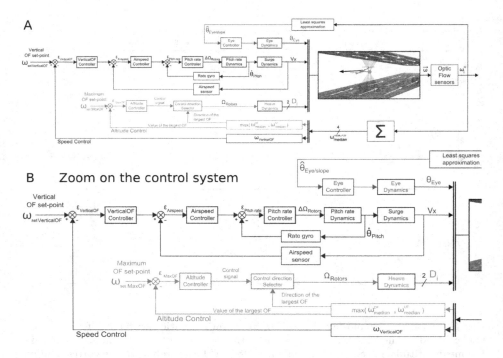

Fig. 4. (A-B) Beerotor autopilot with a speed control loop based on the Vertical optic flow for landing.

it limited the amplitude of the aircraft's altitude undershoot after the changes in the ground profile.

Although not strongly affecting the behavior of the aircraft with our experimental setup, such strategy would allow freely flying aircraft to more robustly avoid obstacles and therefore navigate collision-free in an unknown environment.

6   Regulation of the Vertical OF During Landing

As we have seen in section 4, the Vertical optic flow is proportional to the ratio between the vertical speed V_z and the distance h to the surrounding objects. By means of the altitude control loop regulating the ventral or dorsal optic flow, the distance h to the objects is theoretically always kept constant. By coupling the altitude control loop with a feedback loop controlling the forward speed of the aircraft based on the vertical OF, we can induce an automatic landing. Indeed, by taking the difference between a positive setpoint value $\omega_{setVerticalOF}$ and the measured Vertical OF and use it to drive a lead phase regulator controlling the robot forward speed by means of the two already presented nested feedback loops acting on the airspeed and the pitch rate, the aircraft will automatically reduce its forward speed in order to generate a non null vertical speed by means

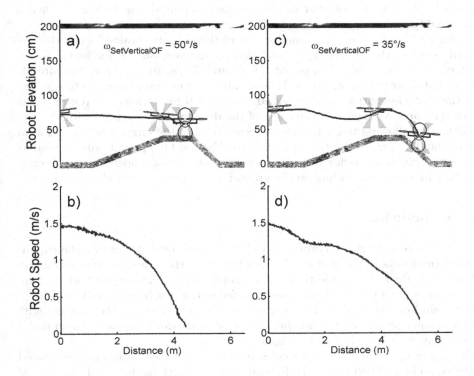

Fig. 5. Automatic landing of the aircraft on the ground achieved by controlling the forward speed feedback loop with the Vertical optic flow. **a-c)** Altitude of the Beerotor robot during automatic landing with $\omega_{setMaxOF} = 150°/s$ and two different values of the Vertical OF setpoint $\omega_{setVerticalOF} = 50°/s$ and $\omega_{setVerticalOF} = 35°/s$. In both cases, the aircraft successfully landed and the duration of the landing increased when the Vertical OF setpoint $\omega_{setVerticalOF}$ decreased. **b-d)** To reach the setpoint value $\omega_{setVerticalOF}$, the rotorcraft reduced its forward speed leading to a loss of altitude induced by the feedback loop regulating the ventral optic flow and hence a smooth landing with almost no speed at touchdown.

of the altitude control loop. Such strategy leads to a smooth landing with almost no speed at touchdown where the duration of the landing depends on the chosen setpoint $\omega_{setVerticalOF}$. The proposed autopilot is presented in figure 4 with the altitude control loop (in red) using the maximum value between the ventral or the dorsal optic flow compared with a setpoint value to act on the vertical lift of the aircraft and therefore its altitude. Contrary to the previous experiments, the speed control loop (in blue) is here based on the difference between a fixed setpoint and the measured Vertical OF $\omega_{VerticalOF}^{meas}$ which is used to control the forward speed of the Beerotor robot. The Vertical OF controller is in that case a simple phase lead controller increasing the damping of the system and hence its stability.

Figure 5 shows the automatic landing of the aircraft obtained when the forward speed of the aircraft was controlled by the vertical OF for two different values of the Vertical OF setpoint: 50 /s and 35 /s. In any case, the aircraft immediately decreased its forward speed and therefore came nearer to the ground as the altitude control loop decreased the vertical lift to keep its ventral optic flow constant. As expected, the slope of the deceleration increased with the setpoint value allowing to control the descent speed of the aircraft or the duration of the landing procedure. In conclusion, the Vertical OF can not only be used as a ventral expansion reflex but also to control the vertical speed of the aircraft leading to a smooth landing on the ground at a forward speed close to $0m/s$.

7   Conclusion

We suggested new visuo-motor feedback loops that exploit suitably the expansion components in the vertical OF in addition to the translational ventral and dorsal OF. Beerotor is able to fly in a steeply sloping environment without an accelerometer and without any need to refer to the absolute vertical.

The main limitation of the Beerotor robot is its mechanical tether which limits its dynamics range and the number of degree of freedom compared to an aerial robot that would freely fly in 3D space.

The advantages of not using an accelerometer are that the strategies we presented here could be embedded into the lightest of robots, such as the insect-scale aerial robot weighing only a few hundred milligrams recently developed by [15].

Acknowledgments. We are very grateful to M. Boyron for his involvement in the electronic design, F. Paganucci, Y. Luparini and J. Diperi for their help with the mechanical design, J. Blanc for improving the English manuscript and A. Manecy, G. Sabiron, G. Portelli, J. Serres, N. Franceschini and S. Viollet for their fruitful comments and suggestions during this research. This work was supported partly by CNRS Institutes (Life Science; Information Science; Engineering Science and Technology), the Aix-Marseille University, the French National Research Agency -ANR- (EVA project under ANR-ContInt grant number: ANR-08-CORD-007-04) and the European Commission via the CURVACE project. The CURVACE project acknowledges the financial support of the Future and Emerging Technologies (FET) programme within the Seventh Framework Programme for Research of the European Commission, under FET-Open grant number: 237940.

References

1. Baird, E., Srinivasan, M.V., Zhang, S., Lamont, R., Cowling, A.: Visual control of flight speed and height in the honeybee. In: Nolfi, S., Baldassarre, G., Calabretta, R., Hallam, J.C.T., Marocco, D., Meyer, J.-A., Miglino, O., Parisi, D. (eds.) SAB 2006. LNCS (LNAI), vol. 4095, pp. 40–51. Springer, Heidelberg (2006)

2. Barron, A., Srinivasan, M.: Visual regulation of ground speed and headwind compensation in freely flying honey bees (apis mellifera l.). Journal of Experimental Biology **209**(5), 978–984 (2006)
3. Barrows, G., Neely, C.: Mixed-mode VLSI optic flow sensors for in-flight control of a micro air vehicle. In: SPIE : Critical Technologies for the Future of Computing, San Diego, USA, vol. 4109, pp. 52–63 (2000)
4. Beyeler, A., Zufferey, J.C., Floreano, D.: Optipilot: control of take-off and landing using optic flow. In: European Micro Aerial Vehicle Conference (EMAV), Delft, Netherlands, pp. 1–8 (2009)
5. Beyeler, A., Zufferey, J.C., Floreano, D.: Vision-based control of near-obstacle flight. Autonomous Robots **27**, 201–219 (2009)
6. Chahl, J., Srinivasan, M., Zhang, S.: Landing strategies in honeybees and applications to uninhabited airborne vehicles. The International Journal of Robotics Research **23**, 101–110 (2004)
7. de Croon, G., Ho, H., de Wagter, C., van Kampen, E., Remes, B., Chu, Q.: Optic-flow based slope estimation for autonomous landing. International Journal of Micro Air Vehicles **5**(4), 287–297 (2013)
8. Expert, F., Ruffier, F.: Controlling docking, altitude and speed in a circular high-roofed tunnel thanks to the optic flow. In: IEEE Int. Conf. on Robots and Systems (IROS), Vilamoura, Portugal, pp. 1125–1132, October 2012
9. Expert, F., Ruffier, F.: Flying over uneven moving terrain based on optic-flow cues without any need for reference frames or accelerometers. Bioinspiration & Biomimetics **10**(2), 026003 (2015)
10. Green, W., Oh, P., Barrows, G.: Flying insect inspired vision for autonomous aerial robot maneuvers in near-earth environments. In: IEEE Int. Conf. on Robotics and Automation (ICRA) (2004)
11. Herisse, B., Hamel, T., Mahony, R., Russotto, F.X.: The landing problem of a VTOL unmanned aerial vehicle on a moving platform using optical flow. In: IEEE International Conference on Intelligent Robots and Systems (IROS), Taipei, Taiwan, pp. 77–89 (2010)
12. Hérissé, B., Hamel, T., Mahony, R., Russotto, F.X.: A terrain-following control approach for a vtol unmanned aerial vehicle using average optical flow. Autonomous Robots **29**(3–4), 381–399 (2010)
13. Herisse, B., Hamel, T., Mahony, R., Russotto, F.X.: Landing a VTOL Unmanned Aerial Vehicle on a moving platform using optical flow. IEEE Transactions on Robotics **28**(1), 77–89 (2012)
14. Kendoul, F., Yu, Z., Nonami, K.: Guidance and nonlinear control system for autonomous flight of minirotorcraft unmanned aerial vehicles. Journal of Field Robotics **27**(3), 311–334 (2010)
15. Ma, K., Chirarattananon, P., Fuller, S., Wood, R.: Controlled flight of a biologically inspired, insect-scale robot. Science **340**(6132), 603–607 (2013)
16. Mellinger, D., Michael, N., Kumar, V.: Trajectory generation and control for precise aggressive maneuvers with quadrotors. International Journal of Robotics Research **31**(5), 664–674 (2012)
17. Moore, R., Thurrowgood, S., Bland, D., Soccol, D., Srinivasan, M.: Uav altitude and attitude stabilisation using a coaxial stereo vision system. In: IEEE Int. Conf. on Robotics and Automation (ICRA), Anchorage, USA, pp. 29–34 (2010)
18. Portelli, G., Ruffier, F., Franceschini, N.: Honeybees change their height to restore their optic flow. Journal of Comparative Physiology **196**(4), 307–313 (2010)
19. Portelli, G., Ruffier, F., Roubieu, F., Franceschini, N.: Honeybees' speed depends on dorsal as well as lateral, ventral and frontal optic flows. PLOS ONE **6**(5) (2011)

20. Ruffier, F., Franceschini, N.: Visually guided micro-aerial vehicle: automatic take off, terrain following, landing and wind reaction. In: Proceeding of IEEE Int. Conf. on Robotics and Automation (ICRA), New Orleans, USA, pp. 2339–2346 (2004)

21. Ruffier, F., Franceschini, N.: Optic flow regulation in unsteady environments : A tethered mav achieves terrain following and targeted landing over a moving platform. Journal of Intelligent and Robotic Systems (2014)

22. Ruffier, F., Franceschini, N.: Optic flow regulation: the key to aircraft automatic guidance. Robotics and Autonomous Systems 50(4), 177–194 (2005)

23. Shen, S., Mulgaonkar, Y., Michael, N., Kumar, V.: Vision-based state estimation for autonomous rotorcraft mavs in complex environments. In: IEEE International Conference on Robotics and Automation (ICRA), Karlsruhe, Germany, pp. 1758–1764, May 2013

24. Srinivasan, M., Zhang, S., Chahl, J., Barth, E., Venkatesh, S.: How honeybees make grazing landings on flat surfaces. Biological Cybernetics 83, 171–183 (2000)

25. Straw, A., Lee, S., Dickinson, M.: Visual control of altitude in flying Drosophila. Current Biology 20(17), 1550–1556 (2010)

26. Strydom, R., Thurrowgood, S., Srinivasan, M.: Visual odometry: autonomous uav navigation using optic flow and stereo. In: Australasian Conf. on Robotics and Automation (ACRA), Melbourne, Australia, December 2014

27. Zufferey, J.C., Beyeler, A., Floreano, D.: Autonomous flight at low altitude using light sensors and little computational power. Journal of Micro Air Vehicles 2(2), 107–117 (2010)

28. Zufferey, J.C., Floreano, D.: Fly-inspired visual steering of ultralight indoor aircraft. IEEE Transactions on Robotics 22(1), 137–146 (2006)

Route Following Without Scanning

Aleksandar Kodzhabashev and Michael Mangan[✉]

School of Informatics, University of Edinburgh, Informatics Forum,
10 Crichton Street, Edinburgh, UK
mmangan@staffmail.ed.ac.uk

Abstract. Desert ants are expert navigators, foraging over large distances using visually guided routes. Recent models of route following can reproduce aspects of route guidance, yet the underlying motor patterns do not reflect those of foraging ants. Specifically, these models select the direction of movement by rotating to find the most familiar view. Yet scanning patterns are only occasionally observed in ants. We propose a novel route following strategy inspired by klinokinesis. By using familiarity of the view to modulate the magnitude of alternating left and right turns, and the size of forward steps, this strategy is able to continually correct the heading of a simulated ant to maintain its course along a route. Route following by klinokinesis and visual compass are evaluated against real ant routes in a simulation study and on a mobile robot in the real ant habitat. We report that in unfamiliar surroundings the proposed method can also generate ant-like scanning behaviours.

Keywords: Navigation · Ant · Robot · Route following · Klinokinesis

1 Introduction

Desert ants are impressive insect navigators, foraging over large distances (sometimes 100s of metres) in extremely hostile environments without the use of pheromone trails. Instead, desert ants navigate through their environment using primarily visual cues [1]. This capability is all the more impressive when one considers the ant's relatively small brain and their low resolution eyes (in the order of 4° angular acuity). As such they have become a model system for bio-roboticists seeking computationally efficient strategies for navigation in natural environments. One specific behaviour will be the focus of this paper; visual route following [2,3] (see figure 3 for an example ant route).

Recently a new family of visual navigation models have been shown sufficient to recreate route following in simulated environments with characteristics similar to those of real ants [4,5]. Fundamental to the success of these algorithms is a methodology termed the "visual compass"; first described by Zeil et al [6]. The visual compass relies on the observation that any stored image inherently encodes information about the viewing direction from which it was taken. Hence, it is possible to later recover that direction by simply rotating on the spot until the best match is found between the current view and the visual memory [6]

© Springer International Publishing Switzerland 2015
S.P. Wilson et al. (Eds.): Living Machines 2015, LNAI 9222, pp. 199–210, 2015.
DOI: 10.1007/978-3-319-22979-9_20

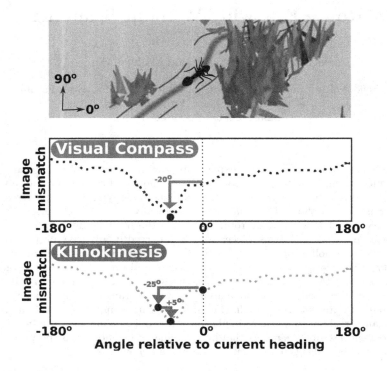

Fig. 1. *Route following using the rotational image difference.* The top panel shows the scenario of an ant coming to a 20° left bend in its route (ant moving from lower left to top right along the blue line). The central panel shows the full rotational image difference function (rIDF) that is generated using a scan. rIDFs are characterised by a minimum where current and stored views match, flanked by increasing values of image mismatch [6]. By moving in the direction giving the best match (in this case -20°) the ant will progress along it's route. The bottom panel shows how a klinokinesis method might work by modulating a turn using only the image mismatch between the current view and all route memories. As the ant heading has deviated from the route a corrective turn is made (in this case -25°). The alternating turns allow the minima to be continually optimised as shown by the second, smaller +5° turn.

(see Fig. 1). This principle can be extended to a bank of memories sampled densely across a previously travelled path. By looking for the best match across all memories, the direction of the travel at any particular route position can be recovered. If this process is repeated at each step, it is sufficient to retrace a previously travelled path [4,5].

Yet, a crucial discrepancy remains between these models and observations of real ants. Specifically, while ants have been observed rotating on the spot before selecting a direction of travel (in a behaviour termed 'scanning') [7], consistent with aligning their current visual input with their memory, this behaviour is only expressed in specific circumstances (e.g. when an ant is released in an unfamiliar environment). Under normal conditions ants travelling along a familiar route scan infrequently, if at all [7]. This suggests that ants might use an alternative strategy to maintain their progress along the route.

In this work we present a novel method by which ants might continuously correct their heading along a learned route without the need to repeatedly scan the environment. Inspiration comes from the klinotaxis and klinokinesis behaviours observed in Drosophila larvae [8] and bacteria [9]. These animals can approach a favourable odour by directing turns relative to a pair of temporally and spatially distinct sensor readings (klinotaxis) or through a simpler undirected strategy by increasing the frequency of turns relative to the sensory input (klinokinesis) [9]. The algorithm described in this work is most similar to klinokinesis given the absence of direct comparison between paired sensory readings. However rather than modulating the likelihood of initiating a turn, the sensory input (in this case the instantaneous image-difference between current view and route memories) modulates the size of alternating left and right turns and also forward step size (or speed). This schema essentially sub-samples the rotational image difference function and exploits its characteristic shape to correct for deviations in orientations along a previously travelled path (see Fig. 1). We show that this simple sensory-motor routine is capable of route following in a simulated ant world and on a custom wheeled robot in the ant habitat. Further, in specific scenarios this algorithm can generate "scan-like" behaviours and we thus propose that scans may be an emergent property of a simpler route following strategy.

2   Methods

2.1   Navigation Algorithms

Two approaches to route following were implemented and tested on a robot and in a simulation environment: visual compass and klinokinesis. The memory bank for both models was identical: comprising densely sampled images (every 1cm for simulation, and every 10cm for the robot) facing along the route to be retraced. Both algorithms use the sum squared per-pixel difference of two grayscale images as a measure of difference between them:

$$SSD(I_1, I_2) = \log \sum_i (I_1(i) - I_2(i))^2 \qquad (1)$$

where I_1 and I_2 are images and i is the index of corresponding pixels within both images. The logarithm of the image difference was used to lineralise the SSD which is proportional to the turn angle and step size of the klinokinesis algorithm.

Visual Compass. The visual compass algorithm works by measuring the difference between the current view as the agent rotates, and all views comprising the route memory (see Fig. 1 central panel). The orientation of the agent when the minimum image difference is observed is chosen as the direction of travel:

$$Best\ Direction = \arg\min_{r} SSD(M(x), C(r))\ for\ all\ x \qquad (2)$$

where M is the bank of images making up the route memory and x is index of each memory; C is the current image taken as the agent rotates through $r°$. In our case we limited scanning to $+/-60°$ around the current heading in $2°$ steps for the virtual environment and in approximately $12°$ steps for robot.

Klinokinesis. Our Klinokinesis-like algorithm works by alternately turning left and right between forward steps. The magnitude of each turn is modulated by the size of the minimum image difference between the current view and all route memories. Thus, while on the route and facing in the correct direction a low image difference would be computed and the agent will turn very little. But when the route bends, and the view mismatch increases, the agent will perform a larger turn. If this turn happens to be in the correct direction a familiar view will be restored, and the agent will recover the path and make a smaller subsequent turn. If, the first turn happens to be away from the route, the view mismatch should increase further so a larger corrective turn in the opposite direction will be implemented and the route recovered (see Fig. 1 lower panel).

At each step, the unfamiliarity is computed as described for the visual compass method above:

$$Unfamiliarity = \min_{x} SSD(M(x), C) \qquad (3)$$

M, C and x follow the convention defined above.

We then normalize unfamiliarity to be between 0 and 1:

$$Normalized\ unfamiliarity = Unfamiliarity/Unfamiliarity_{max} \qquad (4)$$

where $Unfamiliarity_{max}$ is the biggest image difference between the first training image and the rest of the training images:

$$Unfamiliarity_{max} = \max_{x} SSD(M(1), M(x)) \qquad (5)$$

In our method the turn angle is given by:

$$Turn\ angle = \alpha * Normalized\ unfamiliarity \qquad (6)$$

α is the scaling factor which is optimised for route following (described below).

We also scale the forward step size between turns. This allows the agent to take larger steps when the view is very well matched (i.e. on the route), but reduce step size (or slow down) when the mismatch increases, before the route is lost completely. The step size is modulated by the image difference value as shown in equation 7 where β is the scaling parameter.

$$stepsize = (1 - Normalized\ unfamiliarity) * \beta \qquad (7)$$

Parameter Tuning. For simulation tests, we performed a parameter sweep for α and β, in the range 60° to 160° in 20° steps for max turn angle α and 0.5cm to 3.5cm in 0.5cm steps for max step size β. We selected the parameters that resulted in the smallest median number of route following errors over the 15 routes (see 2.3 for definition of error).

2.2   Robot Study

We first assessed the performance of both navigation methods in a robot study in the real ant habitat, as a proof of principle.

Hardware Configuration. The main hardware component of the robot is an Android device (Nexus 5 smartphone), providing an easy to program system supported by dual integrated cameras, a powerful processor (2.26Ghz quad-core; 2Gb RAM; 32Gb of internal storage), 2300mAh battery, and numerous connectivity and sensor options (e.g. GPS, accelerometers) [10]. A panoramic mirror (Kogeto Dot) was attached to the front facing camera giving unobscured 360° vision, only 8cm from the ground. The drive unit is a Pololu Zumo Robot base, providing a small yet semi-rugged, self-powered, platform on which to mount the main device. An Arduino Uno relays motor and sensory commands from base to android device and vice versa via a USB On-The-Go (OTG) cable. A complete guide to constructing the roboant Edinburgh (see Fig. 2) is provided at our blog (https://blog.inf.ed.ac.uk/insectrobotics/roboant/).

Software. A stand-alone app was written implementing both visual compass and klinokinesis algorithms described above. Grayscale panoramic images were sampled using the front facing camera with panoramic lens, before being normalised using OpenCV (adaptive histogram normalisation) and downsampled to approximate ant-eye resolution of 1200 pixels. Example Roboant videos: full resolution (http://youtu.be/W9K1G3XiSAg) and downsampled (http://youtu.be/jWCk40O8jWg).

The entire software project has been made available as an open source package (https://github.com/d3kod/roboant) and the navigation algorithm are available as a downloadable plug-and-play package app on the Google Play Store (https://play.google.com/store/apps/details?id=uk.co.ed.insectlab.ant).

Robot Trials. The robot pilot study was conducted at our current ant field site on the outskirts of Sevilla, Spain, between 18th and 22nd June, 2014. A test track 8m long (similar in length to ant routes in simulation) was constructed using laminated wooden boards which provide a base on which the robot could easily move (persistent toppling occurred on the natural terrain). Two training routes were recorded: one turning left at a junction point and the other turning right (see Fig. 2C); thereby ensuring that any positive route recapitulations were not simply a consequence of an environmental bias. To assess the ability

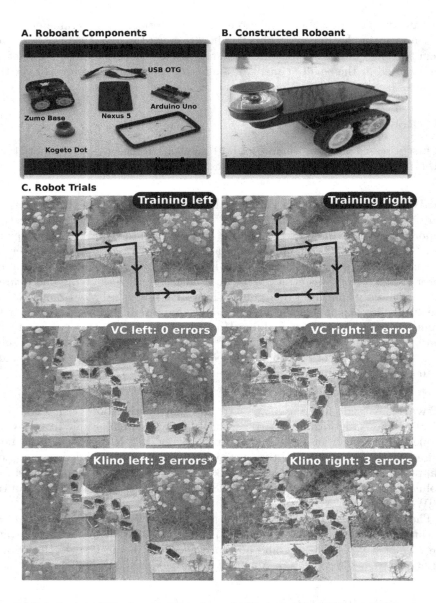

Fig. 2. *Robot route following in the real ant habitat. A.* The off-the-shelf components which make up Roboant are shown prior to assembly. *B.* The completed robot being tested in the lab. *C.* Robot trials at our field site in Seville, Spain. Training images were sampled making either a left turn (top left panel) or right turn (top right panel) at the junction point. The paths taken by the robot in each condition and when driven by both algorithms (VC: visual compass, Klino: klinokinesis) are shown using still frames captured from an overhead camera (GoPro 3) mounted on a pole. *Note that all errors for the klinokinesis on the left trial were clustered at the same location.

of the methods to retrace the learned path we recorded the instances when the robot path deviated more than 20cm from the route (in this case when it fell off the track), whether each method managed to make the correct turn at the crossroads, and also the total time taken to complete each route.

2.3   Simulation Study

Given the positive results from the robot trials we then undertook a detailed simulation study allowing full parameter tuning and performance to be compared to several real ant paths.

Real Ant Routes. Our training and verification data comes directly from the real routes of 15 ant of the species *Cataglyphis velox* as they travelled to and from an experimental feeder located 7.5m from the nest in their natural habitat of scrub near Seville, Spain [3]. Each ant developed an idiosyncratic route leading it to the food and another unique route leading it home again (see Fig. 3 for example ant route).

3D World. Both models were assessed in a corresponding virtual environment of the ant habitat, consisting of a 10x10 metre area in which each tussock is represented as a collection of triangular grass blades of appropriate size and height, with a distribution of shading taken randomly from the intensity range in the panoramic pictures ([11]). Within this virtual environment, we can generate images true to the visual field (296° horizontally, 76° vertically) and resolution (4°) [12] of the ant. The 3D world and image creation software are available at http://www.insectvision.org/. Images were converted to grayscale using the standard Matlab function (Matworks, Inc).

Simulation Trials. Both visual compass and klinokinesis algorithms were supplied with the same route memory database consisting of images sampled every 1cm along the inward routes of 15 ants (route lengths approx 8m each). The performance of the models is again assessed by computing the number of deviations greater than 20cm from the route (more than 20cm from the route the rIDF disappears and all models become lost), the total number of route errors and the number of computations amassed across a route are used to assess performance.

3   Results

3.1   Successful Route Following by a Robot in the Real Ant Habitat

We first tested the visual compass and klinokinesis algorithms on our custom Roboant platform in the real ant environment. The robot faced challenges not experienced in the simulated world. Environmental disturbances such as changing illumination as clouds passed in front of the sun and bushes moving in the

Fig. 3. *Visual compass vs klinokinesis in 3D environment. Top Left Panel.* Example of the real ant path, plotted in our 3D world. The feeder location (start of the route) is labelled F, and the nest (end of the route) marked N. *Top Right & Bottom Left Panels.* Example routes using the visual compass and klinokinesis algorithms when trained with images sampled along real ant route route shown. The black spaced line indicates the boundary 20cm from the route where an error would be generated. An example of an error is shown for the klinokinesis homing by the red cross. *Bottom Right Panel* The number of errors produced across all 15 routes (top), and the number of image comparisons required to those complete simulations (bottom) are shown for both visual compass and klinokinesis algorithms (red and blue boxes respectively).

wind were present. Furthermore, the robot base had no means of accurately measuring its heading angle or distance travelled and as such relied on timing the duration of the motor signal which proved extremely variable in tests.

Both visual compass and klinokinesis methods succeeded in retracing the stored routes, and in both cases made the correct turns at the junction point (see Fig. 2(c-f)). The visual compass algorithm produced only 1 error in total whereas the klinokinesis algorithm accumulated 6 errors (4 if the cluster of errors for the left route are counted as one). This pilot study offers compelling evidence that route following by klinokinesis of the rIDF is possible. Additional benefits of bypassing full scans is also clear from the times taken to complete trials (visual compass: 285s and 375s; klinokinesis: 225s and 190s). The speed up is a modest 63% when comparing the entire time taken to complete the trial, however we note that much of that time was spent by the experimenter emerging from his hidden location before finding the stricken robot and returning it to the route. We expect both a reduction in errors and further speed up when the robot base is fitted with more accurate distance and turning measurement (e.g. wheel encoders).

3.2   Klinokinesis Produces Accurate and Rapid Route Following in Simulation

Following the success of the route following algorithms on the robot we analysed performance of the models further in our simulation, verifying results against real ant routes in an accurate 3D simulation of their environment. Fig. 3 shows an example of the positive route following performance of both visual compass and klinokinesis algorithms. The consistency across the 15 different routes tested are shown in the low number of route deviations reported in the for both models (Fig. 3 bottom right). The klinokinesis method thus appears as effective as full scanning for retracing a previously travelled path.

Klinokinesis significantly outperforms the visual compass in computational workload as shown in Fig. 3 lower right. The median number of image comparisons required by the visual compass algorithm to complete a route is 49980 (IQR=4065) whereas the klinokinesis algorithm requires only 4039 comparisons (IQR=117). This 10-fold reduction in computational load allows routes to be traversed in a fraction of the time required by previous models.

We note that best performance of the klinokinesis algorithm was found for a max turning angle of 80° and max stepsize of 3cm.

4   Discussion

Recently models of route following in ants have concentrated on how ants might use visual cues alone to navigate long range paths [4,5]. While these methods can generate ant-like routes through complex environments, the underlying behaviour (i.e. scanning) does not fit with real ant data [7]. In this work we implemented a novel route following strategy inspired by klinokinesis, as used by Drosophila larvae and bacteria to ascend odour gradients [8,9]. By modulating the size of alternating turns and forward step size in proportion to the instantaneous image difference between current and stored views our algorithm descends the rotational image gradient, and thus maintains heading alignment with a previously travelled path.

We firstly probed the ability of a klinokinesis-like algorithm to perform route guidance on a wheeled robot in the ant's natural environment. Both klinokinesis and visual compass strategies were shown to be capable of at least partial route following along an 8m. In all cases the models followed the training route correctly when faced with a choice of directions. The positive outcome of the pilot study is all the more impressive given the low resolution visual input (approx 1200 pixels) and environmental factors including movement of vegetation by wind and dynamic lighting.

A subsequent simulation study validated models more rigorously by comparing performance against 15 real ant routes in our 3D reconstruction of an ant habitat. The klinokinesis algorithm produced similar error scores as the visual compass methodology but used only 8% of the total number of image comparisons resulting in a marked improvement in speed in both robot and simulation trials. Further, the paths produced by the algorithm show a natural zig-zag pattern, which closely matches the fine-scale movement patterns reported in homing ants which are sinusoidal rather than straight [13–15].

The positive results of both simulation and robot studies support the plausibility of the use of a klinokinesis-like mechanism by ants and also provides inspiration for a computationally cheap visual route following strategy for use on mobile robots. There are a number of immediate amendments that could be made to improve the performance of the robot. Firstly, applying rigorous parameter tuning methods (e.g. allowing non-linear weighting) to optimise turn angle and step size, coupled with improved turn and speed measurements (e.g. by adding wheel encoders) would immediately improve robot performance. Further improvements might be possible by modulating turns not by the instantaneous image difference but by the change in image difference, giving a natural normalisation in readings and the ability to invert the turn direction (e.g. when the change in sensory input becomes negative). We note that an amendment to the algorithm would change the underlying mechanism from a klinokinesis-like to a klinotaxis-like behaviour. Similar klinotaxis inspired methods have recently been implemented in both simulation and robot studies to ascend light gradients [16].

Finally, we report on an interesting behaviour that emerged during both studies. During robot trials, when the robot path diverged from the route, it would start to perform large turns that were exaggerated by its rotational momentum. As the forward step size would be small this behaviour looked very similar to ant scans (see Fig. 4a), and in some cases allowed the robot to recover the route. We occasionally observed similar results during parameter tuning in our simulated study, whereby our simulated ant performed a series of large turns before aligning itself with the route. It is possible to trigger such behaviours by placing the simulation ant at specific locations on the route but at the wrong heading,

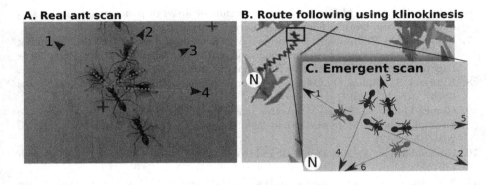

Fig. 4. *Emergent scans. A* Example of an ant scan. Frames in which the ant has paused are overlaid, showing 4 distinct viewing directions (labelled arrows) before a direction of movement was selected. Video adapted from [7] with permission. *B* The motions resulting from klinokinetic homing when the virtual ant is placed on the route but with the wrong heading (red ant, with direction shown by arrow 1). Note, the nest direction to the lower left labelled N. Shown are the 5 subsequent movements of the simulated ant before finally moving towards the nest (green ant, with direction shown by arrow 6). Note that the parameters used are larger than those optimized for homing - $\alpha = 8$ and $\beta = 600$.

with parameters settings that allowed large turns (see Fig. 4b). We thus suggest that scanning may be an emergent property of a simpler route following strategy, rather than a fundamental behavioural routine required for route following.

Acknowledgments. The authors would like to thank our reviewers for constructive comments; Xim Cerda, Antoine Wystrach, Paul Graham, Alex Dewar for assistance with robot trials; and Antoine Wystrach and Barbara Webb for feedback on the manuscript.

References

1. Wehner, R.: The desert ant's navigational toolkit: procedural rather than positional knowledge. Navigation **55**(2), 101–114 (2008)
2. Kohler, M., Wehner, R.: Idiosyncratic route-based memories in desert ants, Melophorus bagoti: How do they interact with path-integration vectors? Neurobiology of Learning and Memory **83**(1), 1–12 (2005)
3. Mangan, M., Webb, B.: Spontaneous formation of multiple routes in individual desert ants (Cataglyphis velox). Behavioral Ecology (2012)
4. Baddeley, B., Graham, P., Husbands, P., Philippides, A.: A model of ant route navigation driven by scene familiarity. PLoS Computational Biology **8**(1), e1002336 (2012)
5. Baddeley, B., Graham, P., Philippides, A., Husbands, P.: Holistic visual encoding of ant-like routes: Navigation without waypoints. Adaptive Behavior **19**(1), 3–15 (2011)

6. Zeil, J., Hofmann, M.I., Chahl, J.S.: Catchment areas of panoramic snapshots in outdoor scenes. Optical Society of America Journal **200**, 450–469 (2003)
7. Wystrach, A., Philippides, A., Aurejac, A., Cheng, K., Graham, P.: Visual scanning behaviours and their role in the navigation of the Australian desert ant Melophorus bagoti. Journal of Comparative Physiology A **200**(7), 615–626 (2014)
8. Khurana, S., Siddiqi, O.: Olfactory responses of Drosophila larvae. Chemical Senses **38**(4), 315–323 (2013)
9. Gomez-Marin, A., Louis, M.: Active sensation during orientation behavior in the Drosophila larva: more sense than luck. Current Opinion in Neurobiology **22**(2), 208–215 (2012)
10. Kodzhabashev, A.: Ant Trackball and Robot. MSc Thesis, University of Edinburgh (2014)
11. Mangan, M.: Visual homing in field crickets and desert ants: A comparative behavioural and modelling study. PhD Thesis, University of Edinburgh (2011)
12. Schwarz, S., Narendra, A., Zeil, J.: Arthropod Structure & Development The properties of the visual system in the Australian desert ant Melophorus bagoti. Arthropod Structure and Development **40**(2), 128–134 (2011)
13. Nicholson, D.J., Judd, S.P.D., Cartwright, B.A., Collett, T.S.: Learning walks and landmark guidance in wood ants (Formica rufa). Journal of Experimental Biology **202**(13), 1831–1838 (1999)
14. Lent, D.D., Graham, P., Collett, T.S.: Phase-dependent visual control of the zigzag paths of navigating wood ants. Current Biology **23**(23), 2393–2399 (2013)
15. Collett, T.S., Lent, D.D., Graham, P.: Scene perception and the visual control of travel direction in navigating wood ants. Philosophical Transactions of the Royal Society of London. Series B, Biological Sciences **369**, 20130035 (2014)
16. Schmickl, T., Hamann, H., Stradner, J., Mayet, R., Crailsheim, K.: Complex Taxis-Behaviour in a Novel Bio-Inspired Robot Controller. ALIFE, 648–658 (2010)

Using Animal Data and Neural Dynamics to Reverse Engineer a Neuromechanical Rat Model

Alexander J. Hunt[1]([✉]), Nicholas S. Szczecinski[1], Emanuel Andrada[2], Martin Fischer[2], and Roger D. Quinn[1]

[1] Case Western Reserve University, Cleveland, OH 44106, USA
[2] Friedrich-Schiller-Universität Jena, 07743 Jena, Germany

Abstract. A baseline model for testing how afferent muscle feedback affects both timing and activation levels of muscle contractions has been constructed. We present an improved version of the neuromechanical model from our previous work [6]. This updated model has carefully tuned muscles, feedback pathways, and central pattern generators (CPGs). Kinematics and force plate data from trotting rats were used to better design muscles for the legs. A recent pattern generator topology [15] is implemented to better mimic the rhythm generation and pattern formation networks in the animal. Phase-space and numerical phase response analyses reveal the dynamics underlying CPG behavior, resulting in an oscillator that produces both robust cycles and favorable perturbation responses. Training methods were used to tune synapse properties to shape desired motor neuron activation patterns. The result is a model which is capable of self-propelled hind leg stepping and will serve as a baseline as we investigate the effects changes in afferent feedback have on muscle activation patterns.

Keywords: Neural controller · Rat · Mammal · Central pattern generator · Intra-leg coordination

1 Introduction

Maneuvering through unstructured environments is a challenging task that animals successfully perform on a variety of terrains. Feedback loops within the nervous system act at many levels, changing the timing and magnitude of motion in a continuous fashion, complicating studies of the interaction between the animal's body and nervous system. Understanding these feedback loops in the context of walking can lead to a better understanding of how behavior is modified to produce moderate changes in output. For example, if the animal wishes to run faster, is a change in motor output timing or magnitude alone sufficient, or must both be commanded?

A.J. Hunt—This work was supported by a NASA Office of the Chief Technologists Space Technology Research Fellowship (Grant Number NNX12AN24H).

© Springer International Publishing Switzerland 2015
S.P. Wilson et al. (Eds.): Living Machines 2015, LNAI 9222, pp. 211–222, 2015.
DOI: 10.1007/978-3-319-22979-9_21

Rats, mice, and cats are often studied in this context for their convenient size. Many models of these animals have been developed to study what makes them successful, including those of muscles [14][7], kinematic and dynamic movements [1][13], the behavior and connectivity of the neural systems involved in locomotion [15], and even network pathways which influence stepping transitions of a neuromechanical model [4]. However, animals are both neural and mechanical walking systems, and separating one component from the other can significantly alter the behavior of the system [11].

Faster computers and the development of more sophisticated modeling systems have lead to models which integrate these components and increase our understanding of the interaction between them [2][6][12]. However, when working with these larger, more comprehensive models, the number of parameters that must be identified can become as large as the data set itself. Additionally, these models are highly non-linear. Adjusting these parameters manually until desired behaviors are achieved is extremely difficult, time consuming, and unreliable. For example, small changes of a single variable may have cascading effects that drastically change or even abolish coordinated walking motion.

We have developed tools and methods for systematic numeric training of multiple parameters simultaneously to achieve rhythmic stepping and we have used these tools to improve our previously published neuromechanical rat model [6]. The network topology in our model has been improved based on recent findings in the mouse [15]. Thorough analysis of central pattern generator (CPG) dynamics has revealed how network parameters can affect oscillation and its sensitivity to inputs. Kinematic and force-plate data from the rat have been used directly to tune feedback parameters and produce propulsive steps in the rat model.

2   Model Description

The rat model was developed in AnimatLab 2 [2] and is shown in Fig. 1. The model consists of both a physical model actuated by linear Hill muscle models and a neural model with leaky conductance-based neurons. The physical model consists of two hind legs, each with three degrees of freedom in the sagittal plane: hip, knee, and ankle. The model was constructed using bone scans of a dissected rat. The body, head, tail, and upper limbs are fixed with respect to the pelvis. The entire mass of the model is approximately 300 g divided between the body components according to lab measurements [1]. The front of the model is supported with a single bar placed approximately 5 cm directly underneath the shoulders and with a coefficient of friction of 0.1. This places approximately half the body weight on the hind legs and requires the legs to overcome some friction to propel the body forward.

The neurons in the model are leaky conductance-based models. Each neuron possesses three states: the membrane voltage V, and sodium channel activation m and deactivation h. The dynamics of each neuron are as follows:

$$C_{mem}\frac{dV}{dt} = G_{mem} \cdot (E_{rest} - V) + G_{syn} \cdot (E_{syn} - V) + G_{Na} \cdot m \cdot h \cdot (E_{Na} - V) \quad (1)$$

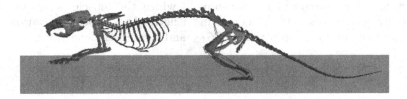

Fig. 1. Model of the rat in AnimatLab 2. The front of the body is supported and hind legs are actuated at the hip, knee, and ankle.

$$\frac{dm}{dt} = \frac{m_\infty(V) - m}{\tau_m(V)} \tag{2}$$

$$\frac{dh}{dt} = \frac{h_\infty(V) - h}{\tau_h(V)} \tag{3}$$

in which G is a conductivity, E is a constant reversal potential (reference voltage), and τ is a time constant. The subscript mem stands for "membrane," $rest$ stands for the resting potential of the neuron, syn stands for "synapse," and ∞ stands for the equilibrium value. $m_\infty(V)$ and $h_\infty(V)$ are sigmoidal functions of the membrane voltage, expressed as:

$$u_\infty(V) = \frac{1}{1 + exp(-S_u \cdot (E_u^{mid} - V))} \tag{4}$$

in which u is either m or h, S_u is the slope of the sigmoid at its midpoint, and E_u^{mid} shifts the sigmoid along the V axis. $m_\infty(V)$ increases monotonically with the voltage and $h_\infty(V)$ decreases monotonically, due to $S_m > 0$ and $S_h < 0$.

Neurons communicate via synapses. The presynaptic (i.e. sending) neuron changes the synaptic conductivity (G_{syn} in Eq. 1) proportional to its membrane voltage in a piecewise-linear fashion:

$$G_{syn} = G_{max} \cdot min\left(max\left(\frac{V_{pre} - E_{lo}}{E_{hi} - E_{lo}}, 0\right), 1\right) \tag{5}$$

in which G_{max} is the maximum conductivity that a synapse can achieve, and E_{lo} and E_{hi} are the lower and upper thresholds.

Muscles are modeled as linear Hill muscles:

$$\frac{dT}{dt} = \frac{k_{se}}{c}\left(k_{pe}x + c\dot{x} - \left(1 + \frac{k_{pe}}{k_{se}}\right) \cdot T + A\right) \tag{6}$$

in which T is the tension in the muscle, x is the length of the muscle, and A is the muscle activation, which is a sigmoid function of the motor neuron voltage. The model also includes passive properties: k_{se} is the series element stiffness, which models the tendon, and k_{pe} and c, the parallel stiffness and damping, which model the muscle fibers' resistance to stretch. The contraction force is subject to a length-tension relationship, in which the muscle activation A is limited by its deviation from the resting length. The muscle activation as a function of the motor neuron voltage V_{MN} and the muscle length x is:

$$A = \left(1 - \frac{(x - x_{rest})^2}{x_{width}^2}\right) \cdot \frac{A_{max}}{1 + exp(-S \cdot (E_{mid} - V_{MN}))} \tag{7}$$

in which A_{max} is the maximum active force the muscle can apply, S is the maximal slope of the activation curve, and E_{mid} shifts the sigmoid along the V axis. More details about all of these models can be found in [6].

Muscle parameters are approximated based on data from rats when available, and otherwise from other mammals. Muscle attachments are based on the main antagonistic muscles that actuate each joint [7]. The resting length x_{rest} is set to the maximum length of each muscle, causing muscles to get stronger as they elongate [14], and x_{width} is set such that the muscles exert 70% of their maximum force when fully contracted [11]. Maximum muscle force A_{max} is set by taking maximum foot forces for several positions of the foot during a step cycle from the full muscled Johnson model [7], and performing inverse analysis to find maximum joint torques. Muscle attachments in our model are then used to find the muscle forces needed to apply these maximum joint torques. Muscle damping c is set to the ratio between maximum muscle force and maximum muscle velocity [7]. Series stiffness is set according to data from [11], and parallel stiffness is set such that the muscle deflects 4% of its total length under maximum load [9].

3   Designing and Training Neural Output

Developing a neurologically derived spinal network capable of producing self-supported hind leg stepping in the rat model was divided into two tasks. The first task was determining the necessary motor neuron activations required to produce walking. The second task was training the network to output these motor neuron activations.

3.1   Calculating Motor Neuron Activations

First, joint torques for steady state walking were calculated. Stance joint torques were calculated using inverse dynamics based on force plate and kinematic data of rats walking on level ground [1]. Kinematic data in conjunction with a full hind leg inertial model were used to solve for the swing phase joint torques. These vectors were concatenated and smoothed to form a full step cycle (bottom left of Fig. 2).

Desired muscle tension was calculated using joint torques and the kinematic model. A unique solution was obtained by assuming only one muscle per joint actively contracts at any time [5]. The active muscle must overcome the previously calculated torques as well as torques created by the passive muscle forces of the opposing antagonistic muscle (top right Fig. 2).

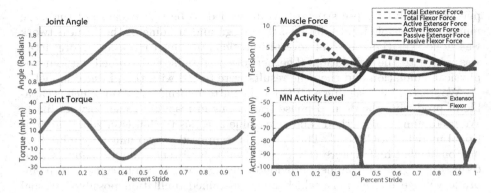

Fig. 2. Intermediate steps in the development of desired motor neuron activations for the hip extensor and flexor muscles. X-axis is stride percent; stance is 0%-50% and swing is 50%-100%. X-ray video reveals the joint angle (top left), which when combined with force plate data and a dynamic model of the rat give joint torque (bottom left). Passive and active muscle forces were then calculated (top right). Motor Neuron activations were calculated using Eq. 6 (bottom right).

Passive forces were calculated using Eq. 6 with $A = 0$. Muscle length (x) and muscle velocity ($\dot{x}$) were calculated using model geometry and kinematic data. The derivative was discretized and T was solved for at the next time step based on the previous tension:

$$T_{i+1} = T_i + \Delta t \cdot \frac{k_{se}}{c} \left(k_{pe} x_i + c\dot{x}_i - \left(1 + \frac{k_{pe}}{k_{se}}\right) \cdot T_i \right) \qquad (8)$$

Starting with $T = 0$ and repeating this process for several step cycles produces a periodic tension profile which counteracts the ground-force and dynamic torques. A bisection root-finder solves for the active muscle tension required to produce the necessary torques and each muscle's tension was numerically differentiated to find its rate of change. Equations 6 and 7 were solved with a bisection root-finder to find the motor neuron voltage (bottom right Fig. 2).

3.2 Designing CPGs

The connectivity of the Zhong locomotor model [15] was chosen as the basis for the neural control system of our model. The implemented model was simplified to a single network for each joint with a pattern formation layer and lower level afferent feedback networks. This allows us to concentrate our efforts on how feedback can be used to shape motor neuron activation level and timing.

Each leg network consists of 82 neurons with 12 parameters each, and 134 synapse connections with 4 parameters each. Many parameters were set using basic heuristics such as resting voltage (-60 mV), time constant (5 ms), and relative size (1). Even after these simplifications, approximately 90 parameters per leg, mostly synapse strengths, still needed to be set. Because of the large number of possible local solutions, the design and training of the CPG network was done over the course of several iterations in which more and more detail was added to the simulation. The network was built and trained without the physical simulation and then the results were tested with the physical model.

We started by designing a CPG for the pattern formation layer which is capable of producing the desired phase transitions in response to sensory feedback. We used numerical methods and eigenvalue analysis to characterize a computational model of a CPG [3]. The system is composed of two mutually inhibitory neurons called half-centers (HCs), each with persistent sodium channels (last term, Eq. 1). These channels provide decaying positive reinforcement to membrane voltage fluctuations, which make sustained oscillation possible. Mutual inhibition is implemented via nonspiking interneurons (INs). Each HC excites an IN, which inhibits the other HC, as shown in Fig. 3.

Our analysis finds nullclines and equilibrium points for the system, characterizes their stability, and examines how oscillation is achieved. To simplify analysis, we assume that the parameters of each HC, its IN, and the synapses that connect them are symmetrical. In addition, since $\tau_m < \tau_V << \tau_h$, we assume $m = m_\infty$, which lets us visualize one neuron's phase portrait in two-dimensional space.

This model may either have one or two equilibrium points, depending on the applied drive. When there are two, we categorize these points by their state relative to the other, yielding an excited neuron equilibrium voltage (E_{exc}) and an inhibited neuron equilibrium voltage (E_{inh}). E_{exc} is the steady state voltage of the neuron when it has no inputs, and E_{inh} is the steady state voltage of the opposite HC when inhibited by its counterpart resting at E_{exc}. E_{inh}'s value relative to synaptic threshold E_{lo} (Eq. 5) was revealed to be important for determining the existence and nature of oscillation. For example, if E_{inh} is below the synaptic threshold, no endogenous oscillation will occur, because the inhibited HC has no means of escape. Applying excitation via descending commands increases E_{inh} and E_{exc}, and when $E_{inh} > E_{lo}$, the system oscillates. However, in this case E_{inh} and E_{exc} no longer correspond to actual equilibrium states of the system; with enough applied drive, the system will possess only one unstable equilibrium point, about which a limit cycle forms. Fig. 3 shows a full sweep of applied drives, revealing a number of bifurcations. However, high-drive scenarios were not applied to our model.

Perturbation response was also studied by numerically generating phase response surfaces (PRSs). These PRSs are different from the typical phase response curve in that the duration of a stimulus (a square pulse) is plotted along with the phase of stimulus onset. We chose this method because our inputs are in the form of sustained signals from afferent receptors, not impulses.

PRSs for inhibitory stimuli to the HC and the IN, for the first or second cycle following the stimulus are shown in Fig. 4. The PRS for stimulus to the IN shows that it is comprised of three planar regions. If inhibition arrives while that IN is active, the CPG advances phase proportional only to the phase of its

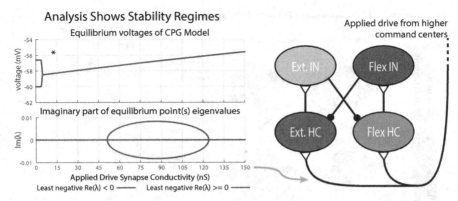

Fig. 3. Plots showing the equilibrium voltages of the neurons in a CPG and their eigenvalues as the conductivity of an excitatory pathway to both HCs is varied. This changes the INEV, resulting in a variety of eigenvalues and thus behaviors. The * denotes the system configuration used in this study.

arrival. If it arrives while the IN is inactive, the CPG phase is delayed only if the stimulus duration is longer than the remaining time in that cycle.

The CPG's response to stimuli applied to the HC is more complicated. The response is linear along the stimulus phase for stimulus durations longer than 10% of the period. It is also linear along the stimulus duration, which suggests that the phase shift depends on the timing of both the application and removal of the stimulus. When the stimulus is removed, a rapid upward change in V following release is reinforced by m, causing the neuron to overshoot E_{lo}, inhibit the other HC, and create a secondary transition. This rebound effect can also be achieved by exciting the presynaptic IN, which we choose to do in our model.

A CPG was designed by setting the equilibrium point of the low neuron to 0.001 mV below threshold, adding a synaptic drive of 13 nS towards 0 mV, and adjusting the slope of $m_\infty(V)$, h_∞, and G_{Na} until a CPG was formed with a peak height approximately 20% above the high equilibrium point and a period of two times the final desired oscillation period.

3.3   CPG Entrainment and Output

The next step in getting the network to produce the desired output was to entrain the CPGs to the sensory feedback such that it oscillates in conjunction with the calculated alternating motor neuron activations. Training the network to produce these results necessitates that feedback be provided as though the correct stepping motions are taking place. Types Ia, Ib and II feedback are calculated as follows in AnimatLab 2:

$$\text{Ia} = k_a x_{series} \qquad \text{Ib} = k_b T \qquad \text{II} = k_c x_{parallel} \qquad (9)$$

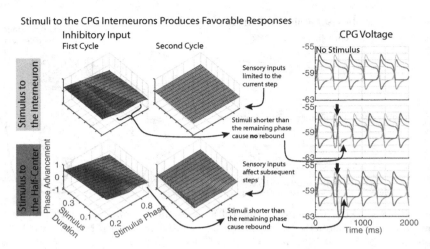

Fig. 4. Phase response surfaces from inhibitory inputs, and their interpretation explained by simulation output. All axis labels for the surfaces are the same, in units of % of the period.

where k_a, k_b, and k_c are gain parameters set by the user, and the expressions in Eq. 9 specify the current injected into sensory neurons. Ia feedback encodes both length and velocity of muscle movement. Ib encodes the muscle tension, and II encodes the length of the parallel elastic element $x_{parallel}$. The gains were set such that the feedback injects between 0 and 20 nA of current into a designated sensory neuron over the full possible range of tension and lengths for each muscle. Using the kinematics and expected muscle tensions calculated earlier, we are then able to calculate the afferent feedback of the AnimatLab 2 simulation for walking with the projected kinematics.

This feedback is used as input into the CPG according to rules discovered in vertebrates. Hip flexor stretch encourages a transition from stance to swing in decerebate cats [10], and hip extensor stretch could contribute to termination of swing [8]. To obtain the proper CPG response, these feedback pathways are implemented as inhibitory synapses to the CPG INs. Research also shows that stance can be prolonged in spinal cats with continued stimulation of ankle Ib feedback [10]. This is implemented as an excitatory connection to the extensor IN to obtain the rebound effect described earlier.

The strength, synaptic thresholds, and driving current of these synapses onto the CPG were first trained using a genetic algorithm with a population of 1500 samples for 5 generations, and then refined using a Nelder-Mead simplex method. The objective of training was period and CPG rise times in relation to desired MN rise times.

Subsequently, the CPG output synapse strength was trained to produce activations of the motor neurons with a peak height that matches each desired motor neuron activation using similar methods to those described in the previous step.

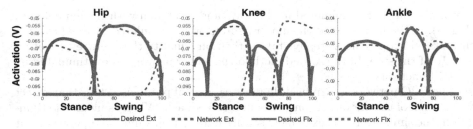

Fig. 5. Trained Network Output compared with desired motor neuron activations. The hip and ankle fit very well, while the bi-phasic nature of the knee posed some problems. Extensor output is much higher than desired at the end of swing, and does not see a full dip down at the beginning of stance. Flexor output at the beginning of stance was ignored during training and later produced through a 'hamstring' connection in which hip extensor activation also activates knee flexion.

3.4   Afferent Influence of MN Activation

Afferent feedback was trained to help shape the MN output and provide additional force if necessary to overcome changes in foot placement (excitatory Ib feedback), or reduced force if the leg is moving too quickly (inhibitory Ia feedback). All neurons and pathways involved in these networks were designed to be completely continuous over all possible ranges. Though sensory feedback could be modulated by thresholds in the animal, we do not train thresholds because our input is only expected input and training thresholds causes the system to become overly dependent on exact threshold points and small changes in feedback strength had significant effects on behavior. The final fit of the training can be seen in Fig. 5.

3.5   Additional Modifications

Two additional modifications were made to the network to produce stable stepping. We added a connection from the hip extensor motor neuron to the knee flexor motor neuron because in the animal, hamstring muscles both extend the hip and flex the knee. This implemented connection produces activation of the knee flexor at the beginning of stance and maintains dual activation of the muscles during stance, stiffening the joint.

The second change modifies the ankle flexor and extensor Ib feedback. Ankle movement was significantly too small and limited propulsive forces. To increase movement, Ib feedback was increased for both joints. The result of this increase can be seen in Fig. 6 and shows higher network motor neuron activations than the calculated desired activations.

4   Walking Comparison

Motor neurons were stimulated in two ways and compared with the animal data (Fig. 6). In the first method the motor neurons were stimulated by a sequence of currents which produces steady state activation equal to the average activation of 1/10 of the step. In the second method the motor neurons were stimulated by the CPG network.

The CPG network is significantly more capable of producing steps than direct stimulation of the motor neurons without any feedback. When using direct stimulation, the simulation was extremely sensitive to friction values and pelvis height and did not produce significant kinematic motions despite having close to the calculated desired muscle activations.

A comparison of the data in Fig. 6 shows that the neuron network is capable of maintaining the necessary rhythms despite not having been trained with the full

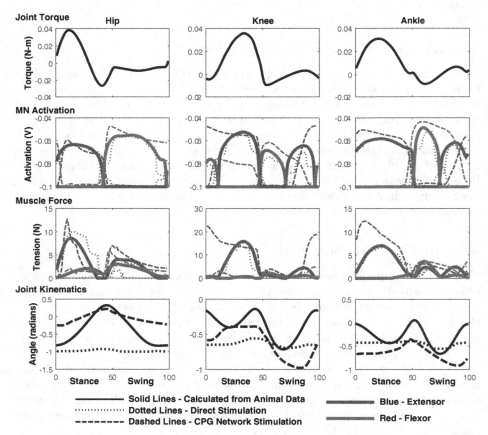

Fig. 6. Comparison of desired motor neuron activations, muscle tensions, and joint kinematics with those produced by direct stimulation of the motor neurons and stimulation of the motor neurons by the CPG network

physical simulation. The careful construction of the CPG model and subsequent training of the coordinating currents is successful in producing the transition from stance to swing. The timing of this transition for all three joints is properly coordinated with the end of ground contact.

The CPG network does not have any connections between the legs, and is not capable of producing sustained alternating stepping. Because the leg is not catching the weight of the rat, the first half of stance for the knee and ankle do not undergo early flexion followed by extension; however the power stroke of the leg raises the pelvis from the ground during each step, indicating the muscles are receiving activations capable of supporting and propelling the rat.

The foot contacts the ground more quickly in simulation than in the animal. This earlier contact could be because the hip does not flex as fast as the other joints and does not raise the knee high enough. Another cause could be higher activations of the knee extensor during swing cause the tibia to extend faster and create early ground contact.

The Ib and Ia feedback pathways that modulate motor neuron output add significant control to the simulation, however there is not enough available data on how these pathways affect walking to properly train them offline. The model is now set up such that systematic adjustments can be made to these pathways with the physical simulation in the loop, and changes in behavior can be recorded and analyzed.

5   Conclusions

A biologically based neuron network has been successfully trained without a physical simulation in the loop to produce hind leg stepping in a neuromechanical rat model. This model is more carefully constructed both in physical and neuronal simulation than our previous model through the use of data, analysis of the base CPG system, and naive training methods. This model is ready to test how changes in afferent feedback signals can affect both timing and activity level of motor neurons in the context of a full physical system.

References

1. Andrada, E., Mämpel, J., Schmidt, A., Fischer, M.S., Karguth, A., Witte, H.: From biomechanics of rats' inclined locomotion to a climbing robot. International Journal of Design and Nature and Ecodynamics **8**(3), 191–212 (2013)
2. Cofer, D., Cymbalyuk, G., Reid, J., Zhu, Y., Heitler, W.J., Edwards, D.H.: AnimatLab: a 3D graphics environment for neuromechanical simulations. Journal of Neuroscience Methods **187**(2), 280–288 (2010). http://www.ncbi.nlm.nih.gov/pubmed/20074588
3. Daun-Gruhn, S., Rubin, J.E., Rybak, I.A.: Control of oscillation periods and phase durations in half-center central pattern generators: a comparative mechanistic analysis. Journal of Computational Neuroscience **27**(1), 3–36 (2009). http://www.pubmedcentral.nih.gov/articlerender.fcgi?artid=2844522&tool=pmcentrez&rendertype=abstract

4. Ekeberg, O., Pearson, K.G.: Computer simulation of stepping in the hind legs of the cat: an examination of mechanisms regulating the stance-to-swing transition. Journal of Neurophysiology **94**(6), 4256–4268 (2005). http://www.ncbi.nlm.nih.gov/pubmed/16049149

5. Hooper, S.L., Guschlbauer, C., Blümel, M., Rosenbaum, P., Gruhn, M., Akay, T., Büschges, A.: Neural control of unloaded leg posture and of leg swing in stick insect, cockroach, and mouse differs from that in larger animals. The Journal of Neuroscience : The Official Journal of the Society for Neuroscience **29**(13), 4109–4119 (2009). http://www.ncbi.nlm.nih.gov/pubmed/19339606

6. Hunt, A., Schmidt, M., Fischer, M., Quinn, R.D.: Neuromechanical simulation of an inter-leg controller for tetrapod coordination. In: Duff, A., Lepora, N.F., Mura, A., Prescott, T.J., Verschure, P.F.M.J. (eds.) Living Machines 2014. LNCS (LNAI), vol. 8608, pp. 142–153. Springer, Heidelberg (2014). http://link.springer.com/chapter/10.1007/978-3-319-09435-9_13

7. Johnson, W.L., Jindrich, D.L., Roland, R.R., Edgerton, V.R.: A three-dimensional model of the rat hindlimb: musculoskeletal geometry and muscle moment arms. Journal of Biomechanics **41**(3), 610–619 (2008). http://www.sciencedirect.com/science/article/pii/S0021929007004344

8. McVea, D.A., Donelan, J.M., Tachibana, A., Pearson, K.G.: A role for hip position in initiating the swing-to-stance transition in walking cats. Journal of Neurophysiology **94**(5), 3497–3508 (2005). http://www.ncbi.nlm.nih.gov/pubmed/16093331

9. Meijer, K., Grootenboer, H., Koopman, H., van der Linden, B., Huijing, P.: A Hill type model of rat medial gastrocnemius muscle that accounts for shortening history effects. Journal of Biomechanics **31**(6), 555–563 (1998). http://linkinghub.elsevier.com/retrieve/pii/S0021929098000487

10. Pearson, K.G.: Role of sensory feedback in the control of stance duration in walking cats. Brain Research Reviews **57**(1), 222–227 (2008). http://www.ncbi.nlm.nih.gov/pubmed/17761295

11. Pearson, K.G., Ekeberg, O., Büschges, A.: Assessing sensory function in locomotor systems using neuro-mechanical simulations. Trends in Neurosciences **29**(11), 625–631 (2006). http://www.ncbi.nlm.nih.gov/pubmed/16956675

12. Szczecinski, N.S., Brown, A.E., Bender, J.A., Quinn, R.D., Ritzmann, R.E.: A Neuromechanical Simulation of Insect Walking and Transition to Turning of the Cockroach Blaberus discoidalis. Biological Cybernetics (2013)

13. Witte, H., Hackert, R., Lilje, K., Schilling, N., Voges, D., Klauer, G., Ilg, W., Albiez, J., Seyfarth, A., Germann, D., Hiller, M., Dillmann, R., Fischer, M.: Transfer of biological principles into the construction of quadruped walking machines. In: Proceedings of the Second International Workshop on Robot Motion and Control, RoMoCo 2001 (IEEE Cat. No.01EX535), pp. 245–249 (2001)

14. Zajac, F.E.: Muscle and tendon: properties, models, scaling, and application to biomechanics and motor control. Critical Reviews in Biomedical Engineering (1989). http://ukpmc.ac.uk/abstract/MED/2676342

15. Zhong, G., Shevtsova, N.A., Rybak, I.A., Harris-Warrick, R.M.: Neuronal activity in the isolated mouse spinal cord during spontaneous deletions in fictive locomotion: insights into locomotor central pattern generator organization. The Journal of Physiology **590**(Pt. 19), 4735–4759 (2012). http://www.pubmedcentral.nih.gov/articlerender.fcgi?artid=3487034&tool=pmcentrez&rendertype=abstract

Entraining and Copying of Temporal Correlations in Dissociated Cultured Neurons

Terri Roberts[1], Kevin Staras[1], Philip Husbands[2], and Andrew Philippides[2(✉)]

[1] School of Life Sciences, University of Sussex, Brighton, UK
{T.P.Roberts,K.Staras}@sussex.ac.uk
[2] Centre for Computational Neuroscience and Robotics,
Department of Informatics, University of Sussex, Brighton, UK
{philh,andrewop}@sussex.ac.uk

Abstract. Here we used multi-electrode array technology to examine the encoding of temporal information in dissociated hippocampal networks. We demonstrate that two connected populations of neurons can be trained to encode a defined time interval, and this memory trace persists for several hours. We also investigate whether the spontaneous firing activity of a trained network, can act as a template for copying the encoded time interval to a naïve network. Such findings are of general significance for understanding fundamental principles of information storage and replication.

Keywords: Temporally correlated activity · Microelectrode array · Neural network · Dissociated cultured neurons

There is substantial current interest in understanding the way that temporal information is encoded in neural networks but the underlying mechanisms remain poorly defined. One emerging idea is that the temporal architecture of a circuit may be readily shaped by the timing of the input that the network receives [1]. If this is the case, how temporal correlations can be both entrained within, and copied between, networks of neurons could be significant with regards to the transfer of information between neural circuits. Here we set out to investigate whether such entraining and copying is possible in dissociated hippocampal networks.

To do this, we exploited multi-electrode array (MEA) technology. MEAs are a micro-engineered grid of extracellular electrodes providing a robust solution for sampling network spiking activity - effectively 'nodes' representing collective output of one or several neurons local to an electrode - as well as allowing the application of targeted stimulation. We have developed a method that allows us to grow neuronal cultures from dissociated hippocampal neurons which form mature highly inter-connected networks in 10-12 days in vitro on which we can effectively record and impose activity patterns (Fig. 1 a-c). Using an MEA also demonstrates the possibility of being able to entrain temporal correlations in neural networks using a biohybrid device.

© Springer International Publishing Switzerland 2015
S.P. Wilson et al. (Eds.): Living Machines 2015, LNAI 9222, pp. 223–226, 2015.
DOI: 10.1007/978-3-319-22979-9_22

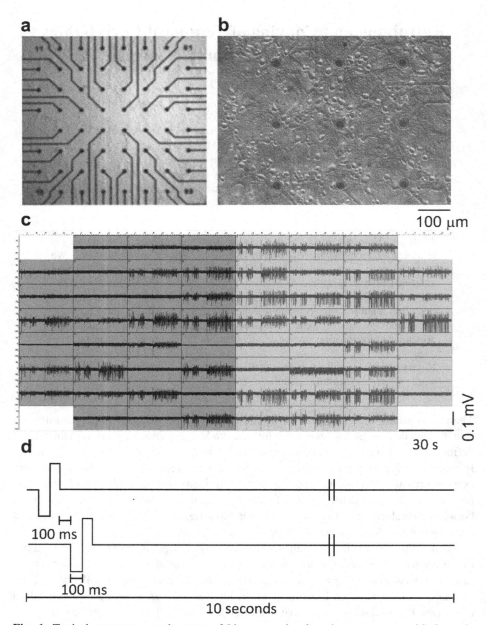

Fig. 1. Typical appearance and output of hippocampal cultured neurons on multi-electrode array (MEA). (a) Image of 64 electrode MEA. (b) High-density neuronal cultures (~14 DIV) used for experiments. (c) Typical readout of activity in 64 electrode recording. A nodes are shaded in dark, B nodes shaded in light. (d) Training protocol. Schematic shows timing of stimulation for A nodes (top) and B nodes (bottom).

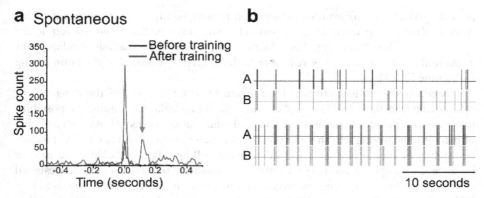

Fig. 2. Neural activity before and after training. (a) Cross-correlogram (A->B) before (green) and after (red) training for an example experiment. Training results in a robust peak at ~120 ms (arrow). (b) Sample of node activity before (blue) and after (red) training showing training leads to synchronized activity. Spikes are seen as vertical lines in the subplots to the right.

We used this preparation to test whether it is possible to train two connected populations of neurons to encode a defined time interval. To do this, we initially divide the population of nodes into two sub-groups, 'A' and 'B' (Fig. 1c). The training paradigm (Fig. 1d) consists of electrical stimulation imposed onto A nodes followed, at a set time-interval (100 ms), by stimulation applied to B nodes and repeated 360 times over 1 h. Using this paradigm, a defined time interval can be readily encoded in the firing relationship of the neural sub-populations stimulated (Fig. 2). In further experiments, we have seen that the encoded change can be persistent for several hours and that the fidelity of encoding changes when the training time interval is modulated.

In a second type of experiment, we are starting to explore the replicative capabilities of the stored information. Specifically, the objective is to use the post-training spontaneous activity recorded in A and B (which encodes the entrained A->B temporal relationship) as the basis for a training paradigm imposed on a new 'naïve' neuronal culture. Our preliminary experiments indicate that a trained network can indeed act as a template for replicating the same timing relationship in a new network. To date we have found that the effectiveness of this training paradigm is substantially lower than the original A->B training. This may relate to the choice of trained data sequences used as the basis for the secondary training. We are currently optimizing the training length and other parameters to improve the fidelity of the copying.

Here we exploit MEA technology to explore features of temporal encoding in hippocampal neural networks. We show robust storage of a defined time interval between the activity of two neuronal populations that becomes a feature of their spontaneous firing patterns after training. Such findings have broad similarities to those reported using large cortical networks in brain slice [1] but our results suggest that, remarkably, this capability is preserved in small reconstituted circuits. Moreover, initial results support the idea that, under appropriate conditions, this temporal relationship can act as a template for replication in independent networks. As such, this information replication mechanism offers a potential substrate on which complex copying properties - for example Darwinian neurodynamics [2] - may operate. This work is thus a step towards the empirical verification of the Neuronal Replicator Hypothesis (NRH) that states that a Darwinian process of production of adaptations by natural selection can run in real-time in the neuronal networks of the brain during its lifetime [2,3,4].

Indeed, this work is motivated by a desire to better understand the dynamics of information transfer between neural circuits and in particular to examine the possibility of neural replicator dynamics in biological neuronal networks [3]. As well as being of benefit to neuroscience, an understanding of such dynamics points the way towards new kinds of biomorphic architectures for controlling adaptive behavior [5,6]. For instance, biomorphic neural devices and brain-machine interfaces could be designed to exploit the kinds of replicator dynamics uncovered in these studies. The use of an MEA to entrain information within a real neuronal network also raises the possibility of hybrid adaptive systems that incorporate cultures as dynamic information processing elements. However, the current work is in its infancy and much more fundamental science is needed to understand, let alone exploit, such information transfer.

Acknowledgements. This project has received funding from the European Union's Seventh Framework Programme for research, technological development and demonstration under grant agreement no 308943.

References

1. Johnson, H.A., Goel, A., Buonomano, D.V.: Neural Dynamics of in Vitro Cortical Networks Reflects Experienced Temporal Patterns. Nat. Neurosci. **13**, 917–919 (2010)
2. Fernando, C., Goldstein, R., Szathmáry, E.: The neuronal replicator hypothesis. Neural Comput. **22**, 2809–2857 (2010)
3. Fernando, C., Karishma, K.K., Szathmáry, E.: Copying and Evolution of Neuronal Topology. PLoS ONE **3**, e3775 (2008)
4. Fernando, C., Szathmáry, E., Husbands, P.: Selectionist and Evolutionary Approaches to Brain Function: A Critical Appraisal. Front. Comput. Neurosci. **6**, 24 (2012)
5. Fernando, C., Szathmáry, E.: Chemical, neuronal and linguistic replicators. Towards an extended evolutionary synthesis. In: Pigliucci, M., Müller, G. (eds.) Evolution: The Extended Synthesis, pp. 209–249. MIT Press, Cambridge (2010)
6. Shim, Y., Husbands, P.: Incremental Embodied Chaotic Exploration of Self-Organised Motor Behaviours with Proprioceptor Adaptation. Front. Robot. AI **2**, 7 (2015)

Remodeling Muscle Cells by Inducing Mechanical Stimulus

Kazuaki Mori[1]([✉]), Masahiro Shimizu[1], Kota Miyasaka[2],
Toshihiko Ogura[2], and Koh Hosoda[1]

[1] Graduate School of Engineering Science, Osaka University,
1-3, Machikaneyama, Toyonaka 560-8531, Japan
{mori.kazuaki,shimizu,hosoda}@arl.sys.es.osaka-u.ac.jp
[2] Institute of Development, Aging, and Cancer, Tohoku University, 4-1, Seiryo,
Aoba, Sendai, Miyagi 980-8575, Japan
{k.miyasako,ogura}@idac.tohoku.ac.jp

Abstract. A muscle cell actuator has been attracting a lot of attention since it is a key technology for realizing bio-machine hybrid systems. This study especially intend to deal with micro robot driven by real muscle cell actuators. To fabricate the actuator, we should study how to control cell aggregation for efficient power generation. This paper proposes a method for remodeling muscle cells by exploiting mechanical stimulus so that we can get an appropriate structure of the actuator. The experimental results demonstrate that the three factors, cell density, cell-matrix adhesion, and mechanical stimulation period, largely contribute to the remodeling of the muscle cells.

Keywords: Muscle cells actuator · Remodeling · Mechanical stimulus · Bio-machine hybrid system

1 Introduction

Living organisms can cope with noise, uncertainties and complexity of the real world. For realizing such abilities, elastic mechanical property of muscles plays an important role. Some researchers have developed bio-machine hybrid systems consisting of muscle cell actuators and mechanical parts so that they can utilize elasticity of the muscles[1–3]. Nawroth[1] et al. developed a jellyfish-like bio-machine hybrid system Medusoid driven by rat cardiomyocytes. Each cell is spatially-arranged manually based on the real jellyfish. In the real biological system, distribution and orientation of cells are adaptively determined based on its body and environment. We believe that also muscle cell actuators should be arrangement adaptively for appropriate actuation of a micro robot. We have to consider how the muscle cells is arranged reflecting environmental changes, particularly mechanical stimulus changes. Previous reports have shown how we can control cell orientation and differentiation by mechanical stimulus[4–6], but the mechanism of remodeling and mechanical stimulus response of muscle cells is still not clear. These reports have paid attention to orientation of single cell. We have

© Springer International Publishing Switzerland 2015
S.P. Wilson et al. (Eds.): Living Machines 2015, LNAI 9222, pp. 227–230, 2015.
DOI: 10.1007/978-3-319-22979-9_23

to observe remodeling cells carefully. This report experimentally investigates the effect of mechanical stimulus to cell aggregation.

2   Method

A stretching machine is developed to give mechanical stimulus to muscle cells systematically. It has a stretchable polydimethylsiloxane(PDMS) silicone chamber(STB-CH-04, STREX Inc. Japan), and muscle cells are cultured on it (see Fig.1). We employed C2C12, which is a cell line derived from mouse skeletal muscles. We investigated several conditions of cell density, cell-matrix adhesion, and mechanical stimulation period as important factors that affect mechanical reaction of the cells. Cell density is expected to affect cell differentiation, cell-cell adhesion, and proliferation; cell-matrix adhesion is expected to affect degrees of cell movement; mechanical stimulation period is expected to affect cell orientation and differentiation.

To change cell-matrix adhesion, we regulated quantity of fibronectin(F1141, Sigma-Aldrich Co., USA) used for adhesive. To change mechanical stimulus period, we set the parameters of the stretch machine; The machine was running in 24, 48 and 72hours; Here, we stretched muscle cells at a frequency of 10cycle/min and a length of 1.2 times of cell length. Mechanical stimulation experiments had been conducted as follows: Muscle cells had cultured on a chamber in Dulbecco's modification of Eagle's medium(DMEM, D5796, Sigma-Aldrich Co., USA) containing 10% fetal bovine serum(FBS, SFBM30, Equitech-Bio Inc., USA) and 1% PS(06168-34, Nacalai Tesque Inc., Japan) for 48 hours. After we replaced medium with DMEM containing 2% FBS, periodic stretch were given to muscle cells.

Fig. 1. A stretch machine and PDMS silicone chambers. Before we cultured muscle cells on the chamber, we need fibronectin coating on PDMS silicone. PDMS silicone has hydrophobicity, therefore, cells can't adhere to it. PDMS silicone is elastic, transparent and harmless to living cells.

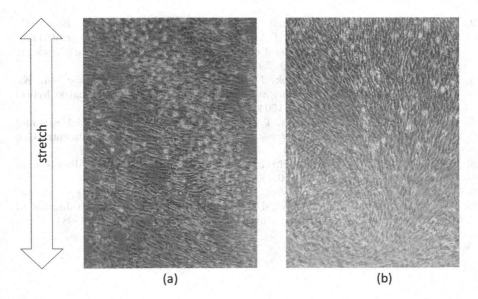

(a)   (b)

Fig. 2. (a)Result1(cell density: low, cell-matrix adhesion: weak, mechanical stimulation period: 72hours) Muscle cells spread in a direction of perpendicular to stretching. Cell distribution was heterogeneous. (b)Result2(cell density: high, cell-matrix adhesion: strong, mechanical stimulation period: 72hours) Muscle cells spread at random and cell distribution is homogeneous.

3   Results

In Fig.2, we show two typical patterns of cell aggregation: in result (1), the cells are growing in the perpendicular direction to the stretch, whereas they are growing randomly in result (2). Experimental conditions are illustrated in the caption of Fig.2. Cell density effects on a orientation. At low cell density, muscle cells were aligned in a direction perpendicular to stretching. At high cell density, muscle cells were not aligned. When cell-matrix adhesion was weak, cell distribution was heterogeneous. Mechanical stimulus period related to a degree of differentiation. We expect these results contribute to basic technique for robotic muscle cells actuators that self-improve.

Acknowledgments. This work was supported partially by Grant-in-Aid for Scientific Research on 24680023, 23220004 and 25540117 from the Ministry of Education, Culture, Sports, Science and Technology of Japan.

References

1. Nawroth, J.C., Lee, H., Feinberg, A.W., Ripplinger, C.M., McCain, M.L., Grosberg, A., Dabiri, J.O., Parker, K.K.: A tissue-engineered jellyfish with biomimetic propulsion. Nature Biotechnology **30**(8), 792–797 (2012)

2. Cvetkovic, C., Raman, R., Chan, V., Williams, B.J., Tolish, M., Bajaj, P., Sakar, M.S., Asada, H.H., Saif, M.T.A., Bashir, R.: Three-dimensionally printed biological machines powered by skeletal muscle. Proceedings of the National Academy of Sciences **111**(28), 10125–10130 (2014)
3. Kim, J., Park, J., Yang, S., Baek, J., Kim, B., Lee, S.H., Yoon, E.S., Chun, K., Park, S.: Establishment of a fabrication method for a long-term actuated hybrid cell robot. Lab on a Chip **7**(11), 1504–1508 (2007)
4. Hayakawa, K., Sato, N., Obinata, T.: Dynamic reorientation of cultured cells and stress fibers under mechanical stress from periodic stretching. Experimental Cell Research **268**(1), 104–114 (2001)
5. Engler, A.J., Sen, S., Sweeney, H.L., Discher, D.E.: Matrix elasticity directs stem cell lineage specification. Cell **126**(4), 677–689 (2006)
6. Naruse, K., Yamada, T., Sokabe, M.: Involvement of sa channels in orienting response of cultured endothelial cells to cyclic stretch. American Journal of Physiology-Heart and Circulatory Physiology **274**(5), H1532–H1538 (1998)

Integration of Biological Neural Models for the Control of Eye Movements in a Robotic Head

Marcello Mulas[(✉)], Manxiu Zhan, and Jörg Conradt[(✉)]

Neuroscientific System Theory Group Department of Electric
and Computer Engineering, Technische Universität München,
Karlsrasse 45, 80333 Munich, Germany
{marcello.mulas,conradt}@tum.de
https://www.nst.ei.tum.de

Abstract. We developed a biologically plausible control algorithm to move the eyes of a six degrees of freedom robotic head in a human-like manner. Our neurocontroller, written with the neural simulator *Nengo*, integrates different biological neural models of eye movements, such as microsaccades, saccades, vestibular-ocular reflex, smooth pursuit and vergence. The coordination of the movements depends on the stream of sensory information acquired by two silicon retinas used as eyes and by an inertial measurement unit, which serves as a vestibular system. The eye movements generated by our neurocontroller resemble those of humans when exposed to the same visual input. This robotic platform can be used to investigate the most efficient exploration strategies used to extract salient features from either a static or dynamic visual scene. Future research should focus on technical enhancements and model refinements of the system.

Keywords: Neural control · Eye movements · Robotic head

1 Introduction

Traditional artificial vision algorithms analyze static visual scenes in order to extract the most salient features of an image. Unfortunately, state-of-the-art artificial vision is still far from reaching the performance of biological vision systems. A possible reason for such a gap is that biological systems might explicitly take advantage of information about the motion of the observer [1]. However, despite decades of research, it is not clear exactly how the human brain control eye movements in order to explore a visual scene.

Neuroscientists have developed several biologically plausible models of eye movement control. The models implemented for robotic control can reproduce human behavior but are usually limited to few specific eye movements (e.g., gaze fixation [2] and saccades [3]). A robotic control that integrates multiple

© Springer International Publishing Switzerland 2015
S.P. Wilson et al. (Eds.): Living Machines 2015, LNAI 9222, pp. 231–242, 2015.
DOI: 10.1007/978-3-319-22979-9_24

movement-specific models in a cohesive way is still lacking. This directly reflects the lack of neuroscientific knowledge about how top-down cognitive influences and bottom-up streams of sensory information interact in order to decide which movement should be executed at a given time.

It is plausible that the active exploration strategy adopted by the human brain might be tuned by evolution to speed up the process of information acquisition about the surrounding environment. An intelligent strategy that can quickly localize relevant features would also contribute to saving computational power. Aside from artificial vision, anthropomorphic robotics would also greatly benefit from a control algorithm that reproduces human-like behavior. Movement is one of the key factors taken into consideration by humans to judge the intelligence of an artificial system [4].

Several stereoscopic vision systems controlled by bio-inspired algorithms have been proposed for different purposes. However, to the best of our knowledge, a robotic head controlled by biologically realistic models that can reproduce human-like behavior is still lacking. For instance, Berthouze and Kuniyoshi [5] studied the emergence of sensorimotor coordination in a four DoF (degrees of freedom) robot head as it interacts with its surrounding environment. In order to provide both a wide field of view in the periphery and high resolution in the fovea, they used special camera lenses [6]. The learning of visual abilities such as gaze fixation and saccadic motion was based on Kohonen self-organizing maps (SOM) [7]. Bjorkman and Kragic proposed (2004) a robot head for object recognition, tracking and pose estimation [8]. They implemented foveal vision by using two sets of binocular cameras mounted on a four DoF robotic head. However, the authors did not aim at implementing neural plausible strategies, so the control program integrated different low level computer vision algorithms. Similarly to Berthouze et al., Asuni et al. used (2005) self-organizing maps to solve the inverse kinematics problem [2]. Once having learned sensory-motor relations, their seven DoF robotic head was able to direct its gaze towards a given target arbitrarily located in a 3D space. According to the authors, the robotic head was able to behave more similarly to humans thanks to additional constraints on the joint angles.

In this paper, we propose a biologically plausible control algorithm, implemented in *Nengo* [9], to move the eyes of a six DoF robotic head in a human-like manner. Our control integrates different biological neural models of specific eye movements, such as saccades, vestibular-ocular reflex, smooth pursuit and vergence. Coordination of these movements is based on the current internal and external context, dependent on the stream of sensory information continuously acquired by on-board sensors. The neural simulator *Nengo* was instrumental in easily defining information processing algorithms based on neural computation. The use of biological plausible neural circuitry can provide a more useful platform to investigate the basic mechanisms underlying human vision albeit at the expense of computational efficiency and movement control precision.

2 Robotic Setup

The robotic head that we designed to develop and validate our neurocontroller incorporates six DoF (Figure 1 to the left). Two DoF are at the neck joint for the pan and tilt of the whole head; the others are two for each eye, one for horizontal and one for vertical rotation. All the components of the frame are made by a 3D printer using a rigid opaque white material (*Objet VeroWhitePlus* photopolymer) except for two T-shaped elements made out of metal to enhance robustness. Due to technical considerations we designed our robot head differently in comparison to the anatomy of a human head. A major difference is the interpupillary distance (IPD) that is equal to 230 mm in our robot head against the IPD of an average human head that is around 63 mm [10].

We chose Dynamixel servomotors as a good compromise between the high speed performance required to match biological performance and the need to keep the cost of the robotic system low. We used two Dynamixel MX-64 servomotors to actuate the joint of the neck and four Dynamixel MX-28 servomotors to actuate the eyes [11]. The servomotors are connected in chain formation, and they are controlled by a computer via serial communication. A power supply provides energy to all servomotors at a constant voltage of 12 V. The electric current necessary to actuate all the servomotors is approximately 300 mA in standby conditions.

We built the eyes of the robot head using two dynamic vision sensors (DVS) [12]. These cameras, also called silicon retinas, have a resolution of 128x128 pixels. Contrarily to standard cameras which record the intensity of light, silicon retinas are sensitive to light temporal contrast, i.e. each pixel signals a change in the intensity of light. Every time the incident light increases or decreases its intensity by certain amount an ON or OFF event is asynchronously generated. This working mechanism greatly reduces the redundancy of visual information providing a data rate that typically is orders of magnitude lower in comparison with the data rate of conventional frame-based cameras [12]. The large dynamic range (> 120 dB) and the asynchronous event-based output (similar to the spiking activity of retinal ganglion cells) make silicon retinas suitable to model human eyes. In addition, despite very different transduction mechanisms, the sensitivity to temporal contrast makes silicon retinas behave similarly to human retinas in presence of static scenes. In fact, during constant light conditions silicon retinas do not signal any changes in light intensity and likewise retinal cells become silent due to neural adaptation [13]. We exploited this property of DVS sensors for the control of eye movements as it seems that the functional role of microsaccades is to prevent the image from fading away [14].

Next, an inertial measurement unit (IMU) sensor is mounted in the center of the robot head to serve as an artificial vestibular system. The IMU provides a continuous stream of sensory information about rotational velocities and accelerations in all directions of the 3D space. However, in order to implement the vestibular-ocular reflex, we took into account only horizontal (i.e., yaw) and vertical (i.e., pitch) rotations of the robot head.

In order to validate our neurocontroller, we recorded the movements of the robotic head using a motion capture system [15] recording at a sampling rate of 120 Hz. To estimate the positions of reflective markers, the tracking software analyzes the recordings of eight cameras mounted close to the ceiling of the recording room and pointing downwards. The positions and orientations of both the robotic head and the two eyes can be accurately estimated from the positions of the reflective markers (1 cm in diameter) appropriately attached to the robotic frame (Figure 1, right panel). Then, based on simple geometric calculations, we computed for each eye the saccadic traces projected on a monitor screen placed in front of the robotic head. We then compared these traces with those generated by humans while watching the same video.

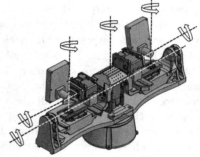

Fig. 1. Robotic head. Left: the robotic head is equipped with two silicon retinas, an inertial measurement unit and six servomotors. Right: the axes of rotation for each of the six DoF (two for each of the two eyes and two for the pan and tilt of the whole head).

3   Neurocontroller

We integrated different biological models of eye movement control using the neural simulator *Nengo* [9]. We chose to model neurons as Leaky Integrate and Fire (LIF) as this model represents a good compromise between biological plausibility and simulation performance. In *Nengo* LIF neurons are described by their tuning curves. Given a certain tuning curve, the spiking rate of a LIF neuron depends only on the input current. In order to cope with the massively parallel computation required by our neural simulation, we took advantage of *Nengo* support for GPU computing. Our neurocontroller runs on a Nvidia GeForce GTX Titan graphic card installed on a computer with an Intel Core i7-4770K processor and 16 Gb of RAM. The GPU we used has 2688 processing cores working at a base clock of 837 MHz.

Despite the available computational power, we had to reduce the spatial resolution of the silicon retinas in order to achieve acceptable real-time performance for the control of the movements. First, accumulated the events coming from the two retinas with a decay time constant of 20 ms without distinguishing between their polarities (ON/OFF). Second, we summed up the accumulated events in contiguous regions of 14x14 pixels and we applied a binary threshold. The final input image to *Nengo* is a 9x9 binary matrix of visual activation.

In this work we integrated five models of eye movements in a single neural simulation. We considered three gaze-shifting movements (i.e., saccade, smooth pursuit and vergence), one gaze-stabilizing movement (i.e., the vestibular-ocular reflex) and one fixational movement (i.e., microsaccade). We implemented microsaccadic and saccadic motion with different models because of their different functional roles. However, it is still not clear if these models correspond to two distinct biological neural control modules [16].

3.1   Eye Movements Integration

A key feature of our neurocontroller is the coordination of several eye movements depending on external and internal contexts. The occurrence of a specific context is determined based on sensory streams from the silicon retinas and from the IMU that serves as a vestibular system. Figure 2A illustrates how information from these sensors flows through different neural modules to generate motor commands for the servomotors. Three main information processes run in parallel. One of them receives visual information from the two silicon retinas in order to control the vergence of the eyes. A second process receives information only from the IMU (*gyroscope*) to compensate for head movements. A third process is dedicated to the selection of one of three mutually exclusive eye movements (i.e., saccades, microsaccades and smooth pursuit).

The action selection working mechanism is further detailed in Figure 2B where excitatory connections are shown in red and inhibitory ones in black. Three neural modules (*saccades, microsaccades* and *smooth pursuit*) receive information about the level of retinal activity, the detection of moving objects in the scene and the motion of the eyes in order to determine how likely the activation of the corresponding eye movement is given the current context. More precisely, a saccadic movement is likely to occur if there is high activity from the retinas but no detection of motion and the eyes are not moving. Microsaccades are activated when the activity from the retinas is below a certain threshold and smooth pursuit can occur if a moving object is detected and the eyes are moving. From a biological perspective, the boundaries of a context are not precisely defined. As a consequence, it is necessary to use a neural mechanism that can decide at any given time which context is the most likely to occur. In the human brain a plausible candidate to perform action selection is the basal ganglia (BG) [17,18]. In our neurocontroller we used the *Nengo* implementation of BG developed by Eliasmith and collaborators [9], which is based on the work of Gurney et al. [19] in order to decide the current context. The BG model, which consists of five neural sub-modules receives excitatory inputs from the context modules. BG computes a *winner-takes-all* function determining the strongest input and selects one of three possible eye movements.

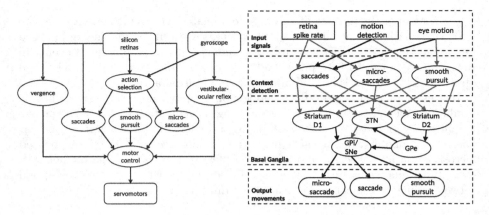

Fig. 2. Neural processing diagrams. Left: the arrows show the information flow between hardware components (rectangular boxes) and functional neural modules (rounded boxes). Right: the BG model receives information about the likelihood of different contexts and select the action associated with the strongest activation.

3.2   Saccades

Saccades are movements of the eye that direct the fovea to a new point of interest in the visual field. Saccades are ballistic movements, that is the final target position cannot change during the execution of the motion. In case the target is missed, smaller corrective saccadic movements are possible, but only after the current saccade is concluded. The velocity of a saccade increases with the amplitude of the movement and can reach up to 900 degrees per second. In our controller we determine the next fixation target on the basis of a bottom-up saliency map model [20], computed at the lowest level on retinal windows of 3x3 pixels. The level of saliency of such a window depends on the detection of basic geometric primitives like points, edges or corners. The neural mechanism that implements this function is based on a standard model of V1 simple cells [21,22]. A generic simple cell receives a strong excitatory connection from the photoreceptor in the center of its receptive field and weaker inhibitory connections from the photoreceptors in the surrounding periphery. The outputs of contiguous simple cells are then combined to estimate the saliency of larger areas of the visual field. In addition, we introduced a global saliency map with a memory effect. This prevents the saliency map from generating fixations lasting too long and also the return of the gaze to previous locations (i.e., inhibition of return). Memory is implemented with recurrent inhibitory connections with a time constant of 50 ms. The saliency map of the current visual field is inhibited by the corresponding area of the global saliency map.

3.3 Smooth Pursuit

Smooth pursuit allows an observer to track a moving object by constantly keeping its image centered in the fovea. This movement is voluntary but the presence of a moving stimulus in the visual field is necessary for untrained people in order to accomplish this motion. Our implementation of smooth pursuit is based on the detection of motion due to direction selective (DS) ganglion cells. These cells are modeled as Elaborated Reichardt Detectors [23]. The presence of the delay in only one of the two input connections allows for the coincident arrival of two excitatory stimuli to the DS ganglion cell only if the upstream photoreceptors are activated in a specific temporal order. In order to solve the motion ambiguity inherent in a local motion detector (known as the aperture problem), the outputs of multiple DS ganglion cells sensitive to different directions of motion at different speeds are combined. The motion detected in the fovea of the silicon retinas is then used to estimate the speed of the eyes required to keep tracking the moving object.

3.4 Vergence

Vergence is a coordinated movement of both eyes in opposite directions in order to converge on a common fixation point. A biologically-inspired technical solution to allow vergence is to compute matching landmarks in stereo images [24]; a biologically plausible way to determine the right velocities of the eyes is to compute a disparity map [25]. Following the model described in Patel et al. [26] we implemented a population of neurons for each specific degree of disparity. Each population receives excitatory connections from the right retinal photoreceptors and inhibitory connections from the left retinal photoreceptors. A *winner-takes-all* mechanism selects the population corresponding to the most likely disparity. Depending on the estimated disparity d, the eyes converge (if $d > 0$) or diverge (if $d < 0$).

3.5 Vestibular-Ocular Reflex (VOR)

VOR is a very important mechanism to stabilize vision based only on the sensory information provided by the vestibular system. During either the translation or rotation of the head, this reflex induces the eyes to move in the opposite direction in order to compensate for the motion of the head. In our neural simulation four vestibular nuclei act as integrators of acceleration signals coming from the vestibular system. Two of the nuclei control the angular velocity of the eyes on the horizontal plane and the other two in the vertical plane.

3.6 Microsaccades

Microsaccades are small involuntary fixational eye movements. Their functional role is still a matter of debate. However, a common hypothesis assumes that they prevent the retinal image from fading in presence of a perfectly static visual scene [27]. Because silicon retinas behave similarly to human retinas in this regard, our neurocontroller triggers microsaccadic movements every time the stream of retinal events decreases below a certain threshold as occurs in the absence of movement. Moreover, it is unclear how microsaccades are generated [28]. A possibility is that they are generated from the noise in the Superior Colliculus (SC) [29]. In our implementation we took advantage of spiking rate fluctuations in the LIF neurons with constant input in order to generate noise sources. We then connected two of these sources with the neural populations that control the horizontal and vertical eye movements.

4   Validation

In order to validate the similarity between our neurocontroller-generated behavior and human behavior, we compared the movements of the robotic eyes to those generated by humans when exposed to the same visual input. We placed a monitor in front of the robotic head and tracked the saccadic movements while showing a video (2′29″ of duration) selected from the *DIEM* database [30]. The video is part of a documentary about monkeys living in the tropical rainforest jungle and was selected amongst others because of the presence of distinct moving elements in the scene such as monkeys jumping from one tree to another.

For a very preliminary assessment of the accuracy of our neurocontroller in reproducing human behavior we directly compared the eye traces on the screen recorded from the eyes of the robot and from those of a randomly chosen human subject as shown in Figure 3. On the one hand, similar to the human counterpart, each of the eyes of the robotic head seems to follow a random trajectory that explores a large area of the screen close to the center. In addition, microsaccadic and saccadic movements alternate in a similar way to human eye motion. On the other hand, the trajectories of the robotic eyes do not converge on close fixation points as observed in human behavior. This is likely due to the blur of the image

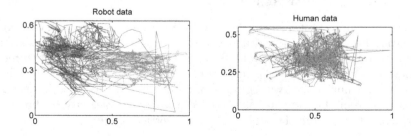

Fig. 3. Saccadic traces on the monitor screen for the left (red) and right (green) eye. The monitor screen size is normalized in reference to its width. The trajectory patterns for each robotic eye resemble the corresponding human one, but the traces do not overlap due to an insufficient convergence of the eyes.

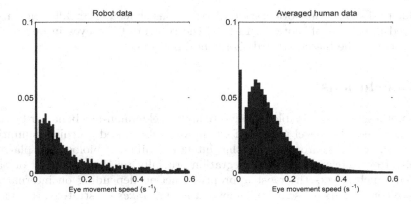

Fig. 4. Velocity distribution of eye motion. The distribution of velocities of robotic (left) and human (right) saccades and microsaccades is similar except for a scale factor.

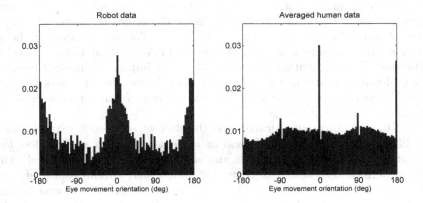

Fig. 5. Orientation distribution of eye motion. The eye movements of the robot head are biased towards horizontal directions in contrast to the behavior of human eyes.

captured by the silicon retinas that is responsible for the lack of a clear common point of fixation.

Figure 4 shows a comparison between the distribution of velocities of robotic and human microsaccades and saccades. To more accurately estimate the velocity distribution, we analyzed data from the DIEM database recorded from 50 human subjects watching the same selected video. As it can be seen, the profiles of velocities are similar except for a scale factor. However, if we consider the whole range of velocities, that is including saccadic motion as well, the average and standard deviation of the distribution match the human data particularly well (0.32 ± 0.80 for the robot head, 0.36 ± 0.99 for humans).

Figure 5 shows a comparison between the distribution of orientations of robotic and human eye movements (averaged across 50 subjects). Contrarily to velocity distributions, orientation distributions show a clear difference in the preferred directions of motions. In fact, the control of the eyes in our robotic setup seems to be biased towards horizontal movements.

5   Conclusions

We hypothesized it is possible to generate high level human-like behavior by coordinating several low level models of eye movements. Based on this assumption, we implemented a neurocontroller that integrates different biologically plausible models of eye movements. The integration crucially relies on a model of basal ganglia, which selects the most appropriate action depending on internal and external contexts. Because it implements a *winner-takes-all* strategy, this model of BG can resolve situations where different competing contexts occur at the same time.

Unfortunately, the performance of our neurocontroller is still far from accurately matching human behavior. One of the major limitations of our current implementation is the severe down-sampling of the silicon retina resolution necessary to make the system operate in real-time. The exploitation of the full resolution would require a number of neurons in *Nengo* that is far beyond the capabilities of our computing hardware. The most important limiting factor for our neural simulations was memory rather than the number of processing cores. A possible solution to this technical limitation is to further preprocess visual information to input in Nengo.

Overall, our neurocontroller still lacks the ability to efficiently explore a visual scene like a human. Besides technical enhancements, which would allow for a richer stream of visual information and more powerful computational neural modules, performance would likely benefit from a more careful tuning of neural model parameters. In order to achieve this, more comprehensive and accurate behavioral tests need to be done. Once our robotic system will reproduce in a more accurate way human-like behavior it could be used to study the importance of active motion for object recognition tasks. In particular, the possibility to disable specific neural modules in our neurocontroller might be helpful for validating current models of visual perception based on lesion studies.

References

1. Rothkopf, C.A., Ballard, D.H., Hayhoe, M.M.: Task and context determine where you look. J Vis. Dec. 19 **7**(14), 16.1–20 (2007). doi:10.1167/7.14.16
2. Asuni, G., Teti, G., Laschi, C., Guglielmelli, E., Dario, P.: A robotic head neurocontroller based on biologically-inspired neural models. In: Proceedings of the 2005 IEEE International Conference on Robotics and Automation, ICRA 2005, pp. 2362–2367, 18–22 (2005)
3. He, H., Ge, S.S., Zhang, Z.: A saliency-driven robotic head with bio-inspired saccadic behaviors for social robotics. Autonomous Robots **36**(3), 225–240 (2013). doi:10.1007/s10514-013-9346-z

4. Shibata T.: An overview of human interactive robots for psychological enrichment. In: Proceedings of the IEEE, vol. 91, no. 11 (2004)
5. Berthouze, L., Kuniyoshi, Y.: Emergence and Categorization of Coordinated Visual Behavior Through Embodied Interaction. Machine Learning **31**, 187–200 (1998)
6. Kuniyoshi, Y., Kita, N., Sugimoto, K., Nakamura, S., Suehiro, T.: A foveated wide angle lens for active vision. In: Proceedings 1995 IEEE International Conference on Robotics and Automation, vol. 3, pp. 2982–2988, 21–27 (1995)
7. Kohonen, T.: Self-Organized Formation of Topologically Correct Feature Maps. Biological Cybernetics **49**(1), 59–69 (1982)
8. Bjorkman, M., Kragic, D.: Combination of foveal and peripheral vision for object recognition and pose estimation. In: Proceedings 2004 IEEE International Conference on Robotics and Automation, ICRA 2004, vol. 5, pp. 5135–5140 (2004)
9. Eliasmith, C., Stewart, T.C., Choo, X., Bekolay, T., DeWolf, T., Tang, Y., Rasmussen, D.: A Large-Scale Model of the Functioning Brain. Science **338**(6111), 1202–1205 (2012)
10. Dodgson N.A.: Variation and extrema of human interpupillary distance. Electronic Imaging 36–46 (2004)
11. Robotis Inc.: http://www.robotis.com
12. Lichtsteiner, P., Posch, C., Delbruck, T.: A 128×128 120 dB 15 μs Latency Asynchronous Temporal Contrast Vision Sensor. IEEE Journal of Solid-State Circuits **43**(2), 566–576 (2008)
13. Riggs, L.A., Ratliff, F.: The effects of counteracting the normal movements of the eye. Journal of the Optical Society of America **42**, 872–873 (1952)
14. Martinez-Conde, S., Macknik, S.L., Troncoso, X.G., Dyar, T.A.: Microsaccades counteract visual fading during fixation. Neuron. **49**(2), 297–305 (2006)
15. OptiTrak Tracking System: https://www.optitrack.com
16. Martinez-Conde, S., Otero-Millan, J., Macknik, S.L.: The impact of microsaccades on vision: towards a unified theory of saccadic function. Nature Reviews Neuroscience **14**, 83–96 (2013)
17. Mink, J.: The basal ganglia: focused selection and inhibition of competing motor programs. Progress in Neurobiology **50**(4), 381–425 (1996)
18. Redgrave, P., Prescott, T., Gurney, K.N.: The basal ganglia: A vertebrate solution to the selection problem? Neuroscience **89**, 1009–1023 (1999)
19. Gurney, K., Prescott, T.J., Redgrave, P.: A computational model of action selection in the basal ganglia. II. Analysis and simulation of behaviour. Biological Cybernetics **84**(6), 411–423 (2001)
20. Itti, L., Koch, C.: Computational modelling of visual attention. Nature Reviews Neuroscience **2**, 194–203 (2001). doi:10.1038/35058500
21. Hubel D.H., Wiesel T.N.: Receptive fields, binocular interaction and functional architecture in the cat's visual cortex. J. Physiol. (1962)
22. Li, Z.: A saliency map in primary visual cortex. Trends in cognitive sciences **6**(1), 9–16 (2002)
23. Van Santen, J.P.H., Sperling, G.: Elaborated Reichardt detectors. JOSA A **2**(2), 300 (1985)
24. Conradt, J., Simon, P., Pescatore, M., Verschure, P.F.M.J.: Saliency maps operating on stereo images detect landmarks and their distance. In: Dorronsoro, J.R. (ed.) ICANN 2002. LNCS, vol. 2415, pp. 795–800. Springer, Heidelberg (2002)
25. Franz A., Triesch J.: Emergence of disparity tuning during the development of vergence eye movements. In: IEEE 6th International Conference on Development and Learning, pp. 31–36, pp. 11–13 (2007) doi:10.1109/DEVLRN.2007.4354029

26. Patel, S.S., Ogmen, H., White, J.M., Jiang, B.C.: Neural network model of short-term horizontal disparity vergence dynamics. Vision Research **37**(10), 1383–1399 (1997)
27. Pritchard, R.M.: Stabilized images on the retina. Sci. Am. **204**, 72–78 (1961)
28. Rolfs, M., Kliegl, R., Engbert, R.: Toward a model of microsaccade generation: the case of microsaccadic inhibition. J. Vis. **8**(11), 5.1–23 (2008)
29. Krekelberg, B.: Microsaccades. Current Biology **21**(11), 416 (2011)
30. Mital, P.K., Smith, T.J., Hill, R., Henderson, J.M.: Clustering of Gaze during Dynamic Scene Viewing is Predicted by Motion. Cognitive Computation **3**(1), 5–24 (2011)

Saying It with Light: A Pilot Study of Affective Communication Using the MIRO Robot

Emily C. Collins[✉], Tony J. Prescott, and Ben Mitchinson

The University Of Sheffield,Sheffield, UK
e.c.collins@sheffield.ac.uk

Abstract. Recently, the concept of a 'companion robot' as a healthcare tool has been popularised, and even commercialised. We present MIRO, a robot that is biomimetic in aesthetics, morphology, behaviour, and control architecture. In this paper, we review how these design choices affect its suitability for a companionship role. In particular, we consider how emulation of the familiar body language and other emotional expressions of mammals may facilitate effective communication with naïve users through the reliable evocation of intended perceptions of emotional state and intent. We go on to present a brief pilot study addressing the question of whether shared cultural signals can be relied upon, similarly, as components of communication systems for companion robots. Such studies form part of our ongoing effort to understand and quantify human responses to robot expressive behaviour and, thereby, develop a methodology for optimising the design of social robots by accounting for individual and cultural differences.

1 Introduction

The 2013 film 'Robot and Frank' (Figure 1) was an exploration of the role that a robot might play in the life and care of a patient with in-home care needs, in this case owing to old age and the onset of dementia. The eponymous robot is remarkable because the most impactful role it plays is as a companion to Frank, inbetween performing physical assistance tasks such as transporting food. The importance of this role to healthcare scenarios in real life is exemplified by the

Fig. 1. Left panel: 'Robot and Frank' (from the 2013 film). Right panel: PARO the robot seal (left) and MIRO the robot mammal (right) are examples of 'social' robots.

© Springer International Publishing Switzerland 2015
S.P. Wilson et al. (Eds.): Living Machines 2015, LNAI 9222, pp. 243–255, 2015.
DOI: 10.1007/978-3-319-22979-9_25

successes already achieved with 'simple' companion robots such as PARO [1] and Kaspar [2]. These robots are defined as 'social' robots, robots designed to interact and communicate with humans—usually, in a naturalistic way by using biological communication channels (e.g. body language or vocalisation rather than a keypad). What marks them out as companion robots is that they not only communicate, but play a role in their user's emotional life through these interactions (other examples of commercially available companion robots include Sony's AIBO [3], Omron's NeCoRo [4], and MobileRobots Inc.'s PeopleBot).

Robots like Frank's may be some way off yet, but on what principles is a contemporary companion robot built? One starting point is the large body of research that exists on the benefits of animal therapy for lowering stress [5], reducing heart rates [6], elevating mood, and social facilitation [7]. Robot therapy borrows from this branch of healthcare by creating robots with the capacity to act as pet surrogates for those who do not have access to animals [8]. PARO is one of the most active commercial examples, and is marketed as a therapeutic tool for use in nursing home settings. It is sold on the premise that it will interact with human beings to make them feel emotional attachment to the robot [9]. It does this by engaging its user with basic capabilities: sensing touch, recognising a limited amount of speech, expressing small utterances and moving its head, flippers and tail. The relationship that develops between a user and the robot is built upon the limited reactions the robot makes to the user's spoken and physical actions [10]. PARO is designed for, amongst other things, use in therapy sessions attended by individuals suffering from dementia and other conditions of cognitive decline. In such individuals emotional capability does not decline in a one-to-one fashion with cognition [11] allowing for meaningful application of psychological and emotional therapy. PARO does not locomote, and is designed to be held and fussed over.

The relevance of biomimetics to human-robot interactions, more generally, is widely attested. Robots that are biomimetic in their morphology, in the way they move, and that have expressive faces are immediately and intuitively engaging, owing to our familiarity with mammalian channels for conveying emotion and intent [12]. Naïve 'users', for example, choose to interact to a greater degree with robots that include naturalistic body language in their interactions [13], and robots can emit powerful social signals simply by following rules long-established by animals [14]. Neither, it appears, does knowledge that a robot is not biological eliminate the impact of these design strategies [15]; anecdotally, our own experience with biomimetic platforms has indicated that even an explicit statement from a robot's 'handler' that there is 'nobody home' leaves engagement more-or-less intact [16]. Meanwhile, these aspects of robotic design are beginning to creep into industrial robots, also [17]. Thus, it seems that a biomimetic component to engagement will remain a design principle for coming generations of companion robots. By good fortune, this is synergistic with the increasing role that biomimetics is playing in functional design [18], so that biomimetics promises to drive forward both functional and relational aspects of performance.

In the remainder of this paper, we introduce a new robot platform, 'MIRO', which follows biomimetic design principles aesthetically, morphologically, and behaviourally, as well as with respect to control architecture. MIRO is intended to act as an accessible biomimetic research platform, providing an opportunity to explore all aspects of that functional/relational synergy. One research role MIRO plays for us, then, is as an engaging robot companion. Below, we describe how we are beginning to use empirical data from human interaction studies to contribute to the iterative design process of channels for emotional expression. Specifically, we present a pilot for a study assessing the performance of pulsating patterns of coloured lights as intuitive signals of affective state. Biological analogues for such a signalling modality are less easy to identify, but it is commonplace to use lights for signalling conditions across human cultures. This study will address the twin hypotheses: (i) Signals with culturally-agreed meaning can communicate affect between a robot and a naïve human and (ii) The effectiveness of that communication will depend on the tailoring of signals to the interactee, individually and/or culturally. Results from our pilot tend to support hypothesis (i); results from the larger study will be required to begin to address hypothesis (ii).

2   MIRO

The MIRO robot was commissioned as a commercial pedagogical and leisure product, targeted particularly at the domestic and school markets. Through the encouragement of exploration of its construction and operation (the flagship configuration has 'build-it-yourself' form and is accompanied by an extensive series of magazines). MIRO is also intended as a artefact to drive public engagement with science, robotics in particular, and biomimetic robotics most of all (this agenda being reflected also in the magazine).

2.1   Aesthetics and Morphology

MIRO's aesthetics and morphology (Figure 2) were chosen to be engaging through evocation of a mammalian identity. Design choices explicitly avoided

Fig. 2. Concept art for MIRO expression of emotion through biomimetic body language (imagery from Sebastian Conran Associates, Kensington, London, UK).

targeting a particular mammal so that the end result is intended to be somewhat of a 'generic mammal', though some specificity is naturally unavoidable. The platform is equipped with some of the same expressive appendages available to many mammals (ears, tail, eyelids) allowing mammal-like direct signalling of emotional state and responses to stimuli.

2.2 Platform

The MIRO platform is built around a core of a differential drive (plus caster) base and a three-DOF (lift, pitch, yaw) neck. Additional DOFs include two for each ear (curl, rotate), two for the tail (droop, wag), one for the caster (raise/lower), and one for the eyelids (open/close). Whilst these latter DOFs target only communication, the movements of the neck and body that serve locomotion and active sensing play a significant role in communication as well. Finally, the platform is equipped for sound production.

All DOFs are equipped with proprioceptive sensors (potentiometers for absolute positions and optical shaft encoders for wheel speed). Four light level sensors are placed at each corner of the base, two task-specific 'cliff sensors' point down from its front face, and four capacitive sensors are arrayed along the inside of the body shell providing sensing of direct human contact. In the head, stereo microphones (in the base of the ears) and stereo cameras (in the eyes) are complemented by a sonar ranger in the nose and an additional four capacitive sensors over the top and back of the head (behind the ears).

Peripherals are reached on an I2C bus from the 'spinal processor' (ARM Cortex M0), which communicates via SPI with the 'brainstem processor' (ARM Cortex M0/M4 dual core), which in turn communicates via USB with the 'forebrain processor' (ARM Cortex A8). Division of the processing in this way is partly pedagogic and partly aesthetic, in service of the product's standard configuration, and plays no direct functional role. Nonetheless, it does align closely with the layered control architecture design (see below). All peripherals and a level of control over processing are accessible from off-board through WiFi connectivity, and the forebrain processor is open if lower-level access is required (all processors can be re-programmed if desired, though with more onerous requirements to respect the specifics of the platform).

Owing to its origins in a commercial project aimed at the general public, the MIRO platform has excellent affordability: the current configuration can be manufactured for around USD250. Whilst a MIRO-like platform would need some development for the healthcare market, maintaining affordability will make companion robots accessible in very considerable volumes, with a consequent impact on their relevance as a healthcare tool.

2.3 Control Architecture and Gross Behaviour

MIRO's control system is a brain model with a layered architecture [19]. That is, its most fundamental organising feature is the presence of sensorimotor loops layered on top of one another, so that lower loops function without the help of higher loops, but higher loops can modulate the behaviour of those lower down. Low-level loops implement reflex-like behaviours, immediate responses to sensory information that make use of neither memory nor signal analysis and can be implemented simply (for instance, soft threshold units respond to cliff sensor signals to inhibit forward wheel motion). Mid-level loops make use of short-term memory and within- and cross-modal signal relationships to implement 'hard-wired' behaviours that require co-ordination across motor systems (a major centre is a model of superior colliculus that represents recent salient events in a multi-modal map of egocentric space and responds to specific 'innate' stimuli with directed action [20]). High-level loops use arbitrarily deep memory and inter-signal relationships to implement cognitive competences (reinforcement learning provides the ability to 'train' MIRO to perform simple stimulus-response tasks, for example).

Whilst this three-level break-down is simplified, it conveys well the architectural principle of layers of increasingly sophisticated processing, with each layer making an important contribution to overall behaviour rather than being obsoleted by higher processing. In order to arbitrate between behavioural subsystems at mid and high levels we implement a model of the basal ganglia [21] in an abstract form as used in several of our previous robots [22]. Thus, MIRO's gross behaviour emerges from the competition between various sub-systems to explore locations with high sensory salience, escape from stimuli that are perceived as threatening, seek out goals (such as a charging station), have social exchanges with an interacting human, and so on.

2.4   Modelling and Expressing Affect

MIRO represents affective state using the circumflex model [23]. This model represents emotions (as well as, on the longer term, moods and temperaments) as points in a space having dimensions of valence and arousal (Figure 3). These dimensions are purported to have neural correlates whilst terms used to describe emotions (such as 'excited') are cast as locations in this space. This stands in contrast to 'basic emotions' theory which considers individual emotions (such as 'excitement') to correspond to discrete neural systems. Whilst continuum models of this sort have overwhelmingly received attention in human studies, recently they have begun to be transposed into the domain of non-human animals [24]. These models are also remarkable for their clarity and accessibility for the non-psychologist, as well as for their light computational weight, and have, accordingly, received some attention from roboticists [12,25,26].

MIRO displays affective state through its behaviour. Affect is fundamental to MIRO's functional behaviour because gross behaviours (such as approach, or flight) have unambiguous emotional correspondences and are, correspondingly, facilitated or suppressed by affective state. Affect is also communicated directly and explicitly through its encoding in MIRO's non-locomotory movements. MIRO has mobile ears, eyelids, and tail expressly for the communication of affect, but body configuration movements are also driven by emotions (activation tending to lead to raised posture, for instance). Body language has been

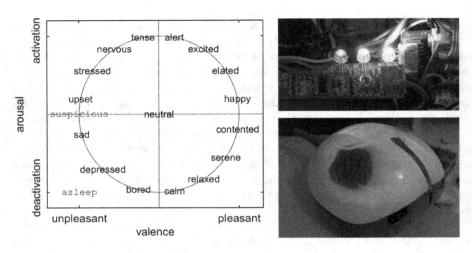

Fig. 3. (Left) Circumplex model of affective state is a space with valence and arousal dimensions. Names for states (sans serif font) are taken from Posner et al. (2005), except for two suggested by experiment participants (typewriter font, described below). (Top right) One way in which MIRO expresses affect is through a changing pattern of coloured lights. (Bottom right) False colour image of one of the lights as it appears through MIRO's body shell.

shown to be effective for the communication of emotions between humans [27] and consistent interpretation of the body language of animals by humans has been demonstrated [28], though there is considerable variation between species in expression [29]. Moreover, the use of human-like body language in humanoid robots is effective for communication of emotion to naïve humans [25].

In addition, MIRO is equipped with six RGB LEDs (three on each side) under its body shell that can be controlled dynamically (at up to 50Hz). Through these, MIRO can display arbitrary light patterns that change in parameters such as colour and rate in a bid to communicate affect. Whilst light displays offer rich expression and low cost, changing patterns of lights—in contrast to body language—do not have a direct biological analogue. Certainly, cultural associations exist for parameters such as colour—red/green for traffic lights is an almost universal contemporary code, for example—but reports have been presented of variability in these associations based on culture [30], gender [31], and context [32]. There is a considerable literature reviewing the effect of colour on physiology, behaviour, and emotion, and individual and cultural differences in colour responses; some population relationships are present, but a clear picture has not emerged [33,34]. Moreover, it is not clear in what way such associations would translate to perception of the affective state of a robot, nor whether these perceptions would be reliable in a naïve interactee. Work addressing this question to date has been somewhat informal and results variable [35]. Below, we report a pilot of a methodology to address this question.

3   Experimental Study

3.1   Methods

In many cultures, red signals danger and green safety; we therefore proposed red/white/green for encoding negative/neutral/positive valence. Red is also a signal for sexuality, and for the ripeness of fruit, and green for nausea and decay (the degree to which these associations are biological or cultural is not always clear), so we could equally well have proposed the opposite encoding; such observations underline the uncertainty in these associations and the need for empirical study. The rate of change of a light pattern may be intuitively linked to arousal—both breathing and heartrate, for example, increase in frequency with increasing physiological arousal—so we proposed slow/medium/fast to encode deactivation/neutral/activation (specifically, 0.25/0.5/2.5Hz, reflecting the frequency range of human breathing/heartrate). Thus, nine points in affect space could be encoded, in total.

We arbitrarily selected the remaining parameters of a pulsating light pattern that could physically be presented through the three RGB LEDs available on each side of MIRO. Specifically, the pattern at each parameter point was monochromatic, with sinusoidal intensity, and with a fixed phase offset between adjacent LEDs of $\pi/2$ radians. Whilst the pattern was chosen to be deliverable through MIRO's LED arrays, patterns were actually delivered to participants through a simulation of one of the arrays on a computer monitor. This choice reflects the more general nature of our experimental question, and was intended to eliminate possible sources of confound stemming from participants' perceptions of other aspects of MIRO's design and presentation (its shape, positioning, etc.). The actual colours delivered ranged, in each case, from zero intensity (black) to maximum intensity of either pure red (i.e. $[255, 0, 0]$), pure green, or white.

Our methodology for measuring the effectiveness of these encodings for evoking emotional perceptions was similar to that established by Beck et al. (2010) [25]. Naïve participants ($n = 5$, 2 female; M age $= 30$, $SD = 5$) were recruited informally from The University Of Sheffield Robotics Laboratory. Prior to study participation written informed consent was obtained from each participant. Participants were then asked to view simulated light patterns and indicate their perceptions on nominal and interval scales.

Participants were seated one at a time in front of a laptop computer. The experimenter gave them initial directions, and then left them to follow on-screen instructions. The computer displayed simulations of one of MIRO's light arrays (Figure 4) at the nine points in affect space comprising each possible combination of negative, neutral, and positive valence and arousal (for analysis, negative/neutral/positive were assigned the values -1/0/+1). Participants were first exposed, over the course of thirty seconds, to all nine points, with instructions to watch the patterns. They were then presented with each of the nine points again, in random order—these we refer to as the 'presented' affect values. Participants were asked to fill a response sheet for each presentation, comprising:

Fig. 4. Stimulus presentation tool. Stimuli ($N = 9$) were presented in random order for each participant, who clicked NEXT when ready to move on.

1. Which of the following words best describes your perception of the emotional state represented by the pattern of light? Please circle one:
   Happy – Depressed – Calm – Stressed – Relaxed – Sad – Alert – Upset – Elated – Nervous – Contented – Bored – Serene – Excited – Neutral – Tense
2. If you think another word or phrase better describes your perception of the emotional state represented by the pattern of lights please write it here: —
3. Place a vertical mark on the line to indicate your perception of the level of arousal represented by the pattern of lights, from relaxed to aroused:
   Relaxed ——————————————————— Aroused
4. Place a vertical mark on the line to indicate your perception of the level of happiness represented by the pattern of lights, from unhappy to happy:
   Unhappy ——————————————————— Happy

The terms used in question 1 were taken from Posner et al. 2005 [23], with the addition of 'neutral', following Beck et al. (2010) [25], and presented in a randomised order. At the end of the response phase the experimenter conducted a short informal interview in which participants were asked whether they found the question 1 word list adequate. If the participant had answered any question two with a word or phrase of their own this was also discussed. The interview was conducted to establish whether the participants had perceived the patterns in emotional terms at all and, if so, whether the word list had allowed them to express their perception. At the end of the interview participants were debriefed.

For numerical analysis, we associated numerical values in $[-1, +1]$ for valence and arousal with each of the terms used in question 1 (each taking a position in affect space on the unit circle, as indicated by their location in Figure 3) and with each of the marks in questions 3 and 4 (with the left/right extrema on the scales being transposed to -1/+1). These values, recovered from participants' responses, we refer to as the 'reported' affect values. Analyses of the reliability of the relationships between presented and reported affect values were conducted independently for valence and arousal.

3.2   Results

We first analysed the results pooled across participants; our results are graphed in Figure 5. We identified positive correlations between presented and reported values for both parameters when using both approaches to reporting. The relationship was apparently robust in all four cases, with between 25% and 70% of the variance in reporting explained by a simple linear predictive model.

We then exploratively reviewed the relationships identified above on an individual basis (see also Figure 5). Data from each participant displayed relationships of the same polarity as those displayed by the pooled data, indicating that pooled results reflected the responses of all participants in this sense.

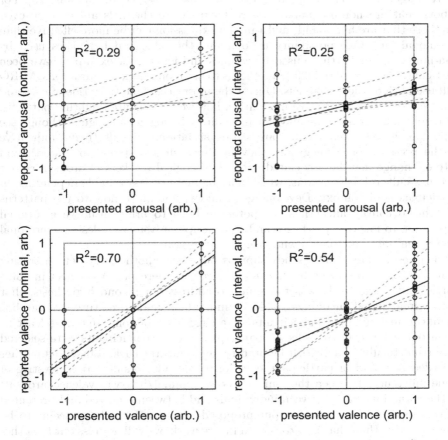

Fig. 5. Reported affect values against presented affect values. Top/bottom: arousal/valence. Left/right: nominal/interval reporting. All units arbitrary (arb.). Individual trials (circles, $N = 45$). Trend line (solid) and R^2 values are from simple linear regression of pooled data ($N = 45$). Trend lines over samples from each participant ($N = 9$ per participant) are also shown (dashed grey).

In response to question 2, only two responses were received (of a possible 45). These are the terms indicated in typewriter font in Figure 3, and they are placed in the affect space at the location of the presented stimulus for each of those 2 trials. Informal interviews generally indicated a high level of satisfaction with the word list for expressing participants' perceptions.

4   Discussion

To be an effective companion, a robot must be able to convey affect [12]. Work with humanoid robots has shown that affect can be communicated well using body language (gesture) based directly on that observed in humans [25]. For robots that are non-humanoid, different expressive channels are needed; even for those that are humanoid, multi-modal expression can be more effective than uni-modal [26]. One possibility is to mimic the biological languages used by non-human animals (those used by canines and felines, for example, have been particularly well explored [29]), another is to use biomimetic vocalisations; MIRO will use both of these channels. One of the most accessible (in terms of cost and practicality) of all expressive modalities, however, is coloured lighting patterns. Dynamic lighting patterns may not have direct biological analogues (though see cephalopods [36]), but colours are strong situational signals (being indicative of the presence of ethologically-relevant items such as blood and food), and rate of change may be associated with physiological markers of arousal; colour also has cultural associations, which may be more or less reliable depending on participant and context. Developing an understanding of how to use patterns of light to convey affect has the potential both to bring intuitive emotional expression to low cost platforms and to firm up our ability to design 'emotional expression' into our robots, whatever form they take.

The results of our pilot study support our first hypothesis by demonstrating that patterns of pulsating lights can evoke reliable perceptions of affect in naïve participants. The study was too small to address our second hypothesis, that the optimal signal encodings would be individual- and/or culture-specific, but results from individual participants were suggestive of consistency, at least at the grossest level, in the selected participant group (participants were selected opportunistically in a British laboratory, and cultural background was neither recorded nor used in participant selection). The pilot results are suggestive of some differences between the four analyses (nominal/interval, valence/arousal) in the variability both between individuals and between reported and presented affective states—in particular, our proposed encoding for valence seems to be more effective than that for arousal. In future work, we will address our hypotheses formally using larger studies in varying cultural contexts and exploring pattern space in more detail to allow the identification of encodings that were not, as here, preconceived. In addition, we will investigate the degree to which perceptions formed in response to a simulated light display under test conditions translate to the case of signalling through the light displays of MIRO, as an example of an interacting robot.

Emotional expression is so deeply a function of the response of human inter-
actees that deriving design principles is not a trivial process. Simply copying
known examples (such as human body language, vocal patterns) is, no doubt,
an excellent starting point. However, broadening the gamut of possible expressive modalities is only one way in which we can benefit from empirical studies
of the communication of affective states. It is our intent, therefore, to develop a
methodology for distilling descriptions of effective expression channels through
empirical study, accounting for individual and cultural differences between inter-
actees. One of the long term aims of this work has to be adaptation of the communication strategy based on the responses of the interactee; that is, to adapt
to the individual differences specific to a person with whom the robot must
interact [37].

We also hope to make MIRO, the platform, widely available. With low cost
and extensive suites of sensory and motor peripherals, MIRO is an attractive
research platform for many investigations and at all levels.

References

1. Shibata, T., Yoshida, M., Yamato, J.: Artificial emotional creature for human-
   machine interaction. In: 1997 IEEE International Conference on Computational
   Cybernetics and Simulation Systems, Man, and Cybernetics, vol. 3, pp. 2269–2274.
   IEEE (1997)
2. Dautenhahn, K., Nehaniv, C.L., Walters, M.L., Robins, B., Kose-Bagci, H.,
   Mirza, N.A., Blow, M.: Kaspar-a minimally expressive humanoid robot for human-
   robot interaction research. Applied Bionics and Biomechanics 6(3–4), 369–397
   (2009)
3. Tamura, T., Yonemitsu, S., Itoh, A., Oikawa, D., Kawakami, A., Higashi, Y.,
   Fujimooto, T., Nakajima, K.: Is an entertainment robot useful in the care of elderly
   people with severe dementia? The Journals of Gerontology Series A: Biological Sciences and Medical Sciences 59(1), M83–M85 (2004)
4. Libin, A.V., Libin, E.V.: Person-robot interactions from the robopsychologists'
   point of view: the robotic psychology and robotherapy approach. Proceedings of
   the IEEE 92(11), 1789–1803 (2004)
5. Allen, K.M., Blascovich, J., Tomaka, J., Kelsey, R.M.: Presence of human friends
   and pet dogs as moderators of autonomic responses to stress in women. Journal of
   personality and social psychology 61(4), 582 (1991)
6. Ballarini, G.: Pet therapy. animals in human therapy. Acta Bio Medica Atenei
   Parmensis 74(2), 97–100 (2003)
7. Collis, G., McNicholas, J.: A theoretical basis for health benefits of pet ownership.
   Companion Animals in Human Health, pp. 105–22 (1998)
8. Stiehl, W.D., Lieberman, J., Breazeal, C., Basel, L., Lalla, L., Wolf, M.: Design
   of a therapeutic robotic companion for relational, affective touch. In: IEEE International Workshop on Robot and Human Interactive Communication, ROMAN
   2005, pp. 408–415. IEEE (2005)
9. Shibata, T.: Paro's Goal (Purpose) and Effects (2015). http://paro.jp/?page_
   id=336 (Accessed July 9, 2015)
10. Kidd, C.D., Taggart, W., Turkle, S.: A sociable robot to encourage social inter-
    action among the elderly. In: Proceedings 2006 IEEE International Conference on
    Robotics and Automation, ICRA 2006, pp. 3972–3976. IEEE (2006)

11. Magai, C., Cohen, C., Gomberg, D., Malatesta, C., Culver, C.: Emotional expression during mid-to late-stage dementia. International Psychogeriatrics **8**(03), 383–395 (1996)

12. Breazeal, C., Scassellati, B.: How to build robots that make friends and influence people. In: Proceedings 1999 IEEE/RSJ International Conference on Intelligent Robots and Systems, IROS 1999, vol. 2, pp. 858–863. IEEE (1999)

13. Bruce, A., Nourbakhsh, I., Simmons, R.: The role of expressiveness and attention in human-robot interaction. In: Proceedings IEEE International Conference on Robotics and Automation, ICRA 2002, vol. 4, pp. 4138–4142. IEEE (2002)

14. Mutlu, B., Shiwa, T., Kanda, T., Ishiguro, H., Hagita, N.: Footing in human-robot conversations: how robots might shape participant roles using gaze cues. In: Proceedings of the 4th ACM/IEEE international conference on Human robot interaction, pp. 61–68. ACM (2009)

15. Banks, M.R., Willoughby, L.M., Banks, W.A.: Animal-assisted therapy and loneliness in nursing homes: use of robotic versus living dogs. Journal of the American Medical Directors Association **9**(3), 173–177 (2008)

16. Pearson, M.J., Mitchinson, B., Sullivan, J.C., Pipe, A.G., Prescott, T.J.: Biomimetic vibrissal sensing for robots. Philosophical Transactions of the Royal Society of London B: Biological Sciences **366**(1581), 3085–3096 (2011)

17. Guizzo, E., Ackerman, E.: The rise of the robot worker. IEEE Spectrum **49**(10), 34–41 (2012)

18. Lepora, N.F., Verschure, P., Prescott, T.J.: The state of the art in biomimetics. Bioinspiration & biomimetics **8**(1), 013001 (2013)

19. Prescott, T.J., Redgrave, P., Gurney, K.: Layered control architectures in robots and vertebrates. Adaptive Behavior **7**(1), 99–127 (1999)

20. Dean, P., Redgrave, P., Westby, G.: Event or emergency? two response systems in the mammalian superior colliculus. Trends in Neurosciences **12**(4), 137–147 (1989)

21. Gurney, K., Prescott, T.J., Redgrave, P.: A computational model of action selection in the basal ganglia. i. a new functional anatomy. Biological Cybernetics **84**(6), 401–410 (2001)

22. Pearson, M.J., Pipe, A.G., Melhuish, C., Mitchinson, B., Prescott, T.J.: Whiskerbot: a robotic active touch system modeled on the rat whisker sensory system. Adaptive Behavior **15**(3), 223–240 (2007)

23. Posner, J., Russell, J.A., Peterson, B.S.: The circumplex model of affect: An integrative approach to affective neuroscience, cognitive development, and psychopathology. Development and Psychopathology **17**(03), 715–734 (2005)

24. Mendl, M., Burman, O.H., Paul, E.S.: An integrative and functional framework for the study of animal emotion and mood. Proceedings of the Royal Society B: Biological Sciences **277**(1696), 2895–2904 (2010)

25. Beck, A., Hiolle, A., Mazel, A., Cañamero, L.: Interpretation of emotional body language displayed by robots. In: Proceedings of the 3rd International Workshop on Affective Interaction in Natural Environments, pp. 37–42. ACM (2010)

26. Yilmazyildiz, S., Henderickx, D., Vanderborght, B., Verhelst, W., Soetens, E., Lefeber, D.: Multi-modal emotion expression for affective human-robot interaction. In: Proceedings of the Workshop on Affective Social Speech Signals (WASSS 2013), Grenoble, France (2013)

27. Wallbott, H.G.: Bodily expression of emotion. European journal of social psychology **28**(6), 879–896 (1998)

28. Wemelsfelder, F., Hunter, A., Paul, E., Lawrence, A.: Assessing pig body language: Agreement and consistency between pig farmers, veterinarians, and animal activists. Journal of animal science **90**(10), 3652–3665 (2012)

29. Darwin, C.: The expression of the emotions in man and animals. Oxford University Press (2002)
30. Courtney, A.J.: Chinese population stereotypes: color associations. Human Factors: The Journal of the Human Factors and Ergonomics Society **28**(1), 97–99 (1986)
31. Hurlbert, A.C., Ling, Y.: Biological components of sex differences in color preference. Current Biology **17**(16), R623–R625 (2007)
32. Maier, M.A., Barchfeld, P., Elliot, A.J., Pekrun, R.: Context specificity of implicit preferences: the case of human preference for red. Emotion **9**(5), 734 (2009)
33. Valdez, P., Mehrabian, A.: Effects of color on emotions. Journal of Experimental Psychology: General **123**(4), 394 (1994)
34. Manav, B.: Color-emotion associations and color preferences: A case study for residences. Color Research & Application **32**(2), 144–150 (2007)
35. Haring, M., Bee, N., André, E.: Creation and evaluation of emotion expression with body movement, sound and eye color for humanoid robots. In: 2011 IEEE RO-MAN, pp. 204–209. IEEE (2011)
36. Mäthger, L.M., Hanlon, R.T.: Malleable skin coloration in cephalopods: selective reflectance, transmission and absorbance of light by chromatophores and iridophores. Cell and tissue research **329**(1), 179–186 (2007)
37. Collins, E.C., Prescott, T.J.: Individual differences and biohybrid societies. In: Duff, A., Lepora, N.F., Mura, A., Prescott, T.J., Verschure, P.F.M.J. (eds.) Living Machines 2014. LNCS, vol. 8608, pp. 374–376. Springer, Heidelberg (2014)

Integrating Feedback and Predictive Control in a Bio-inspired Model of Visual Pursuit Implemented on a Humanoid Robot

Lorenzo Vannucci$^{(\boxtimes)}$, Egidio Falotico, Nicola Di Lecce, Paolo Dario, and Cecilia Laschi

The BioRobotics Institute, Scuola Superiore Sant'Anna,
Viale Rinaldo Piaggio 34, 56025 Pontedera (Pisa), Italy
{lorenzo.vannucci,egidio.falotico,nicola.dilecce,
paolo.dario,cecilia.laschi}@sssup.it
http://sssa.biaroboticsinstitute.it

Abstract. In order to follow a moving visual target, humans generate voluntary smooth pursuit eye movements. The purpose of smooth pursuit eye movements is to minimize the retinal slip, i.e. the target velocity projected onto the retina. In this paper we propose a model able to integrate the major characteristics of visually guided and predictive control of the smooth pursuit. The model is composed of an Inverse Dynamics Controller (IDC) for the feedback control, a neural predictor for the anticipation of the target motion and a Weighted Sum module that is able to combine the previous systems in a proper way. In order to validate the general model, two implementations with two different IDC controllers have been carried out. The first one uses a backstepping-based controller to generate velocity motor commands for the eye movements and the other one uses a bio-inspired neurocontroller to generate position motor commands for eye-neck coordinated movements. Our results, tested on the iCub robot simulator, show that both implementations can use prediction for a zero-lag visual tracking, a feedback based control for "unpredictable" target pursuit and can combine these two approaches by properly switching from one to the other, guaranteeing a stable visual pursuit.

Keywords: Eye movements · Smooth-pursuit · Predictive control · Neurocontroller

1 Introduction

Following a moving target with a foveal vision is one of the essential tasks of humans and humanoid robots. Humans accomplish this task through a combination of two forms of eye movements: saccade and smooth pursuit. Saccades are high velocity gaze shifts that bring the image of an object of interest onto fovea. The purpose of smooth pursuit eye movements is to minimize the retinal

© Springer International Publishing Switzerland 2015
S.P. Wilson et al. (Eds.): Living Machines 2015, LNAI 9222, pp. 256–267, 2015.
DOI: 10.1007/978-3-319-22979-9_26

slip, i.e. the target velocity projected onto the retina, stabilizing the image of the moving object on the fovea. This cannot be achieved by a simple visual negative feedback controller due to the long delays (around 100 ms in the human brain), most of which are caused by visual information processing. During maintained smooth pursuit, the lag in eye movement can be reduced or even cancelled if the target trajectory can be predicted [1].

One of the first models of pursuit systems was built by Robinson and colleagues [2]. It included three feedback loops: the outer one computed a velocity error and the other two (a positive one and a negative one) ensure that the model could reproduce adequately the ringing settling to a steady state.Later, Krauzlis and Lisberger [3] prosed a model conposed of three pathways that simulate pursuit behaviour when tracking targets with constant velocity. An important biologically plausible smooth pursuit controller has been proposed by Shibata and Schaal [4]. This controller learns to predict the visual target velocity in head coordinates, based on on-line learning of the target dynamics. The models proposed by Shibata and Schaal and improved in [5] results in a good solution to represent predictive behaviour in the smooth pursuit eye movement.

Although the presented systems may consistently represent specific features of the smooth pursuit eye movement, they do not consider to integrate the major characteristics of visually guided and predictive control of the smooth pursuit. The first instance of a model integrating these aspects of the pursuit dynamics in a coherent model has been proposed by Orban de Xivry and colleagues [6]. Their model aims at reproducing the eye dynamics during the switching between feedback and predicting control and it has a memory based on previous values of the target velocity. The eye controller, designed specifically for the human eye and the memory approach makes this model not reproducible on a robotic platform. The model we propose, inspired in part by their observations [6], aims at producing a model, suitable for a robotic implementation, able to combine feedback and predictive control.

From a robotic point of view, few works focus on the implementation of the smooth pursuit eye movement and none accounts for integrating the characteristics previously described. Shibata and colleagues suggested a control circuits for the integration of the most basic oculomotor behaviours [7] including the smooth pursuit eye movement. This model, based on the prediction of target dynamics did not consider a feedback control. The same consideration can be applied to the model of smooth pursuit and catch-up saccade [8] or occlusions [9] implemented on the iCub robot. Also the model proposed by Vannucci and colleagues [10] based on a neurocontroller did not include a feedback control, but only prediction based on a Kalman filter.

The objective of this work is to investigate the applicability of smooth pursuit functional principles derived from neuroscience research on humanoid robots [11], in order to achieve a human-like visual pursuit able to: (i) use prediction for a zero-lag visual tracking, (ii) use a feedback based control for "unpredictable" target pursuit and (iii) combine these two approaches in order to guarantee a stable visual pursuit. In the next sections we will present the model and its components in detail together with two different implementations for the iCub humanoid robot. In order to validate the general model, two different inverse dynamics controllers (IDC) were implemented. In the past two approaches have been used to implement IDCs for oculomotor control: one that uses classic techniques to identify the eye plant parameters and build suitable controllers[12] and one that uses neurocontrollers capable of adapting to the eye plant without having to identify its parameters[13][14]. Thus, we propose two different implementations: the first one, working in the velocity space, uses a backstepping-based controller to generate velocity motor commands for the eye movements and the other one uses a bio-inspired neurocontroller to generate position motor commands for eye-neck coordinated movements.

2 Model

The overall schema of the controller can be seen in Figure 1. The model consist of two main pathways: a sensory pathway (*Visual Processing*), where the camera image is processed to get sensory information about the target, and a predictive pathway (*Predictor*) where the sensory information is used to predict future states of the target.

Input coming from the sensory pathway (a target reference TR^{sens} and an error reference ER^{sens}) can be used to implement a classic feedback controller that is able to make the robot follow a moving target but with a certain delay.

On the other hand, input coming from the predictive pathway (a predicted target reference TR^{pred} and a predicted error reference ER^{pred}) can be used to anticipate the movement of the target and overcome the said delay, making the robot follow the target with zero lag. But the predictor will not perform well when the target velocity is unpredictable or when there is a change in the target motion. In such cases, the feedback controller will perform better. Thus, a proper switching mechanism is introduced to smoothly change between the input coming from the two pathways at the appropriate time (*Weighted Sum*).

In the end, the smoothed error reference ER is given to a controller capable of generating the appropriate motor commands for the Robot.

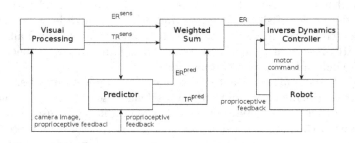

Fig. 1. Model of the controller

2.1 Predictor

In order to implement a model capable of predicting the target reference at future steps, a simple linear model has been chosen: Rosenblatt's single layer perceptron[15]. To use this kind of model to predict future values of the signal a temporal sliding window of size d containing past values of the signal are given as an input along with the current value [16]. The activation function of such unit is linear.

Thus, the output of the network at step t, for a given input signal x is:

$$out(t) = \sum_{i=0}^{d} x(t - d) \cdot w_d \tag{1}$$

The model was trained by an on-line version of the Widrow-Hoff rule[17], where the target value is the signal anticipated by p and η is the learning rate.

$$\Delta w(t) = \eta \cdot (x(t + p) - out(t)) \cdot x(t) \tag{2}$$

Using the on-line version of the learning algorithm gives the possibility of using the model without doing any previous training phase, as the perceptron will adapt the weights for the prediction during the execution phase. In order to move the signal towards 0 and thus facilitate learning, the mean observed value up to time t was subtracted from the signal.

After choosing the parameters η and d, the model proved capable of predicting periodic signals, even with noise. The model was also able to quickly adapt to variations in frequency and amplitude of the signal and decay. In the overall control scheme the predictive model is used to predict future values of the target reference (TR^{pred}) starting from the current value of TR^{sens}. Then, the predicted error reference is computed as follows, where is g is an appropriate function of the sensors:

$$ER^{pred}(t) = TR^{pred}(t) - g(sensors(t)) \tag{3}$$

2.2 Weighted Sum

The control system must be able to automatically switch between the sensory and predictive pathways. In order to do so in a smooth manner, a weighted sum of the error references coming from the two pathways is performed:

$$ER(t) = (1 - \alpha(t))ER^{sens}(t) + \alpha(t)ER^{pred}(t) \tag{4}$$

where, for each t, $\alpha(t) \leq 1$. At each control step, α is chosen as a measure of how much the prediction has been accurate in the last 100 steps:

$$\alpha(t) = f(\max\{err(t), ..., err(t - 100)\}) \tag{5}$$

where, $err(t)$ is the error of the predictor at time t and f is a threshold function. In order to make the function α more smooth, f was implemented as

a spline function, interpolating between these points:
$\{(0,1),(0.04,0.9),(0.08,0.5),(0.184,0.1),(0.4,0)\}$.

Finally, to compute the error of the predictor at each step, its output is delayed by the same number of steps of the ahead prediction and then confronted with the input signal:

$$err(t) = \frac{|TR^{sens}(t) - TR^{pred}(t-p)|}{maxTR - meanTR} \tag{6}$$

where $maxTR$ and $meanTR$ are, respectively, the maximum observed value and the observed mean value of TR^{sens}.

As such, if the prediction is accurate enough the system will smoothly transition to take advantage of it, while, on the other hand, if the signal becomes unpredictable, it will revert to using only the sensory pathway.

3   Implementation

In order to validate the proposed model two different robotic implementations were tested on the iCub robot, using the simulator given with the iCub libraries. The iCub head [18] contains a total of 6 DOFs: 3 for the neck (pan, tilt and swing) and 3 for the eyes (an independent pan for each eye and a common tilt). The visual stereo system consists of 2 cameras with a maximal resolution of 640X480 pixels.

The first implementation has a backstepping-based controller as IDC receiving a velocity target reference to control the pan of the eyes through velocity motor commands, while the second has a neurocontroller receiving position target reference and to perform a full eye-head coordination through a joint position control.

3.1   Backstepping-Based Controller as IDC

In this implementation the incoming reference is the velocity of the target on the horizontal axis and the error signal is the retinal slip, that is the speed of the target velocity projected in the eye reference frame. These values are computed by the *Visual Processing* module as follows:

$$TR^{sens}(t) = ER^{sens}(t) + \dot{\theta}(t) \tag{7}$$

where $\dot{\theta}$ is the current velocity of the pan joint. In the predictive pathway the retinal slip is computed from the predicted target velocity:

$$ER^{pred}(t) = TR^{pred}(t) - \dot{\theta}(t) \tag{8}$$

The smoothed error reference is then given to a controller, the objective of which is to cancel the dynamics of the eye plant. The system model of the eye plant is a second order transfer function given by

$$G(s) = \frac{1}{Js^2 + Bs} = \frac{k}{s^2 + as} \tag{9}$$

where J is the inertia of the system and B is the damp, with $k = 1/J$ and $a = B/J$. The system input is a velocity reference and its output is a position measurement. In particular, the velocity reference is the target velocity (TR) and its output is the position of the eye. For the system parameters we used those estimated through an identification methodology on the simulated robot. The system can be expressed by the state-space function such as:

$$\begin{cases} \dot{\theta} = x_2 \\ \dot{x}_2 = ku - ax_2 \\ \theta = x_1 \end{cases} \tag{10}$$

The IDC that can negate the delay of such a plant is implemented using the *backstepping* technique, a design methodology for the construction of a robust feedback control law through a recursive construction of a Control Lyapunov Function (CLF). The proposed backstepping is based on [19] and involves the inclusion of an integrator to the input of linear servo-motor model with a known feedback-stabilizing control law, and so the stabilizing approach is known as *integrator backstepping*. In particular, the *integrator backstepping* law provides suitable velocity commands and the possibility to exploit "good" non-linearities while "bad" non-linearities can be dominated by adding non-linear damping. The standard SISO model is:

$$\begin{cases} \dot{x} = f(x) + g(x)u = F(x, u) \\ f(0) = 0 \end{cases} \tag{11}$$

where $x \in \mathbb{R}^n$ and $u \in \mathbb{R}$. The existence of a differentiable feedback control law $u = \Gamma(x)$, with $\Gamma_i(0) = 0$ and of a CLF $\mathbf{V}(x) : \mathbb{R}^n \to \mathbb{R}$ such that $L_F\mathbf{V}(x) \le 0$ are assumed. Hence, the augmented system with an integrator is given by:

$$\begin{cases} \dot{x} = f(x) + g(x)\nu = F(x, \nu) \\ \dot{\nu} = u \end{cases} \tag{12}$$

where $f(x)$ is the drift function of the model and u is the input command. If $L_F\mathbf{V}(x)$ is negative definite, then the CLF becomes:

$$\mathbf{V}_e(x, \nu) = \mathbf{V}(x) + \frac{1}{2}[\nu - \Gamma(x)]^2 \tag{13}$$

where the function $\mathbf{V}(x)$ is the possible CLF that fulfils other constraints. A feedback control $u = \Gamma(x, \nu)$ exists such that the equilibrium point $(x, \nu) = (0, 0)$ is globally asymptotically stable. The possible control that satisfies the assumption (13) is:

$$u = -\lambda(\nu - \Gamma(x)) + \frac{\partial \Gamma(x)}{\partial x}g(x)\nu - \frac{\partial \mathbf{V}(x)}{\partial x}g(x) \tag{14}$$

with $\lambda > 0$. This control can be replaced in the CLF directional derivative obtaining:

$$L_F\mathbf{V}_e(x) = L_F\mathbf{V}(x) - \lambda[\nu - \Gamma(x)]^2 \tag{15}$$

where $L_F \mathbf{V}(x)$ is the directional derivative of the possible other constraints. Then, the overall CLF is defined negative. Therefore, the expanded Lyapunov candidate is:

$$\mathbf{V} = \frac{1}{2}(ER)^2 \tag{16}$$

where in this case the retinal slip is computed as $ER = (TR - x_2)$. The time derivative of Lyapunov candidate (16) can be derived as:

$$\dot{\mathbf{V}} = (ER)(\dot{ER}) = ER(\dot{TR} - ku + ax_2) \tag{17}$$

and after some easy algebraic computations, the control input for the motor can be rewritten as listed below:

$$u = \frac{1}{k}[\dot{TR} + ax_2 + \lambda(ER)] \tag{18}$$

The values of λ is strictly positive and is defined on the basis of the desired convergence velocity of the algorithm. In this case, the control law does not include the derivative terms as in (14) as it could represent a possible noise source. The absence of derivative terms is balanced by the presence of the constant λ. Furthermore, replacing (18) in the CLF derivative (17) the asymptotic stability of Lyapunov candidate is demonstrated:

$$\dot{\mathbf{V}} = -\lambda(ER)^2 \leq 0. \tag{19}$$

3.2 Neurocontroller as IDC

In this implementation the pieces of information gathered from the *Visual Processing* are the target position in the 3D space as the target reference and the difference between the target position and the gaze fixation point. The current gaze fixation point was computed by using direct kinematics, so, if θ are the values of the head joints, then, for both the sensory and predictive pathways:

$$ER(t) = TR(t) - K(\theta(t)) \tag{20}$$

The smoothed ER signal is given to a neurocontroller capable of moving the current gaze fixation point to compensate for the error in the 3D space with a coordinated movement of both the robot eyes and neck joints. This controller, fully described in [10], makes use of two different bio-inspired neural models: a Growing Neural Gas for associative learning of motion patterns and a biologically recurrent neural network that generates the position motor commands. The neurocontroller is trained offline through a motor babbling algorithm.

4 Results

In order to test the effectiveness of the proposed model, two kinds of tests have been performed on both implementations.

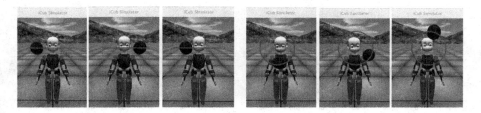

Fig. 2. Trial executions showing the iCub robot pursuing a target. Left images show the horizontal task for the first implementation, while the right ones show the circular motion used in the second implementation. The red line shows the target motion.

In the first test, the reference signal switches from a sinusoidal wave to another, after a certain period of time (Figure 2). This test is performed in order to show the capability of the model to adapt to a change in frequency and amplitude of the signal.

In the second test, the reference signal switches to a random, unpredictable one. This test is performed in order to demonstrate that the model is still able to follow an unpredictable signal even if with some delay.

For both implementations the tests were performed on the iCub simulator where a ball was moved inside the simulated environment and then tracked from the camera images. It should be noticed that the objective of this work is not to provide a comparison among the two controllers performance, but to asses the effectiveness of the model through these controller implementations.

4.1   Results for the Backstepping-Based Controller as IDC

Using this implementation, the two kinds of tests were performed by moving the ball on the horizontal axis, with a sinusoidal motion. The retinal slip and the joint velocity are computed by applying a least-squares algorithm [20]. The parameters used for the predictor in these tests were: prediction step $p = 5$, tap delay $d = 50$ and learning rate $\eta = 0.5$. The parameters for the controller were found empirically as: $k = 25.14$, $a = 30.43$, $\lambda = 60$.

Results for the signal switching task can be seen in Figure 3. The initial frequency of the movement was 0.25Hz with an amplitude of 0.1m, and it can be observed that the value of α increases, meaning that the prediction becomes more accurate. The maximum peak-to-peak value of the retinal slip decreases from 8 deg/s to 3 deg/s. When the motion changes in both frequency and amplitude (0.125Hz and 0.15m), after 26 seconds, the value of α suddenly decreases towards zero, resulting in an higher retinal slip. After a certain period of time, the prediction is again accurate, therefore the value of α increase again and the eye velocity aligns again with the target one, resulting in an maximum retinal slip peak-to-peak value of 3 deg/s.

During the initial phase of the second task, the motion has the same frequency and amplitude as the previous one, but after 26 seconds it changes to a random one (Figure 4). After the change, the value of α suddenly decrease towards zero,

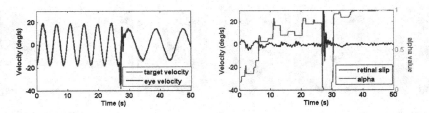

Fig. 3. Results for a frequency and amplitude changing task, in the backstepping-based implementation. The left plot shows the reference velocity and the eye velocity, while the right plot shows the velocity error and the value of α. Values of α close to 1 correspond to a strong contribution of the prediction in the control loop, values close to 0 correspond to a pure feedback control.

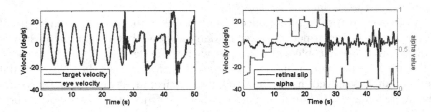

Fig. 4. Results for a task with a switch to a random signal, in the backstepping-based implementation. The left plot shows the reference velocity and the eye velocity, while the right plot shows the velocity error and the value of α.

but the eye is still able to reach the target velocity, even with some delay. During this second phase, the mean peak-to-peak value of the error is 6.86 deg/s.

4.2 Results for the Neurocontroller as IDC

For the neural implementation the test were performed by moving the ball on two axes, both the horizontal and vertical, with a sinusoidal motion on each. The parameters used for the predictor in these tests were: prediction step $p = 5$, tap delay $d = 50$ and learning rate $\eta = 0.5$.

The same values for the frequency and amplitude of the movements used for the other implementation were used for this one.

Figure 5 shows the results for the first task. During the first phase of the motion the position error decreases from a maximum peak-to-peak amplitude of 0.04m to 0.2m on both axes. After the change of frequency and amplitude at 26s, the value of α decreases and then grows again, meaning that the predictor adjusted for the new signal. The error then decreases again under 0.02m on both axes.

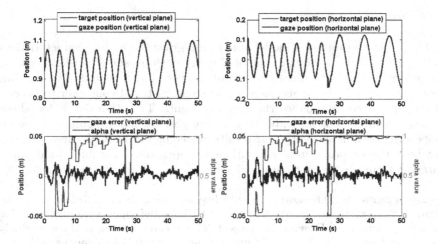

Fig. 5. Results for the neural implementation on a signal switching task. The top plots show the reference position and the gaze position on both the horizontal and the vertical axes, while the bottom plots show the position error and the value of α for both axes.

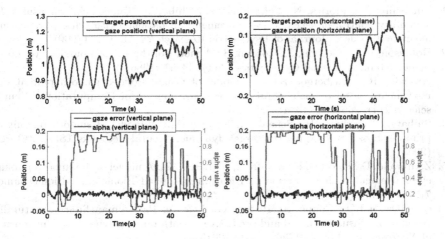

Fig. 6. Results for the neural implementation task with a switch to random motion. The top plots show the reference position and the gaze position on both the horizontal and the vertical axes, while the bottom plots show the position error and the value of α for both axes.

Also in this implementation the second task starts with an initial motion phases equal to the previous one (Figure 6). Again, after the change, the value of α suddenly decreases, but the robot is still able to follow the target, with a maximum peak-to-peak amplitude of the error of 0.05m on each axis.

5 Conclusions

In this paper we propose a model able to integrate the major characteristics of visually guided and predictive control of the smooth pursuit. In order to validate the general model, two different controllers were implemented. The first one, working in the velocity space, uses a backstepping-based controller to generate velocity motor commands for the eye movements. The other one is based on a bio-inspired neurocontroller able to generate position motor commands for eye-neck coordinated movements. Our results, tested on on the iCub simulator, show that both the proposed implementations can guarantee a stable visual pursuit switching from predictive to feedback control.

Acknowledgments. The authors would like to thank Italian Ministry of Foreign Affairs and International Cooperation DGSP-UST for the support through Joint Laboratory on Biorobotics Engeneering project. The research leading to these results has received funding from the European Union Seventh Framework Programme (FP7/2007-2013) under grant agreement no. 604102 (Human Brain Project).

References

1. Fukushima, J., Morita, N., Fukushima, K., Chiba, T., Tanaka, S., Yamashita, I.: Voluntary control of saccadic eye movements in patients with schizophrenic and affective disorders. Journal of Psychiatric Research **24**(1), 9–24 (1990)
2. Robinson, D.A., Gordon, J., Gordon, S.: A model of the smooth pursuit eye movement system. Biological Cybernetics **55**(1), 43–57 (1986)
3. Krauzlis, R.J., Lisberger, S.G.: A model of visually-guided smooth pursuit eye movements based on behavioral observations. Journal of Computational Neuroscience **1**(4), 265–283 (1994)
4. Shibata, T., Tabata, H., Schaal, S., Kawato, M.: A model of smooth pursuit in primates based on learning the target dynamics. Neural Networks **18**(3), 213–224 (2005)
5. Zambrano, D., Falotico, E., Manfredi, L., Laschi, C.: A model of the smooth pursuit eye movement with prediction and learning. Applied Bionics and Biomechanics **7**(2), 109–118 (2010)
6. de Xivry, J.J.O., Coppe, S., Blohm, G., Lefevre, P.: Kalman filtering naturally accounts for visually guided and predictive smooth pursuit dynamics. The Journal of Neuroscience **33**(44), 17301–17313 (2013)
7. Shibata, T., Vijayakumar, S., Conradt, J., Schaal, S.: Biomimetic oculomotor control. Adaptive Behavior **9**(3–4), 189–207 (2001)
8. Falotico, E., Zambrano, D., Muscolo, G.G., Marazzato, L., Dario, P., Laschi, C.: Implementation of a bio-inspired visual tracking model on the icub robot. In: Proc. 19th IEEE International Symposium on Robot and Human Interactive Communication (RO-MAN 2010), pp. 564–569. IEEE (2010)
9. Falotico, E., Taiana, M., Zambrano, D., Bernardino, A., Santos-Victor, J., Dario, P., Laschi, C.: Predictive tracking across occlusions in the icub robot. In: Proceedings of the 9th IEEE-RAS International Conference on Humanoid Robots (Humanoids 2009), pp. 486–491, December 2009

10. Vannucci, L., Cauli, N., Falotico, E., Bernardino, A., Laschi, C.: Adaptive visual pursuit involving eye-head coordination and prediction of the target motion. In: Proceedings of the 14th IEEE-RAS International Conference on Humanoid Robots (Humanoids 2014), pp. 541–546. IEEE (2014)
11. Dario, P., Carrozza, M.C., Guglielmelli, E., Laschi, C., Menciassi, A., Micera, S., Vecchi, F.: Robotics as a future and emerging technology: biomimetics, cybernetics, and neuro-robotics in european projects. IEEE Robotics & Automation Magazine 12(2), 29–45 (2005)
12. Viollet, S., Franceschini, N.: A high speed gaze control system based on the vestibulo-ocular reflex. Robotics and Autonomous systems 50(4), 147–161 (2005)
13. Lenz, A., Balakrishnan, T., Pipe, A.G., Melhuish, C.: An adaptive gaze stabilization controller inspired by the vestibulo-ocular reflex. Bioinspiration & Biomimetics 3(3), 035001 (2008)
14. Franchi, E., Falotico, E., Zambrano, D., Muscolo, G., Marazzato, L., Dario, P., Laschi, C.: A comparison between two bio-inspired adaptive models of vestibulo-ocular reflex (vor) implemented on the icub robot. In: Proceedings of the 10th IEEE-RAS International Conference on Humanoid Robots (Humanoids 2010), pp. 251–256, December 2010
15. Rosenblatt, F.: The perceptron: a probabilistic model for information storage and organization in the brain. Psychological Review 65(6), 386 (1958)
16. Weigend, A.S., Huberman, B.A., Rumelhart, D.E.: Predicting the future: A connectionist approach. International Journal of Neural Systems 1(03), 193–209 (1990)
17. Widrow, B., Hoff, M.E.: Adaptive switching circuits (1960)
18. Beira, R., Lopes, M., Praga, M., Santos-Victor, J., Bernardino, A., Metta, G., Becchi, F., Saltarén, R.: Design of the robot-cub (icub) head. In: Proceedings of the 2006 IEEE International Conference on Robotics and Automation (ICRA 2006), pp. 94–100. IEEE (2006)
19. Kokotovie, P.V.: The joy of feedback: nonlinear and adaptive. IEEE Control Systems Magazine 12(3), 7–17 (1992)
20. Janabi-Sharifi, F., Hayward, V., Chen, C.S.: Discrete-time adaptive windowing for velocity estimation. IEEE Transactions on Control Systems Technology 8(6), 1003–1009 (2000)

Knowledge Transfer in Deep Block-Modular Neural Networks

Alexander V. Terekhov[(✉)], Guglielmo Montone, and J. Kevin O'Regan

Laboratoire Psychologie de la Perception, Université Paris Descartes,
75006 Paris, France
{avterekhov,montone.guglielmo}@gmail.com,
kevin.oregan@parisdescartes.fr
http://lpp.psycho.univ-paris5.fr/feel

Abstract. Although deep neural networks (DNNs) have demonstrated impressive results during the last decade, they remain highly specialized tools, which are trained – often from scratch – to solve each particular task. The human brain, in contrast, significantly re-uses existing capacities when learning to solve new tasks. In the current study we explore a block-modular architecture for DNNs, which allows parts of the existing network to be re-used to solve a new task without a decrease in performance when solving the original task. We show that networks with such architectures can outperform networks trained from scratch, or perform comparably, while having to learn nearly 10 times fewer weights than the networks trained from scratch.

Keywords: Deep learning · Neural networks · Modular · Knowledge transfer

1 Introduction

Deep Neural Networks (DNN) have demonstrated impressive results in the last 10-15 years. They have established new benchmarks in such tasks as the classification of hand-written digits [14], object recognition [17], speech recognition [12], machine translation [23] and many others. These successes can mainly be attributed to the use of pre-training [7,21], sharing [16], and various forms of regularization [8,10], as well as to increases in computational capacity and data availability [22].

In spite of these impressive results, DNNs are still unable to match humans in the diversity of tasks we can solve and the ease which we learn. For example, we are able to progressively accumulate and abstract knowledge from previous experiences and re-use this knowledge to perform new tasks. On the contrary, in DNNs learning a new task tends to erase the knowledge about the previous tasks — a phenomenon known as *catastrophic forgetting* [20].

This specialization contrasts with the operating principles of the human nervous system, as it is well known that the brain re-uses existing structures when

© Springer International Publishing Switzerland 2015
S.P. Wilson et al. (Eds.): Living Machines 2015, LNAI 9222, pp. 268–279, 2015.
DOI: 10.1007/978-3-319-22979-9_27

learning a new task [1,6]. Ideally, we would like neural networks to possess similar capacities and operating principles. Imagine a neural network NN_1 has been trained on task T_1, and we would like to have a new network NN_2 which solves task T_2, where T_1 and T_2 share common features. If we simply take the network NN_1 and train it on T_2, it will most probably perform more poorly when solving T_1 after retraining. Of course, we can make an exact copy of NN_1 and then train it on T_2, but this is computationally cumbersome. Moreover, when using such a paradigm, we will simply obtain a collection of networks which are all very specialized and all work independently from one another. Rather, we would like to have a system that re-uses features learned in task T_1 to solve task T_2, and eventually, after having learned a number of tasks, would re-use relevant features from all (or more likely some) previous tasks to solve a new task.

The idea of exploiting features learned in a previous task when solving a new task is not original in the field of artificial neural networks. Simultaneous learning of multiple related tasks can lead to improves the performance in DNNs [5] especially if the training data is limited in one of the tasks [18]. The studies with sequential learning of multiple tasks are rare. In work by Gutstein and colleagues [13] a multi-layer convolutional neural net was trained to recognize a set of digits. It was then shown that the same network, when trained on a second set of digits, achieved better performance when only the upper layers of the network – and not the entire network – were re-trained on the second task. Clearly the main problem with this kind of architecture is that learning a new task necessarily undermines those previously learned [11]. The problem of solving several tasks has been addressed in modular neural networks. These networks are able to detect different training patterns, corresponding to different tasks, and allocate different sub-networks to learn them [15,19].

In the current paper we explore an alternative approach to training DNNs, partly inspired by modular NNs. This approach will allow the network to learn a new task by exploiting previously learned features. At the same time, the learning procedure is such that training on a new task will not affect the performance of the network on the previously learned tasks. Our procedure consists of training an initial network on a task, and then, instead of copying the network and training it on a new task, adding blocks of neurons to the original network and learning the connections between the neurons in the original network and the neurons in the introduced blocks. We repeat this procedure on multiple tasks, showing that the final architecture is able to learn a new task by adding a rather small number of blocks of neurons and connection weights to the original network, when compared with a number of neurons and connection weights in a network which must learn the new task from scratch.

The paper is organized as follows: in the next section we describe the structure of the neural networks used and the techniques used to add blocks of neurons to such networks. In the Methods section we present the tasks on which we train different networks. We then provide details on the networks and on the learning algorithm. Finally, in the Results section we report the performance of the networks which learned from scratch and those which learned by adding blocks of neurons to existing networks.

1.1   Block-Modular Network Architecture

Consider a neural network with an input layer, multiple hidden layers, and an output layer such as the one presented in Figure 1a. Such a network, after being trained on a certain task, T_1, has definite values of weights and biases. We create a new network by adding neurons to each of the layers, including the output layer, as shown in Figure 1b. We refer to the added neurons as block neurons.

The first layer block neurons receive projections from the input only, and in this aspect are qualitatively similar to the original first-layer neurons. The second-layer hidden block neurons receive inputs from both the first layer original neurons and the first layer block neurons. This pattern is repeated for all hidden layers. The output layer block neurons receive inputs from the last hidden layer of both original and block neurons.

In the current study we explore the applicability of such an architecture to classification tasks. We use softmax units in the output layer, which we refer to as the classification layer. Note that the softmax is computed independently for the original and block classification layer neurons. We train the original network on a task T_1 and then train the weights of the block neurons on a task T_2. The resulting network is then able to perform both tasks T_1 and T_2. Note that such a network has two classifiers: one for T_1 (original classification neurons) and one for T_2 (block classification neurons).

We also consider variations of this type of block network. Most frequently, we add blocks to all layers, except for the first hidden layer, as illustrated in Figure 1c. In this situation the neural network does not receive raw input information; its only inputs are the outputs of the first hidden layer of the original network. As in the previous case (Figure 1b), only the weights to the block neurons are learned.

Additionally, we make use of several original networks, trained for different tasks. An example of adding blocks to a pair of original networks is presented in Figure 1d. Both original networks remain unaffected by the introduction of new neurons. The block neurons receive inputs from both of the original networks, and only connection weights to the block neurons are changed when learning a new task.

2   Methods

2.1   Tasks

The driving force behind our suggested architectures (e.g. Figure 1b-d) is the idea that the added neurons (being trained on task T_2) will re-use the capacities of the original neurons (trained on task T_1) whenever these capacities are relevant to the new task (T_2). Of course, such re-use will be most efficient when the tasks T_1 and T_2 have something in common.

In order to explore the possibility of the re-use of network capacities we designed several tasks which, to a human observer, involve the notions of line and angle. Specifically, we designed six tasks, as illustrated in Figure 2.

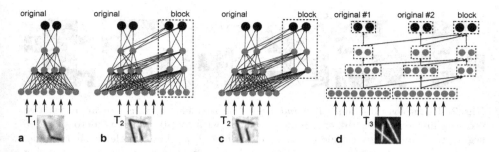

Fig. 1. (a) The architecture of the original network. We use a feedforward network with three hidden layers. (b) An additional block of neurons is added to each layer of the network. This is represented by the neurons in a dashed box. Each layer of the additional block is fully connected with the layer directly above and/or below within the block itself and with the layer below in the original network. (c) One of the architectures most used in the present work. The architecture is based on an original network trained on a first task T_1. A block of two hidden layers was added to the original network. The neurons were added at the second and third hidden layers only. The resulting architecture was trained on a second task T_2. (d) Adding blocks to two original networks. The dashed boxes indicate the layers of the two original networks and the blocks added. An arrow connecting two boxes indicates that all the neurons in the first box are connected to all the neurons in the second box.

In each task the stimuli were gray scale images, 32 x 32 pixels in size. Each image contained two to four line segments, each at least 13 pixels long (30% of the image diagonal). The distance between the end points of each line segment and every other line segment was at least 4 pixels (10% of the image diagonal). In order to obtain anti-aliased images, the lines were first generated on a grid three times larger (96 x 96). The images were then filtered with a Gaussian filter with sigma equal to 3 pixels, and downsampled to the final dimensions of 32 x 32 pixels.

We used the following conditions (see illustrations in Figure 2).

ang_crs: requires classifying the images into those containing an angle (between 20° and 160°) and a pair of crossing line segments (the crossing point must lay between 20% and 80% along each segment's length).

ang_crs_line: the same as *ang_crs*, but has an addition line segment crossing neither of the other line segments.

ang_tri_ln: distinguishes between images containing an angle (between 20° and 160°) and a triangle (with each angle between 20° and 160°); each image also contains a line segment crossing neither angle nor triangle.

blnt_shrp: requires classifying the images into those having blunt (between 100° and 160°) and those having sharp (between 20° and 80°) angles in them.

blnt_shrp_ln: the same as *blnt_shrp*, but has an additional line segment, crossing neither of the line segments forming the angle.

crs_ncrs: distinguishes between a pair of crossing and a pair of non-crossing lines (the crossing point must lay between 20% and 80% of each segment length).

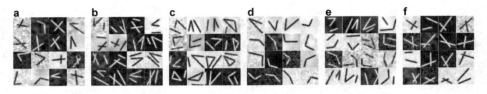

Fig. 2. Examples of stimuli: (a) *ang_crs* – line segments forming an angle vs. two crossing line segments; (b) *ang_crs_line* – same, with an additional non-crossing line segment; (c) *ang_tri_ln* – angle vs. triangle; (d) *blnt_shrp* – blunt angle vs. sharp angle; (e) *blnt_shrp_ln* – same, with a non-crossing line segment; (f) *crs_ncrs* – two crossing line segments vs. two non-crossing line segments.

Each stimulus was generated by randomly selecting an appropriate number of points and verifying that all conditions were satisfied. Each image was then combined with a random background, $I_{background}$. Four different types of random background were generated with four patterns changing with different velocities. In particular a grid of step size s with $s \in \{3, 7, 11, 15\}$ was superimposed onto a 32 x 32 image. The values of the pixels corresponding to the grid nodes were randomly drawn from a uniform distribution between 0.1 and 0.9. The values of the remaining image pixels were obtained by linearly interpolating the randomly drawn ones. We used positive and negative stimuli. The positive stimuli were defined by the formula:

$$I_{positive}(x, y) = \varepsilon I_{background} + (1 - \varepsilon)I_{stimulus},$$

where ε was randomly selected between 0.1 and 0.4 for each stimulus.

The negative stimuli were defined as $I_{negative} = 1 - I_{positive}$

For our experiments we generated 700,000 stimuli for each condition.

2.2 Neural Network Details

Original Neural Networks. The original neural networks had three hidden layers and one classification layer with two softmax neurons:

$$z_i = \frac{e^{x_i}}{e^{x_1} + e^{x_2}},$$

where z_i is the output of the i-th classification neuron ($i = 1, 2$), and x_i is the activation of the corresponding neuron.

Each of the hidden neurons had a rectified linear activation function:

$$y = x_+ = \begin{cases} x, & x \geq 0, \\ 0, & x < 0. \end{cases}$$

where y is the output of this neuron and x is its activation.

The activation of the i-th neuron at the k-th level (with $m^{(k)}$ neurons) was computed as a weighted sum of the outputs of neurons from the previous layer:

$$x_i^{(k)} = \sum_{j=1}^{m^{(k)}} w_{ji}^{(k)} y_j^{(k-1)} + b_i^{(k)}$$

Here $b_i^{(k)}$ is the bias.

For the first hidden layer, the activation was computed as a weighted sum of the inputs.

For every task, the original neural network had three hidden layers with 200, 100 and 50 neurons, respectively, and one classification layer with 2 neurons. The network received a vector of image pixels, which was of dimension 1024.

Block Neural Networks. We used several block neural network configurations. The block neural network always had two softmax output neurons. The structure of the hidden layers is described by a triplet of numbers. For example, the triplet 100-50-25 represents 100 neurons added to the first layer, 50 to the second, and 25 to the third (see Figure 1b). In many cases the block network only received inputs from the first layer of the original network (see Figure 1c), rather than from the stimuli directly. In these cases the first number in the triplet is zero, e.g. 0-50-50.

Cost Function. The network was trained to minimize a cost function J which combined three terms: the quality of the prediction J_1, the sparsity of the neurons' activation J_2, and the values of the weights J_3:

$$J = J_1 + \beta J_2 + \lambda J_3$$

The quality of the prediction was measured using the negative log-likelihood of the prediction given the data:

$$J_1 = -\frac{1}{N} \sum_{n=1}^{N} \log z_{i(n)}(x_n^{(0)})$$

where $x_n^{(0)}$ is the n-th training example, $z_{i(n)}$ is the output of the classifier corresponding to the correct answer.

The sparsity term of the cost function requires that each neuron in the hidden layer be active for ρN samples from the training data and silent otherwise. This measure was evaluated using the KL divergence [4]:

$$J_2 = \sum_{k=1}^{M-1} \sum_{j=1}^{m_k} \rho \log \frac{\rho}{\rho_j^{(k)}} + (1 - \rho) \log \frac{1 - \rho}{1 - \rho_j^{(k)}},$$

where $\rho_j^{(k)}$ is the average activation of the j-th neuron in the k-th layer, M is the number of layers (we do not apply this regularization to the classification layer), and $m^{(k)}$ is the number of neurons in the k-th layer.

The third term in the cost function is simply

$$J_3 = \sum_{k=1}^{M-1} \sum_{j=1}^{m_k} \sum_{i=1}^{m_{k-1}} \left(w_{ij}^{(k)} \right)^2.$$

The target sparseness ρ was set to 0.05, and the coefficient β was equal to 0.01. The weight-limiting coefficient λ was set to 0.0001.

Training. The weights of the k-th hidden level were initialized with random uniformly distributed values in the range $\pm\sqrt{6/(n^{(k-1)} + n^{(k)})}$, where $n^{(k)}$ is the number of neurons in the k-th layer and $n^{(0)}$ is equal to the number of inputs. This initialization has been recommended for networks with tanh activation function [9]; in the current study we used rectified linear units, but we kept the initalization range the same for compatibility with our pilot studies.

The total dataset was split into training (680,000 samples), validation (10,000 samples), and test (10,000 samples) datasets. The neural network was trained on the entire training dataset using mini-batch gradient descent learning with a batch size of 20. The initial update rate for the gradient descent was set to 0.01 and it decreased by a factor of 0.985 after every epoch. We used early stopping of the training process if the error on the validation dataset did not decrease after 5 epochs. The test score corresponding to the minimal validation error is presented as the performance of the network. Every condition was repeated 5 times, and the performance of the network was evaluated by the median value of the error.

Implementation. All code was written in python using Theano [2,3]. Source files are available online: https://github.com/feel-project/abstraction

3   Results

We first present the results for the original networks, which were trained from scratch on each task. Then we present the results for the networks produced by adding blocks to the original networks and training the added weights on new tasks. We compare the performance of such networks to the original ones. Given the large number of permutations of weights and tasks, all possible combinations were not studied. Instead we tried to examine a sample set that gives an understanding of the performance of block-modular networks.

Original Neural Networks. The performance of the original neural networks is presented in Table 1. These results show that adding a line to the image (e.g. Figure 2ad vs. 2be) made the task significantly more complicated. The first layer weights learned by the network are shown in Figure 3. Interestingly, visually, the weights are rather similar for *blnt_shrp* and *blnt_shrp_ln*, as well as for *ang_crs* and *ang_crs_ln*.

a *angle_crossing*   **b** *angle_crossing_line*   **c** *angle_triangle_line*

d *blunt_sharp*   **e** *blunt_sharp_line*   **f** *crossing_noncrossing*

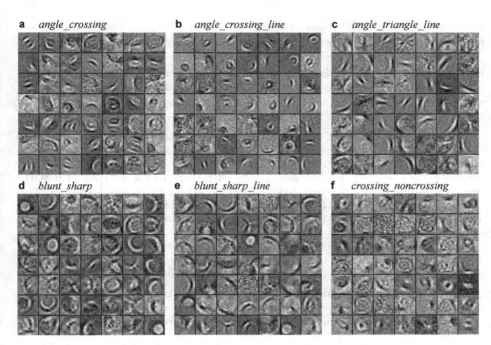

Fig. 3. Examples of weights. Weights (normalized) correspond to randomly selected first layer neurons of original networks trained to perform the corresponding tasks.

Table 1. The performance of the original 200-100-50 networks on different tasks

condition	performance
ang_crs	3.1 (2.4–3.6)
ang_crs_ln	7.8 (6.7–9.0)
ang_tri_ln	3.6 (3.0–4.1)
blnt_shrp	1.2 (0.8–1.4)
blnt_shrp_ln	4.1 (3.6–5.2)
crs_ncrs	1.6 (1.3–2.6)

The numbers correspond to median (min–max) percentage of misclassified examples.

Adding 0-50-50 and 0-100-50 Blocks to Original Networks. We tested whether the networks performing *ang_crs* and *blnt_shrp* tasks could be re-used to perform *ang_crs_ln* and *blnt_shrp_ln* tasks by adding blocks to the original networks. In spite of the apparent similarity of the first-layer weights for both tasks (see Figure 3), adding block networks resulted in rather poor performance on tasks with lines (compare results in Table 1 and 2). The results improved slightly when more neurons were added at the second hidden layer.

Re-using *ang_crs_ln* for *ang_crs* and *blnt_shrp_ln* for *blnt_shrp*, yielded comparable performance on *ang_crs* and improved performance on *blnt_shrp* compared to that of the corresponding original networks. This, however, could be partially attributed to the fact that the tasks are similar and, as such, re-using the original network on a new task in some sense increases the available training data.

We also attempted to re-use the network trained on one task to perform tasks that were more substantially different, such as *ang_crs_ln* for *ang_tri_ln*. In spite of certain similarities in the weights in both tasks (Figure 3bc), the performance was rather poor compared to that of the corresponding original network.

Table 2. Adding blocks to original networks

condition	0-50-50	0-100-50	50-50-50	100-50-50
ang_crs_ln (*ang_crs*)	9.0 (8.5–9.4)	8.2 (7.8–8.5)	**7.7 (7.4–8.7)**	**7.5 (7.0–9.0)**
blnt_shrp_ln (*blnt_shrp*)	6.6 (6.0–7.1)	5.9 (5.2-6.2)	5.5 (5.2–5.9)	5.0 (4.7–5.3)
ang_crs (*ang_crs_ln*)	**2.7 (2.4–3.3)**	**2.7 (2.4–3.2)**	**2.9 (2.5–3.3)**	**2.8 (2.6–3.5)**
blnt_shrp (*blnt_shrp_ln*)	**0.8 (0.7–1.0)**	**0.7 (0.6–0.8)**	**0.8 (0.7–0.9)**	**0.7 (0.7–0.9)**
ang_tri_ln (*ang_crs_ln*)	6.7 (6.3–7.6)	5.6 (5.0–6.6)	4.8 (4.2–5.5)	4.3 (3.8–4.9)
blnt_shrp_ln (*ang_crs_ln*)	7.8 (7.4–8.2)	6.6 (6.2–7.1)	5.0 (4.6–5.1)	4.3 (4.3–4.6)
ang_crs_ln (*ang_tri_ln*)	12.4 (11.8–13.4)	10.4 (10.1–11.2)	8.9 (8.7–10.2)	**7.8 (7.4–8.6)**
ang_crs_ln (*blnt_shrp_ln*)	13.0 (12.8–13.4)	11.7 (10.4–12.8)	10.0 (9.5–10.4)	8.8 (8.5–9.6)

The name of the task on which the block neurons were trained is followed by the name of the task on which the original networks were trained (in brackets). The cases when the block networks outperform the original ones are marked with bold font.

Adding 50-50-50 and 100-50-50 Blocks to Original Networks. One potential reason for the poor performance of the block networks discussed above is that they did not receive stimuli as inputs, but only the outputs of the first layer of the original neural networks. To test this hypothesis we added blocks which had 50 or 100 neurons in the first layer in addition to neurons in the other layers. These results are presented in Table 2. This modification yielded substantial improvement, with some of the conditions performing better than the original networks.

Adding Blocks to Pairs of Networks. The previous observation suggested that having a richer first hidden layer may improve the performance of block networks. We further explored this by creating block networks based on pairs of original networks (see Figure 1d). The results are presented in Table 3. Clearly, block networks with an empty first layer performed comparably to or better than the original networks. It must be noted that the block networks learned an order of magnitude fewer parameters. Each original network had approximately 10^5 parameters, while each block network had about $2 \cdot 10^4$ which were not shared with original networks (for 0-50-50 blocks). Adding neurons to the first hidden layer provided additional improvement, as shown in the last column of Table 3.

Adding Blocks to Triplets of Networks. The results of adding blocks to triplets of networks are presented in Table 4. They show that having three original networks improved the performance of the block network. For example,

Table 3. Adding blocks to pairs of original networks

condition	0-50-50	0-100-50	50-50-50
ang_crs_ln (ang_tri_ln+crs_ncrs)	8.8 (8.5–9.0)	8.3 (7.3–8.4)	**7.7 (7.6–7.8)**
ang_crs_ln (ang_tri_ln+blnt_shrp_ln)	8.3 (7.9–9.2)	**7.5 (7.3–8.4)**	7.9 (7.7–8.5)
blnt_shrp (ang_tri_ln+crs_ncrs)	**1.0 (0.9–1.2)**	0.9 (0.8–1.0)	0.9 (0.8–1.0)
blnt_shrp (ang_tri_ln+ang_crs_ln)	**0.8 (0.7–0.9)**	0.7 (0.7–0.8)	0.8 (0.6–0.9)
blnt_shrp (ang_tri_ln+blnt_shrp_ln)	**0.6 (0.6–0.7)**	0.6 (0.6–0.8)	0.6 (0.5–0.7)
blnt_shrp_ln (ang_tri_ln+ang_crs_ln)	**4.1 (3.9–4.4)**	3.8 (3.2–4.1)	3.6 (3.4–4.0)
blnt_shrp_ln (ang_tri_ln+crs_ncrs)	5.0 (4.3–5.1)	4.4 (4.1–4.6)	**4.0 (3.9–4.3)**
ang_crs (ang_tri_ln+blnt_shrp_ln)	3.2 (3.0–3.9)	**3.1 (2.8–3.3)**	3.1 (2.6–3.6)
ang_crs (ang_tri_ln+ang_crs_ln)	**2.2 (2.1–2.6)**	2.2 (1.8–2.7)	2.3 (2.0–2.7)

using three original networks (ang_tri_ln+crs_ncrs+blnt_shrp_ln) a block which is trained for ang_crs outperformed a similar network learned from scratch. Interestingly, for blnt_shrp_ln, combining multiple original networks (two or three) and adding a block with an empty first layer was more efficient than adding a block with 100 neurons in the first hidden layer to a single original network (compare Tables 1–4). The respective number of trained weights was also smaller in the former case.

Table 4. Adding blocks to triplets of original networks

condition	0-50-50
ang_crs (ang_tri_ln+crs_ncrs+blnt_shrp)	3.2 (3.1–3.5)
ang_crs (ang_tri_ln+ang_crs_ln+crs_ncrs)	**2.3 (2.1–2.7)**
ang_crs (ang_tri_ln+crs_ncrs+blnt_shrp_ln)	**2.9 (2.7–3.2)**
ang_crs_ln (ang_tri_ln+crs_ncrs+blnt_shrp_ln)	**7.8 (7.3–8.2)**
ang_crs_ln (ang_tri_ln+crs_ncrs+blnt_shrp)	8.4 (8.2–8.6)
blnt_shrp (ang_crs+ang_tri_ln+crs_ncrs)	**0.7 (0.7–0.8)**
blnt_shrp (ang_crs_ln+ang_tri_ln+crs_ncrs)	**0.7 (0.7–0.8)**
blnt_shrp_ln (ang_crs_ln+ang_tri_ln+crs_ncrs)	**4.0 (3.5–4.1)**
blnt_shrp_ln (ang_crs+ang_tri_ln+crs_ncrs)	**3.9 (3.4–4.2)**

4 Conclusions

Our results suggest that adding blocks to neural networks can be an efficient way to obtain networks capable of performing several tasks. In certain cases such composite networks outperform the networks trained from scratch, while having almost one order of magnitude fewer weights. Also, we observed a smaller range of variability in performance for block networks when compared to the original networks. Adding fewer new weights offers a significant gain in computational time if the stimuli are to be tested for all tasks. The performance of the block networks can be partly explained by the pre-training effect; the original network trained on its task can be considered as a pre-trained sub-network for a new task, to which the added block is trained. Another reason for better performance could be the increase in available training data. Since the tasks share some similarities, certain features could be learned from both original and new datasets. Pilot computational experiments show that the improvement offered by the block networks becomes more significant when the amount of training data is reduced for both the original and new tasks. This ability to re-use features from different datasets may be beneficial for practical problems where it is often difficult to collect a large amount of training data for a specific task.

Acknowledgments. This work was funded by the European Research Council (FP 7 Program) ERC Advanced Grant "FEEL" to KO'R.

References

1. Anderson, M.L.: Neural reuse: A fundamental organizational principle of the brain. Behavioral and Srain Sciences **33**(04), 245–266 (2010)
2. Bastien, F., Lamblin, P., Pascanu, R., Bergstra, J., Goodfellow, I.J., Bergeron, A., Bouchard, N., Bengio, Y.: Theano: new features and speed improvements. In: Deep Learning and Unsupervised Feature Learning NIPS 2012 Workshop (2012)
3. Bergstra, J., Breuleux, O., Bastien, F., Lamblin, P., Pascanu, R., Desjardins, G., Turian, J., Warde-Farley, D., Bengio, Y.: Theano: a CPU and GPU math expression compiler. In: Proceedings of the Python for Scientific Computing Conference (SciPy), June 2010. oral Presentation
4. Bradley, D.M., Bagnell, J.A.: Differential sparse coding (2008)
5. Collobert, R., Weston, J.: A unified architecture for natural language processing: Deep neural networks with multitask learning. In: Proceedings of the 25th International Conference on Machine Learning, pp. 160–167. ACM (2008)
6. Dehaene, S., Cohen, L.: Cultural recycling of cortical maps. Neuron **56**(2), 384–398 (2007)
7. Erhan, D., Manzagol, P.A., Bengio, Y., Bengio, S., Vincent, P.: The difficulty of training deep architectures and the effect of unsupervised pre-training. In: International Conference on Srtificial Intelligence and Statistics, pp. 153–160 (2009)
8. Girosi, F., Jones, M., Poggio, T.: Regularization theory and neural networks architectures. Neural Computation **7**(2), 219–269 (1995)
9. Glorot, X., Bengio, Y.: Understanding the difficulty of training deep feedforward neural networks. In: International Conference on Artificial Intelligence and Statistics, pp. 249–256 (2010)
10. Glorot, X., Bordes, A., Bengio, Y.: Deep sparse rectifier networks. In: Proceedings of the 14th International Conference on Artificial Intelligence and Statistics, JMLR W&CP, vol. 15, pp. 315–323 (2011)
11. Goodfellow, I.J., Mirza, M., Xiao, D., Courville, A., Bengio, Y.: An empirical investigation of catastrophic forgetting in gradient-based neural networks. arXiv:1312.6211 (2013)
12. Graves, A., Mohamed, A.R., Hinton, G.: Speech recognition with deep recurrent neural networks. In: 2013 IEEE International Conference on Acoustics, Speech and Signal Processing (ICASSP), pp. 6645–6649. IEEE (2013)

13. Gutstein, S., Fuentes, O., Freudenthal, E.: Knowledge transfer in deep convolutional neural nets. International Journal on Artificial Intelligence Tools **17**(03), 555–567 (2008)
14. Hinton, G., Osindero, S., Teh, Y.W.: A fast learning algorithm for deep belief nets. Neural Computation **18**(7), 1527–1554 (2006)
15. Jordan, M.I., Jacobs, R.A.: A competitive modular connectionist architecture. In: Advances in Neural Information Processing Systems, pp. 767–773 (1991)
16. LeCun, Y., Bengio, Y.: Convolutional networks for images, speech, and time series. In: The Handbook of Brain Theory and Neural Networks, vol. 3361, p. 310 (1995)
17. Lee, H., Grosse, R., Ranganath, R., Ng, A.Y.: Convolutional deep belief networks for scalable unsupervised learning of hierarchical representations. In: Proceedings of the 26th Annual International Conference on Machine Learning, pp. 609–616. ACM (2009)
18. Liu, X., Gao, J., He, X., Deng, L., Duh, K., Wang, Y.Y.: Representation learning using multi-task deep neural networks for semantic classification and information retrieval
19. Lu, B.L., Ito, M.: Task decomposition and module combination based on class relations: a modular neural network for pattern classification. IEEE Transactions on Neural Networks **10**(5), 1244–1256 (1999)
20. McCloskey, M., Cohen, N.J.: Catastrophic interference in connectionist networks: The sequential learning problem. Psychology of Learning and Motivation **24**, 109–165 (1989)
21. Salakhutdinov, R., Hinton, G.: An efficient learning procedure for deep boltzmann machines. Neural Computation **24**(8), 1967–2006 (2012)
22. Schmidhuber, J.: Deep learning in neural networks: An overview. Neural Networks **61**, 85–117 (2015)
23. Sutskever, I., Vinyals, O., Le, Q.V.: Sequence to sequence learning with neural networks. In: Advances in Neural Information Processing Systems, pp. 3104–3112 (2014)

A Top-Down Approach for a Synthetic Autobiographical Memory System

Andreas Damianou[1,2]([✉]), Carl Henrik Ek[3], Luke Boorman[1],
Neil D. Lawrence[2], and Tony J. Prescott[1]

[1] Sheffield Centre for Robotics (SCentRo), University of Sheffield,
Sheffield S10 2TN, UK
[2] Department of Computer Science, University of Sheffield, Sheffield S1 4DP, UK
[3] CVAP Lab, KTH, Stockholm, Sweden
andreas.damianou@shef.ac.uk

Abstract. Autobiographical memory (AM) refers to the organisation of
one's experience into a coherent narrative. The exact neural mechanisms
responsible for the manifestation of AM in humans are unknown. On the
other hand, the field of psychology has provided us with useful understanding about the *functionality* of a bio-inspired synthetic AM (SAM)
system, in a higher level of description. This paper is concerned with
a top-down approach to SAM, where known components and organisation guide the architecture but the unknown details of each module are
abstracted. By using Bayesian latent variable models we obtain a transparent SAM system with which we can interact in a structured way. This
allows us to reveal the properties of specific sub-modules and map them
to functionality observed in biological systems. The top-down approach
can cope well with the high performance requirements of a bio-inspired
cognitive system. This is demonstrated in experiments using faces data.

Keywords: Synthetic autobiographical memory · Hippocampus ·
Robotics · Deep Gaussian process · MRD

1 Introduction and Motivation

Autobiographical memory (AM) refers to the ability to recollect episodes from
one's experience, relying on organising events and context (semantics) into a
narrative. A key task for the intersection of cognitive robotics and biomimetics
is to create *Synthetic* Autobiographical Memory (SAM) systems inspired by
the current understanding of brain physiology. However, our current knowledge
of neural connection formation and activity does not go as far as to enable
understanding of how high-level structures, such as semantics, emerge. On the
other hand, experimental psychology has provided us with useful understanding
about how the *functionality* of AM is organised in "modules" and upon which
requirements. Consequently, for practical purposes *top-down* approaches to SAM
are developed. These approaches ensure that the known AM requirements are
respected and focus on implementing the functional (rather than physiological)

© Springer International Publishing Switzerland 2015
S.P. Wilson et al. (Eds.): Living Machines 2015, LNAI 9222, pp. 280–292, 2015.
DOI: 10.1007/978-3-319-22979-9_28

AM modules and their interconnections. Any known physiological information is sought to be incorporated in the low-level components of the approach, which implement specific tasks (e.g. pattern completion).

Machine learning (ML) has been used in the past to implement the higher levels of top-down approaches to synthetic physiological systems. However, if ML is used as a "black-box" and purely out of necessity to improve functionality, then consistency in the overall framework is lost. In other words, it is no longer clear whether certain properties of the artificial system emerge due to the low-level bio-inspired components or due to the high-level ML methods. This hinders subsequent evaluation of hypotheses about the system. This paper studies the requirements for enabling top-down approaches to SAM, in a way that functionality is improved (making it usable in a real robotic system) without sacrificing transparency. Subsequently, an existing ML approach, referred to as deep Gaussian processes (deep GPs) [1], is studied here and linked to the SAM framework as a key ingredient of the top-down SAM approach we present. Finally, the results section demonstrates how deep GPs differ from many ML "black-box" approaches (in the context of biomimetics) by enabling uncertainty quantification and intuitive interaction with the model. It is shown that by using a deep GP, not only do we obtain high-level SAM functionality, but we can also recognise individual low-level components of the model as proxies of known subfunctions of AM, such as compression or pattern completion.

2   Requirements for a Top-Down SAM System

In [2] the authors recognise the following requirements enabling a biologically inspired SAM system to match the functionality of AM:

- **Compression** of perceived signals in a way that *information* emerges from raw data (recognition of patterns).
- **Pattern separation** to encode different contexts separately and ensure that weak but important signals are not overwhelmed by stronger ones.
- **Pattern completion** for reconstructing events from partial information.

In [2], unitary coherent perception was also added to the requirements of the generic SAM system, although it was highlighted that this selection is suboptimal in practice and choices stemming from Bayesian brain hypotheses [3] are an alternative. In practice, the unitary coherent perception is usually used as a means of avoiding the costly computational requirements associated with full Bayesian inference. However, in a top-down approach that targets improved functionality and transparency, the Bayesian component is vital. This is because the model components are only functional approximations to the true underlying physiological system and, therefore, the Bayesian quantification of uncertainty in our (unavoidably imperfect) representation is important. Deterministic inference is also desired to achieve transparency. In [2] this requirement was fulfilled indirectly through a deterministic Gibbs sampler variant within the unitary coherent perception framework. Here, the intractable full Bayesian inference is approximated with a deterministic *variational* approximation. Based on the above, two more requirements are recognised particularly for the top-down SAM approach:

- **Deterministic inference:** The same starting conditions and parameters should always result in the same outcome.
- **Encoding consistency:** Supervised, semi-supervised and unsupervised learning should be handled using the same representations of memory

3   A Top-Down Approach to SAM

3.1   Properties

The top-down SAM approach needs to be robust for embedding in a real robotic system while, at the same time, fulfilling the requirements specified in the previous section, so that connections with expert knowledge from the domain of psychology and neuroscience can be established. For this reason, the approach proposed in this paper comes from a family of models which is:

- **Bayesian probabilistic:** Random variables encode the observables (signal perceived by the robotic agent) and unknowns/latents (internal representation of memories). In a Bayesian framework, prior knowledge (e.g. mammals have legs) can be combined with observations (e.g. past memories) to define *posterior distributions* (e.g. probability that an observed animal is a mammal). The posterior uncertainty is important in practice. For example, an agent operating in a dangerous environment can avoid actions associated with high uncertainty.
- **Latent variable method:** Latent variables correspond to the unknowns in the modelling scenario, and can be inferred from the data. In the approach proposed in this paper, the latent variables are taken to be much simpler and compact (low-dimensional) compared to the high-dimensional observables. By further associating the simple latent variables with the complex, noisy observations we can implement the compression and pattern separation requirements.
- **Generative:** The latent variables are associated with the observables via a *generative mapping function*. This encodes our assumption that the highly compressed latent variables (encoding memory events) should be able to generate fantasy data in the observable domain.
- **Non-parametric:** A non-parametric approach allows to define the generative mapping without having to make crude assumptions about its nature. Memory models built upon artificial neural networks [4] often assume parametric activation functions. Instead, the approach suggested in this paper is based on Gaussian processes (GPs) [5], which learn functional relationships from data with minimal assumptions. For example, fig. 1 shows two functions (posterior processes) learned with the same GP in the presence of different data.

3.2   Gaussian Processes

To formalise the above, let us denote the noisy observables as $\mathbf{y}$ and the latent variables as $\mathbf{x}$. A mapping function f relates latent points to observables, so that $\mathbf{y} = f(\mathbf{x}) + \epsilon$; here, ϵ denotes Gaussian noise, which leads to the Gaussian likelihood $p(\mathbf{y}|\mathbf{f})$, where $\mathbf{f} \triangleq f(\mathbf{x})$ is the collection of mapping function values (noise-free versions of $\mathbf{y}$). To obtain a non-parametric mapping, we place a Gaussian process prior distribution on the mapping function f. Given finite inputs $\mathbf{x}$, this leads to a prior $p(\mathbf{f}|\mathbf{x})$ which is also Gaussian. Thanks to the analytic Bayesian framework, the mapping function values $\mathbf{f}$ can actually be integrated out, to obtain the marginal likelihood (which is again Gaussian):

$$p(\mathbf{y}|\mathbf{x}) = \int_f p(\mathbf{y}|\mathbf{f})p(\mathbf{f}|\mathbf{x}) = \int_f \mathcal{N}\left(\mathbf{y}|\mathbf{f}, \sigma_\epsilon^2\right) \mathcal{N}\left(\mathbf{f}|\mathbf{0}, \mathbf{K}\right). \tag{1}$$

Contrast this with the parametric Bayesian regression approach, which assumes that the mapping function has a fixed form $\mathbf{w}\phi(\mathbf{x})$ parametrised by $\mathbf{w}$, and marginal likelihood $p(\mathbf{y}|\mathbf{x}) = \int_{\mathbf{w}} \mathcal{N}\left(\mathbf{y}|\mathbf{w}\phi(\mathbf{x}), \sigma_\epsilon^2\right) \mathcal{N}\left(\mathbf{w}|\mathbf{0}, \sigma_w^2\right)$. As can be seen, rather than assuming a fixed parametric form and placing a prior on the parameters, in the Gaussian process framework we place the prior directly on the mapping function. More details on GPs can be found in [5].

Notice that so far we have assumed that the latent points $\mathbf{x}$ are known. In the approach taken in this paper, the latent variables are unknown. The Gaussian process latent variable model [6] handles the unknown latent variables by placing a prior on them and optimising the new objective $p(\mathbf{y}|\mathbf{x})p(\mathbf{x})$ also w.r.t $\mathbf{x}$.

Inducing Point Representations. In a non-parametric model, the learned quantities (posteriors over the mapping function process and over the latent points) are conditioned on the training data. Being a generative model, this already achieves compression; indeed, the posterior Gaussian process is able to interpolate between the training points for every point in the input domain. This is demonstrated in fig. 1, where the blue solid line gives an estimate for the function in the whole line of real numbers. To achieve further compression via a fixed set of points, one can use inducing point representations. In this case, all the information in the set of pairs $\{\mathbf{x}, \mathbf{f}\}$ is compressed through a smaller set of pairs $\{\mathbf{z}, \mathbf{u}\}$ that remains constant in size as the training dataset grows. This is achieved by replacing the original Gaussian prior $p(\mathbf{f}|\mathbf{x})$ with a *sparse* prior $p(\mathbf{f}|\mathbf{u}, \mathbf{x}, \mathbf{z})$ which depends on the inducing points. This is demonstrated in fig. 1.

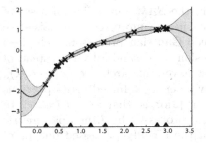
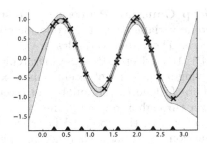

Fig. 1. The same GP prior combined with two different sets of observations (black x's) to obtain the posterior processes (blue solid line) and posterior uncertainty (shaded area). Triangles along the x axis indicate the position of the inducing inputs, $\mathbf{z}$.

3.3   Top-Down SAM Architecture

In this paper the focus is on developing a top-down SAM approach the functionality of which is inspired by physiology. In particular, deep probabilistic approaches have been linked in the past with functionality that can be observed in the human brain [7]. For example, human vision involves a hierarchy of visual cortices (V1 – V5) which process visual signals in a progressive manner [8]. Evidence that this kind of hierarchical learning is performed in the brain areas associated with memory is not yet existent; the exact localisation and functionality of the billions of neural interconnections associated with memory is yet unknown and, therefore, making a neural simulation is currently impossible. Instead, the top-down approach to SAM works on a higher level and seeks to simulate the organisation of *functionality* (rather than the organisation of neurons) into hierarchically structured (sub)modules. As such, the designed architecture involves high level modules grouped into a core and a set which extends beyond the core of the SAM, as can be seen in fig. 2. These modules are explained below.

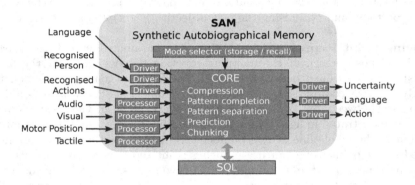

Fig. 2. The developed Synthetic Autobiographical Memory (SAM) system. This paper focuses on the core module, which is implemented using deep Gaussian processes [1].

Deep Gaussian Processes. To start with, the *SAM core* relies on a set of latent variables which encode memory events in a compressed and noise-free space. The latent variables are part of a deep Gaussian process model (deep GP) [1]. A deep GP is the hierarchical extension to a standard GP. Instead of having a single set of latent points, $\mathbf{x}$, we now have a hierarchy: $\mathbf{x}_1, \mathbf{x}_2, \cdots, \mathbf{x}_L$, where L denotes the number of layers. Every layer $\mathbf{x}_\ell$ is linked to its previous layer through a mapping function f_ℓ with a GP prior, so that $\mathbf{x}_\ell = f_\ell(\mathbf{x}_{\ell-1})$. In other words, the observed layer is successively processed by L non-parametric functions, so that each function operates on the already processed output of the previous in the hierarchy. Importantly, the intermediate latent spaces are available for inspection, revealing intuitive features. Fig. 3 demonstrates this process for the task of recognising handwritten digits; samples can be drawn from the latent spaces in each layer, to reveal features that successively encode more abstract information due to the successive processing.

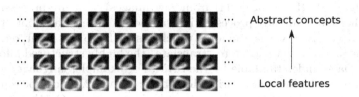

Abstract concepts

Local features

Fig. 3. Samples from the hierarchy of the latent spaces discovered for a collection of handwritten digit images. The lowest layer encodes very local features (e.g. if the circle in a zero is closed or not), but successive processing allows the top layer to encode abstract information, such as general characteristics of different digits.

Inference in deep Gaussian processes is not analytically tractable straight-forwardly. This is because the model is required to marginalise over the latent representation $\mathbf{x}$ so as to obtain a posterior over it through the Bayes rule: $p(\mathbf{x}|\mathbf{y}) = \frac{p(\mathbf{y}|\mathbf{x})p(\mathbf{x})}{\int_{\mathbf{x}} p(\mathbf{y}|\mathbf{x})p(\mathbf{x})}$. The intractability in the denominator requires approximate solutions. In deep GPs, inference proceeds through a variational framework. In contrast to stochastic inference approaches like sampling and MCMC, a variational inference approach is deterministic [9]. This means that the same starting conditions (e.g. initialisation of parameters) will always result in the same approximation of the quantities of interest (posterior distributions, inducing points, latent representation). Therefore, the requirement for deterministic inference is fulfilled when deep GPs are used within the SAM core.

Multiple Modalities. In the SAM framework, multiple representations of the same event must be taken into account consistently. Consider e.g. the separate signals (visual, audio) associated with a memory from watching a theatrical play. However, there is some commonality (specific scenes are associated with specific sounds). Formally, assume that $\mathbf{y}^{(1)}, \mathbf{y}^{(2)}, \cdots, \mathbf{y}^{(M)}$ represent the segmentation of the observables into M different modalities. These can be accounted for in the latent variable framework by maintaining for all modalities a single Q-dimensional latent space (i.e. a single representation compressed in Q features). Subsequently, learning which parts of the whole latent space are relevant for which modality is achieved by optimising a set of *relevance weights* $\mathbf{w}^{(m)} \in \Re^Q$ for each modality. If, for example, $w_5^{(2)}, w_5^{(3)} \neq 0$, this means that the latent dimension 5 encodes information for views 2 and 3, thus avoiding redundancy and achieving compression. This idea was developed in [10]. This approach can be embedded in the deep GP framework of the SAM core, and is demonstrated in the next section.

Modules Outside of the SAM Core. As can be seen in fig. 2, the top-down SAM architecture links inputs to the SAM core through drivers and processors.

A processor is a module that allows raw stimuli to be pre-processed. This is not a compulsory requirement, since the SAM core achieves compression through the deep GP. However, this allows the usage of sophisticated feature extraction methods (e.g. SURF features [11] from raw images) as a pre-processing step. On the other hand, already processed information is incorporated into the SAM through drivers. A driver is a module which is specific to the input/output it is responsible for, and is employed to "translate" the highly structured information into a language understandable by the SAM. For example, language/actions in the context of social interaction with the robotic agent [12,13] can be represented as a set of frequencies of terms from a pre-built dictionary. On the other hand, recalling an action involves another driver which translates the memory into a series of motor commands. The current implementation accompanying this paper only contains the appropriate processors and drivers for visual stimuli.

Tasks and Consistency. The SAM model operates in a supervised as well as unsupervised scenario. In the supervised scenario, the latent representation learned by the internal deep GP model is guided through additional input information, $\mathbf{t}$, expressed through a prior, that is, $p(\mathbf{x}|\mathbf{t})$. For example, $\mathbf{t}$ might be the time-stamp of a particular frame associated with a visual stimulus and would force the latent representation to form a smooth time-series . Unsupervised learning corresponds to the scenario where the latent representation is learned in an unconstrained manner, only from the data. In this case, the latent representation is assigned a fairly uninformative prior $p(\mathbf{x}) = \mathcal{N}(\mathbf{x}|\mathbf{0}, \mathbf{I})$. Semi-supervised learning can also be handled by following a data-imputation approach [14]. Overall, the suggested SAM approach satisfies the encoding consistency requirement.

Related Work. Related work involves methods inspired by low-level neural structures (e.g. temporal codes [17]) and "traditional" bio-inspired but high-level deep learning approaches, e.g. convolutional neural networks [18]. In particular, the latter have achieved remarkable results in many vision tasks, but do not always provide transparent/manipulable representations as required for a SAM.

4   Demonstration Using Human Faces Data

This section demonstrates a selection of representative results from the top-down SAM system. Additional results can be seen at https://youtu.be/rIPX3CIOhKY. Software for reproducing the results is available at: http://git.io/vTBMt.

The employed machine learning approach has been previously demonstrated quantitatively in classification tasks [10]. In contrast, the focus of this paper is to demonstrate the emergence of ABM functionality through our interpretation of the model components, in particular the latent and the inducing points which compress the perceptual information from multiple event modalities.

4.1 Face Rotations Experiment

For the first demonstration, images of 3 subjects were captured using a standard mobile phone's low-resolution camera (140×140 pixels per image). For each subject, multiple images were recorded under different rotations of the face with respect to the camera. To demonstrate the method in imperfect data, the images were collected while the camera was held by hand and the subjects were rotating on their own; no further processing was made on the data (e.g. no cropping). 250 images of each subject were stored in three matrices, $(\mathbf{Y}^{(1)}, \mathbf{Y}^{(2)}, \mathbf{Y}^{(3)})$, so that each triplet of rows, $(\mathbf{y}_n^{(1)}, \mathbf{y}_n^{(2)}, \mathbf{y}_n^{(3)})$, corresponds to the three faces under a similar rotation. These data were presented to a top-down SAM system. The Gaussian processes used 45 inducing inputs (see end of section 3.2) and a $Q = 20-$dimensional latent space, much smaller than the original output dimensionality ($140 \times 140 = 19,600$). The inducing points and latent space together achieve strong compression and chunking of the original signal.

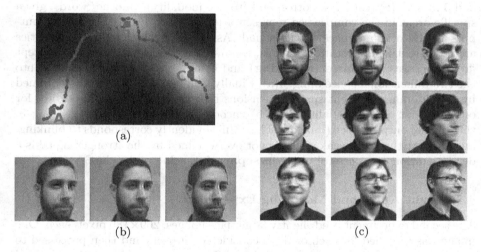

Fig. 4. Results from the rotating faces experiment. Fig. (a) depicts the projection of the internal SAM representation on the two dimensions shared for all modalities. Background intensities correspond to the variance of the distribution for predicting $\mathbf{y}$ from $\mathbf{x}$. Fig. (c) shows the corresponding outputs generated by conditioning on the selected locations shown as A, B, C in fig. (a). Fig. (b) shows outputs generated by conditioning on latent locations which encode weak but highly descriptive signal.

The internal SAM representation of the raw visual signal was obtained after a training phase, required for tuning the parameters of the core's model. Next, this internal representation was investigated, to understand the way in which weak and strong signals are chunked and how low-level (subtle, e.g. blinking) and high-level (e.g. face characteristics) concepts emerge automatically. Out of the 20 features used to compress the observed signal, fig. 4(a) depicts two (one plotted versus the other) which were deemed important by all three modalities. Red crosses correspond to latent points $\mathbf{x}_n$ which, in turn, correspond to observations in each of the three modalities, $(\mathbf{y}_n^{(1)}, \mathbf{y}_n^{(2)}, \mathbf{y}_n^{(3)})$. The SAM system successfully recovers a semi-circular shape, corresponding to the rotation of the faces from 1 to 180 degrees. Importantly, this information is encoded once for all three modalities and is learned automatically from the given dataset. Blue circles represent inducing points which further compress the internal representation. The optimisation procedure spreads the inducing points nicely along the latent path. Notice that the compressed discrete representation obtained with the latent and inducing points can be conditioned upon to perform inference for *any* possible area in the latent space. This demonstrates the real power of the model in terms of performing **compression** and **pattern completion**. Specifically, the background intensity of fig. 4(a) represents the variance associated with the distribution, where bright intensities correspond to areas where the SAM model is confident in its predictions. To demonstrate this, predictions were made for the depicted points A, B, C, obtaining the outputs in column 1, 2 and 3 in fig. 4(c) (each row corresponds to one modality). In other words, given specific areas in the compressed representation, the SAM model generated outputs in the original space of the signal. As can be seen, all three modalities are consistent in the rotation, that is, the memory associated with the concept "rotation" was recognised (**chunking**) and compressed for all three faces into the two-dimensional space of fig. 4(a). Finally, fig. 4(b) depicts outputs obtained by conditioning on latent space dimensions that were a) deemed important for only the first of the modalities and b) encoded signal which was very weak in the original output space (images). This signal evidently corresponds to blinking, and the fact that this weak signal is not overwhelmed by the stronger signal is a demonstration of successful **pattern separation** achieved by the SAM system.

4.2   Light Angle and Morphing Experiment

The second experiment used slightly larger face images, 269×186 pixels each. One image was recorded for each of the 6 considered subjects and then processed to simulate illumination under one out of 42 different light source positions around the face, similarly to [15]. The total set of 6×42 images was then split into two groups (modalities) $\mathbf{Y}^{(1)}$ and $\mathbf{Y}^{(2)}$ where: a) each row $\mathbf{y}_n^{(m)}$ corresponds to an image in modality m; b) group $\mathbf{Y}^{(1)}$ contains images only from subjects 1,2,3 and $\mathbf{Y}^{(2)}$ only from 4,5,6; c) the rows of the two matrices were aligned, so that $\mathbf{y}_n^{(1)}$ and $\mathbf{y}_n^{(2)}$ are matched in the angle of the light source (but the ordering of the subjects is arbitrary, i.e. not matched). In other words, the two modalities were created such that the illumination condition is a common signal and the face identity is signal private (specific) to each of the two modalities. The challenge is for the SAM system to compress the data by also encapsulating this information.

Fig. 5 depicts the results. Fig. 5(a) depicts a bar graph of the relevance weights corresponding to each of the two modalities and are of the same dimensionality as the latent space ($Q = 14$). Thick/blue bars correspond to modality

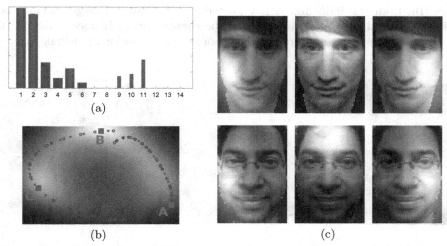

Fig. 5. Second experiment: (a) depicts the relevance weights optimised for the two modalities; (b) shows the compressed representation of the common signal; (c) depicts the outputs obtained by conditioning on the locations depicted as A,B,C in (b).

1 and red/thinner bars to modality 2. Dimensions 1,2,5 encode information for both modalities. To verify this, fig. 5(b) plots dimension 1 versus 2. As can be seen, the SAM system successfully mapped the information for the light source position from the original $50,034$ dimensional space to only two dimensions. Indeed, by conditioning on the latent space locations indicated by A, B and C we obtain the outputs in columns 1,2,3 respectively of fig. 5(c), which depict faces under the same illumination condition with no other features changed.

On the other hand, one can also perform the same procedure for dimensions taken from the sets $(3, 4, 6)$ or $(9, 10, 11)$ which, from fig. 5(a), is obvious that they are relevant to only one of the modalities. In particular, fig. 6(a) depicts the corresponding internal representation of dimensions 3, 4, which encode signal relevant only to the first modality. This plot demonstrates the successful **chunking** achieved by the SAM system, since the three clusters which were automatically discovered correspond to each of the three faces contained in modality 1. Again, the inducing points (blue circles) nicely cover each cluster and do not fall in between clusters, thereby using the full compressing capacity of the model. To verify these intuitions, 8 locations were selected from this space (the rest of the dimensions were kept fixed) along the path depicted as a black dotted line in fig. 6(a). Notice that this scenario is different than the procedure followed so far in the experiments, in that the selected latent locations are interpolations between those corresponding to training points. This procedure results in the images depicted in fig. 6(c). The morphing effect verifies the intuition that this part of the compressed space is responsible for encoding face characteristics, and manifests **pattern completion** by producing *novel outputs*. Finally, fig. 6(b) depicts the "fantasy" memories used as a compressed basis, computed as eigenfaces [16] from the inducing output posterior. For example, large variance is observed around the eye area (reflecting the changes in the images of fig. 6(c)) for the two male faces, and around the eyebrows and nose for the female face.

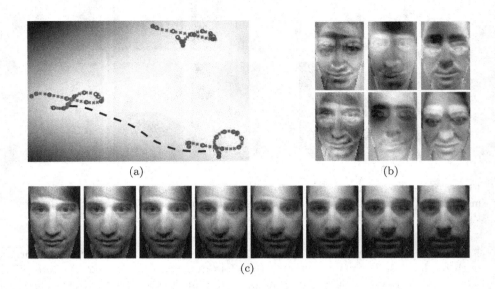

(a)   (b)

(c)

Fig. 6. Morphing effect obtained by sampling outside of the training compressed representations' region (black line, fig. (a)) to obtain novel outputs (c). The top row of fig. (b) depicts three of the "fantasy" memories used as a compressed basis (inducing outputs' eigenfaces). Bottom row is just the color-inverted version of the top row. This reveals that the plotted eigenvectors define the tangent direction for interpolating between faces.

5   Discussion and Future Work

Previous work in psychology and bio-inspired robotics has resulted in extracting high-level descriptions of the organisation of modules in a SAM system. This paper discussed a top-down approach to SAM where the aforementioned high-level descriptions are guiding the system architecture while the specific (unknown) details of each component are abstracted. This is made possible through a flexible representation of memories, based on Bayesian latent variable models [1,10] which filter all functionality through a smaller set of learned variables (inducing points). Experiments on "noisy", real-world faces data revealed the robustness of the method in learning powerful representations of the data (simulating *memory formation*), while structured interaction with the framework allowed for examining its properties with respect to requirements for a biologically inspired SAM.

Future work will aim at integrating the SAM system into the cognitive component of a robot, such as the iCub. Although preliminary experiments with regards to handling auditory streams have been performed, a more complete solution which handles heterogeneous sensory data is planned for the future. Finally, we will work towards achieving stronger connections with biology, by incorporating more detailed bio-inspired structure in the lowest levels of the top-down architecture (e.g. through priors and constraints on the inducing points).

Acknowledgments. This research was funded by the European research project EU FP7-ICT (Project Ref 612139 "WYSIWY"). We thank the following for participating in the creation of the faces dataset: F. Yousefi, A. Saul, J. Gonzalez, M. Zwiessele, M. A. Rahman, Z. Dai, R. A. Pacheco. We also thank our colleagues at Sheffield Robotics and, in particular, Uriel Martinez-Hernandez.

References

1. Damianou, A., Lawrence, N.: Deep Gaussian processes. Proceedings of the 16th International Workshop on A.I. and Statistics (AISTATS), pp. 207–215 (2013)
2. Evans, M.H., Fox, C.W., Prescott, T.J.: Machines learning - towards a new synthetic autobiographical memory. In: Duff, A., Lepora, N.F., Mura, A., Prescott, T.J., Verschure, P.F.M.J. (eds.) Living Machines 2014. LNCS, vol. 8608, pp. 84–96. Springer, Heidelberg (2014)
3. Pouget, A., Beck, J.M., Ma, W.J., Latham, P.E.: Probabilistic brains: knowns and unknowns. Nature Neuroscience **16**(9), 1170–1178 (2013)
4. Rojas, R.: Neural networks: a systematic introduction. Springer Science & Business Media (1996)
5. Rasmussen, C.E., Williams, C.K.I.: Gaussian processes for machine learning. MIT Press, Cambridge (2006)
6. Lawrence, N.D.: Probabilistic non-linear principal component analysis with Gaussian process latent variable models. Journal of Machine Learning Research **6**, 1783–1816 (2005)
7. Bengio, Y., LeCun, Y.: Tutorial on Learning Deep Architectures. Videlectures.net, June 2009. http://videolectures.net/icml09_bengio_lecun_tldar/
8. Nielsen, M. A.: Neural Networks and Deep Learning. Determination Press (2015)
9. Bishop, C. M.: Pattern Recognition and Machine Learning. Springer-Verlag (2006). ISBN 0387310738
10. Damianou, A., Ek, C.H., Titsias, M., Lawrence, N.: Manifold relevance determination. Proceedings of the 29th International Conference on Machine Learning (ICML), pp. 145–152. omnipress, New York (2012)
11. Bay, H., Tuytelaars, T., Van Gool, L.: SURF: speeded up robust features. In: Leonardis, A., Bischof, H., Pinz, A. (eds.) ECCV 2006, Part I. LNCS, vol. 3951, pp. 404–417. Springer, Heidelberg (2006)
12. Pointeau, G., Petit, M., Dominey, P.F.: Embodied simulation based on autobiographical memory. In: Lepora, N.F., Mura, A., Krapp, H.G., Verschure, P.F.M.J., Prescott, T.J. (eds.) Living Machines 2013. LNCS, vol. 8064, pp. 240–250. Springer, Heidelberg (2013)
13. Pointeau, G., Petit, M., Dominey, P.F.: Successive Developmental Levels of Autobiographical Memory for Learning Through Social Interaction. IEEE Transactions on Autonomous Mental Development **6**, 200–212 (2014)

14. Damianou, A., Titsias, M., Lawrence, N.: Variational inference for uncertainty on the inputs of Gaussian process models. arXiv preprint, arXiv:1409.2287 (2014)
15. Georghiades, A.S., Belhumeur, P.N., Kriegman, D.J.: From few to many: Illumination cone models for face recognition under variable lighting and pose. IEEE Trans. Pattern Anal. Mach. Intelligence **23**(6) (2001)
16. Turk, M., Pentland, A.: Eigenfaces for recognition. Journal of Cognitive Neuroscience **3**(1), 71–86 (1991)
17. Luvizotto, A., Renn-Costa, C., Verschure, P.: A Framework for Mobile Robot Navigation Using a Temporal Population Code. Biomimetic & Biohybrid Systems (2012)
18. Fukushima, K.: Neocognitron: A self-organizing neural network model for a mechanism of pattern recognition unaffected by shift in position. Biological Cybernetics **36**(4), 193–202 (1980)

Crowdseeding: A Novel Approach
for Designing Bioinspired Machines

Mark D. Wagy[✉] and Josh C. Bongard

Computer Science Department, University of Vermont, Burlington, VT, USA
{mwagy,jbongard}@uvm.edu

Abstract. Crowdsourcing is a popular technique for distributing tasks to a group of anonymous workers over the web. Similarly, crowdseeding is any mechanism that extracts knowledge from the crowd, and then uses that knowledge to guide an automated process. Here we demonstrate a method that automatically distills features from a set of robot body plans designed by the crowd, and then uses those features to guide the automated design of robot body plans and controllers. This approach outperforms past work in which one feature was detected and distilled manually. This provides evidence that the crowd collectively possesses intuitions about the biomechanical advantages of certain body plans; we hypothesize that these intuitions derive from their experiences with biological organisms.

Keywords: Evolutionary robotics · Biomimetics · Crowdsourcing · Data mining · Machine learning

1 Introduction

Embodied cognition [18] is the view that the intelligent behavior of an animal or human is influenced not just by its nervous system but also by its body plan. Engineers that produce bio-inspired designs are (implicitly or explicitly) adhering to this view. Wings on an airplane strongly suggest the influence of the morphology of birds. Many robots that have been developed are either humanoid in form [20,9,11,17] or resemble other animals, such as the canine *Bigdog* [19], the serpentine *OT-4* [3] or the chelonian *Aqua* [7]. Some biomimetic designs result from an explicit aim to exploit some desirable property of the behavior or feature of animals or of their environment. But in some cases the tendency to bias search toward specific design spaces could be considered an implicit tendency of collective human design behavior.

In [28], web participants collectively designed robot bodies. It was found that, among the successful designs, there was an overrepresentation of symmetric designs. This suggests that contributors have a strong proclivity for locomoting agents that resemble animals found in the physical world. Symmetry and single-component designs were the most explicit biases in robot bodies created by the crowd. However, there could be other, less obvious, traits and relations between

© Springer International Publishing Switzerland 2015
S.P. Wilson et al. (Eds.): Living Machines 2015, LNAI 9222, pp. 293–303, 2015.
DOI: 10.1007/978-3-319-22979-9_29

them that could be exploited. In this study, we describe a novel methodology for extracting latent information from a group of participants in a crowdsourced experiment. Using the design of robot bodies and control as our domain, we use symbolic regression to discover implicit relations between properties of robot designs and an objective, in our case rapid forward locomotion. We then use these latent relationships to seed a new robot design process. This methodology represents a new mode of collaborative interaction between a crowd of human designers and a machine learning algorithm.

2   Related Work

There has been of late a great deal of interest in finding methods to utilize the collective intelligence of crowds to solve complex problems [23,8,10,12]. The field of crowdsourcing has moved from being a convenient way to source simple, separable human intelligence tasks [15] that cannot yet be completed by machine intelligence [26] to being used to combine the efforts of individuals to solve larger problems that might not be amenable to reduction into a divide-and-conquer strategy [16,31,1,25].

Use of human participants in evolutionary algorithms has been common for both selection of individuals in a population [6,24] and in introducing variation in the evolutionary population [13]. But each of these examples involved direct user participation in the evolutionary search. Crowdsourcing in evolutionary robotics has been used to guide search for better robots and robot control [4,2,27]; but in these studies, the use of human intuition was used to actively guide search during the experiment rather than to distill out useful features in a crowdsourced study that were then incorporated into a separate search algorithm. In [28], features were extracted to seed the fitness objective of an evolutionary algorithm. But instead of using automated methods to find latent variables and their relations that contribute to performance – as is the case in this study – obvious characteristics of robot bodies favored by the crowd were distilled manually to seed the objective function.

3   Methods

We conducted an experiment in three stages. In the first stage of the experiment (Section 3.1), we deployed a web-based tool in which participants in a crowdsourced study designed robots collaboratively. In the second stage (Section 3.2), we used symbolic regression via genetic programming [14] to identify a novel relationship between the attributes of the crowdsourced robot designs and the distance that robots were able to move. In the third stage (Section 3.3), we used the relationship found through symbolic regression to augment a stochastic search process from a single-objective search problem to that of a multiobjective search.

3.1 First Stage: Crowdsourcing

We deployed a web-based tool that allowed participants to rapidly design robots using a simple grid-based drawing panel (Figure 1). We invited participants to design robots by recruiting them through the online forum *Reddit* (*www.reddit.com*). Participation was unpaid and voluntary. Participants were only given the instructions to design a robot that could "move farther". They were asked to connect dots to design a robot and click *GO*. They were told that their robot will learn new behaviors if they run the same robot multiple times.

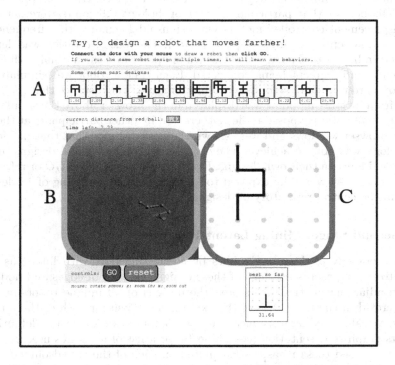

Fig. 1. Screenshot of web-based robot design tool. Users designed robot bodies by "connecting the dots" (Panel **C**). When they clicked "GO", they would see their robot move in a simulation (Panel **B**). They were shown a random sampling of 13 robots designed by others (Panel **A**).

Users connected dots in the design panel by clicking on a dot and dragging their mouse to another dot, which would form a line. Only lines between adjacent dots were allowed. When they clicked *GO*, each line was translated into a $0.1 \times 0.1 \times 0.1$ meter rigid segment in the simulation panel and each dot that was adjacent to a line was translated into a $0.2 \times 0.2 \times 0.2$ rigid cube. Cubes were connected to segments by a one-degree-of-freedom hinge joint. Robots were simulated in the three-dimensional physics simulation engine, Ammo.js[1], and were rendered using Web3D[2] in the participant's web browser.

Each of the robot's joints was assigned to move either in-phase ($0°$) or out-of-phase ($180°$) with other joints. When a particular robot design was run for the first time, it was assigned its own hill-climber algorithm on a central repository, which would determine whether each of its joints would be in-phase or out-of-phase. If the same or another user repeated that design for a run in the simulation, the joint configuration would either be repeated or a joint could be randomly mutated from in-phase to out-of-phase or vice-versa at a 0.1 mutation rate. Thus every time a participant clicked *GO*, it was contributing one run to the hillclimber for that particular robot morphology. All joints were actuated with displacement-controlled motors via a sinusoidal signal with a frequency of 1.5 Hz, and sweeping an angle of $[-45°, +45°]$. The axis of rotation was defined to be perpindicular to the normal of the ground plane and the normal of the faces of the cube and segment being connected. Each time a robot was simulated, it was allowed to run for 15 seconds of simulation time. The distance that the robot moved from its starting point was displayed in the browser above the simulation.

Participants were exposed to designs created by other participants at the top of their browser windows and were shown the best distance that that particular morphology was able to achieve. They were free to ignore these designs or use them as guidance in their own designs. Every time a user clicked *GO* or refreshed the web site, they would be exposed to another random sampling of 13 designs stored in the central repository of designs.

3.2   Second Stage: Mining Latent Features

The methodology of designing robots by connecting dots with lines was conducive to storing representations of these designs as a set of edges and nodes in a graph adjacency matrix. We stored the designs of all unique robots created by the crowd in the first stage of the experiment. Then for each of these robot designs, we calculated network measures (see [5] for network measure definitions) using its graph representation (see Table 1a for a list of measures used).

We then used these measures calculated on each of the crowdsourced robot morphologies as explanatory variables in a symbolic regression model. The best distance that that morphology was able to move was the response variable used to train the model. We used the *Eureqa* [21] symbolic regression package to build the models. The set of functions allowed are listed in Table 1b.

Eureqa is a multiobjective search tool: it maintains non-dominated solutions along a Pareto front with respect to either minimum error or minimal solution complexity [22]. We ran ten trials of symbolic regression until they reached 100% convergence and at least 80% *maturity* (proportion of time since the last improvement to a solution, as defined in the *Eureqa* tool) to find expressions that related distance to the set of explanatory network measures. We then selected

---

[1]   www.github.com/kripken/ammo.js

[2]   www.khronos.org/webgl

Table 1. (a) Variables used as explanatory variables and (b) functions allowed for use in symbolic regression expressions

(a) Variables

Variable	Symbol
Maximal matching	M_{max}
Number of connected components	c
Maximum degree	D_{max}
Minimum degree	D_{min}
Number of limbs	L
Not a chordal graph	C
Symmetry	S
Number of segments	G
Average Degree	D_{ave}
Average degree connectivity	D_{con}
Average clustering	T_{ave}

(b) Functions

Function Name	Symbol
Addition	$+$
Subtraction	$-$
Multiplication	$\cdot$
Division	$/$
Logistic function	$\sigma()$
Indicator function	$I()$
Cosine	$cos()$
Sine	$sin()$
Tangent	$tan()$
Exponential	$exp()$
Natural Logarithm	$log()$
Power	x^y
Square root	$\sqrt{}$
Gaussian	$G()$
Less than or equal	$\leq$
Greater than or equal	$\geq$

the best solution of the ten trials with the minimum fit error value (and thus maximal complexity) for use in the third stage of the study.

3.3   Third Stage: Seeding the Objective

In the third stage of our study, we used the best expression found using symbolic regression as an additional objective in a genetic algorithm to evolve new robots and their controllers. The genomes in this genetic algorithm consisted of bitstrings grouped into sets of three. Each of these groups corresponded to a potential lines position between adjacent dots in the same 5×5 grid that users in the crowdsourced portion of the study were given to design robots, resulting in a bitstring that consisted of $2 \times 5 \times 4 \times 3 = 120$ bits. Within each group of three bits, the center bit determined whether there was a segment present at that location. If a segment was present (a 1 in the center bit location) the bits to the left and right of the center bit determined whether the motor at the left/top or right/bottom hinge joint would move in-phase or out-of-phase with other motors depending on whether the line was vertical or horizontal (as illustrated in Figure 2).

We compared the best distances achieved by two different, *seeded* evolutionary algorithms: both had the primary objective of maximizing the distance that a robot was able to move in a physics simulation, but they were given additional objectives derived from the preferences of the crowdsourced portion of the study. The two treatments differed in this second, seeded, objective. In the first

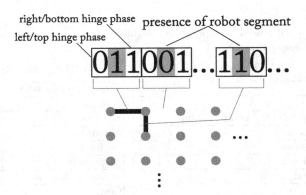

Fig. 2. Genotype to phenotype translation

(control) treatment, the secondary objective was to maximize the ratio of symmetry to number of connected components that made up the robot body. This seeded objective was shown [28] to outperform an evolutionary algorithm with the single objective of maximizing distance that a robot could move. The second (experimental) treatment consisted of using the single best expression found by symbolic regression described in Section 3.2 to minimize the error between the expression and distance that a particular design with those parameters could achieve.

We performed 100 independent trials for each of the control and experimental treatments. Each evolutionary process ran for 100 generations with a population of 50 individuals. We used bit-flip mutation at a rate of 0.1 as well as uniform crossover at a rate of 0.1.

4   Results

In the first, crowdsourced, stage of the experiment, a total of 947 volunteers participated in robot design. They created 2292 unique designs. On average, 5.63 designs were created by each participant with a long-tailed distribution pattern of participation as is common in crowdsourced studies [29,30]. Drawings of the best five designs can be seen in Figure 3.

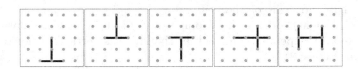

Fig. 3. Top five unique designs in stage one (left to right, top to bottom). Note that some designs that are considered unique are morphologically similar but at different grid coordinates.

Examples of the ten best symbolic regression expressions found in the second stage of the experiment can be seen in Table 2. The expression with the minimum error value (marked with *) was used in the obtaining distance values for the experimental treatment reported in Figure 4.

Table 2. Lowest error solutions found in the ten symbolic regression trials. See Table 1 for variable definitions and the list of functions used in the expressions. Constants are rounded to the nearest thousandth for brevity. Expression 1 was used as the second objective in the experimental treatment.

	Solution	Error
1*	$I(D_{max} \leq L) \cdot min(G(D_{conn}), S) + 0.177 min(c + 6.800 D_{conn}, D_{max}^{4.510 - max(S, log(L))})$	0.771
2	$0.0.041 \cdot c \cdot S \cdot I(L \geq 0.044 \cdot c \cdot D_{max}) + min(c^{0.350}, D_{max})$	0.773
3	$min(D_{max} \cdot \sigma(M_{max}), 2.187) - 0.085 \cdot M_{max}$	0.804
4	$min(3.150, D_{max} + 0.170 \cdot M_{max}) - 0.144 \cdot M_{max}$	0.806
5	$\sqrt{D_{max}} - (0.003 \cdot N \cdot c^2)^S \cdot cos(D_{max})$	0.786
6	$min(3.054, D_{max} \cdot min(M_{max}, 1.325)) - 0.125 \cdot M_{max}$	0.807
7	$4.374 \cdot c \cdot min(0.033, D_{con}) + I(D_{max} \geq 3) + min(S, I(D_{max} \leq L))$	0.781
8	$0.053 \cdot c \cdot S \cdot I(L \geq 0.044 \cdot c \cdot D_{max}) + min(log \, c + S, D_{max})$	0.773
9	$D_{con} + S \cdot I(L \geq D_{max}) + 0.196 \cdot min(3.471^{D_{max}}, c) - I(L \geq G)$	0.771
10	$2.819 \cdot \sigma(S) \cdot min(3.180, D_{max}) - 0.141 \cdot M_{max} \cdot min(M_{max}, D_{max}) - 2.960 \cdot I(3.887 - L \geq M_{max})$	0.779

The best distances achieved in the 100 independent runs of the control and experimental treatments in the third stage of the experiment are compared in Figure 4. The best five designs in this third stage of the study are illustrated in Figure 5.

5   Discussion

The introduction of an additional objective using a symbolic regression expression significantly outperformed the inclusion of manually-derived additional objectives to the fitness in the genetic algorithm as reported in [28] ($p < 0.0001$; independent two group t-test), which was itself shown to significantly outperform the use of the primary objective alone to guide search. We are comparing performance of these methods on the basis of the primary, distance, objective alone. Since adding additional objectives decreases selection pressure on the primary objective, this finding is surprising and encouraging for the method that we introduce here.

The error of the symbolic regression expressions are fairly high. This indicates that either the measures used in the expressions or the expressions themselves were not particularly accurate models of distance traveled by the robots designed by the crowd. Despite this fact, we still achieved a significant improvement by

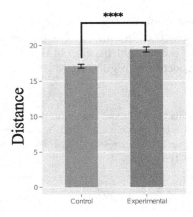

Fig. 4. Distance comparison between treatments. The control case (pink) shows the distance achieved by using the primary distance objective as well as a combined symmetry/number of components objective in 100 independent runs of a genetic algorithm. The experimental case (blue) shows the distance achieved by using the primary distance objective as well as the expression obtained using symbolic regression in 100 independent runs of the genetic algorithm described in Section 3.3. Error bars indicate the 95% confidence intervals around the mean distance value. The difference is significant at the $p < 0.0001$ level (independent two group t-test).

Fig. 5. Top five unique designs in stage three (left to right, top to bottom)

using the expressions as a additional objective in our experiment. This suggests that we might be able to achieve even better results if we were to explore other objective measures of the robot bodies that were discovered by the crowd or if we were to improve the expressions found by symbolic regression by running evolution longer or fine-tuning genetic programming parameters.

The expressions obtained using symbolic regression are quite complex. We took the expressions from the Pareto Front that had the lowest error of all solutions on the front, which necessarily means that these solutions would be the most complex expressions. Despite their complexity, we have seen empirically that they were able to outperform more simple multiobjective fitness values. This suggests that there is some valuable latent information in the symbolic regression expressions distilled from the crowd despite the difficulty of parsing out exactly what the expressions entail at an intuitive level.

6 Conclusion and Future Work

In this work, we demonstrated a new process for automating the distillation of a crowd's design preferences into an additional objective used to seed an evolutionary algorithm. This objective was in addition to the primary objective of maximizing the distance that a robot was able to move from an initial fixed point in a simulation. We showed that this process significantly outperformed a manual method for extraction of information from crowdsourced studies, which itself was shown to outperform the use of the primary objective alone in the evolutionary algorithm.

In future work, we will investigate whether this technique can be incorporated into other domains. In an era of large amounts of data – much of it generated by humans – there is potential for mining features and expressions from that data that can then be used to improve training of machine learning algorithms, in the development of better design requirements for engineering projects or in seeding algorithms for creative algorithmic work.

In the present study, past designs that were presented to participants consisted of a random sampling of 13 historical designs by other users. In future work, we will investigate whether systematic methods of choosing the historical designs that are shown to participants has an impact on the overall design decisions of the crowd.

Additionally, we will investigate methods to obtain intuitive information from the symbolic regression expressions themselves. In the present study, expressions were directly copied into a multiobjective search. However, there may be kernels of information in the expressions that could be distilled out to further to reduce complexity and drill down to the terms within the expression that are the basis for this method's ability to outperform manually-derived objectives.

Acknowledgments. This work was supported by the National Science Foundation (NSF) under project DGE-1144388. This work was also supported by the NSF under grant PECASE-0953837, and by the Defense Advanced Research Projects Agency (DARPA) under grants W911NF-11-1-0076 and FA8650-11-1-7155.

References

1. Ambati, V., Vogel, S., Carbonell, J.G.: Active learning and crowd-sourcing for machine translation. In: LREC, vol. 1, p. 2. Citeseer (2010)
2. Bongard, J.C., Hornby, G.S.: Combining fitness-based search and user modeling in evolutionary robotics. In: Proceeding of the Fifteenth Annual Conference on Genetic and Evolutionary Computation Conference, pp. 159–166. ACM (2013)
3. Borenstein, J., Hansen, M., Borrell, A.: The omnitread ot-4 serpentine robot design and performance. Journal of Field Robotics 24(7), 601–621 (2007)
4. Celis, S., Hornby, G.S., Bongard, J.: Avoiding local optima with user demonstrations and low-level control. In: 2013 IEEE Congress on Evolutionary Computation (CEC), pp. 3403–3410. IEEE (2013)
5. Cohen, R., Havlin, S.: Complex networks: structure, robustness and function. Cambridge University Press (2010)

6. Dawkins, R.: The blind watchmaker: Why the evidence of evolution reveals a universe without design. WW Norton & Company (1996)
7. Dudek, G., Jenkin, M., Prahacs, C., Hogue, A., Sattar, J., Giguere, P., German, A., Liu, H., Saunderson, S., Ripsman, A., et al.: A visually guided swimming robot. In: 2005 IEEE/RSJ International Conference on Intelligent Robots and Systems (IROS 2005), pp. 3604–3609. IEEE (2005)
8. Gao, H., Barbier, G., Goolsby, R., Zeng, D.: Harnessing the crowdsourcing power of social media for disaster relief. Tech. rep, DTIC Document (2011)
9. Hirai, K., Hirose, M., Haikawa, Y., Takenaka, T.: The development of honda humanoid robot. In: Proceedings of 1998 IEEE International Conference on Robotics and Automation, vol. 2, pp. 1321–1326. IEEE (1998)
10. Kamar, E., Hacker, S., Horvitz, E.: Combining human and machine intelligence in large-scale crowdsourcing. In: Proceedings of the 11th International Conference on Autonomous Agents and Multiagent Systems, vol. 1. pp. 467–474. International Foundation for Autonomous Agents and Multiagent Systems (2012)
11. Kaneko, K., Harada, K., Kanehiro, F., Miyamori, G., Akachi, K.: Humanoid robot hrp-3. In: IEEE/RSJ International Conference on Intelligent Robots and Systems, IROS 2008, pp. 2471–2478. IEEE (2008)
12. Khatib, F., Cooper, S., Tyka, M.D., Xu, K., Makedon, I., Popović, Z., Baker, D., Players, F.: Algorithm discovery by protein folding game players. Proceedings of the National Academy of Sciences 108(47), 18949–18953 (2011)
13. Kosorukoff, A.: Human based genetic algorithm. In: 2001 IEEE International Conference on Systems, Man, and Cybernetics, vol. 5, pp. 3464–3469. IEEE (2001)
14. Koza, J.R.: Genetic programming: on the programming of computers by means of natural selection, vol. 1. MIT press (1992)
15. Law, E., Ahn, L.V.: Human computation. Synthesis Lectures on Artificial Intelligence and Machine Learning 5(3), 1–121 (2011)
16. Laws, F., Scheible, C., Schütze, H.: Active learning with amazon mechanical turk. In: Proceedings of the Conference on Empirical Methods in Natural Language Processing, pp. 1546–1556. Association for Computational Linguistics (2011)
17. Lohmeier, S., Buschmann, T., Ulbrich, H.: Humanoid robot lola. In: IEEE International Conference on Robotics and Automation, ICRA 2009, pp. 775–780. IEEE (2009)
18. Pfeifer, R., Bongard, J.: How the body shapes the way we think: a new view of intelligence. MIT Press (2006)
19. Raibert, M., Blankespoor, K., Nelson, G., Playter, R., et al.: Bigdog, the rough-terrain quadruped robot. In: Proceedings of the 17th World Congress, vol. 17, pp. 10822–10825 (2008)
20. Sanders, D., Kusuda, Y.: Toyota's violin-playing robot. Industrial Robot: An International Journal 35(6), 504–506 (2008)
21. Schmidt, M., Lipson, H.: Eureqa (version 0.98 beta)[software]. nutonian inc., cambridge, ma (2014)
22. Schmidt, M., Lipson, H.: Distilling free-form natural laws from experimental data. Science 324(5923), 81–85 (2009)
23. Surowiecki, J., Silverman, M.P., et al.: The wisdom of crowds. American Journal of Physics 75(2), 190–192 (2007)
24. Takagi, H.: Interactive evolutionary computation: Fusion of the capabilities of ec optimization and human evaluation. Proceedings of the IEEE 89(9), 1275–1296 (2001)
25. Von Ahn, L.: Games with a purpose. Computer 39(6), 92–94 (2006)

26. Von Ahn, L.: Human computation. In: 46th ACM/IEEE Design Automation Conference, DAC 2009, pp. 418–419. IEEE (2009)
27. Wagy, M., Bongard, J.: Collective design of robot locomotion. In: ALIFE 14: The Fourteenth Conference on the Synthesis and Simulation of Living Systems, vol. 14, pp. 138–145
28. Wagy, M.D., Bongard, J.: Crowdseeding robot design. In: Proceedings of the 24th Annual Conference on Genetic and Evolutionary Computation. ACM (forthcoming)
29. Wilkinson, D.M.: Strong regularities in online peer production. In: Proceedings of the 9th ACM Conference on Electronic Commerce, pp. 302–309. ACM (2008)
30. Wu, F., Wilkinson, D.M., Huberman, B.A.: Feedback loops of attention in peer production. In: International Conference on Computational Science and Engineering, CSE 2009, vol. 4, pp. 409–415. IEEE (2009)
31. Yan, Y., Fung, G.M., Rosales, R., Dy, J.G.: Active learning from crowds. In: Proceedings of the 28th International Conference on Machine Learning (ICML 2011), pp. 1161–1168 (2011)

Studying the Coupled Learning of Procedural and Declarative Knowledge in Cognitive Robotics

Rodrigo Salgado, Francisco Bellas[✉], and Richard J. Duro

Integrated Group for Engineering Research, Universidade da Coruña, A Coruña, Spain
{rodrigo.salgado,francisco.bellas,richard}@udc.es
http://www.gii.udc.es

Abstract. Procedural and Declarative knowledge play a key role in cognitive architectures for robots. These types of architectures use the human brain as inspiration to design control structures that allow robots to be fully autonomous, in the sense that their development depends only on their own experience in the environment. The two main components that make up cognitive architectures are models (prediction) and action-selection structures (decision). Models represent the declarative knowledge the robot acquires during its lifetime. On the other hand, action-selection structures represent the procedural knowledge, and its autonomous acquisition depends on the quality of the models that are being learned concurrently. The coupled learning of models and action-selection structures is a key aspect in robot development, and it has been rarely studied in the field. This work aims to start filling this gap by analyzing how these concurrent learning processes affect each other using an evolutionary-based cognitive architecture, the Multilevel Darwinist Brain, in a simulated robotic experiment.

1   Introduction

Cognitive robotics is a research field that uses the human brain and its main functionalities as inspiration to design control architectures that allow the robots to be fully autonomous, in the sense that their development depends only on their own experience [2][11]. Consequently, the main objective of researchers in this field is designing and implementing developmental programs that can run during the robots' lifetime instead of creating complex control systems for a set of predefined tasks. These developmental programs make up what is called a cognitive architecture, which controls the acquisition of knowledge, its consolidation and use [13].

Some of the most relevant cognitive functions for robots such as anticipation and planning may be achieved by internally simulating the robot's interaction with the environment through its actions and their consequences [6]. These capabilities, however, require having good enough and updated models of the world and of the robot itself that can be used within processes that have been formalized, for example, in terms of the Simulation Hypothesis [10]. Any type of cognitive mechanism that aims to do this type of cognitive process must have models that can adapt to changing circumstances and that need to be remembered and generalized in order to be reused. Also, for a cognitive mechanism to be meaningful in real life robotic scenarios,

S.P. Wilson et al. (Eds.): Living Machines 2015, LNAI 9222, pp. 304–315, 2015.
DOI: 10.1007/978-3-319-22979-9_30

it must be able to generate actions and sequences of actions in real time so that robots can interact with the environment in a timely manner and perform their tasks. Thus, two main types of knowledge need to be acquired, manipulated and preserved by cognitive architectures: procedural (non-conscious) and declarative (conscious). These play a key role in cognitive architectures for robots [7].

Declarative knowledge is usually presented in the form of models of the external and internal world of the robot, or explicit data related to specific occurrences of things. These models are prediction structures that allow the robot to infer how the world or itself are going to behave in a given situation. Procedural knowledge is related with how to do things "without thinking" (skills), and it is typically represented in robotics through action-selection structures (like behaviors or policies). This knowledge is crucial, and the faster it can be applied when the correct circumstances arise, the better the chance of "survival".

Most existing cognitive architectures include declarative and procedural knowledge using very different types of representations, from symbolic [1][12] to distributed [3][9] ones. What is common to all of them is that, except in the simplified case of establishing a predefined set of models or action-selection structures, the learning of both types of knowledge must be carried out concurrently. The problem that arises is how to perform this coupled learning, that is, how to learn proper action-selection structures from models while the models are also being learned. In this paper, we are going to study this problem for the specific case of the Multilevel Darwinist Brain (MDB) cognitive architecture, but with the aim of providing results that can be generalized to other representations.

The rest of the paper is structured as follows. Section 2 explains in detail how knowledge is acquired and represented within the MDB. Section 3 presents the simulation experiment that has been setup to study the coupled learning of declarative and procedural knowledge using the AIBO robot. Section 4 devoted to the presentation and discussion of the main results of this study. Finally, in section 5 we extract some relevant conclusions trying to generalize them for other cognitive architectures.

2 Knowledge Acquisition in the Multilevel Darwinist Brain

The Multilevel Darwinist Brain (MDB) is an evolutionary cognitive architecture that has been under development since the early 2000s and that allows an artificial agent, real or simulated robot, to learn from its experience in a dynamic and unknown environment in order to fulfill its motivations or objectives [4]. Its main features have been tested in real robotic experiments in the past [4][5]. This architecture is based in 4 basic elements:

- *Episodes*: real world samples that are obtained from the robot sensors and actuators after applying an action. Formally:
    - o *External perception* (*e*): sensory information the robot is capable of acquiring through its sensors from the environment in which it operates.
    - o *Internal perception* or proprioception (*i*): sensory information provided by the internal sensors (i.e. battery, cpu load)
    - o *Satisfaction* (*s*): degree of fulfillment of the motivations
    - o *Action* (*a*): actuation performed by the robot in the environment

The episodes are made up of the sensorial information plus the applied action in time t and the sensorial information derived from the execution of the action in time $t+1$, that is:

$$episode=\{e(t), i(t), a(t), e(t+1), i(t+1), s(t+1)\}$$

- *Models*: prediction structures that conform the *declarative knowledge* a robot has. Three types of models are considered:

  o *World Model* (*W*): function that predicts the external perception in t+1 from the perception in the current instant t after applying an action:

  $$e(t+1) = W(e(t), a(t))$$

  o *Internal Model* (*I*): function that predicts the internal perception in t+1 from the perception in the current instant t after applying an action:

  $$i(t+1) = I(i(t), a(t))$$

  o *Satisfaction Model* (*S*): function that predicts the satisfaction in t+1 from the internal and external perceptions in t+1:

  $$s(t+1) = S(e(t+1), i(t+1))$$

  $$s(t+1) = S(W(e(t), a(t)), I(i(t), a(t)))$$

  These models W, I and S must be learned during the robot's lifetime using the episodes as the known samples that must be predicted and generalized.

- *Behaviors*: they make up the *procedural knowledge* the robot has so they can be assimilated to learned skills. A behavior (*B*) is a decision structure that provides the action to be applied in time t from the sensorial inputs in t:

  $$a(t)=B(e(t),i(t))$$

  As in the case of models, the behaviors should not be predefined in a cognitive mechanism like the MDB, and they must also be learned on line.

- *Memories*: two main kinds of memory elements are considered: Long-Term (LTM) and Short-Term (STM) [5]. The STM is made up of a set of *model and behavior memories*, which contain models and behaviors that are relevant to the current task, and a set of *episodic buffers (EB)* that store the last episodes experienced by the robot. The EB has a very limited capacity according to the temporal nature of the STM. The LTM is made up of a *declarative memory (DM)*, which contains the models that have been consolidated due to their significance and reliability, and a *procedural memory (PM)* that stores the consolidated behaviors. It is important to highlight that both DM and PM are filled during the robot lifetime in a completely autonomous fashion, that is, no predefined library of models or behaviors are considered.

This work will focus on knowledge acquisition within the MDB, that is, model and behavior learning. Details about the overall operation of the architecture can be found in [4][5]. It must be pointed out here that these learning processes are performed in the MDB through evolutionary algorithms, being a key aspect of this architecture. How to achieve an on-line efficient operation using evolution has been one of the main research challenges in this line [4], which has determined several operational details of the architecture.

Fig. 1 displays a functional diagram of the MDB elements involved in knowledge acquisition. As we can see, there are two asynchronous time scales, one devoted to the execution of actions in the environment in real time (*execution scale*) and another that deals with the knowledge acquisition (*learning scale*). The elements present in the execution scale necessarily run in the robot's physical hardware while the elements present in the learning scale can run on it if the computational power allows it, but they can also run in distributed computers.

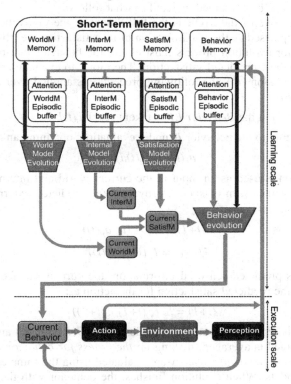

Fig. 1. Diagram of knowledge acquisition in the MDB

In the learning scale, all of the evolutionary processes related to declarative knowledge (models) are executed in parallel (*World Model Evolution, Internal Model Evolution and Satisfaction Model Evolution* blocks). Every time the robot executes an action in the environment, a new episode is acquired and a new MDB *iteration* starts. This episode is a candidate to be stored in the EB of each type of model (*WorldM EB, InterM EB, SatisfM EB*) if it is passes an attention filter (explained in detail in [5]). At this moment, the model evolution starts using as population the models stored in the model memory (*WorldM Memory, InterM Memory, SatisfM Memory*). The fitness function used is the same for all models: the *minimization of the error* between the model prediction and the episodes stored in the EB. To avoid a premature convergence of the model towards a specific EB, and thus to produce gradual learning, the evolution is carried out for a small number of generations each iteration (interaction with the world) and the model memories are preserved through iterations [4].

After the model evolution ends, the best models are selected as the current instance of each model type (*Current InterM, Current WorldM* and *Current SatisfM*), and the behavior evolution can start (*Behavior Evolution* block)[1]. This set of current models in a given iteration make up the best internal representation the robot has in that instant of time, and they are applied, as "simulators" of the real world, in the behavior learning. Behaviors are also evolved starting from a population of behavior candidates (*Behavior Memory* block), which is stored and preserved through iterations. The behavior evolution will be explained in detail in what follows.

The main objective of the MDB is the fulfillment of the motivation of the agent, which may be expressed as the maximization of its satisfaction *s* each instant of time. Thus, the behavior evolution process must obtain the behavior that provides the actions with the highest predicted satisfaction. Therefore, to evaluate an individual *i* of the behavior population (B_i), the following procedure is applied over all the episodes of the *Behavior EB*:

1. An episode *j* of the *Behavior EB* is taken: $\{e_j(t), i_j(t)\}$

2. It is the input to the behavior B_i under evaluation, obtaining an action a_{ji}:

$$a_{ji}(t)=B_i(e_j(t),i_j(t))$$

3. This action is used as an input in the current world and internal models (W_c and I_c) together with episode *j*, providing the predicted external and internal sensing values:

$$e_j(t+1) = W_c\,(e_j(t),\,a_{ji}(t))$$

$$i_j(t+1) = I_c\,(i_j(t),\,a_{ji}(t))$$

4. From this predicted sensorial information, the current satisfaction model (S_c) provides the predicted satisfaction for the action a_{ji}:

$$s_j(t+1) = S_c\,(e_j(t+1),\,i_j(t+1))$$

This process is repeated for all the episodes in the *Behavior EB* and the predicted satisfaction obtained is averaged, making up *the fitness for behavior B_i*. The remaining behaviors in the *Behavior Memory* are evaluated using the same episodes and the same current models. When evolution finishes, the behavior with the highest fitness (highest average satisfaction) is selected as the *Current Behavior* and it is transferred to the execution scale (see Fig. 1). As in the case of models, the evolution of behaviors is also carried out for a small number of generations for every iteration with the aim of preventing convergence towards particular contents of the *Behavior EB*.

With the existence of a behavior in the execution scale, the robot is able to react instantaneously to the task demands. That is, the robot control is decoupled from the "thought" process, which occurs in the learning scale. Moreover, a behavior is able to provide actions or motor commands for unlimited lengths of time, it can be stored for future cases in which this behavior may become useful again, and it can be improved over time with local adaptation processes.

In the current version of the MDB, both models and behaviors have been represented by means of Artificial Neural Networks (ANN), so the learning processes

---

[1] It is performed sequentially to ensure that the most updated models are available.

are neuroevolutionary processes [8]. Moreover, behavior learning is carried out on-line using models that are being learned concurrently[2]. Consequently, there is a coupled neuroevolution behind the success or failure of the MDB operation that must be analyzed in depth. Taking into account that initially both, models and behaviors, are randomly initialized, two main questions arise: how is behavior learning affected by the poor precision of the models in the first iterations? How is model learning affected by the poor actions selected by the quasi-random behaviors in those initial stages? In the following sections we will try to answer them using illustrative example of developmental learning with a simulated robot. The example itself has been chosen to be extremely simple in order to minimize the influence of other factors or parts of the architecture on the objective we seek, that is, to see how the coupling of the two aforementioned learning processes influences the operation of the mechanism.

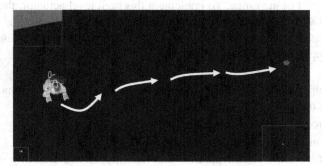

Fig. 2. Snapshot of the simulation experiment

3   Illustrative Example Setup

The experiment has been set up using the Webots simulator with a modified AIBO model in a simple learning example. As shown in Fig. 2, we place an AIBO robot and a pink ball in an empty scenario, and the robot has to learn how to reach the ball. It is a very simple experiment that is easy to evaluate and that includes the bare minimum components of the architecture for this purpose. That is, there is basically nothing other than the elements displayed in Fig. 1, no long-term memory or any other models are present in the architecture. However, it does imply the on-line coupled learning of models and behaviors from scratch, which is what we aim to study in this work.

Each iteration works in the same way: the robot looks for the ball by moving its neck and estimates a distance and an angle. This sensorial data is provided to the MDB, which decides, based on its model predictions, the best action in order to reach the ball. After executing the action, new distance and angle measures are obtained, conforming a new episode. If the robot reaches the ball, it is placed in a random position within the robot's field of view creating a sort of guided learning experiment. The following MDB elements (sensors, actuators…) have been considered:

---

[2] Although the model and behavior learning are performed sequentially within the same iteration, they may be considered as concurrent in the MDB *iteration* time scale, especially taking into account that a very gradual evolution is performed.

- Sensorial information:
    1. *Distance* (*d*): distance from the robot chest to the center of the ball extracted from the camera
    2. *Angle* (*a*): angle from the robot chest to the center of the ball extracted from the camera information and the neck angle (the robot searches for the ball with the body in a fixed position by moving its head until it is in the field of view)
    3. *Visibility* (*v*): binary value extracted from the camera that represents whether the ball is in the robot's field of view or not
- Action: the action represents a high level actuator that performs the final *movement* (*m*) of the AIBO robot. The possible values for *m* were encoded in the range [-1:1], being -1 a left turn of 180° and 1 a right turn of 180°, with continuous intermediate turns in the rest of the range. To implement this high level actuator, the linear speed of the robot was fixed and only the angular speed was varied.
- Episodes: taking into account the sensorial information and the action defined above, the episodes for the model evolution are made up of:
$$e = \{d(t),\ a(t),\ m(t),\ d(t+1),\ a(t+1),\ v(t+1)\}$$
- Models: four models are considered in this experiment, three world models and one satisfaction model
    1. *Distance model (D):* $d(t+1) = D(d(t),\ a(t),\ m(t))$
    2. *Angle model (A):* $a(t+1) = A(d(t),\ a(t),\ m(t))$
    3. *Visibility model (V):* it predicts the visibility or non-visibility of the ball to avoid learning when the robot does not perceive it: $v(t+1) = V(d(t),\ a(t),\ m(t))$
    4. *Satisfaction model (S):* $s(t+1) = S(d(t+1),a(t+1),v(t+1))$

    The satisfaction model has to learn a satisfaction function that has been defined in the simulator and that can be perceived by the robot: its value is higher when the robot is closer to the ball in a frontal direction, and it is zero if the robot does not see the ball.
- Behavior *(B):* one behavior has been defined in this experiment: $m(t) = B(d(t),a(t))$

The models and behaviors have been represented by standard multilayer perceptron ANNs evolved using the Differential Evolution (DE) algorithm with the basic implementation *DE/1/rand/bin*. The ANN parameters that are evolved are the synaptic weights, and the bias and slope of the sigmoid transfer function of each neuron. Table 1 displays the main parameters of the MDB and DE used in this experiment.

Regarding the fitness function, in the case of models it is the mean squared error in the prediction of the episodes in the EB, as commented in the previous section. For example, for the distance world model (*D*), the fitness will be calculated using:

$$f_D = \frac{1}{|EB_D|} \sum_{n=1}^{|EB_D|} \left(D\big(d_j(t), a_j(t), m_j(t)\big) - d_j(t+1) \right)^2$$

where $|EB_D|$ represents the distance model EB size, while (d_j, a_j, m_j) are the distance, angle and movement stored in episode *j*. The expression for the other models is equivalent but using the corresponding EB in each case. For the behavior evolution, as commented above, the average predicted satisfaction is used as fitness value for each behavior in the population. Formally, being m_j the movement predicted by the behavior B for episode j, the fitness function f_b would be in this case:

$$m_j(t) = B(d_j(t), a_j(t)) \qquad d_j(t+1) = D(d_j(t), a_j(t), m_j(t))$$

$$a_j(t+1) = A(d_j(t), a_j(t), m_j(t)) \qquad v_j(t+1) = V(d_j(t), a_j(t), m_j(t))$$

$$s_j(t+1) = S(d_j(t+1), a_j(t+1), v_j(t+1))$$

$$f_b = \frac{1}{|EB_D|} \sum_{j=1}^{|EB_D|} (s_j(t+1))$$

Table 1. MDB parameterization in the AIBO experiment

	World Models	Satisfaction Model	Behavior
EB size	20	20	20
ANN size	3-6-3-1	3-6-3-1	2-5-1
Generations per iteration	2	2	2
Population size (Model and Behavior Memories)	80	80	26
F (Differential Evolution)	rand(0.3,0.6)	rand(0.3,0.6)	rand(0.3,0.6)
CR (Differential Evolution)	rand(0.7,0.9)	rand(0.7,0.9)	rand(0.7,0.9)

4   Results on the Coupled Learning of Models and Behaviors

In a first stage, the experiment was executed using the behavior learning procedure explained in the previous sub-section and an *exhaustive action selection* strategy, in order to evaluate the basic capability of the behaviors to solve the task compared to a reference procedure that always provides the best possible action. Specifically, the exhaustive action selection strategy works as follows: each iteration, all the actions in a predefined set are used as inputs to the world models obtaining predicted sensing values that are used as inputs in the satisfaction model. This way, we have a predicted satisfaction for each possible action, and the one that provides the highest value can be selected and executed by the robot. These actions make up the robot response over time and can be compared to those provided by the learned behaviors. It must be highlighted that the exhaustive action selection is computationally much more costly than behavior execution. The latter implies executing a single ANN while the former implies executing, in this case, four ANNs for the whole set of predefined actions.

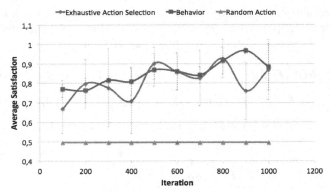

Fig. 3. Evolution of the average satisfaction using the behavior learning procedure (green) and the exhaustive action selection (red). The blue line represents a random selection procedure (the value is constant because the average satisfaction does not depend on the models in this case, so its average value in the set of 189 positions that have been defined is the same).

Fig. 3 shows the evolution of the average satisfaction of ten independent runs using the behavior learning procedure (green) and the exhaustive action selection procedure (red). Moreover, a random action selection has been represented in blue. To obtain these data for the behavior learning case, the current behavior was taken from the execution scale of the MDB every 100 iterations from 100 to 1000, and applied to 189 different combinations of initial robot and ball positions in an independent simulation, thus creating a sort of objective testing procedure[3]. In these 189 cases, the action provided by the behavior was executed and a real satisfaction value was computed and averaged, which is any of the points displayed in red. For the exhaustive action selection case, the current models of the corresponding iterations were taken and used to select the action in the 189 cases, obtaining a comparable satisfaction measure. As shown in the figure, the behaviors provide a very similar satisfaction, albeit more stable, to that of the exhaustive procedure, which is a highly successful result taking into account all the advantages of behaviors in terms of computational cost, reusability and generalization.

Once the behavior learning procedure has been shown to be capable of solving this task, we are going to present the results obtained in the study of the coupled learning of models and behaviors for this experiment. Two different learning policies have been designed: the first one has been called *Developmental*, and it implies that the MDB uses a random action selection in the execution scale during the initial 500 iterations. During this stage, the behavior is being learned in background although it does not take control of the robot. Thus, in the Developmental policy there is a forced exploration phase in order to learn more general models before starting to execute the behavior. The second policy, *Non-Developmental*, corresponds to the normal MDB operation where the behavior is used in the action selection from the beginning. The idea behind these two policies is to compare the model and behavior learning in a case where good general models (learning generality) are sought before executing the behavior to fulfill the motivations, and in another case where the MDB starts learning towards the objective from scratch.

---

[3] The set of 189 positions are consecutive in a squared grid, from long distances to short ones with different angles to the robot chest.

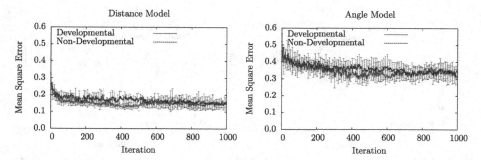

Fig. 4. Evolution of the generalization error for the world models used in the experiment

Fig. 4 displays the average generalization error and standard deviation of the current world models[4] in ten independent runs of the experiment for each policy (in red and blue). To obtain these data, the current model of each iteration was applied to a set of 4000 random perceptions and actions extracted from the simulator, and their outputs were compared to the real ones, obtaining thus a generalization error. As we can see in the plots of Fig. 4, the distance and angle models obtained using the Developmental policy provide a lower and more stable error value than those of the Non-Developmental one in the first 500 iterations. This is a clear consequence of the higher variety in the sensorimotor space in the Developmental case, as expected. The Non-Developmental policy implies an action space that is more task-oriented, specifically, actions that perform small turns and straight movements (the typical sequence of actions is that displayed in Fig. 2, where the robot first turns to be facing the ball, and then moves directly towards it). With this action space, it is easy to understand the learning generalization of distance and angle models displayed in Fig. 4. From iteration 500 onwards, both policies are the same (the behavior provides the action), so the generalization error tends to be the same for the two models.

The main topic of this work is the coupled learning of models and behaviors, that is, whether behavior learning is affected by model learning, because the behaviors are evaluated using the models. To answer this question, we have represented in Fig. 5 the behavior response over the previously used set of 189 different positions in the scenario comparing the behaviors obtained using the two learning policies. The left plot of Fig. 5 corresponds to iteration 500 (when the developmental stage finishes) and the right one to iteration 1000. The blue dotted curve of the plots corresponds to the actions provided by the "hidden" behavior that was learned in background during the random action selection stage. The left plot shows that the Developmental behavior is better in the first positions (far from the ball) while the Non-Developmental one outperforms it in the last positions (short distances). This is a consequence of the higher generality of the Developmental models that lead to a behavior with a general successful response in all the positions. Accordingly, the Non-developmental behavior is more task-specific, and it performs better in short distances because they are more frequent according to the evolution of the learning process.

---

[4] The visibility model is very simple and no relevant conclusions can be extracted from the differences in the error evolution of the two policies. For the satisfaction model, the sensorial space does not depend on the action, so the difference between the policies is not remarkable.

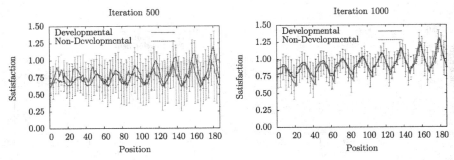

Fig. 5. Average satisfaction and standard deviation of 10 runs provided by the behaviors in the Developmental and Non-Developmental learning policies for the same 189 positions used in Fig. 3. The repetitive pattern that is shown in the curves is a consequence of the predefined order followed to sweep the set of initial positions of the robot related to the variation of the ball angle position respect to the robot chest.

Analyzing now the right plot of Fig. 5 that corresponds to iteration 1000, the main conclusion that can be extracted is that both learning policies lead to similar behaviors as time progresses, and both of them solve the task properly. This is highly relevant for the MDB, because it means that the higher generality of the developmental policy is not a key aspect for behavior learning. In other words, it seems that a set of more task-oriented models does not imply a poorer behavior in terms of how the motivation is fulfilled. As a consequence, if one seeks a more general modeling with the aim of, for example, consolidating more general knowledge in LTM, it would be interesting to perform this initial exploration stage. But, in general, the coupled evolution of models and behaviors from scratch does not represent a problem in order to obtain behaviors that solve the task. The variability in the sensorimotor samples (episodes) that is achieved during the initial iterations as a consequence of the random response of the behaviors, is enough to obtain a set of models that allow learning this task.

5   Conclusions

The coupled learning of declarative and procedural knowledge in the MDB cognitive architecture has been studied through a simulated robotic experiment. The learning processes in this architecture are based on neuroevolution over models and behaviors, so the conclusions that have been extracted could be of application to other architectures that use this type of computational processing. Thus, the main conclusion from a neuroevolutionary perspective would be that the initial stages of development introduce enough randomness in the action-selection to adequately explore the sensorimotor space and to obtain proper ANNs. Therefore, general knowledge can be obtained without requiring a dedicated exploration stage. Moreover, the concurrent neuroevolution converges towards successful models, in the sense that they were adequate to learn behaviors that solve the task, although there is a transient stage where these behaviors are poorly evaluated. Despite the fact that this initial result requires of a deeper analysis with more general cases, it is very relevant for the MDB development because it suggests that coupled learning from the beginning is possible, thus reducing the necessity of carefully designing experiments of incremental complexity.

Acknowledgements. This work has been partially funded by the EU's H2020 research and innovation programme under grant agreement No 640891 (DREAM project) and by the Xunta de Galicia and European Regional Development Funds under grants GRC 2013-050 and redTEIC network (R2014/037).

References

1. Anderson, J.R., Bothell, D., Byrne, M.D., Douglass, S., Lebiere, C., Qin, Y.: An Integrated Theory of the Mind. Psychological Review **111**(4), 1036–1060 (2004)
2. Asada, M., MacDorman, K.F., Ishiguro, H., Kuniyoshi, Y.: Cognitive developmental robotics as a new paradigm for the design of humanoid robots. Robotics and Autonomous Systems **37**, 185–193 (2001)
3. Bach, S.: Principles of Synthetic Intelligence. PSI: An Architecture of Motivated Cognition. Oxford Univ. Press (2009)
4. Bellas, F., Duro, R.J., Faiña, A., Souto, D.: Multilevel Darwinist Brain (MDB): Artificial Evolution in a Cognitive Architecture for Real Robots. IEEE Transactions on Autonomous Mental Development **2**(4), 340–354 (2010)
5. Bellas, F., Caamaño, P., Faiña, A., Duro, R.J.: Dynamic learning in cognitive robotics through a procedural long term memory. Evolving Systems **5**(1), 49–63 (2014)
6. Cotterill, R.: Enchanted looms: Conscious networks in brains and computers. Cambridge University Press (2000)
7. Duro, R.J., Bellas, F., Becerra, J.A.: Brain-Like Robotics. Springer Handbook of Bio-/Neuroinformatics, pp. 1019–1056 (2014)
8. Floreano, D., Dürr, P., Mattiussi, C.: Neuroevolution: from architectures to learning. Evolutionary Intelligence **1**(2008), 47–62 (2008)
9. Goertzel, B., de Garis, H.: XIA-MAN: An extensible, integrative architecture for intelligent humanoid robotics. AAAI Fall Symp. Biol. Inspired Cogn. Archit., 65–74 (2008)
10. Hesslow, G.: The current status of the simulation theory of cognition. Brain Research **1428**, 71–79 (2012)
11. Krichmar, J.L., Edelman, G.M.: Principles underlying the construction of brain-based devices. In: Proceedings of AISB 2006, vol. 2, pp. 37–42 (2006)
12. Laird, J.: The Soar Cognitive Architecture. MIT Press (2012)
13. Weng, J., McClelland, J., Pentland, A., Sporns, O., Stockman, I., Sur, M., Thelen, E.: Autonomous mental development by robots and animals. Science **291**, 599–600 (2001)

Damasio's Somatic Marker for Social Robotics: Preliminary Implementation and Test

Lorenzo Cominelli[✉], Daniele Mazzei, Michael Pieroni, Abolfazl Zaraki, Roberto Garofalo, and Danilo De Rossi

Research Center "E. Piaggio", University of Pisa,
Largo Lucio Lazzarino 1, 56126 Pisa, Italy
lorenzo.cominelli@for.unipi.it
http://www.faceteam.it

Abstract. How experienced emotional states, induced by the events that emerge in our context, influence our behaviour? Are they an obstacle or a helpful assistant for our reasoning process? Antonio Damasio gave exhaustive answers to these questions through his studies on patients with brain injuries. He demonstrated how the emotions guide decision-making and he has identified a region of the brain which has a fundamental role in this process. Antoine Bechara devised a test to validate the proper functioning of that cortical region of the brain. Inspired from Damasio's theories we developed a mechanism in an artificial agent that enables it to represent emotional states and to exploit them for biasing its decisions. We also implement the card gambling task that Bechara used on his patients as a validating test. Finally we put our artificial agent through this test for 100 trials. The results of this experiment are analysed and discussed highlighting the demonstrated efficiency of the implemented somatic marker mechanism and the potential impact of this system in the field of social robotics.

Keywords: Somatic marker hypothesis · Human-inspired · Cognitive systems · Artificial intelligence · Emotion · Decision-making · Reinforcement learning

1 Introduction

The theory that in human beings emotions not only affect but are mandatory for advantageous decision-making, social world comprehension and behaviour modulation is increasingly affirmed in scientific literature [6,13,16]. In particular, the neurologist Antonio Damasio developed and investigated the *Somatic Marker Hypothesis* (SMH) reported in [9] and more in detail in the book "Descartes' Error" published in 1994 [8]. Damasio's theory derives from the examination of patients with bilateral lesions who had a particular profile of difficulties in decision-making. Thanks to these studies Damasio discovered a specific region of the brain, in the prefrontal cortex, which is involved in the coupling of entities (or memories of entities) with somatic states. He demonstrated that this labeling

© Springer International Publishing Switzerland 2015
S.P. Wilson et al. (Eds.): Living Machines 2015, LNAI 9222, pp. 316–328, 2015.
DOI: 10.1007/978-3-319-22979-9_31

mechanism is fundamental for the human reasoning process and it is a pillar of the process that allows us to take quick and crucial decisions. In the light of these considerations it is necessary to go beyond the development of human-inspired cognitive systems conceived as disembodied and purely computational minds. Accordingly, taking inspiration from well-known behavioural and cognitive frameworks for social robotics [7,10,19], we developed an artificial intelligence which includes artificial 'emotions' and an emotionally-based marking process. In this work we describe the implementation of artificial emotions and of the somatic marker modules for the I-CLIPS Brain cognitive architecture [15]. In addition, the implemented system has been tested for a standard decision-making task, also used by Damasio in his studies, that is the Iowa Gambling Task [2]. Such psychological test was conceived by Antoine Bechara to evaluate the loss of sensitivity to future consequences in patients with prefrontal cortex lesions when exposed to a task on which important and risky decision have to be taken in short time. We have replicated the game with all its rules and tested it on the artificial agent with many trials, comparing the results of the experiment in case of presence or absence of the somatic marker mechanism, in line with Bechara's test [2].

In the first part of this paper we report a brief introduction to the somatic marker mechanism and a presentation of the mentioned experiment. In the second part we introduce the method used to simulate this process in the cognitive system of an artificial agent and the implementation of the gambling task. In the third and last part of the work, since we have an artificial intelligence endowed with the somatic maker and the simulation of the game, we make the agent play the card game and we present statistical analysis of the results, comparing the outcome that we obtain with and without the implemented somatic marker mechanism.

2   The Somatic Marker Hypothesis

Human emotions guide human decision-making and behaviour through the activation of somatic states as asserted by Damasio in [8,9] and discussed in detail by Suzuki et al. in [18]. According to this theory, every time we experience an event, or interact with an external entity, that induces a relevant somatic response (including neural responses), we associate that entity with the experienced emotional consequence. This emotional response is stored in our brain as a somatic state, or *Somatic Marker* (SM). We exploit these markers to optimize our decision-making in case of a second exposure to the same, or similar, event. Somatic markers are triggered also by the mere thought of an option encouraging our appetitive (approach) or aversive (withdrawal) attitudes to the world. The somatic marker mechanism has been investigated with the Iowa Gambling Task designed by Bechara [2]. Damasio and Bechara were interested in the investigation of two regions of the brain which they considered fundamental for the SM mechanism. The experiment demonstrated that patients with damage of the Ventromedial Prefrontal Cortex (VMPFc) or with amygdala lesions show high propensity for risky activities. Therefore, Damasio and Bechara demonstrated that amygdala damage determines an inability to evoke somatic responses, which are fundamental for the evaluation of an affective situation, while VMPFc damage precludes the integration of somatic representations. Since SMs are acquired through the integration of somatic representations, both defects impede the acquisition of somatic markers and thus lead to decision-making impairment.

3   The Iowa Gambling Task

In the Iowa Gambling Task participants usually play a card game over a computer monitor where four decks comprising 40 cards are presented [1]. All the decks are face down and labelled as deck A, B, C, and D. Two of the four decks (the *conservative decks*) were associated with small rewards and small punishments, whereas the other two decks (the *risky decks*) yielded large rewards and large punishments. The player start the game with a specific amount of money, usually $1000. The participants can sequentially select one of the four decks from which to pick a card. Each card is associated with a reward, a penalty, or both. For instance, one card might yield a reward of $100, a penalty of $150, and thus a net loss of $50. The overall money amount is then presented to the player highlighting the difference from the outcome of the previous deal. Participants do not know how many deals will have to play and this number is not specified, nevertheless one trial is usually made of 100 deals. Participants do not even know the probability of a reward or penalty as well as how the distribution of these rewards and penalties varies across the decks.

Typically, at the first deals of the game, most participants will select the disadvantageous decks, attracted by their higher profits. Over time, usually within the first 40 deals, healthy participants learn to choose the advantageous decks, whereas participants with lesions in the mentioned regions of the brain, such as VMPFc [4], do not learn to select the advantageous decks, incurring relevant losses of money. This difference between the two populations of participants was highlighted in Damasio's experiments [8] as shown in figure 1. Moreover, healthy subjects and patients with VMPFc lesion, after receiving the monetary outcome, shown changes in skin conductance responses (SCRs). This somatic reaction was not measured in amygdala-damaged patients [3]. Therefore, Damasio and Bechara demonstrated that amygdala damage determines an inability to evoke somatic responses, which are fundamental for the evaluation of an affective situation, while VMPFc damage precludes the integration of somatic representations. Since SMs are acquired through the integration of somatic representations, both defects impede the acquisition of somatic markers and thus lead to decision-making impairment [1,3,5].

4   The Somatic Marker integrated in an Artificial Agent

"When the choice of option X, which leads to bad outcome Y, is followed by punishment and thus painful body states, the somaticmarker system acquires

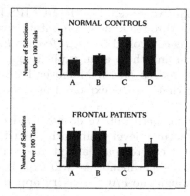

Fig. 1. Significant differences in decks selection between normal controls and frontal patients. Here, A and B are *risky decks*, while C and D are *conservative decks*. Figure from [8].

the hidden, dispositional representation of this experience-driven, non inherited, arbitrary connection. Reexposure of the organism to option X, or thoughts about outcome Y, will now have the power to re enact the painful body state and thus serve as an automated reminder of bad consequences to come." [8]

With this essential sentence, extracted from "Descartes' Error", Antonio Damasio explains the somatic marker mechanism in the human brain. From a system developer point of view, these words seem to suggest a possible implementation for an artificial intelligence. In order to implement such a mechanism, we identified the following requirements: a method by which represent the emotional states that the agent 'feels' ; the rules which determine how certain external *entities* and *events* affect the emotional state of the agent; an algorithm that makes a somatic marker (SM) trigger whenever an induced emotional state has to be considered particularly relevant; a storage in which to save every triggered somatic marker associated with the related inducer; several *basic behavioural rules* that determine a neutral approach of the agent to the novel entities and some *alternative behavioural rules* that bias the course of an action in case of marked entity recognition.

In the Iowa Gambling Task, the *entities* with which the agent has to face are the four decks of cards; the *events* are the rewards and punishments associated with each card drawn by the selected deck during the game; the *behaviour* of the agent is expressed by its decisions about card decks selection deal after deal.

Referring to the above described requirements it is evident that we are dealing with a highly dynamic system that needs to be solved with a forward-chaining approach. This kind of problem solving can be implemented on top of a cognitive system we previously developed: the I-CLIPS Brain [15]. I-CLIPS Brain is an innovative hybrid deliberative/reactive cognitive system for controlling social humanoid robots. It has been developed embedding CLIPS, a programming language that provides support for rule-based non-procedural programming of expert systems [11]. A cognitive system like the I-CLIPS Brain allows to deliver a solution to a problem not through a mere computational algorithm but thanks to *rules* and *inference*. It is controlled by an *Inference Engine* provided with a powerful conflict resolution and there is an explicit separation between control (*rules*) and data (represented as *facts*). Moreover I-CLIPS Brain is provided with a *Working Memory* that can be exploited as a database for the somatic markers.

For the purpose of this experiment we do not need the overall system presented in [15] but only the high-level deliberating part. Therefore, the FACE robot [14] and its low-level control part aimed at driving the physical robot are not necessary. In future experiments, we will exploit the influence of external stimuli, perceived through the FACE sensory apparatus, on the SM-based decision-making process.

A schema of the reasoning process developed on top of the I-CLIPS brain cognitive architecture is reported in figure 2. In the next sections we explain the components of the system following this figure.

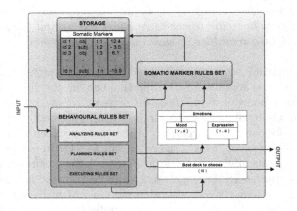

Fig. 2. Framework of the cognitive system

4.1   The Behavioural Rules Set

The *Behavioural Rules Set* (BRS) is a set of rules that process the incoming information to determine which is the best decision to take according to the emerging events in the context and which are the emotional responses. BRS is divided in three subset of rules:

Analyzing Rules Set (**A-RS**) analyses the incoming data and the list of stored somatic markers identifying if a sensory input is already marked in the storage (e.g. checking if a deck is already labelled with a SM);

Planning Rules Set (**P-RS**) on the basis of A-RS investigation, activates a reasoning chain intended to determine the emotional immediate reaction, the consequence in the emotional state and which will be the best choice for the next deal; the conclusions reached in the P-RS decision-making process result in a modification of internal data (i.e. *mood*, *expression* and *best deck to choose*);

Executing Rules Set (E-RS) reads the internal data upgraded by the P-RS and disposes the sending of this information outside of the cognitive system.

4.2 How the Events Affect the Emotional States

For the representation of the emotional states we follow the Russell theory of Circumplex Model of Affect [17]. Therefore, every emotional state is defined by two values: *valence* (v), which describes the pleasantness of an emotion, and *arousal* (a), indicative of the level of activation that an emotion leads. This method is particularly useful as it allows to associate the emotions with numerical values which can be processed. The variation of these values leads to a dynamic of the emotions 'felt' by the artificial agent. As briefly mentioned in the explanation of the BRS processing, two different kind of internal data have been created to represent emotional states: *mood* and *expression*. Both are defined in (v, a) coordinates that are modified by rules of the BRS but they are used for different purposes. The *expression* is a kind of fast emotional reaction that occurs immediately after the reward or loss of money. This kind of emotional state is shown by the agent every time it plays a deal when it perceives the consequence of his last choice of deck, but have no influence on the game or the decision-making process. For instance, a social robot with facial expressiveness would execute expressions in a discrete way, according to values of valence and arousal that are proportional to the amount of money won or lost, without taking into account the previous executed expressions. On the other hand, the *mood* has a continuous trend because its (v, a) values are modified incrementally by the BRS. Practically, every event that involves the agent induces an emotional consequence that corresponds to an increase or decrease of the (v, a) mood values. This variation is always referred to the previous state in order to create a continuous dynamic of the agent mood. The *mood* is also used by the *Somatic Marker Rules Set* since it works as a threshold trigger for markers, as detailed in the next section. The difference between an emotional immediate reaction *expression* and the mood is extensively discussed in [12].

4.3 How Somatic Markers are Assigned to the Entities

In order to detect when an emotional state becomes relevant, the *Somatic Marker Rules Set* (SMRS) continuously monitor the (v, a) values of the *mood* and the *id* of the entity on which the attention is focused (i.e. one of the four decks) before the action of BRS can make it change. We assume a threshold to distinguish a non emotionally-relevant event rather than a relevant one. Considering the circumplex model of affect used to represent the mood of the robot, and v and a as spatial coordinates, this threshold is graphically represented by an internal circle. The radius of this internal circle can be considered as the *sensibility* of the robot to external events and it is defined as s. The *mood* is a vector, determined by the (v, a) coordinates. The *emotional state* (es) is the magnitude of the *mood* vector, the intensity of an emotion. If es_t is higher than s, the SMRS extracts the *id* of the entity on which the attention of the agent is currently focused, and associates a SM to that entity. The sign of the marker is decided by the sign of the mood's valence (v_t), while the numerical value is proportional to the peak of the aroused stimulus (a_t).

All this information is stored in the *Storage* and each SM reports the following attributes: the *id* of the marked entity; the *type* of the marked entity (i.e. *subject* or *object*); the *time* in which the SM has been stored or upgraded; the current SM *numerical value*, a real number that can be increased and decreased.

The SM assignment mechanism is shown schematically in Fig 3.

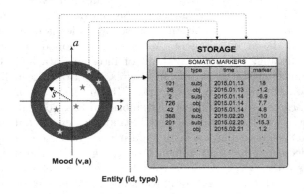

Fig. 3. Schema of the somatic marker assignment mechanism

4.4 How the SM is Exploited by the Cognitive System to Affect Behaviour

In the cognitive system framework represented in Fig. 2 it can be noticed a connection that goes from the *Storage* to the BRS. This link represents the way in which the SM affects the behaviour of the robot.

The working principle is the following: while several rules of the BRS receive and process the information about the external world determining what we call the *basic behaviour* of the agent, some internal rules of the A-RS continuously check the *Storage* to match the *id*s of external detected entities with the *id*s of the internal stored SMs. Every time a marker is recognized by successful *id* matching, a secondary fact is asserted by the A-RS and this will bias the decision-making process that takes place in the P-RS. This deviation of the reasoning chain results in a different behaviour of the agent.

4.5   A Brief Consideration about the Implemented Somatic Marker Mechanism

This implementation of the SM leads to a labeling process that may be considered controversial. Indeed, every entity and every event can modify the mood of the robot, and the *mood* is the main responsible for the SM triggering. Therefore, the causal connection between the triggering entity and the agent judgement of the entity itself, seems to be lost. Nonetheless, we claim that this is the added value of the presented cognitive system rather than a weak point. We believe that this mechanism is alike the way humans behave and perceive the external world. Many are the factors that affect our emotional state and our judgement. We are influenced by past experienced interactions as much as the entities that are around us in this specific moment and environment. This is the process we include in our system.

5   Test and Results: Playing the Game

The interaction between the card game and the robot has been automated to get a significant sample of tasks. Therefore we implemented the Iowa gabling Task as a separated program which autonomously dialogues with the I-CLIPS Brain. The simulation accurately replicates all the game rules of the task proposed by Bechara in his experiments, taking into account the number of deals, the amounts of money used for punishments and rewards associated to each deck as well as the probability to run into a penalty depending on the selected deck. Every deal, the information received as input by the I-CLIPS Brain is the following: the deck that has been selected at the previous deal (d_{n-1}), the card that has been drawn from that deck (c_{n-1}) and the related monetary punishment or reward ($\$_{n-1}$). The A-RS compare these data with the current state of the system and other information from the working memory (e.g. potential SMs, current mood values) and instantiates the reasoning chain of the P-RS that leads to the modification of the inner emotional state, the execution of an immediate reaction (i.e. facial expression), and the decision about the next deck to choose (d_n). The E-RS communicates the outcome of this decisional process back to the game simulator and a new deal starts. Being *entities* that influence the mood of the agent, the four decks can be marked and associated to somatic markers. Therefore, during the trials, the SMRS actively exerts its dual function: upgrading the decks SMs according to the consequence of their selection during the game, and biasing the selection of the deck polarizing the choice towards the deck with the higher somatic marker value. In figure 4 is reported a schema of this closed loop between the agent and the simulated Iowa Gambling Task.

In order to investigate whether the results of the experiment are or not dependent by the SM mechanism, we have implemented two rules set for the BRS: with and without SM. The SM rules yields the selection of the deck associated with the highest marker value while if the SM mechanism is removed decks are randomly selected (basic behaviour). For instance, at the beginning of a game when no markers have been instantiated and the emotional state never went out

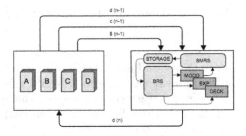

Fig. 4. The closed loop between the I-CLIPS Brain and the Iowa Gambling Task Simulator

of its sensibility threshold, the basic behaviour is followed. When the emotional state, following subsequent monetary winnings or losses, exceeds the threshold then the decks SMs begin to be stored and modified by the SMRS and the decision-making process starts to be influenced by the robot mood through the SM mechanism. One trial is made up of 100 deals. We run 100 trials with the SM mechanism activated by a sensibility threshold of 0.7 and 100 trials without the SM mechanism. In order to ensure the disabling of the SM mechanism, it has been assigned a sensibility threshold that the emotional state can never reach ($s > \sqrt{2}$). During the test several type of data per each deal have been acquired, such as total cash, monetary rewards and penalties, expression values (v, a) , mood values (v, a) , number of selections of every deck and all the decks SMs. These data have been averaged among trials of the same kind creating two populations of data (with SM versus without SM).

As shown in the top-left chart of figure 5 the trend of the cash, averaged over the 100 trials conducted without the implemented SM mechanism, is disastrous. The artificial player, in just 100 hands, choosing randomly the decks, loses money almost linearly and ends the game with a mean debt of about −$6,000. The standard deviation increases with the progress of the game as a random selection of decks can lead to scattered results. On the contrary, the chart at the top-right of figure 5 shows that the performance of the agent when using the mechanism of somatic marker is drastically different. After an initial phase in which the agent loses the provided cash, following the frustrating emotional states induced by the risky decks penalties he starts to avoid them preferring to seek the little but more frequent and comfortable feelings induced by the conservative decks. In this case the agent is not only able to contain the losses but in most of the trials the game ends with a benefit in cash (61% of the trials versus 2% in case of no SM). The overall winnings and losses are reduced in the amount of money but the trend is clear and the reduced dispersion of the standard deviation demonstrates that it is not due to chance. This has been confirmed by the statistical analysis of the decks selection distribution.

The total number of times that the decks were chosen during each game has been averaged over 100 trials conducted without the SM mechanism and over the 100 trials performed with it. A two-way ANOVA test with a 99% of CI reveals that there is no significant difference between the selection of decks in the no-SM

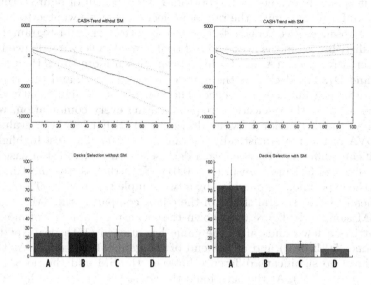

Fig. 5. Plots of the cash trend and histograms of decks selection distribution with and without the SM mechanism

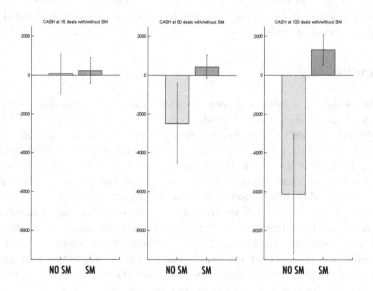

Fig. 6. Cash at the 15th, 50th and 100th deal with (yellow bar) and without (cyan bar) SM mechanism

case. As it is clearly evident in the bottom-left histogram of figure 5 the distribution is uniform confirming the random selection that we wanted to achieve in case of SM absence. Vice versa, referring to the bottom-right histogram, the deck selection distribution in the averaged SM trials reveals a very unbalanced profile. The conservative decks (A and C) have been chosen more times than the risky ones (B and D). The deck A is the more conservative deck and this is reflected in the results as a marked preference. The four decks are significantly different from each other in their selection frequency using every combination, with the exception of the distinction between decks B and D that has been evaluated by the ANOVA test as not statistically relevant ($p > 0.01$). The test highlights also the significant difference between each deck selection preference in case of activated or disabled SM mechanism ($p < 0.01$). Nevertheless, the most important result of the experiment is provided by a two sample t-test with 99% of CI made on the mean cash at several stages of the trials, comparing the SM sample with the no-SM sample. In figure 6 is shown the mean total amount of money that the agent has a few hands after the game has started (15th deal), at the half of the game (50th deal) and at the end of the game (100th deal). At the 15th deal we have no significant difference between the SM and the no-SM sample ($p > 0.01, p = 0.2431$). At the 50th hand the t-test rejects the null hypothesis at the 1% significance level ($p < 0.01$) and at the end of the game we obtain an even stronger significant difference ($p < 0.0001$). Testing the comparison between SM and no-SM samples of the mean cash deal by deal we found a significant difference between the two samples at the 18th deal. This means that after about twenty hands the SM marker mechanism start to influence the decision-making process comporting a behaviour change that can be clearly identified with a confidence interval of 99 %.

6   Conclusions

In this work an emotionally-driven decision-making process has been implemented on top of the I-CLIPS Brain cognitive architecture. The system has been endowed with an in-silico implementation of the Damasio's somatic marker mechanism and validated by simulating an artificial agent involved in the Iowa Gambling Task. The game simulator and the cognitive system have been connected together in a closed-loop without connecting the physical FACE robot to the testing setup. The implemented system has been tested running 100 games of 100 deals each. The test has been conducted with both the experimental condition: with and without SM. The analysis of collected data has confirmed that the presence of the somatic marker mechanism influences the agent decision-making process with differences that are statistically evident since the first (from the 20th) steps of the decision-making process. In addition, the results of our experiment reflect the outcome of the Damasio and Bechara tests [8]. Therefore we demonstrated that the I-CLIPS Brain has adequately included artificial emotions and it can exploit them not only for reinforcement learning to successfully cope an option selection task, but also as a key means to create a diversified scenario, a personal perspective in the mind of a robot about the entities of the world in which is involved. This outcome paves the way for future experiments in which the presented system will be integrated in the FACE robot. Internal and external stimuli will be perceived from the body and from the environment through the sensory apparatus of the physical robot. Decks will be replaced with human interlocutors and verbal and non-verbal cues will influence the mood of the robot. Furthermore, artificial emotions will be represented through the body of the android becoming a meaningful expressive component of the human-robot social interaction.

Acknowledgement. This work was partially funded by the European Commission under the 7th Framework Program projects EASEL, Expressive Agents for Symbiotic Education and Learning, under Grant 611971-FP7- ICT-2013-10.

References

1. Bechara, A., Damasio, H., Tranel, D., Damasio, A.: Deciding conservatively before knowing the conservative strategy. Science **275**, 1293–1294 (1997)
2. Bechara, A., Damasio, A.R., Damasio, H., Anderson, S.W.: Insensitivity to future consequences following damage to human prefrontal cortex. Cognition **50**(1), 7–15 (1994)
3. Bechara, A., Damasio, H., Damasio, A.R., Lee, G.P.: Different contributions of the human amygdala and ventromedial prefrontal cortex to decision-making. The Journal of Neuroscience **19**(13), 5473–5481 (1999)
4. Bechara, A., Damasio, H., Tranel, D., Damasio, A.R.: Deciding advantageously before knowing the advantageous strategy. Science **275**(5304), 1293–1295 (1997)
5. Bechara, A., Tranel, D., Damasio, H., Damasio, A.R.: Failure to respond autonomically to anticipated future outcomes following damage to prefrontal cortex. Cerebral cortex **6**(2), 215–225 (1996)
6. Borst, C.V.: The Mind-Brain Identity Theory: A Collection of Papers. St Martin's P., New York (1970)
7. Breazeal, C., Velasquez, J.: Robot in society: friend or appliance. In: Proceedings of the 1999 Autonomous Agents Workshop on Emotion-Based Agent Architectures, pp. 18–26 (1999)
8. Damasio, A.: Descartes' Error: Emotion, Reason, and the Human Brain. Grosset/Putnam, New York (1994)
9. Damasio, A.R., Tranel, D., Damasio, H.: Somatic markers and the guidance of behavior: Theory and preliminary testing. Frontal Lobe Function and Dysfunction, 217–229 (1991)
10. Gadanho, S., Hallam, J.: Emotion triggered learning for autonomous robots. DAI Research Paper (1998)
11. Giarratano, J.C., Riley, G.D.: Expert Systems: Principles and Programming. Brooks/Cole Publishing Co., Pacific Grove (2005)
12. Gray, E.K., Watson, D., Payne, R., Cooper, C.: Emotion, mood, and temperament: Similarities, differences, and a synthesis. Emotions at work: Theory, research and applications for management, 21–43 (2001)

13. Kiverstein, J., Miller, M.: The embodied brain: towards a radical embodied cognitive neuroscience. Frontiers in Human Neuroscience **9**, 237 (2015)
14. Mazzei, D., Billeci, L., Armato, A., Lazzeri, N., Cisternino, A., Pioggia, G., Igliozzi, R., Muratori, F., Ahluwalia, A., De Rossi, D.: The face of autism. In: The 19th IEEE International Symposium on Robot and Human Interactive Communication, RO-MAN 2010, pp. 791–796. IEEE Computer Society publisher (2010)
15. Mazzei, D., Cominelli, L., Lazzeri, N., Zaraki, A., De Rossi, D.: I-CLIPS brain: a hybrid cognitive system for social robots. In: Duff, A., Lepora, N.F., Mura, A., Prescott, T.J., Verschure, P.F.M.J. (eds.) Living Machines 2014. LNCS, vol. 8608, pp. 213–224. Springer, Heidelberg (2014)
16. Pally, R.: Emotional processing: The mind-body connection. The International Journal of Psychoanalysis (1998)
17. Russell, J.A.: The circumplex model of affect. Journal of Personality and Social Psychology **39**, 1161–1178 (1980)
18. Suzuki, A., Hirota, A., Takasawa, N., Shigemasu, K.: Application of the somatic marker hypothesis to individual differences in decision making. Biological Psychology **65**(1), 81–88 (2003)
19. Velasquez, J.: Modeling emotion-based decision-making. Emotional and intelligent: The tangled knot of cognition, 164–169 (1998)

Learning Sensory Correlations
for 3D Egomotion Estimation

Cristian Axenie[✉] and Jörg Conradt[✉]

Neuroscientific System Theory Group, Department of Electric
and Computer Engineering, Technische Universität München,
Karlstrasse 45, 80333 Munich, Germany
{cristian.axenie,conradt}@tum.de

Abstract. Learning processes which take place during the development of a biological nervous system enable it to extract mappings between external stimuli and its internal state. Precise egomotion estimation is essential to keep these external and internal cues coherent given the rich multisensory environment. In this paper we present a learning model which, given various sensory inputs, converges to a state providing a coherent representation of the sensory space and the cross-sensory relations. The developed model, implemented for 3D egomotion estimation on a quadrotor, provides precise estimates for roll, pitch and yaw angles.

Keywords: Egomotion estimation · Cross-modal learning · Multisensory fusion · Mobile robots

1 Introduction

Human perception improves through exposure to the environment. A wealth of sensory streams which provide a rich experience continuously refine the internal representations of the environment and own state. Furthermore, these representations determine more precise motor planning [1]. An essential component in motor planning and navigation, in both real and artificial systems, is egomotion estimation. Given the multimodal nature of the sensory cues, learning cross-modal correlations improves the precision and flexibility of motion estimates.

Various methods, ranging from neural circuitry implementations to statistical correlation analysis, have been developed to extract correlational structure in sensory data. Related work [2] used a combination of simple biologically plausible mechanisms, like WTA circuitry, Hebbian learning, and homeostatic activity regulation, to extract relations in artificially generated sensory data. After learning, the model was able to infer missing quantities given the learned relations and available sensors. Moreover, due to recurrent connectivity, the sensory representations were continuously refined, de-noising the encoded real-world variable. Finally, due to the constraints imposed by the learned relations, the model was able to combine consistent and correlated data (i.e. cue-integration) and discriminate and penalize inconsistent data contributions (i.e. decision making).

© Springer International Publishing Switzerland 2015
S.P. Wilson et al. (Eds.): Living Machines 2015, LNAI 9222, pp. 329–338, 2015.
DOI: 10.1007/978-3-319-22979-9_32

Using a different neurally inspired substrate, [3] combined competition and cooperation in a self-organizing network of simple processing units to extract coordinate transformations in a robotic scenario. Inspired by sensorimotor transformations in the prefrontal cortex, the algorithm produced invariant representations and a topographic map representation of the scene guiding a robot's behaviour.

Going away from biological inspiration, [4] used nonlinear canonical correlation analysis to extract relations between sets of multi-dimensional random variables. The model implemented a change of representation from the variables input space to a new space of canonical variants. Subsequently, the model mapped the representations back to the initial space minimising the relative mismatch between the original data and the mapping. The extracted relation was encoded in the weights configuration maximising the correlation between the canonical variants.

Although these methods provide good results for dedicated scenarios, they lack the capability to handle non-uniform sensory data distributions and, at the design stage, need judicious parametrisation and prior information about sensory data. The proposed model tries to address these aspects. It uses relatively simple mechanisms to provide a flexible way to learn sensory correlations for precise egomotion estimation.

2   A Neurally Inspired Model for Learning Sensory Correlations

During development, the biological nervous system must constantly combine various sources of information and moreover track and anticipate changes in one or more of the cues. Furthermore, the adaptive development of the functional organisation of the cortical areas seems to depend strongly on the available sensory inputs, which gradually sharpen their response, given the constraints imposed by the cross-sensory relations [5].

Following this principle, we propose a model based on Self-Organizing Map (SOM) and Hebbian Learning (HL) as main ingredients for extracting underlying relations in sensory data. In order to introduce the proposed model, we provide a simple example in Figure 1. In the basic dual-modality scenario, the relation between sources of sensory data is extracted between the pair of sensory modalities, as shown in Figure 1b. In a multisensory scenario, all-to-all connections between modalities are considered, and similar dynamics applies to each possible modality pair.

The input SOMs are responsible for extracting the statistics of the incoming data, depicted in Figure 1a, and encoding sensory samples in a distributed activity pattern, as shown in Figure 1c. This activity pattern is generated such that the closest preferred value of a neuron to the input sample will be strongly activated and will decay, proportional with distance, for neighbouring units. Figure 2 provides a detailed depiction of processing stages which take place when sensory input samples are presented to the network. Using the SOM distributed

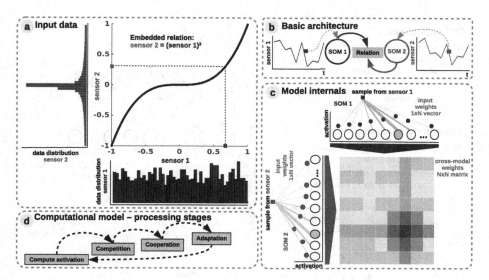

Fig. 1. Model architecture. a) Input data resembling a nonlinear relation and its distribution. b) Basic architecture; c) Model internal structure. d) Processing stages.

representation, the model learns the boundaries of the input data, such that, after relaxation, the SOMs provide a topological representation of the input space. We extend the basic SOM in such a way that each neuron not only specialises in representing a certain (preferred) value in the input space, but also learns its own sensitivity (i.e. tuning curve shape). Given an input sample, $s^p(k)$ at time step k, the network follows the processing stages depicted in Figure 1d and explicitly presented in Figure 2. For each $i - th$ neuron in the $p - th$ input SOM, with the preferred value $w_{in,i}^p$ and $\xi_i^p(k)$ tuning curve size, the sensory elicited activation is given by

$$a_i^p(k) = \frac{1}{\sqrt{2\pi}\xi_i^p(k)} e^{\frac{-(s^p(k)-w_{in,i}^p(k))^2}{2\xi_i^p(k)^2}}. \tag{1}$$

The winner neuron of the $p - th$ population, $b^p(k)$, is the one which elicits the highest activation given the sensory input at time k

$$b^p(k) = \underset{i}{argmax}\ a^p(k). \tag{2}$$

During self-organisation, at the input level, competition for highest activation is followed by cooperation in representing the input space (second and third step in Figure 1d). Given the winner neuron, $b^p(k)$, the interaction kernel,

$$h_{b,i}^p(k) = e^{\frac{-||r_i-r_b||^2}{2\sigma(k)^2}}. \tag{3}$$

allows neighbouring cells (found at position r_i in the network) to precisely represent the sensory input sample given their location in the neighbourhood $\sigma(k)$.

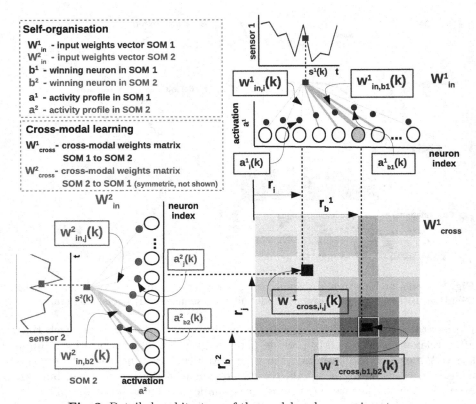

Fig. 2. Detailed architecture of the model and processing stages

The interaction kernel in Equation 3, ensures that specific neurons in the network specialise on different areas in the sensory space, such that the input weights (i.e. preferred values) of the neurons are pulled closer to the input sample,

$$\Delta w_{in,i}^{p}(k) = \alpha(k)h_{b,i}^{p}(k)(s^{p}(k) - w_{in,i}^{p}(k)). \tag{4}$$

This corresponds to the adaptation stage in Figure 1d and ends with updating the tuning curves. Each neuron's tuning curve is modulated by the spatial location of the neuron, the distance to the input sample, the interaction kernel size, and a decaying learning rate $\alpha(k)$,

$$\Delta \xi_{i}^{p}(k) = \alpha(k)h_{b,i}^{p}(k)((s^{p}(k) - w_{in,i}^{p}(k))^{2} - \xi_{i}^{p}(k)^{2}). \tag{5}$$

If we consider learned tuning curves shapes for 5 neurons in the input SOMs (i.e. neurons 1, 6, 13, 40, 45), depicted in Figure 3, we notice that higher input probability distributions are represented by dense and sharp tuning curves. Whereas lower or uniform probability distributions are represented by more sparse and wide tuning curves. Using this mechanism, the network optimally allocates resources (i.e. neurons): a higher amount to areas in the input space,

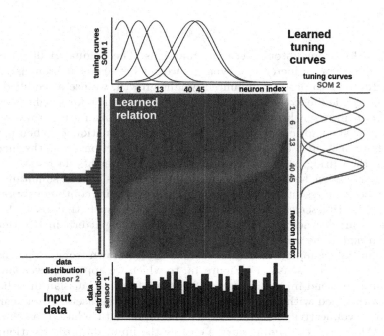

Fig. 3. Extracted sensory relation and data statistics using the proposed model

which need a finer representation; and a lower amount for more coarsely represented areas. This feature emerging from the model is consistent with recent work on optimal sensory encoding in neural populations [6]. This claims that, in order to maximise the information extracted from the sensory streams, the prior distribution of sensory data must be embedded in the neural representation.

The second component of the proposed model is the Hebbian linkage. This consists of a fully connected matrix of synaptic connections between neurons in each input SOM. Using an all-to-all connectivity pattern each SOM unit activation is projected to the Hebbian matrix. The Hebbian learning process is responsible for extracting the co-activation pattern between the input layers (i.e. SOMs), as shown in Figure 1c, and for eventually encoding the learned relation between the sensors, as shown in Figure 3. As one can see in Figure 3 - central panel, the connections between uncorrelated (or weakly correlated) neurons in each population (i.e. w_{cross}) is suppressed (i.e. darker color) while correlated neurons connections are enhanced (i.e. brighter color). The effective correlation pattern encoded in the w_{cross} matrix, imposes constraints upon possible sensory values. Moreover, after the network converges, the learned sensory dependency will make sure that values are "pulled" towards the correct (i.e. learned) corresponding values, will detect outliers, and will allow inferring missing sensory quantities. Formally, Hebbian connection weights, $w^p_{cross,i,j}$, between neurons i, j in each of the input SOM population are updated using

$$\Delta w^p_{cross,i,j}(k) = \eta(k)(a^p_i(k) - \overline{a}^p_i(k))(a^q_j(k) - \overline{a}^q_j(k)), \tag{6}$$

where

$$\overline{a}_i^p(k) = (1 - \beta(k))\overline{a}_i^p(k - 1) + \beta(k)a_i^p(k), \qquad (7)$$

and $\eta(k)$, $\beta(k)$ are monotonic decaying functions. The original Hebbian postulate only allows for an increase in synaptic weight between synchronously firing neurons. In order to prevent unlimited weight growth, we use a modified Hebbian learning rule (i.e. covariance rule, Equation 6) to allow for weight decreases when neurons fire asynchronously. The proposed mechanism uses a time average of pre- and postsynaptic activities, $\overline{a}_i^p(k)$, defined in Equation 7. When neurons fire synchronously in a correlated manner their connection strengths increase, whereas if their firing patterns are anticorrelated the weights decrease.

Self-organisation and correlation learning processes evolve simultaneously, such that both representation and correlation pattern are continuously refined. Moreover, the timescales of the two processes align, such that once the representations are learned in the SOMs the correlation pattern in the Hebbian connection matrix becomes sharper.

In the initial example we consider a set of values drawn from a uniform random distribution (i.e. sensor 1), Figure 1a, to which we apply a power-law, and we compute a second input (i.e. sensor 2) drawn from a Gaussian distribution. The network is fed with random pairs from the two datasets. After learning, the Hebbian connectivity matrix encodes the input data relation, as shown in Figure 3. Moreover, the tuning curves encode the input data distribution: narrower spaced for higher probability distributions and widely spaced for lower (or uniform) distributions of the input data.

In order to develop and test the specific hypotheses of the proposed model, we apply it for a quadrotor 3D egomotion estimation, briefly depicted in Figure 4.

3   Instantiating the Model

For the basic testing scenario, a quadrotor hovers (remote controlled) in an uncluttered environment, while an overhead camera system keeps track of its position and orientation.

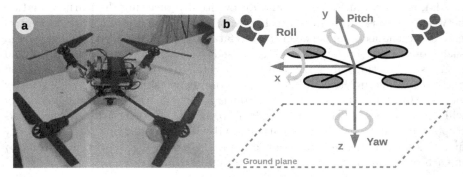

Fig. 4. Experimental setup: a) Quadrotor platform; b) Reference system alignment and ground truth camera tracking system

After the flight, preprocessed data from the available sensors (i.e. gyroscope, accelerometer and a magnetic sensor, Figure 5a) is fed to the model to extract the relations between the sensors for each of the three degrees of freedom (i.e. roll, pitch and yaw). The reference system setup is depicted in Figure 4b. Although initially the model considers all sensory contributions for the estimation of all motion components, as shown in Figure 5b, it will enforce only those connections providing a coherent relation for each degree of freedom, as shown in Figure 5c. As mentioned earlier, all-to-all connections between sensors are considered, but only those encoding coherent relations (i.e. contributions to same degree of freedom estimate) are enforced and considered for subsequent fusion. For roll and

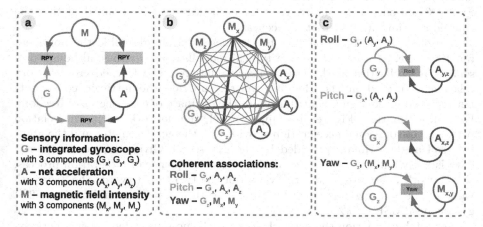

Fig. 5. Network instantiation for 3D egomotion estimation: network structure and sensory associations. a) Sensors configuration; b) Network connectivity; c) Sensory associations for learning.

pitch angles (i.e. rotation around the x and y reference frame axes), the network learns the relation between the roll and pitch angle estimates from integrated gyroscope data and rotational acceleration components (i.e. orthogonal x and y with respect to z reference frame axes). Similarly, the yaw angle is extracted by learning the relation between the yaw angle estimate from integrated gyroscope data (i.e. absolute angle) and aligned magnetic field components from the magnetic sensor (i.e. projected magnetic field vectors on orthogonal x and y reference frame axes). The sensory associations are not arbitrary, but rather represent the dynamics of the system and are consistent with recently developed modelling and control approaches for quadrotors [7], [8]. To make use of the learned relations we decode the Hebbian connectivity matrix using a relatively simple optimisation method [9]. After learning, we apply sensory data from one source and compute the sensory elicited activation in its corresponding (presynaptic) SOM neural population. Furthermore, using the learned cross-modal Hebbian weights and the presynaptic activation, we can compute the postsynaptic activation.

Given that the neural populations encoding the sensory data are topologically organised (i.e. adjacent values coding for similar places in the input space), we can precisely extract (through optimisation) the sensory value for the second sensor given the postsynaptic activation pattern. Without using an explicit function to optimise, but rather the correlation in activation patterns in the input SOMs, the network can extract the relation between the sensors.

4   Experimental Results

In order to validate the extracted relations, we use the aforementioned mechanism to extract the roll, pitch, and yaw estimates for the quadrotor scenario. Figure 6 presents a decoupled view for each degree of freedom, depicting the learned relations and estimation accuracy.

We observe in Figure 6a that the learned relations resemble the nonlinear functions (i.e. arctangent) used in typical modelling approaches, although preserving irregularities in the cross-sensory relations. The learned cross-sensory relations, encoded in the Hebbian matrix, provide the intrinsic constraints between the sensory cues contributing to the estimate of each degree of freedom. For roll estimation, Figure 6b - upper panel, the network learns the relation between net rotational acceleration provided by the accelerometer and the absolute roll angle estimate provided by the gyroscope. Given that accelerometer data is noisy and gyroscope data drifts, as a consequence of integration process, the network is able to "pull" the values of the two cues towards the "correct" value of the roll angle as given by ground truth (accelerometer RMSE: $< 2\%$, gyroscope RMSE: $< 3\%$).

For pitch estimation the network extracts the nonlinear dependency between the accelerometer data and the gyroscope data. Although both cues follow the trend of change in angle, as shown in Figure 6b - middle panel, the accelerometer is overestimating, due to the noisy signal and the overall limited motion of the drone on this axis. The gyroscope contribution was able to modulate the accelerometer contribution such that the overall estimates are acceptable (accelerometer RMSE: $< 7\%$, gyroscope RMSE: $< 3\%$).

Finally, for yaw estimation the network uses the gyroscope absolute angle and the magnetometer contribution, based on magnetic field readings on the other two axes. Interestingly, albeit the fact that the yaw estimate of the magnetometer follows the trend, Figure 6b - lower panel, there is an intrinsic offset (RMSE:$\sim$ 15%) visible from $t = 5s$. Investigating during many test flights, we noticed that the current change generated when arming the rotors introduced a significant modification in magnetic field distribution, subsequently reflected in the magnetometer readings. In the current setup, the network is not able to explicitly compensate for the offset, as one can see in Figure 6a - lower panel, where the co-activation pattern is not sharp like for roll and pitch.

As our results show, the model is able to extract the underlying data statistics without any prior information, as depicted in Figure 3, such that the sensory data distribution was learned directly from the input data. Moreover, following

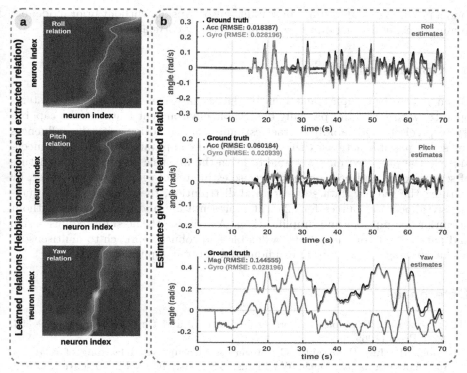

Fig. 6. Network instantiation for 3D egomotion estimation: a decoupled view analysis. a) Learned relations; b) Estimation quality using learned relations.

the statistics of the data, the network allocates more neurons to represent areas in the sensory space with a higher density such that the cross-sensory relations are sharpened, visible in Figure 6a.

As shown in Section 2, there is no specific parameter tuning routine (Figure 1d) to handle different kinds of input data for different scenarios. The generic processing elements (i.e. SOM, Hebbian learning) and their extensions (i.e. tuning curve adaptation, covariance update) ensure that the network first learns (in an unsupervised manner) the structure of the data, and then uses this representation to sharpen its correlational structure. Moreover, given the learned relations, the network is able to infer missing quantities in the case of sensor failures. As the relation is encoded as a synaptic weight, after learning, it is enough to provide samples from one sensor, encode them in the SOM, and project the activity pattern through the Hebbian matrix. The resulting activity pattern, subsequently decoded, will provide the missing real-world sensory value.

The proposed learning scheme extends [10], in which given various sensory inputs and simple relations defining inter-sensory dependencies, the model infers a precise estimate of the perceived motion. Now, by alleviating the need to explicitly encode sensory relations in the network dynamics, we propose a model providing flexible and robust multisensory fusion, without prior modelling assumptions, and using only the intrinsic sensory correlation pattern.

5   Conclusions

Given relatively complex and multimodal scenarios in which robotic systems operate, with noisy and partially observable environment features, the capability to precisely and rapidly extract estimates of egomotion critically influences the set of possible actions. Utilising simple and computationally effective mechanisms, the proposed model is able to learn the intrinsic correlational structure of sensory data and provide more precise estimates of egomotion. Moreover, by learning the sensory data statistics and distribution, the model is able to judiciously allocate resources for efficient representation and computation without any prior assumptions and simplifications. Alleviating the need for tedious design and parametrisation, it provides a flexible and robust approach to multisensory fusion, making it a promising candidate for robotic applications.

References

1. Gibson, E.J.: Principles of Perceptual Learning and Development, pp. 369–394. ACC Press (1969)
2. Cook, M., Jug, F., Krautz, C., Steger, A.: Unsupervised learning of relations. In: Diamantaras, K., Duch, W., Iliadis, L.S. (eds.) ICANN 2010, Part I. LNCS, vol. 6352, pp. 164–173. Springer, Heidelberg (2010)
3. Weber, C., Wermter, S.: A self-organizing map of sigma-pi units. Neurocomputing **50**, 2552–2560 (2007)
4. Mandal, A., Cichoki, A.: Non-Linear Canonical Correlation Analysis Using Alpha-Beta Divergence. Entropy **15**, 2788–2804 (2013)
5. Westermann, G., Mareschal, D., Johnson, M.H., Sirois, S., Spratling, M.W., Thomas, M.S.: Neuroconstructivism. Dev. Sci. **10**, 75–83 (2007)
6. Ganguli, D., Simoncelli, E.P.: Efficient Sensory Encoding and Bayesian Inference with Heterogeneous Neural Populations. Neural Computation **26**, 2103–2134 (2014)
7. Hyon, L., Park, J., Lee, D., Kim, H.J.: Build your own quadrotor. IEEE Robotics and Automation Magazine, 33–45 (2012)
8. Lee, J.K., Park, E.J., Robinovich, S.N.: Estimation of attitude and external acceleration using inertial sensor measurement during various dynamic conditions. IEEE Transactions on Instrumentation and Measurement **61**, 2262–2273 (2012)
9. Brent, R.P.: An Algorithm with Guaranteed Convergence for Finding a Zero of a Function. Algorithms for Minimization without Derivatives. Dover Books on Mathematics, pp. 47–58 (2013)
10. Axenie, C., Conradt, J.: Cortically inspired sensor fusion network for mobile robot egomotion estimation. Robotics and Autonomous Systems (2014)

How iCub Learns to Imitate Use of a Tool Quickly by Recycling the Past Knowledge Learnt During Drawing

Ajaz Ahmad Bhat[✉] and Vishwanathan Mohan

Robotics, Brain and Cognitive Sciences Department,
Istituto Italiano di Tecnologia, Via Morego 30, Genova, Italy
{ajaz.bhat,vishwanathan.mohan}@iit.it

Abstract. Using a skill learning architecture being developed for the humanoid iCub, in this article we show through experiments how a cognitive robot can learn to imitate quickly the use of a tool after a teacher's demonstration by recycling previously learnt knowledge from drawing experiments. The employed architecture incorporates novel principles for constructing a growing motor vocabulary in cognitive robots enabling "cumulative" and "swift" learning of skills through integration of multiple streams of learning like imitation and mental simulation. A central notion emphasized in the architecture is that movements can be represented in the form of 'shapes' instead of trajectories per se in order to liberate them from task specific details. The idea is to abstract out critical features/knowledge in the movement trajectory, which can later be reused in a context-independent manner.

Keywords: Imitation · Skill learning · Shape · Passive motion paradigm · Tool use · iCub

1 Introduction

Once upon a time, barter economy prevailed. Goods or services were exchanged for other goods or services. Then someone invented the concept of 'currency'. With this, humans started conducting trade and economics at one further level of 'Abstraction'. The core idea was to exploit the flexibility resulting from the establishment of a 'common measure' of value (and of course ease of storage too!). Simply, based on ones requirements (i.e. the goal), the right amount of currency could be transformed into any substance or service. What is the brains 'currency' for generating skilled goal directed 'movement'? Can we arrive at a small set of abstract motor vocabulary that when combined, sequenced, and shaped to 'context' (or the goal), allows the emergence of the staggering flexibility, dexterity and range that human actions possess? Consider for example, that, while learning to scribble a circle, draw a face or a flower, the brain of a child is also capable of learning something in general about 'circularity'. From a perception-action viewpoint, by the word 'circularity', we basically mean learning the essential ingredients' that will allow the infant in future to both 'perceive and generate' circular movements of any scale, in any location, using any body effector or a tool coupled to the body (of itself or someone else). The point is that the

© Springer International Publishing Switzerland 2015
S.P. Wilson et al. (Eds.): Living Machines 2015, LNAI 9222, pp. 339–347, 2015.
DOI: 10.1007/978-3-319-22979-9_33

straightforward advantage of learning a motor skill (even movement) in an 'abstract' fashion is that it immediately unlocks the implicit potential to 'perceive, mime and begin to perform' several other skills (that share a similar structure). For example, consider actions like turning a steering wheel, uncorking a bottle of water, paddling a bicycle, unwinding a fishing rod, using a screwdriver, among others that also share an essence of "circularity" in the task space. Of course, it is applied in different contexts, they result in different environmental consequences and utilities (that also needs to be learnt). But does the capability to perceive the structure in these similar actions and "spontaneously" imitate someone performing them with a good enough first prototype becomes possible because the 'seeds' already exist in the form of abstract motor knowledge (learnt previously). What are the seeds, how are they learnt, how are they transformed into task-specific movements based on the goal?

The ability to imitate is one of the pivotal behavioral traits of human beings for learning of motor skills. Imitation occurs through a complex set of mechanisms that map the observed movement of the teacher to the motor apparatus of the learner [1]. However, humans can swiftly employ skills learnt during imitating one task in a totally different context. How do humans 'recycle' the knowledge of movement across skills? And for a cognitive robot, how should motor knowledge be represented such that learning one motor skill also evokes the implicit potential to 'perceive, mime and begin to perform' several other skills that share a similar structure? In addition, for the robots entering the social domain for household support or industrial domain for automation, painting, manufacturing etc., or healthcare for surgery and elderly support, the capability to maneuver the range of tools which humans use (requiring an overlapping set of motor skills) is of prime importance.

In this article, we attempt to address this problem by showing how iCub utilizes motor knowledge learnt in one task in a different task swiftly. This work builds on general framework for skill learning and tool use in iCub proposed in an earlier work on drawing shapes [2]. A shape is a high level invariant representation of movement which is independent of the scale, location, orientation, time and also the end-effector/body chain that creates it. Building on the notion of shapes, the goal of this article is to elucidate how teacher's demonstration, practice and past experience of learning to draw can be meaningfully exploited and integrated to both aid the learning process and perform goal directed "body+tool" actions required to complete the loop of imitation. The sections that follow gradually build up the discussion starting from the necessary primitive building blocks (section 2), followed by details on learning to imitate use of a new tool by a combination of physical and social interactions (section 3) and ending with a few concluding remarks (section 4).

2   Building Blocks

This section briefly describes two subsystems that are necessary for robot's interaction with the world through perception and action to facilitate the learning process of imitation. The first is a visual perception system for acquiring and processing of visual information from the environment. The other is an action coordination system for the robot that is necessary to simulate actions in mental space as well as generate motor commands for controlling iCub's body.

2.1   Visual Perception System

iCub's eyes (two stereo cameras) form the main source of sensory input for the perception modules. The image frames (320 x 240 pixels) from the left and right cameras are first passed to a color segmentation module which recognizes objects in the images based on their color information (for example in our case the tool handle was recognized in green color). A motion detection module uses the information about detected objects from the color segmentation module to capture the trajectory of a moving tool in the visual workspace of iCub. The result is a time series of 2D coordinates, for the left and right eye, respectively of the detected tool:

$$\begin{Bmatrix} U_{left}(t) \\ V_{left}(t) \end{Bmatrix}, \quad \begin{Bmatrix} U_{right}(t) \\ V_{right}(t) \end{Bmatrix}, \quad t \in [t_{init}, t_{fin}]$$

The time interval $[t_{init}, t_{fin}]$ specifies the duration of teacher's demonstration and the U,V are the coordinates of the centroid of the detected color blob/tool-handle in left and right eye image frames (in the horizontal and vertical directions respectively). In the first part of imitation loop, this subsystem is used to track the movement of the tool end-effector when the teacher is demonstrating tool use. The third component within visual perception system is for obtaining 3D coordinates of a desired point (U(t),V(t)) in the time series of 2D coordinates recorded from both eyes. We re-use an existing 3D- reconstruction system [3] on iCub that basically controls the iCub gaze by steering the neck and the eyes to perform saccades towards the point of interest in 2D image frames such that the desired pixel is in the centre of fovea of both eyes. Then using a low-level kinematic solver and tri-angulation method, 3D location of the desired point is calculated in iCub's egocentric frame of reference. The approach is fairly accurate for our purposes to carry out the desired experiments. In summary, the perception subsystem recognizes the tool, extracts the movement of the tool during teacher demonstration and calculates out 3D locations of desired salient points in the acquired movement trajectory.

2.2   Action Generation System: Neural PMP, A Forward-Inverse Model

Passive Motion Paradigm (PMP) is a forward inverse model for action generation/ simulation in robots implemented using non–linear attractor dynamics on a plastic, learnt neural implementation of the body-schema (see Figure 1). The basic idea can be expressed in qualitative terms by suggesting that the process by which the robot can determine the distribution of work across a redundant set of joints, when the end-effector is assigned the task of reaching a target point in space, can be represented as an 'internal simulation process on the internal body model that calculates how much each joint would relax if an externally induced force (i.e. the goal) pulls the end-effector by a small amount towards the target. This process of relaxation is similar to coordinating the movements of a puppet by means of attached strings, as the puppeteer pulls the task relevant tip of the body to a target, the rest of its body elastically reconfigures so as to allow the tip to reach the target. In the context of this work, two features of PMP need to be highlighted. The forward model of PMP can predict the sensory consequences of the imagined actions. This allows for learning the sensorimotor knowledge about shapes through mental simulation without any need for execution of real movements.

PMP owing to its plasticity allows coupling of tools to the body in a seamless manner allowing to perform any goal directed 'body+tool' actions. In our experiments, this subsystem is used to reach, grasp the tool (turn-disk tool) and then further generate motor commands for torso-arm movements both during learning phase and goal directed execution of actions using 'body+tool' chain. We refer the interested readers to a review [4] that describes the PMP framework in detail.

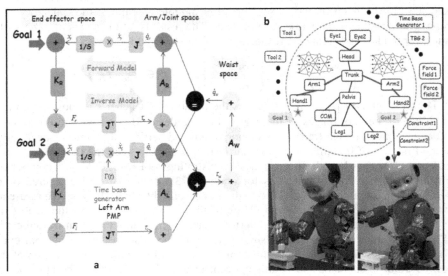

Fig. 1. Panel (a) PMP network for iCub upper body coordination. As shown the network is grouped into multiple motor spaces (end effector, arm joint, waist), each motor space consisting of a displacement node (blue) and a force node (pink). Vertical connections (purple) denote impedances (K: Stiffness, A: Admittance) in the respective motor spaces and horizontal connections (green) denote the geometric relation between the two motor spaces represented by the Jacobian (J). (b) Shows the task specific configurability of the body schema. Based on the goal and end-effector, only the concerned motor parts are activated (as shown by red lines). Panel (c) shows iCub using the right arm-torso combination in one case and the left-arm-torso one in the other based on the task.

3   Learning to Imitate the Use of Tool

Building on the existing skill learning architecture in iCub [2], in this article we show how iCub recycles the past knowledge of creating different kinds of shapes learnt while drawing, in the context of a new goal to imitate a teacher's demonstration of using a tool in order to bring an unreachable object in the reachable workspace of the robot. To give an overview, the loop of imitation begins with iCub visually tracking the trajectory generated by the moving tool-end effector as the teacher demonstrates operating the tool. The tracked trajectory is analyzed in line with formalism of the Catastrophe Theory [5] to detect the 'shape' of the generated trajectory. The 'shape' is an abstract representation in terms of a set of critical points that define the essential local features within the trajectory. From the critical points, iCub generates a set of virtual trajectories by recycling the past knowledge about shapes and exploring a bit in the parameter space to best match the desired trajectory. After a series of self-evaluations of the similarity of the 'shape counterparts' of the iCub generated trajectories (in the mental space) and the desired observed trajectory, parameters for the appropriate virtual trajectory can be learnt and corresponding movement patterns can be executed by iCub. In the following text, we describe in sequence the steps carried out to realize this loop of imitation.

Observing the Teacher's Demonstration: The loop begins with the teacher's demonstration of tool use. The tool (see Figure 2 Panel A) is a 'turn-disk' tool similar to the ones used in experiments on imitation and tool-use in animals [6],(see Figure 2 Panel C) where animals learn to 'rotate' the tool after observing a human or another animal (which already has learnt) using the tool to get food. After holding the tool end-effector, as the teacher rotates it in the workspace of the robot, the visual perception system tracks the tool end-effector and records trajectories as pairs of (U(t),V(t)) coordinates of the detected tool end-effector in the image planes of two eyes from the beginning till the end of the demonstration.

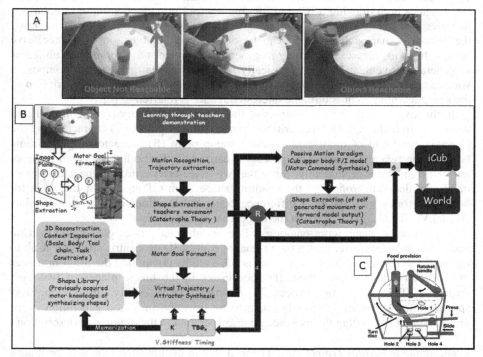

Fig. 2. Panel A shows the tool used in our experiments with iCub. The figure on left shows the target object is not reachable to the robot. The figure in the middle shows the tool as captured through the left camera during a demonstration with a curved arrow drawn over it showing the flow of trajectory. The green blob represents the identified tool end-effector. The figure in right shows the target object becomes reachable after rotating the tool. Panel B illustrates the computational architecture of the imitation learning system, the building blocks and the flow of information. Panel C shows the turn-disk tool used in experiments on imitation, tool-use and cognition in animals.

From the Observed Trajectory to its Shape: The introduction of a level of abstraction in the form of shapes in the domain of imitation against the explicit point-to-point imitation of observed trajectory is a central feature in this architecture. A global shape of the recorded trajectory is figured out using the catastrophe analysis approach [5]. This approach describes the shape of a smooth trajectory in terms of a set of 12 primitive shapes or critical points which basically characterize special local features like "bumps", "cusps" etc. in the trajectory. The set is sufficient enough to represent the shape of any trajectory in general [5].

By introducing shape representation, learning in iCub goes far beyond the reproduction of mere 3D trajectory to the ability to generate a vast repertoire of shapes independent of task-specific details like scale, position, coordinate frames and body effectors that underlie the creation of a trajectory. The motor knowledge of synthesized shapes once learnt can be 'reused' in a range of seemingly unrelated motor tasks by imposing a suitable context to the learnt abstract action representation. For example, the knowledge of creating a bump like shape learnt while drawing the letter 'C' can be efficiently exploited when iCub learns to rotate the turn-disk tool. This is possible because both drawing a 'C' and rotating the tool rely on creation of circular trajectories in different contexts. More complex shapes can be composed by combining the primitive shapes; like an 'S' can be drawn as a combination of 4 successive bumps. In short, instead of learning to generate movement trajectories, if iCub learns to generate movement shapes, it basically has the graphical grammar to compose a wide range of context-based spatiotemporal trajectories. For a detailed description of the shape learning architecture, the interested reader is referred to [2].

In this experiment, the CT analysis of the observed trajectory (see Figure 3 Panel A) results in a shape with three critical points; two end-points (E) characterizing the beginning and end of the trajectory and a bump point (B) characterizing a minima between the two end-points. This process is carried out for both the trajectories from left and right eye independently. Since the two trajectories represent the same movement recorded synchronously, the resulting shape from CT analysis is the same for both trajectories with the only difference being the locations of the critical points $(U_{left}(t), V_{left}(t), U_{right}(t), V_{right}(t))$ in the respective image planes.

The extracted shape representation is like an 'abstract' visual goal in 2D to iCub and needs to be converted into a motor goal in iCub's 3D egocentric space for iCub to reproduce the observed movement. This is achieved by computing the 3D coordinates of the control points from their 2D coordinates in the image planes of two eyes through a 3D reconstruction process [3] described above briefly. In addition, task-specific information like the body chain to be involved in action generation, wrist orientation while holding the tool etc. are also specified in the motor goal description.

From Shape to Virtual Trajectories: The next step in the imitation loop is to generate a trajectory from the motor goal whose shape matches the shape of the demonstrated trajectory as closely as possible. A Virtual Trajectory Generation System (VTGS) transforms the discrete set of shape critical points into a continuous set of 'equilibrium points', i.e., a virtual trajectory that acts as an attractor to the PMP based internal body model of iCub. Since a range of trajectories can be shaped through the sequence of discrete shape points, VTGS is characterized by two parameters i.e., the virtual stiffness 'K' and timing τ (Figure 2 Panel B, green boxes) to shape the attractor landscape in a desired manner.

By exploring this parameter space, a range of virtual trajectories can be generated and using a self-evaluation procedure correct parameters can be learnt that result in a virtual trajectory of the same shape as that of the demonstration.

Recycling Past Knowledge about Shapes: Starting with a Good Approximation:
However, since iCub has learnt to generate many primitive shapes while learning to draw letters [2] that included learning to generate a bump to draw the letter 'C', this knowledge comes of use here providing a good starting approximation of the right parameters needed to synthesize the correct trajectory for tool use (see Figure 3 Panel B). The recycling of previously learnt knowledge drastically reduces the exploration needed in the parameter space. What more is needed now is just a small amount of fine tuning in the parameter space to get a virtual trajectory of the same shape as that of the demonstration.

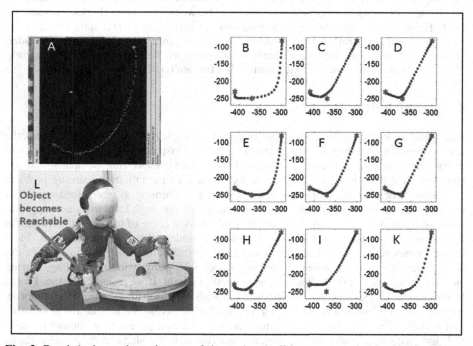

Fig. 3. Panel A shows the trajectory of the tool-end effector as recorded by iCub's left eye camera when the teacher has demonstrated use of the tool. The control points detected after the CT analysis are highlighted in red color i.e., the two end points and a bump point. Panels B-K illustrate an exemplary subset of the mentally simulated trajectories by iCub while learning the correct trajectory for using the tool. These trajectories are in the X-Y plane of iCub's egocentric frame of reference in which the tool rotates, the Z coordinate value equals the height of the tool (50 mm above the origin of the egocentric frame of reference). Panel B is the trajectory generated by recycling the parameters learnt while drawing the letter 'C'. Panels C-K are trajectories generated after small changes in the stiffness parameter 'K'. Panel K shows the trajectory that matches the similarity criterion and the motor equivalents of which are sent to the iCub's actuators. Panel L shows iCub replicating the trajectory completing the loop of imitation; with the consequence the object becomes reachable to the robot.

From Virtual Trajectory to Motor Actions: Learning the Best Possible Imitative Action: The resulting trajectory from VTGS acts as a moving attractor to PMP network which derives the corresponding motor commands to be sent to the actuators. In the end, when parameter learning has achieved the desired level of accuracy, the time course of motor commands is actively sent to the robots. However while learning the appropriate parameters for virtual trajectory; the forward model of PMP predicts the end-effector trajectory created as a consequence of the motor commands without executing any real action on iCub. Now in order to measure the similarity between the PMP generated trajectory (yet in mental space only) and the teacher demonstrated trajectory, the PMP generated trajectory is converted to the same shape format using CT analysis. The shapes of the desired motor goal and the generated motor action are then compared for equivalence i.e., they should contain the same set of shape critical points, in the same sequence and approximately in the same locations. iCub executes the trajectory if it meets the equivalence criterion thus completing the loop of imitation; otherwise, the relative distance between the two shapes is used to guide further exploration of the parameter space. Figure 3 (B) – (H) show some of the mentally simulated trajectories by iCub while fine tuning the parameters. A video recording of the described experiment on iCub is available at the following link: https://www.dropbox.com/s/d5xic7oinmz8dzn/ImitateToolUse.mp4?dl=0.

4   Concluding Remarks

The scenario of iCub using a tool after observing a teacher's demonstration embeds the core loop of the imitation, i.e. transformation from the visual perception of the teacher to motor commands of a learner and back. Though many interesting computational approaches have been proposed recently to tackle parts of the imitation learning problem (for review see [7]), our approach goes beyond these aiming at generalization across learning streams and multiple task contexts through a rather high level invariant representation of the motor skills in the form of 'Shapes'. A wide range of human actions result in formation of trajectories that ultimately result in similar 'shape' representations. Most important among them are line, bump and cusp critical points described in [5]. For example, drawing a circle, driving a steering wheel, uncorking, winding, cycling, stirring etc., are actions that have 'circularity' as invariant in them. If we teach a humanoid robot to synthesize 'Shapes' [6], we can endow them with the powerful capability to 'compose, recycle' the previously acquired motor knowledge to swiftly learn wide range of other motor skills. The employed skill learning architecture is a preliminary attempt to unify task independent knowledge acquisition/reuse (through the shape perception/synthesis hypothesis) and task specific compositionality (both at the level of shapes and at the level of force fields i.e. PMP) in a single computational framework. Is 'shape' the invariant through which humans imitate/mime effortlessly and instantaneously? Future research in this direction will be aimed at answering/validating such questions. Work described in this article and further efforts in this direction will guide research towards skillful learning in autonomous robots which can acquire a rich, growing 'action' vocabulary reusable in a wide range of tasks to assist humans meaningfully in daily life.

Acknowledgments. The research presented in this article is supported by the EU FP7 project DARWIN (www.darwin-project.eu, Grant No: FP7-270138).

References

1. Zentall, T.R.: Imitation: definitions, evidence, and mechanisms. Animal Cognition **9**, 335–353 (2006)
2. Mohan, V., Morasso, P., Zenzeri, J., Metta, G., Chakravarthy, V.S., Sandini, G.: Teaching a humanoid robot to draw `Shapes'. Autonomous Robots **31**(1), 21–53 (2011)
3. Pattacini, U.: Modular Cartesian Controllers for Humanoid Robots: Design and Implementation on the iCub, Ph.D. Dissertation, RBCS, Istituto Italiano di Tecnologia. (2011)
4. Mohan, V., Morasso, P.: Passive motion paradigm: an alternative to optimal control. Front. Neurorobot. **5**, 4 (2011). doi:10.3389/fnbot.2011.00004
5. Chakravarthy, V.S., Kompella, B.: The shape of handwritten characters. Pattern Recognition Letters **24**, 1901–1913 (2003)
6. Whiten, A., Spiteri, A., Horner, V., Bonnie, K.E., Lambeth, S.P., Schapiro, S.J., de Wall, B.M.: Transmission of Multiple Traditions within and between Chimpanzee Groups. Current Biology **17**, 1038–1043 (2007)
7. Lopes, M., Melo, F., Montesano, L., Santos-Victor, J.: Abstraction levels for robotic imitation: overview and computational approaches. In: Sigaud, O., Peters, J. (eds.) From Motor Learning to Interaction Learning in Robots. SCI, vol. 264, pp. 313–355. Springer, Heidelberg (2010)

Children's Age Influences Their Perceptions
of a Humanoid Robot as Being Like a Person or Machine

David Cameron[(⊠)], Samuel Fernando, Abigail Millings, Roger Moore,
Amanda Sharkey, and Tony Prescott

Sheffield Robotics, The University of Sheffield, Sheffield, UK
{d.s.cameron,s.fernando,a.millings,r.k.moore,
a.sharkey,t.j.prescott}@sheffield.ac.uk

Abstract. Models of children's cognitive development indicate that as children
grow, they transition from using behavioral cues to knowledge of biology to determine a target's animacy. This paper explores the impact of children's' ages
and a humanoid robot's expressive behavior on their perceptions of the robot,
using a simple, low-demand measure. Results indicate that children's ages have
influence on their perceptions in terms of the robot's status being a person, a
machine, or a composite. Younger children (aged 6) tended to rate the robot as
being like a person to a substantially greater extent than older children (aged 7)
did. However, additional facially-expressive cues from the robot did not substantively impact on children's responses. Implications for future HRI studies are
discussed.

Keywords: HRI · Design

1 Introduction

1.1 Background

With the increasing development of robots as toys, social companions, and tutors
targeted towards children as users in human-robot interaction (HRI), understanding
the means by which children perceive and evaluate robots is critical for HRI. Determining whether HRI is a special boundary-case for children or simply part of their
existing interactions with non-living objects allows the use and testing of established
developmental-psychology models of children's beliefs and behaviors. One such
model considers children to develop naïve theories of animacy, transitioning from
reliance on observing a target's behavior to knowledge of its biology [1]. Bio-inspired
robotics can be used to explore such models, altering robots' appearances and behaviors to determine influence on children's perceptions of their animacy. It is important
to consider the influence on HRI that humanoid robots' appearance and behavior has
for child users, given that young children use their animistic intuition to attribute intelligence, biology, and goals to encountered objects [1].

Current research offers mixed indications of children's perceptions of robots as ani-
mate or machine. Children aged 3-5 have a piecemeal understanding of animacy in their

© Springer International Publishing Switzerland 2015
S.P. Wilson et al. (Eds.): Living Machines 2015, LNAI 9222, pp. 348–353, 2015.
DOI: 10.1007/978-3-319-22979-9_34

beliefs concerning a robot dog; while they show refinements with age to their attribution of animacy, they still keep mixed beliefs regarding its agency and biology [2]. With age, children are more likely to classify pictures of humanoid robots as being pictures of machines rather than living [3]. However, further HRI work indicates children of that age do not evaluate a robotic dog as different from a stuffed dog toy in terms of animacy, biology, and mental states [4]. In sum, factors influencing children's perceptions of humanoid robots as persons or machines still remain to be uncovered.

1.2   Use Case

The Expressive Agents for Symbiotic Education and Learning (EASEL) project explores human robot symbiotic interaction (HRSI) with a view to understand the development of symbiosis over long-term robot and child tutoring interactions.

Symbiosis is the capacity for both the robot and the human user to mutually influence each other's behavior both within and across repeated encounters. To suitably explore HRSI, a robot needs to be responsive to the behavior and affective states of the human user and adapt its own behavior in ways that have meaningful and measurable influence on the person. This responsiveness in a Synthetic Tutoring Assistant (STA) may range from broad changes, such as tailoring tutoring style to meet a student's learning requirements, to focused changes, such as a robot's use of simulated affect expression when giving feedback.

Early research in the EASEL project [5] indicates that the presence of life-like facial expressions from a humanoid robot (Robokind Zeno R50 [6]) during feedback from the robot regarding children's game performance has a differential impact based on demographics. One mechanism proposed for this effect is that the presence of facial expressions might encourage the user to respond to the robot as a social agent or person rather than a machine or object. Perceptions of the robot as a social agent or a machine may in turn have important influence on users' attitudes and behaviors during HRI. As outlined above, age is a key factor likely to influence the target demographics' perceptions of the STA as a social agent.

This paper explores the influence that a user's age and the presence of life-like facial expressions by the robot impact on user perceptions of the robot as a social agent. We anticipate that younger children will rate the humanoid Zeno R25 as being significantly more like a person than older children will. This difference is anticipated to be strengthened by the presence of life-like robotic facial expressions.

2   Method

2.1   Design

A repeated measures design was employed so that differences in responses to the facially-expressive and non-expressive states of the robot could be explored. Allocation to condition was counterbalanced so that any order effects of the repeated exposure to the robot could be accounted for. The study took place at a local primary school across two days, allowing for approximately a day's break for participants between conditions.

2.2   Participants

Children from UK school years two and three were recruited from a local primary school to take part in the study by invite to play a game with Zeno the robot. 44 children volunteered to take part in the study and 39 completed both conditions. Of completers, there were 20 female and 19 male participants; 15 participants were from year 2 (age M = 6.38, S.E. = .07) and 24 were from year 3 (age M = 7.40, S.E. = .05). These age groups were identified in previous work as being engaged with the interaction and capable of completing self-report measures [5].

2.3   Measures

The primary measure for assessing the children's perceptions of the robot was a single item 100 point thermometer scale on which children pick a point for the robot ranging from 'Zeno is more like a machine' at the 0 point mark and 'Zeno is more like a person' at the 100 point mark. This measure had been pre-tested in a pilot study on adults and children and had been identified as suitable for children to articulate the differences between the two end points. Zeno had been described as a robot in the recruitment phase of the study and so use of the word 'robot' as an anchor point might have unduly primed children's responses.

This paper details part of a broader study concerning children's feelings, thoughts, and behavior in HRI, so it was necessary to use minimal items for measures to reduce overburdening participants with questions. Additional items used to control for children's interaction and understanding of the game consisted of: children's number of correct actions in the game and recognition of the robot's expression "Zeno was pleased / was disappointed / didn't mind, when I got an answer right / wrong".

2.4   Procedure

This study was developed to replicate procedure from a prior field-study [5] in a new, controlled testing environment. The experiment took place in a local primary school, where participants completed the game under the supervision of the research staff and one member of school staff. Information regarding participation was sent before recruitment and informed consent was obtained from parents.

During the game, children were free to position themselves within a designated 'play zone' (so that movements could be correctly detected by the system) marked out by tape starting 1.80m from the robot and extending to 3.66m away.

Interaction with Zeno took the form of the well-known game of *Simon Says*, which initiated once children stepped into the play zone and was autonomously delivered by the robot, including instructions for the children, commands to obey (or not), and feedback on the children's performance – reported as their score[1]. Children played a maximum of ten rounds and the game was repeated to present the alternate condition for the child on the next day of testing.

After each session, children completed a brief self-report questionnaire, including the critical measures described above. Participant experimenter interaction consistency was maintained across sessions for all tasks and the experimenter remained blind

---

[1]   Full details of the game's procedure can be found in [5], which was unchanged for this study.

to condition throughout. The sole experimental manipulation coincided with Zeno's vocal feedback after each game turn by including happy or sad facial animations corresponding to the vocal feedback on the children's performance. By contrast, in the control condition, Zeno's facial expression remained static when delivering feedback.

3   Results

There was a significant main effect for age $F(1,37) = 7.92$, $p < .01$. Children in the younger age group rate Zeno as being more like being a person (M = 77.27, S.E. = 8.83) to a substantially greater extent than those in the older age group (M = 41.58, S.E. = 6.98). This is a large effect observed (d = .95), which post-hoc tests indicate to be sufficiently powered (.88).

There was no main effect for gender $F(1,37) = .71$ $p = .41$, as both boys and girls tended to rate Zeno as being a mix of a machine and a person (Ms = 48.61 & 58.68, S.E.s = 8.57 & 8.35 respectively).

Results did not materially change when controlling for number of correct responses, nor children's perceptions of Zeno's responses to their actions. There was a strong correlation in children's responses between the two conditions $r(39) = .77$, p < .01.

There was no main effect seen for the presence of robotic facial expressions $F(1,37) = .87$, p = .36 as children reported similar scores for rating Zeno as a person or machine in the facially-expressive condition (M = 59.49, S.E. = 6.07) and the non-expressive condition (M = 55.36, S.E. 6.03). Similarly, there was no observed interaction effects between the condition and age groups of participants $F(1,37) = .61$, p = .44 (See Figure 1), gender $F(1,37) < .01$, p = .99, nor order in which conditions were presented $F(1,37) < .01$, $p = .94$.

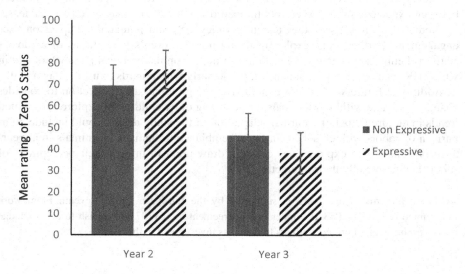

Fig. 1. Mean ratings of children's perceptions of Zeno as being like a machine (lower values) or like a person (higher values)

4   Discussion

In line with existing literature [3, 4], on average older children considered the Zeno robot to be significantly more like a machine than the younger children did. Critically, this large effect was demonstrated using a simple and accessible measure that could be included in future studies without overburdening young participants. Consistency in ratings between conditions was high; although the inclusion of expressions did not substantively affect children's ratings for either age group, this early work might have seen ceiling effects rating Zeno as like a person due to its apparent autonomous movement and response to the children. In particular for young children, these cues could be instrumental in their regarding a humanoid robot as being like a person [1].

While not formally recorded in the current study, think-aloud reasoning by the children on making a judgment included statements such as "He talks like a person but he's got oil and gears", (when rating as more like a machine) and "He's like a person because he knows when I move" (when rating as a person). These suggest possible use of behavioral cues and biological/mechanical knowledge to inform their judgements and further work exploring why children are making particular ratings is recommended.

The results have implications for the use of Zeno in the planned tutoring role as an STA [7] in both a social, tutoring capacity and a user-acceptance capacity. Firstly, it may impact on the potential for Zeno to act as a co-learner or tutor for scenarios concerning biology or health education. For example, user engagement in inquiry learning through comparative work on the differences between machines and people may be influenced by a child's perception of Zeno as like a machine or person. Second, a child's perspective of whether Zeno is like a machine or person may impact on the type of behaviors the user expects form the robot. Sufficient differences between expected and observed robotic features or behaviors may give rise to unease, as predicted by models of the 'uncanny valley' [8] and potentially impact on user engagement. Further work exploring if and how children's perceptions of a robot's status as being like a person or machine can impact on the practical parameters within which HRI can occur is recommended, particularly with regards to user engagement.

Additional future work could benefit from longitudinal data collection to consider changes with age within individuals, or including older children to explore age-related trends in children's beliefs. Further adaptation of the study design could include comparison of robots with closer or further resemblance to humans or animals in appearance or behavior to explore cues children draw upon to inform their perceptions of robots and potentially their interactions.

Acknowledgments. This work is supported by the European Union Seventh Framework Programme (FP7-ICT-2013-10) under grant agreement no. 611971. We wish to acknowledge the contribution of all project partners to the ideas investigated in this study.

References

1. Carey, S.: Conceptual change in childhood. MIT press, Cambridge (1985)
2. Okita, S.Y., Schwartz, D.L.: Young children's understanding of animacy and entertainment robots. International Journal of Humanoid Robotics 3(3), 393–412 (2006)
3. Saylor, M.M., Somanader, M., Levin, D.T., Kawamura, K.: How do young children deal with hybrids of living and non-living things: The case of humanoid robots. British Journal of Developmental Psychology 28(4), 835–851 (2010)
4. Kahn Jr, P.H., Friedman, B., Perez-Granados, D.R., Freier, N.G.: Robotic pets in the lives of preschool children. In: CHI 2004 Extended Abstracts on Human Factors in Computing Systems, pp. 1449–1452. ACM (2004)
5. Cameron, D., Fernando, S., Collins, E., Millings, A., Moore, R.K., Sharkey, A., Evers, V., Prescott, T.: Presence of life-like robot expressions influences children's enjoyment of human-robot interactions in the field. In: AISB2015: Proceedings of the Symposium on New Frontiers in Human-Robot Interaction (to appear 2015)
6. Hanson, D., Baurmann, S., Riccio, T., Margolin, R., Dockins, T., Tavares, M., Carpenter, K.: Zeno: A cognitive character, pp. 9–11. AI Magazine (2009)
7. Fernando, S., Collins, E.C., Duff, A., Moore, R.K., Verschure, P.F., Prescott, T.J.: Optimising robot personalities for symbiotic interaction. In: Duff, A., Lepora, N.F., Mura, A., Prescott, T.J., Verschure, P.F.M.J. (eds.) Living Machines 2014. LNCS, vol. 8608, pp. 392–395. Springer, Heidelberg (2014)
8. Moore, R.K.: A Bayesian explanation of the 'Uncanny Valley' effect and related psychological phenomena. Scientific Reports 2(864) (2012)

Help! I Can't Reach the Buttons: Facilitating Helping Behaviors Towards Robots

David Cameron[✉], Emily C. Collins, Adriel Chua, Samuel Fernando,
Owen McAree, Uriel Martinez-Hernandez, Jonathan M. Aitken,
Luke Boorman, and James Law

Sheffield Robotics, The University of Sheffield, Sheffield, UK
{d.s.cameron,e.c.collins,dxachua1,s.fernando,o.mcaree,
uriel.martinez,jonathan.aitken,l.boorman,j.law}@sheffield.ac.uk

Abstract. Human-Robot-Interaction (HRI) research is often built around the premise that the robot is serving to assist a human in achieving a human-led goal or shared task. However, there are many circumstances during HRI in which a robot may need the assistance of a human in shared tasks or to achieve goals. We use the ROBO-GUIDE model as a case study, and insights from social psychology, to examine two factors of user trust and situational ambiguity which may impact promote human user assistance towards a robot. These factors are argued to determine the likelihood of human assistance arriving, individuals' perceived competence of the robot, and individuals' trust towards the robot. We outline an experimental approach to test these proposals.

Keywords: HRI · Design · Guidance · Mapping

1 Background

Human robot interaction (HRI) research typically explores interactions based around a robot in a supportive or assistive role for the human user [1]. However, there are circumstances in HRI that could require the user to support a robot in shared tasks or to achieve its aims. Current research in this area identifies means for robots to determine *when* to request help [2] and where to *seek* help [3] from humans. However effective means of *how* socially adaptive robots ask for help to encourage user response and helping behavior is still a challenge in HRI.

In this paper, we discuss two factors identified in social psychology, trust and ambiguity, which have impact on human-human cooperative and helping behavior, and their relevance to HRI. Insights from social psychology can be useful in exploring HRI, given that HRI is a somewhat novel and developing area of research [4]. We explore models of cooperation and helping from social psychology in an applied HRI scenario of humans helping robots and outline an experimental proposal to test these.

1.1 Promoting Cooperation

A prominent social psychological model of interpersonal cooperation, including helping behavior towards an individual, identifies trust as its foundation [5].

© Springer International Publishing Switzerland 2015
S.P. Wilson et al. (Eds.): Living Machines 2015, LNAI 9222, pp. 354–358, 2015.
DOI: 10.1007/978-3-319-22979-9_35

McAllister argues that for a task that necessitates two individuals working together to be achieved successfully and efficiently, both individuals need to trust each other. More specifically, cooperation is argued to benefit from both affective trust (arising from personable interactions by one's peers) and cognitive trust (evidence that one's peers carry out their responsibilities reliably) [5]. Analogues for both forms are seen in, for example, 1) a meta-analysis of factors promoting trust in HRI [1]; 2) user's perceptions of robots' performance (analogous to cognitive trust); and 3) a robots' attributes such as personality (analogous to affective trust), which are found to positively contribute towards user trust in robots.

1.2   Promoting Helping Behavior

Key studies from social psychology indicate that ambiguity in helping scenarios result in substantial detriment to pro-active helping behavior in individuals [6]. Interaction with robots in cooperative environments may present a novel scenario for many people, as a result human agents in this situation face potential ambiguity in how to behave towards the robot, and uncertainty regarding *whether to help* the robot. Similarly, with respect to cognitive trust, a lack of a clear plan or intention communicated by the robot may create further ambiguity in *how* the individual may help a robot, limiting their action taken[1].

2   ROBO-GUIDE Helping Scenario

To explore the suggested social factors influencing cooperation with, and helping behavior displayed towards a robot by a human agent it is useful to consider an interactive scenario in which these circumstances arise. The ROBOtic GUidance and Interaction DEvelopment (ROBO-GUIDE) project [8] is an ideal scenario to consider the impact of such social factors as it requires humans to place trust in a robot, whilst also assisting in the robot overcoming obstacles or barriers to progress.

ROBO-GUIDE is implemented on the Pioneer LX mobile platform. The platform is able to autonomously navigate a multistory building and lead users from their arrival to their destination. Our focus here is the point at which the platform changes between floors as it navigates the multistory building. As the ROBO-GUIDE is unable to call for an elevator itself it must rely on human support to press buttons to call the elevator and select the required floor[2]. For both the user and the robot this scenario presents a simple and low-risk circumstance in which a human user can act in order to meet a robot's needs.

---

[1]   Existing robot design may have indirectly addressed user unceratinty to some extent by using animal-like or non-threatening [7] morpholiges.

[2]   The wheeled ROBO-GUIDE is also unable to use stairs to navigate between floors – or rather, would only able to travel down stairs and likely only the once.

3   Helping Behavior and Experimental Proposal

Factors potentially affecting user helping behavior in HRI, identified in section 1, can be applied to the ROBO-GUIDE usage of elevators. We outline a brief experimental proposal to test their impact for the user in the helping scenario, described in section 2. Requests for help by the robot are planned to be manipulated in a 2x2 experimental design: inclusion or absence of competency-oriented, intentional statements and the inclusion or absence of friendly-oriented, relatable statements. Statements will be communicated using the on-board speech synthesizer, (although the viable alternatives of pre-recorded spoken phrases or an on-screen display are acknowledged and considered in further work). These statements are predicted to impact through four channels on helping-related affect and cognitions, along with perceptions of the situational ambiguity, robot's task-capability, and user-liking of the robot (see figure 1). The following subsections show proposed control-condition statements or requests in quotes and normal font, whereas additional, experimental statements are in the same quotes and italics.

3.1   Influencing Trust and Cooperation

Affective Trust
ROBO-GUIDE's primary purpose is to guide new visitors to the robotics laboratory. It is anticipated that many guests would be unfamiliar with everyday interaction with social and responsive robots and could find the experience unusual. To promote affective trust, we aim to include friendly and relatable references by the robot to the situation that is helping the user as a tour guide, "Please follow me; *I am here as your tour guide*". It is anticipated that perceptions of the robot's likability and affective trustworthiness would be greater with additional phrases regarding its offering 'face-to-face' direct assistance to the human user.

Cognitive Trust
Cognitive trust is developed through clear demonstration of an agent's ability to capably meet its intended and/or required responsibilities [5]. In this case the robot may imply its capability of finding the target destination by clearly communicating its aims. "Please follow me *to get to the robot labs*," gives indications that ROBO-GUIDE has a goal and will work towards it. It is anticipated that perceptions of the robot's competence and cognitive trustworthiness would be greater with the inclusion of additional phrases that directly address its intended aims.

3.2   Influencing Ambiguity and Helping

Global Circumstances
As discussed for affective trust, this scenario and HRI *in general* may be an unusual circumstance for many users. To reduce ambiguity in the case of interacting with a novel robot, especially a robot that needs help, we aim to include friendly and

relatable references by the robot to HRI. Essentially, we aim to explore the impact of ROBO-GUIDE stating that it is a robot with limitations, which the user can help with: "Please press the down button; *I can't quite reach the buttons*", and "Please press Ground floor; *good thing you're here to do that for me*". It is anticipated that this could diffuse the ambiguity in whether the robot needs help and encourage user liking of, and relating to, the robot.

Local Circumstances

It is anticipated that ROBO-GUIDE's statements of intentions alongside requests for help will promote user helping through a reduction of ambiguity *specific* to the interaction. In this case, demonstrate the robot has a clear goal it is trying to achieve but it is now facing an obstacle and so asks for help: "Please press the down button *to call the lift*" and "Please press Ground floor *for the Robot Labs*". Without declaring *why* the tasks are to be completed, requests for help are ambiguous in their purpose. It is anticipated that perceptions of the robot's competence would be greater and interaction ambiguity be lower with the additional phrases regarding its aims.

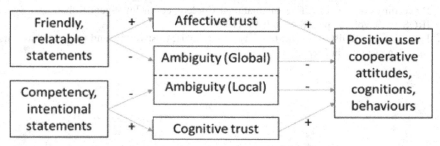

Fig. 1. Structural model of hypothesized impact of friendly and competency statements by the robot on user cooperative behaviors and cognitions

3.3   Experimental Summary

We anticipate that the use of two forms of phrases: friendly-oriented, relatable statements and competency-oriented, intentional statements will both encourage user helping affect, cognitions, and behaviors. We further anticipate that these statements will act through separate channels of raising trust and lowering ambiguity, independently targeting distinct elements of both (Figure 1). This paper offers a novel proposal for exploring HRI in terms of how robots may effectively encourage user helping.

Acknowledgments. This work is supported by the European Union Seventh Framework Programme (FP7-ICT-2013-10) under grant agreement no. 611971. We wish to acknowledge the contribution of all project partners to the ideas investigated in this study.

References

1. Hancock, P.A., Billings, D.R., Schaefer, K.E., Chen, J., De Visser, E., Parasuraman, R.: A meta-analysis of factors affecting trust in human-robot interaction. Human Factors: The Journal of the Human Factors and Ergonomics Society 53(5), 517–527 (2011)
2. Rosenthal, S.; Biswas, J.; Veloso, M.: An effective personal mobile robot agent through a symbiotic human-robot interaction. In: AAMAS 2010, pp. 915–922 (2010)
3. Rosenthal, S., Veloso, M.: Mobile robot planning to seek help with spatially-situated tasks. In: Proceedings of the Twenty-Sixth Conference on Artificial Intelligence (AAAI 2012), Toronto, Canada, pp. 886–891, July 22–26, 2012
4. Collins, C., Millings, A, Prescott. T.J.: Attachment to assistive technology: a new conceptualisation. In: Proceedings of the 12th European AAATE Conference Association for the Advancement of Assistive Technology in Europe (2013)
5. McAllister, D.J.: Affect-and cognition-based trust as foundations for interpersonal cooperation in organizations. Academy of Management Journal 38(1), 24–59 (1995)
6. Clark, D., Word, L.E.: Why don't bystanders help? Because of ambiguity? Journal of Personality and Social Psychology 24(3), 392–400 (1972)
7. Breazeal, C., Scassellati, B.: How to build robots that make friends and influence people. In: Proceedings of the IEEE/RSJ International Conference on Intelligent Robots and Systems IROS 1999, vol. 2, pp. 858–863. IEEE (1999)
8. Law, J., et al.: ROBO-GUIDE: towards safe, reliable, trustworthy, and natural behaviours in robotic assistants (in Prep.)

Tactile Language for a Head-Mounted Sensory Augmentation Device

Hamideh Kerdegari[1(✉)], Yeongmi Kim[2], and Tony Prescott[1]

[1] Sheffield Robotics, University of Sheffield, Sheffield, UK
{h.kerdegari,t.j.prescott}@sheffield.ac.uk
[2] Department of Mechatronics, MCI, Innsbruck, Austria
yeongmi.kim@mi.edu

Abstract. Sensory augmentation is one of the most exciting domains for research in human-machine biohybridicity. The current paper presents the design of a 2nd generation vibrotactile helmet as a sensory augmentation prototype that is being developed to help users to navigate in low visibility environments. The paper outlines a study in which the user navigates along a virtual wall whilst the position and orientation of the user's head is tracked by a motion capture system. Vibrotactile feedback is presented according to the user's distance from the virtual wall and their head orientation. The research builds on our previous work by developing a simplified "tactile language" for communicating navigation commands. A key goal is to identify language tokens suitable to a head-mounted tactile interface that are maximally informative, minimize information overload, intuitive, and that have the potential to become 'experientially transparent'.

Keywords: Sensory augmentation · Vibrotactile feedback · Tactile language

1 Introduction

Sensory substitution (translating one sensory modality into another [1]) was one of the first domains for research in human-machine biohybrid systems [2]. The development of devices for both sensory substitution and sensory augmentation (synthesizing new information to an existing sensory channel) remains an exciting prospect for biohybrid technology. For example, whilst sensory substitution can help people with impaired sensing systems, the additional senses provided by sensory augmentation can be used to augment the spatial awareness of people operating in hazardous environments such as smoked-filled buildings, on construction sites, or on the battlefield [3, 4].

Research in this area has been strongly influenced by the enactive view of cognition (see e.g. [5, 6, 7, 8]). Here, a key design aim is to make the device 'experientially transparent' such that the goal-directed behavior of the user naturally incorporates properties of the artifact including its capacity to transform from one sensory modality to another. Another influential approach has been from research on active perception—the view that sensing in animals including humans is purposeful and information-seeking. That

© Springer International Publishing Switzerland 2015
S.P. Wilson et al. (Eds.): Living Machines 2015, LNAI 9222, pp. 359–365, 2015.
DOI: 10.1007/978-3-319-22979-9_36

approach, together with bio-inspiration from mammalian sensing systems, informed our earlier efforts to develop a sensory augmentation device that incorporated a haptic interface for remote touch [3]. In the current contribution we describe our research on a second generation device that seeks to overcome some of the limitations of the earlier system. Here we describe the motivation for the approach and the design of a new prototype. Pilot results from the experiment outlined below will be presented at the conference.

2   A Sensory Augmentation System Inspired by the Mammalian Vibrissal System

Many mammals have a sensitive tactile sensing capacity provided by their facial whiskers (or vibrissae) that allows them to acquire detailed information about local environment useful for local navigation and object detection and recognition. Similar information could be provided to humans using a sensory augmentation system that combines active distance sensing of nearby surfaces with a head-mounted tactile display [3, 9]. Two such devices have been investigated to date: the Haptic Radar [9] and the Tactile Helmet [3].

The *Haptic Radar* [9] linked infrared sensors to head-mounted vibrotactile displays allowing users to perceive and respond simultaneously to multiple spatial information sources. Here, several sense-act modules were mounted together on a band wrapped around the head, each module measured distance from the user to nearby surfaces, in the direction of the sensor, and transduced this information into a vibrotactile signal presented to the skin directly beneath the module. Users intuitively responded to nearby objects, for example, by tilting away from the direction of an object moving close to the head, indicating that the device could be useful for detecting and avoiding collisions.

The *Tactile Helmet* [3] was a prototype sensory augmentation device developed in Sheffield in collaboration with *South Yorkshire Fire and Rescue* (SYFR) services. We selected a head-based tactile display as this allows rapid reactions to unexpected obstacles, is intuitive for navigation, can easily fit inside the helmet, and leaves the fire fighter's hands free for tactile exploration of objects and surfaces [9]. The first generation device (see Figure 1) comprised a ring of eight ultrasound sensors on the outside of a fire-fighter's safety helmet with four vibrotactile actuators fitted to the inside headband. Ultrasound distance signals from the sensors were converted into a pattern of vibrotactile stimulation across all four actuators. Thus, unlike Haptic Radar, the Tactile Helmet was non-modular, allowing direction signals from the array of sensing elements to be combined into an appropriate display pattern to be presented to the new user. One of the goals of this approach was to have greater control over the information displayed to the user, and, in particular, to avoid overloading tactile sensory channels by displaying too much information at once. This is particularly important in the case of head-mounted tactile displays, as vibration against the forehead is also detected as a sound signal (buzzing) in the ears; too much vibrotactile information can therefore be confusing and irritating and could mask important auditory stimuli. Despite seeking to provide better control over the signal display, however, field tests with the Tactile Helmet, con ducted at SYFR's training facility, showed that tuning the device to suit the user needs and situation was problematic. Specifically, a design that directly converted local distance information into vibration on multiple actuators generated far too much vibrotactile stimuli in confined situations such as a narrow corridor.

Fig. 1. Ist generation Tactile Helmet design undergoing field testing. In right-hand picture the fire-fighter is in a confined smoke-filled space in the South Yorkshire Fire and Rescue training facility.

The above tests established the need to better regulate the tactile display of information to ensure clear signals and to minimize distracting or uninformative signals. Through a series of psychophysical studies (e.g. [10]) we are investigating how to best optimize signals to relay information to the user. For instance, we want quantify people's ability to localize tactile stimuli on the forehead and to understand, and make use of, sensory phenomena such as the "funneling illusion" whereby nearby concurrent tactile stimuli are experienced as a single stimulus at a central point. Based on the outcome of these studies, we are currently developing a "tactile language" for testing with a new Tactile Helmet prototype. Specifically, using our new device we are seeking to understand what are the minimal haptic signals—the tokens of the command language—that can be used to relay useful navigational information. In the current study we wished to have full control over the information provided to the user and therefore we imagine a virtual wall, and used a motion capture system, to directly calculate the user's distance and orientation to that wall. The actuators on the helmet are then used to relay navigation commands to help the user move in a trajectory parallel to the wall. We evaluate the effectiveness of the commands according to speed of movement and the smoothness of the user's trajectory. In future studies we will also examine how the language could be used to convey navigational signals calculated directly from active distance sensors for real-world obstacles. The eventual aim is to identify a tactile command language that can be used with a map of local surface positions, estimated with ultrasound or ladar, and that is maximally informative, minimizes information overload, and intuitive; hopefully with the potential to become experientially transparent. The remainder of the paper explains the design of our new prototype and the experiment we are conducting to evaluate some of the possible tokens of the tactile language.

3   System Overview

3.1   Vibrotactile Helmet

The second-generation Tactile Helmet (fig. 2) consists of an array of twelve ultra-sound sensors mounted with approximately 30 degrees separation to the outside of a skiing helmet (2d), and a tactile display composed of 7 tactors (2b) [10].

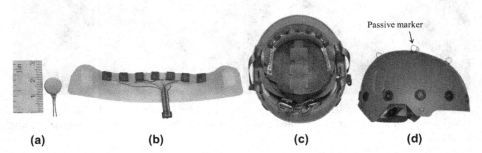

Fig. 2. (a): Eccentric rotating mass vibration motor (Model 310-113 by Precision Microdrives). (b): Tactile display interface. (c): Tactile display position inside the helmet. (d): Vibrotactile helmet.

The tactile display consists of seven eccentric rotating mass (ERM) vibration motors (2a) with 3V operating voltage and 220Hz operating frequency at 3V. These vibration motors are mounted on a neoprene fabric and attached on a plastic sheet (2b) with 2.5 cm inter-tactor spacing which can easily be adjusted inside the helmet. The helmet also incorporates an inertial measurement unit (IMU), a microcontroller unit and two small lithium polymer batteries (7.4 V) to provide the system power. As shown in Figure 3, the ultrasound sensors and IMU data are sent to the microcontroller through I2C BUS. The microcontroller in the helmet reads the sensors values and sends them to the PC wirelessly using its built-in WiFi support. The PC receives the sensor values and generates commands for the tactile actuators sending them back to microcontroller wirelessly for onward transmission to the tactile display. For the experiment described below we disable the direct generation of actuator commands and substitute signals based on information from the motion-capture system.

3.2   Tracking System

We used Vicon motion capture system as a precise optical marker tracking system to track the user's position and orientation. It consists of 10 cameras and reflective markers. The vibrotactile helmet, whose motion is to be captured by cameras, has five reflective passive markers attached to its surface (Fig. 2.d). Data generated by the Vicon software is streamed in real time to a PC via TCP/IP. Finally, the proper tactile command is generated and sent wirelessly to the helmet to navigate the user in the capture room.

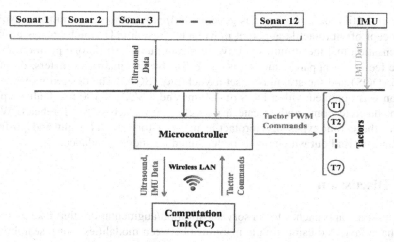

Fig. 3. Data flow diagram of vibrotactile helmet

4   Procedure

The aim of our experiment is to investigate the optimal vibrotactile commands and effectiveness of the proposed tactile commands for navigation along a virtual wall. The experiment is performed in the motion capture room (4 * 5m^2). The user's distance from the virtual wall is calculated continuously, based on this distance, and on the helmet orientation measured by motion capture system, the proper tactile command is produced. For our initial experiment we are evaluating different ways of communicating three simple tactile commands: *turn-right, turn-left* and *go-forward*. Turn right/left command induce a rotation around self (right/left rotation) which is used to control the human orientation; while go-forward command is intended to induce a motion toward forward direction. Fig. 4 illustrates the vibrotactile patterns for presenting turn left/turn right and go-forward commands in the tactile display.

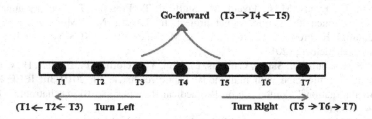

Fig. 4. Vibrotactile patterns for turn left, turn right and go-forward commands

We present these commands in four different modes: *recurring apparent motion, single apparent motion, recurring discrete* and *single discrete*. In *recurring* cues the tactile command is presented to user' forehead repeatedly until a new command is received, in the single cue case, the system presents the tactile command once and then waits until a new command is generated. *Apparent motion* commands exploit on the concept of vibrotactile apparent movement illusion [11] which creates an illusionary sensation that the stimulus is travelling continuously from one position to another. The feeling of apparent motion is controlled by two main parameters: duration of stimuli (DOS) and the stimulus onset asynchrony (SOA). The desired movement impression was obtained with a DOS of 400 ms and a SOA of 100 ms. Unlike apparent motion, discrete commands create a discrete motion across the forehead. We will evaluate these four types of vibrotactile patterns for turn left/right and go-forward commands to find out which one is better suited for indoor guidance.

5   Discussion

Whereas some approaches to sensory substitution/augmentation, that take an enactive view, have favoured using simple mappings between modalities, our research is moving in the direction of more complex mappings. One reason is that the sensorimotor contingencies [12] are often very different in the modalities we are mapping from (here ultrasound for distance sensing) and to (here cutaneous touch). In particular, our project aims to investigate the hypothesis that the transparency of the device depends primarily on having a clear and timely mapping between the environmental affordances (e.g. surfaces for navigational guidance) and the display presented on the sensory surface. We suggest that to achieve this may require significant processing of the primary sensory data to identify the relevant affordances before re-coding them for the new modality.

References

1. Bach-y-Rita, P., Tyler, M.E., Kaczmarek, K.A.: Seeing with the brain. International Journal Of Human-Computer Interaction **15**(2), 285–295 (2003)
2. Bach-Y-Rita, P., Collins, C.C., Saunders, F.A., White, B., Scadden, L.: Vision Substitution by Tactile Image Projection. Nature **221**(5184), 963–964 (1969)
3. Bertram, C., Evans, M.H., Javaid, M., Stafford, T., Prescott, T.: Sensory augmentation with distal touch: the tactile helmet project. In: Lepora, N.F., Mura, A., Krapp, H.G., Verschure, P.F., Prescott, T.J. (eds.) Living Machines 2013. LNCS, vol. 8064, pp. 24–35. Springer, Heidelberg (2013)
4. Gallo, S., Chapuis, D., Santos-Carreras, L., Kim, Y., Retornaz, P., Bleuler, H., Gassert, R.: Augmented white cane with multimodal haptic feedback. In: 2010 3rd IEEE RAS and EMBS International Conference on Biomedical Robotics and Biomechatronics (BioRob), pp. 149–155 (2010)
5. Engel, A.K., Maye, A., Kurthen, M., König, P.: Where's the action? The pragmatic turn in cognitive science. Trends Cogn. Sci. **17**(5), 202–209 (2013)
6. Froese, T., McGann, M., Bigge, W., Spiers, A., Seth, A.K.: The Enactive Torch: A New Tool for the Science of Perception. IEEE Transactions on Haptics **5**(4), 363–375 (2012)
7. Nagel, S.K., Carl, C., Kringe, T., Märtin, R., König, P.: Beyond sensory substitution-learning the sixth sense. J. Neural Eng. **2**(4), R13–R26 (2005)

8. Auvray, M., Hanneton, S., Regan, J.K.O.: Learning to perceive with a visuo-auditory substitution system: Localisation and object recognition withThe vOICe. Perception-London- **36**(3), 416 (2007)

9. Cassinelli, A., Reynolds, C., Ishikawa, M.: Augmenting spatial awareness with haptic radar. In: 2006 10th IEEE International Symposium on Wearable Computers, pp. 61–64 (2006)

10. Kerdegari, H., Kim, Y., Stafford, T., Prescott, T.J.: Centralizing bias and the vibrotactile funneling illusion on the forehead. In: Auvray, M., Duriez, C. (eds.) EuroHaptics 2014, Part II. LNCS, vol. 8619, pp. 55–62. Springer, Heidelberg (2014)

11. Sherrick, C.E., Rogers, R.: Apparent haptic movement. Percept. Psychophys. **1**(6), 175–180 (1966)

12. O'Regan, J.K., Noë, A.: A sensorimotor account of vision and visual consciousness. Behav. Brain Sci. **24**(5), 939–973 (2001). discussion 973–1031

An Energetically-Autonomous Robotic Tadpole with Single Membrane Stomach and Tail

Hemma Philamore[1,3]($\boxtimes$), Jonathan Rossiter[1,3], and Ioannis Ieropoulos[2,3]

[1] Department of Engineering Mathematics, University of Bristol, Bristol, UK
hemma.philamore@bristol.ac.uk
[2] Bristol BioEnergy Centre, University of the West of England, Bristol, UK
[3] Bristol Robotics Laboratory, Bristol, UK

Abstract. We present an energetically autonomous robotic tadpole that uses a single membrane component for both electrical energy generation and propulsive actuation. The coupling of this small bio-inspired power source to a bio-inspired actuator demonstrates the first generation design for an energetically autonomous swimming robot consisting of a single membrane. An ionic polymer metal composite (IPMC) with a Nafion polymer layer is demonstrated in a novel application as the ion exchange membrane and anode and cathode electrode of a microbial fuel cell (MFC), whilst being used concurrently as an artificial muscle tail. In contrast to previous work using stacked units for increased voltage, a single MFC with novel, 0.88ml anode chamber architecture is used to generate suitable voltages for driving artificial muscle actuation, with minimal step up. This shows the potential of the small forces generated by IPMCs for propulsion of a bio-energy source. The work demonstrates great potential for reducing the mass and complexity of bio-inspired autonomous robots. The performance of the IPMC as an ion exchange membrane is compared to two conventional ion exchange membranes, Nafion and cation exchange membrane (CEM). The MFC anode and cathode show increased resistance following inclusion within the MFC environment.

Keywords: Energetic-Autonomy · Ionic polymer metal composite · Microbial fuel cell · Soft robots

1 Introduction

When compared to natural organisms, conventional rigid-body robots are notably inferior in their ability to operate in varied and unpredictable environments for prolonged periods of time. Complex multi-component systems with low mechanical compliance are ill-equipped to deal with the irregularities that characterise most real world environments. Furthermore, the dependency of conventional power sources severely limits the range of operation of autonomous robots. Consequently, biomimickry has become a driving feature in the design of autonomous systems. The goal is to develop robots which are more robust

© Springer International Publishing Switzerland 2015
S.P. Wilson et al. (Eds.): Living Machines 2015, LNAI 9222, pp. 366–378, 2015.
DOI: 10.1007/978-3-319-22979-9_37

and adaptable, while being simpler in their construction than current multi-component systems.

We present a robot that uses a single, soft component for both propulsive actuation and power generation fuelled by raw, natural substrates. The work demonstrates a major step towards the development of artificial animals by bridging the crucial gap between a bio-inspired energy source and a bio-inspired actuator. Furthermore, the novel use of soft materials as multifunctional components shows great potential for systems capable of robust operation in unstructured environments through low system complexity.

The use of soft artificial muscles, including those comprising electro-active polymers (EAPs), to emulate the soft physical mechanisms of natural organisms, allows greater adaptability to irregular environments, low mass to power ratios and good thermodynamic efficiency compared to conventional electromechanical actuators [13]. Among these, ionic polymer metal composites (IPMCs) are capable of significant actuation at low voltages (1-3V) due to induced ionic migration within a polymer layer when a potential is applied across it. Previous work has documented their use in applications such as biomimetic propulsion in small, soft robots [7] & [16], soft compliant mechanical grippers and stents [9] due to high biocompatibility and as a diaphragm in a micro-pump for medical applications [2]. Nafion is widely used as the polymer layer of an IPMC due to its high ionic conductivity. It is commercially available as both thin sheets and as a castable liquid, and offers geometric versatility in fabrication.

Previous work has also documented the use of Nafion as an ion exchange membrane in microbial fuel cells (MFCs) [11], a bio-inspired means of electricity generation using the redox reaction during microbial anaerobic respiration. MFCs convert the chemical energy stored in raw decomposing organic matter to electrical energy. This technology presents a promising option for self powering autonomous robots as well as wearable energetically-autonomous devices, and emulates the foraging behaviour of natural organisms. Past work has used MFCs as 'artificial stomachs' to power the sensor and actuator systems of autonomous robots [5]. However, these robots have previously comprised large stacks of MFCs to multiply the low redox potential of MFCs. The theoretical maximum open circuit voltage is 1.14V [8], with real systems producing significantly lower operating voltages. While the low voltages associated with MFCs are ideal for the actuation of IPMCs, past work showing the combination of these technologies uses energy generation hardware far greater in physical size than the actuator stage of the system[4]. Therefore, this work seeks to explore the combined potential of these technologies for bio-inspired, bio-compatible, environmentally robust, systems such as biologically interfacing devices and autonomous robots by implementing them at a more complementary scale, in an artificial organism.

This investigation considers the ability of a single Nafion membrane to function simultaneously as an ion exchange membrane in an MFC and as a soft robotic actuator in a first generation design for an autonomous robotic tadpole (Figure 1). We evaluate the efficacy for ion exchange in an MFC of a Nafion 112 (Dupont) membrane, compared to an IPMC with a Nafion 112 polymer

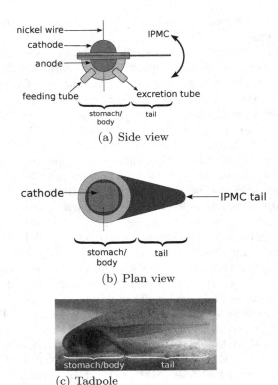

(a) Side view

(b) Plan view

(c) Tadpole
LiquidGhoul, 'Haswell's Frog - Paracrinia haswelli tadpole', February 15, 2008,
Licensed under CC BY-SA 3.0 via Wikimedia Commons

Fig. 1. Schematic: the tadpole robot comprising an MFC with IPMC membrane for
ion exchange in artificial stomach and as tail actuator

layer and gold surface electrodes and CMI-7000 cation exchange membrane
(CEM)(Membranes International Inc.) conventionally used in MFCs. The performance of the three membrane types is considered in both continuous flow and
batch feeding modes in a novel, 0.88 ml anode chamber, MFC design. Furthermore, we investigate the functionality of the noble metal surface electrodes of
the IPMC as the anode and cathode in an MFC, contributing to the design of a
single-component system for both energy generation and robotic actuation.

The effect of prolonged inclusion in the MFC environment on IPMC actuation
and the power consumed per actuation are also considered. Additionally the
effect on MFC performance of applying this electrical stimulus is documented.
Furthermore, IPMC actuation, using the MFC anode and cathode as a means to
deliver electrical charge for actuation is compared to the use of electrodes which
are external to the MFC environment. The study validates the combined use of
EAP technologies for robotic systems combining bio-inspired energy generation
and actuation.

This work will further the development of systems such as small swimming robots by providing energetic autonomy and propulsion from a single IPMC component. This could greatly improve the range of operation and lifetime of small autonomous robots. Furthermore, the use of a novel fabrication process for the construction of the anode, cathode and separator in an MFC as a single component will contribute to miniaturisation and lower system mass, which are key challenges in the use of MFCs as a power source for autonomous mobile robots.

2 Methods and Processes

2.1 Energy Generation from MFC Artificial Stomachs

A microbial fuel cell is a biological fuel cell which uses the redox reactions that takes place during bacterial metabolism to generate electrical charge. An MFC comprises anode and cathode electrodes, usually with a separator to provide the electrical insulation and conduction of protons between the two electrodes required to generate cell potential. The reaction is fuelled by bacterial degradation of bio-matter in the anode chamber.

Table 1. MFC configurations investigated in the study

	Membrane	Carbon veil MFC anode and cathode	Protruding tail	Carbon veil electrodes at base of IPMC tail
Type 1	CEM	✓	X	X
Type 2	Nafion	✓	X	X
Type 3	IPMC with protruding tail	✓	✓	✓
Type 4	IPMC with protruding tail	X	✓	✓

The anode chamber of each MFC had a 0.88mL capacity. A circular ion exchange membrane of 15mm diameter separated the anode and cathode. Three membrane types were investigated (Table 1); cation exchange membrane (Type 1), Nafion (Type 2), and ionic polymer metal composite made from Nafion 112 with gold electrodes fabricated using electroless plating using the method described in [12](Type 3). MFC anode and cathode electrodes were made from carbon fibre veil, with surface areas of $1800mm^2$ and $4500mm^2$, respectively (Figures 2(a) and 2(b)). The open to air cathode of each system was coated with a conductive latex made using a method derived from [15], to maintain a continuous redox reaction without the need to hydrate the cathode electrode.

The performance of IPMC surface electrodes to function as the MFC anode and cathode, with the polymer layer used as the ion exchange membrane, was investigated in an additional configuration without carbon veil and conductive rubber MFC electrodes (Type 4). To maintain a consistent volume of anolyte, Type 4 MFCs included a piece of cellulose (85%) blended with a bonding polymer (15%)(Dri-fresh cellulose, Siriane), of the same volume as the carbon veil anode in Type 1, 2 and 3 MFCs. The cellulose material was used due to its bio-compatibility and non-biodegradability in anaerobic environments.

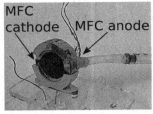 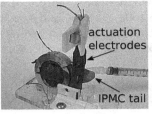 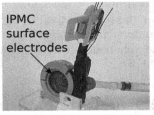

(a) Type 1 (CEM) and Type 2 (Nafion) MFCs   (b) Type 3 (IPMC) MFCs   (c) Type 4 (IPMC) MFCs

Fig. 2. MFC configurations using different membrane types

The open circuit voltage of the MFCs was recorded using a Pico Technology ADC-24 data logger. The MFCs were inoculated with 2mL sludge (Wessex Water, Cam Valley UK 13/01/14) mixed w/ 2.5% nutrient broth, and were fed weekly by syringe using 2mL of the same mixture for a period of five weeks. A peristaltic pump was subsequently used to deliver anolyte (5mM acetate solution w/ 0.2% tryptone, 0.1% yeast) at a constant flow rate of 0.002 mL/min to each MFC through an individual feed tube from a reservoir of anolyte. This mode of feeding required regular intervention to prevent the blockage of supply micro-tubes and was terminated at 174 days.

Each Type 3 and 4 IPMC membrane featuring a tab that protruded from the MFC forming the tadpole tail (Figures 2(b) and 2(c)). Two additional carbon veil actuation electrodes (surface area 1500mm^2) for actuation of the IPMC tail (surface area 140mm^2, length 14mm) were held either side of the tab at its base and pressure was applied using a small hinged clip to press them to the gold surface in order to supply electrical charge for actuation of the IPMC.

Previous work has examined the resistance of MFCs to prolonged periods of anolyte starvation [15]. Such conditions reflect the likely fluctations in available fuel that an autonomous system would have to survive in an unstructured environment. Hence, termination of continuous flow feeding at 174 days was used as an opportunity to evaluate the capability of the MFCs to withstand starvation. From 174 days the MFCs were starved for 168 days. At 342 days, the MFCs were revived by batch feeding with 2 mL of anolyte through a syringe, at 2 day intervals.

2.2   Actuation of IPMC Tail

Actuation in water of the area of the IPMC tail was recorded with the complete system submerged in deionised water, mimicking the undulatory swimming mechanism of a tadpole [10]. Actuation was recorded before inoculation of the MFC and, subsequently, at 349 days, while under batch feeding conditions, to study the effect of the MFC environment on the actuation of the artificial muscle. Tip displacement was recorded using a laser displacement sensor (Keyence). The neutral plane of the IPMC was orthogonal to the laser displacement sensor.

Power consumption during delivery of a 1 Hz square wave for a total time of 10 s using a potentiostat (Hokuto Denko) was recorded using a National Instruments PCI-6229 data acquisition board. Displacement relative to the laser source was documented at voltages of amplitude +/-1 V and +/-3 V. The effect of driving the IPMC using carbon veil electrodes positioned at the base of the tab was compared to using the anode and cathode electrode inside the MFC, as well as using both sets of electrodes simultaneously.

3   Results

3.1   MFC Performance

During inoculation, large variation was shown in the behaviour of different microbial fuel cells of same membrane type 3. However, the conventionally used membranes, Nafion and cation exchange membrane, in general, showed a significantly higher open circuit voltage than the MFCs with ionic polymer metal composite membranes demonstrated by higher peak voltage production (average of 5 batches) (Table 2). The behaviour of MFCs with conventional membranes (Type 1 and 2) was notably more consistent between like systems than under continuous flow feeding conditions, and showed superior mean continuous voltage output to MFCs with IPMC membranes (Type 3 and 4)(Table 2).

During the 168 days that the MFCs were left unfed, negligible voltage was produced from all MFCs. The MFCs were revived at day 342 with a single batch of anolyte. Initially, this resulted in lower performance than the previous feed conditions (Figure 4). Reversal of MFC cell polarity indicated an increase in the internal resistance of the MFCs. However, improved open circuit voltage was shown by the second batch, demonstrating the ability of the MFCs to recover quickly from periods of starvation. By the third batch, peak voltage per batch for Type 1 (CEM) and and Type 2 (Nafion) MFCs was comparable to average continuous voltage shown during continuous flow feeding (Table 2), showing the resilience of MFC performance to the effect of long term anolyte starvation. While, performance of MFCs with IPMC membranes (Type 3 and 4) was inferior to those with conventional membranes (Type 1 and 2), under the new batch feeding mode, both MFCs with IPMC membranes of Type 4 showed a significant improvement in open circuit voltage 4(b). This showed the potential for significantly downsizing the combined MFC-actuator system by removal of the carbon veil anode and cathode electrode, showing the viability of a single component IPMC system for use in an energetically autonomous tadpole robot, or diaphragm pump fed MFC. Furthermore, MFCs with IPMC membranes of Type 3 showed negligible voltages suggesting that this conventional electrode configuration was detrimental to the performance of the IPMC for ion exchange in an MFC.

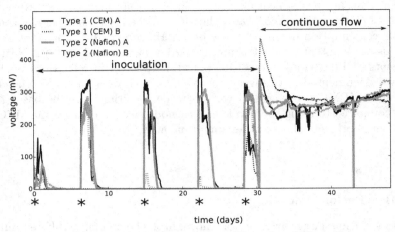

(a) Conventional membranes (Type 1 and Type 2)

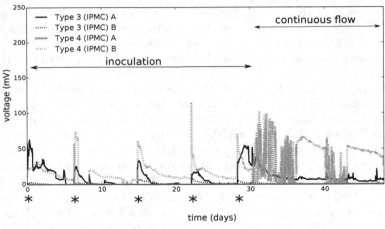

(b) IPMC membranes (Type 3 and Type 4)

Fig. 3. Temporal open circuit voltage of MFCs during inoculation (day 0-29)and continuous flow feeding (day 29-48). Stars indicate batch feed times during inoculation.

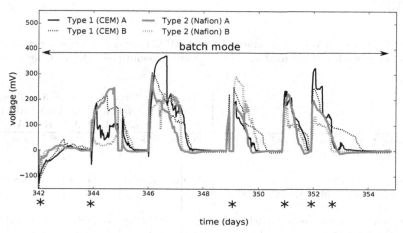

(a) Conventional membranes (Type 1 and Type 2)

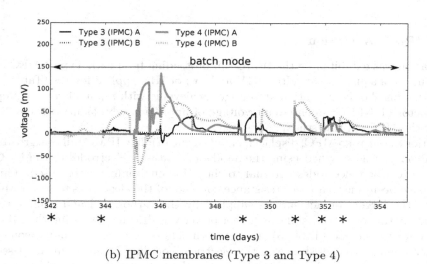

(b) IPMC membranes (Type 3 and Type 4)

Fig. 4. Temporal open circuit voltage of MFCs during batch feeding mode (day 342-355). Stars indicate batch feed times.

During the second batch, all MFCs were removed from the data logging hardware to allow actuation of the IPMC using external hardware to be recorded. On completion of actuation tests, the open circuit voltage of all MFCs with IPMC membranes had not deteriorated, showing the ability of the MFC bacterial culture to withstand significantly larger applied voltages than those generated by the MFC.

Table 2. Open circuit voltage produced by each MFC type under inoculation, continuous flow and batch feeding modes. MFC A and B denote replicate MFC systems of same type

MFC	Type 1 (CEM)		Type 2 (Nafion)		Type 3 (IPMC)		Type 4 (IPMC)	
	A	B	A	B	A	B	A	B
Inoculation, peak voltage per batch (mV) (average of 5 batches, day 1-29)	249.1	143.5	307.2	56.5	41.0	6.6	1.3	81.9
Continuous flow (mV) (average continuous voltage, day 29-48)	260.9	263.4	261.1	283.7	9.5	1.0	0.05	37.6
Batch feeding, peak voltage per batch (mV) (average of 6 batches, day 342-355)	217.2	170.3	225.0	225.9	46.5	10.94	75.2	54.5

3.2 IPMC Actuation

Actuation was exhibited by the IPMC tail protruding from each Type 4 MFC in response to a square wave voltage (1V and 3V) of 1Hz, applied for 10s (Tables 3 and 4). This showed actuation at voltages achievable with minimal voltage step up (factor of 4-12 for open circuit voltage of approximately 250mV (Table 2)) using the MFCs in this study.

Prior to use as an MFC, displacement of the protruding IPMC tail was greater when charge was supplied using the anode and cathode electrodes of the MFC relative to when electrodes external to the MFC environment were used. This may have been due to a lower resistance coupling of the electrodes to the IPMC due to the larger contact surface supplied by the carbon veil and conductive rubber coating on the cathode, as higher power was drawn when using the MFC electrodes to actuate (Table 3). Use of both sets of electrodes simultaneously further reduced the resistance of the electrical coupling resulting in increased power consumption and larger displacement. Hence, future work should seek to further reduce the resistance of the electrical coupling to the IPMC actuator in order to achieve actuation at a suitable amplitude for propulsion.

When actuated at 349 days, the period of MFC batch feeding, lower power was drawn by all electrode configurations during IPMC actuation. This suggested an increase in the resistance of the MFC. This may have been caused by biofouling of the IPMC. The displacement and power recorded when using the MFC electrodes was greater then when using the actuation electrodes, but showed a relative decrease in performance suggesting an increased resistance of the electrodes due to inclusion within the MFC environment, potentially due to biofouling. The order of magnitude of IPMC displacement was the same before

Table 3. Amplitude of IPMC tail displacement, maximum power per stroke and energy per stroke, average of two replicate systems (A and B) over 20 actuations, prior to inoculation of MFCs. Response using actuation electrodes considered for Type 3 and 4 IPMC membranes. Response using carbon veil MFC electrodes and combining both sets of electrodes considered for Type 3 IPMC membranes.

	Displacement (mm)		Power (mW)		Energy (mJ)	
	1V	3V	1V	3V	1V	3V
actuation electrodes	0.01	0.02	5	58	2	97
MFC electrodes	0.01	0.09	10	166	3	260
both electrodes	0.07	0.21	13	184	5	308

Table 4. Amplitude of cantilever displacement, maximum power per stroke and energy per stroke, average of two replicate systems (A and B) over 20 actuations, while MFCs active. Response using actuation electrodes considered for Type 3 and 4 IPMC membranes. Response using carbon veil MFC electrodes and combining both sets of electrodes considered for Type 3 IPMC membranes.

	Displacement (mm)		Power (mW)		Energy (mJ)	
	1V	3V	1V	3V	1V	3V
actuation electrodes	0.02	0.06	4	21	4	30
MFC electrodes	0.02	0.07	6	28	6	23
both electrodes	0.02	0.08	6	20	1	32

and after long term exposure to the MFC environment showing resilience of the IPMC actuator to biodegradation.

4 Discussion

Soft robotic actuators have been applied in the design of mechanical pumps similar to those widely used in continuous flow feeding of microbial fuel cells. Previous studies have shown the use of an MFC stack to power an artificial muscle shape memory alloy actuator, which is used to drive a diaphragm pump constructed from an electrically passive membrane for biomimetic, heart-beat-like actuation [14]. A system using an ionically active ionic polymer metal composite diaphragm demonstrates a similar pump with reduced system dimensions and complexity [2]. The results of the current study show that such an IPMC micropump could be powered by a small MFC, with the same component used for both energy generation and actuation to feed the system. Applications of this may include supply the low flow rate of anolyte that is required by small MFCs, which is limited by current off the shelf technology and is therefore a significant challenge to miniaturisation of MFCs [6].

Mechanical agitation of anolyte within the MFC has been shown to increase mass transfer of the anolyte to the biofilm, stimulating increased power production [3]. Previous work demonstrates combined energy generation and actuation using a single component concurrently as the electrode of a chemical fuel cell and a cantilever actuator driven by the chemical gradients present within the

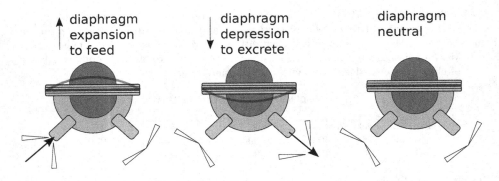

diaphragm
expansion
to feed

diaphragm
depression
to excrete

diaphragm
neutral

Fig. 5. Schematic of design for a self-feeding MFC with IPMC diaphragm pump with one way valves conrolling direction of fluid flow through feeding and excretion tubes

cell [1]. The results of the current study show the potential of IPMCs for use in similar cilia-like mixing driven by power from an MFC (Figure 6). Suitability for this application is indicated by the low power actuation of IPMCs and their resistance to biodegradation by the bacterial environment.

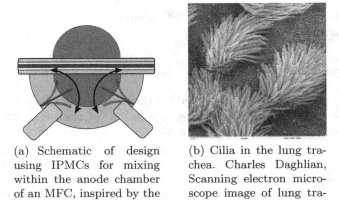

(a) Schematic of design using IPMCs for mixing within the anode chamber of an MFC, inspired by the motion of cilia

(b) Cilia in the lung trachea. Charles Daghlian, Scanning electron microscope image of lung trachea epithelium, October 7, 2006, Licensed under PDM via Wikipedia.

Fig. 6. Cilia-inspired use of IPMCs for mixing within the anode chamber of an MFC

Future work will employ simple voltage step up hardware to reach the IPMC actuation voltages necessary for the aforementioned systems. Capacitors could be charged in parallel from an MFC, then configured in series to supply a multiplied voltage to an actuator. Only four capacitors would need to be charged from Type 1 and 2 MFCs to achieve voltages over 1V. This demonstrates the potential for a relatively simple system comprising few components to use the charge generated by the MFCs for actuation of the IPMC tail. Furthermore, the superior performance of Nafion membranes compared to IPMC membranes could result in Nafion membranes, selectively plated to optimise particular areas for either power generation or actuation.

By considering the energy required per IPMC actuation stroke (Tables 3 and 4) capacitors could be charged from the MFCs to the total energy required,

$$E = CV^2 \tag{1}$$

where V is the voltage to which the capacitor of capacitance, C is charged. Further work will therefore compare the charge times provided by the different MFC membranes considered and hence their suitability for pulsed, charge-actuation cycles. Additionally, voltage step up hardware will be investigated in future studies.

5 Conclusion

This work shows an ionic polymer metal composite membrane functioning as both an actuator and the anode, cathode and separator in a microbial fuel cell. This study demonstrates the feasibility of an autonomous robot fabricated from a single, electroless plated membrane, for both electrical energy generation using an MFC and actuation using minimal multiplication of the generated voltages. Further investigation will develop the power management needed to drive IPMC actuation directly from an MFC which was outside the scope of the current study. Future work will allow the use of this multi functional component to enable energetic autonomy in small soft, IPMC-based robots. The report shows performance of an IPMC as an ion exchange membrane in an MFC. A reduction in IPMC actuation is associated with development of the biofilm. Future work will exploit the multi-functional behaviour of the IPMC to reduce the mass and complexity of the current system. Additional studies will further characterise the behaviour of the novel MFCs under load.

References

1. Ebron, V.H., Yang, Z., Seyer, D.J., Kozlov, M.E., Oh, J., Xie, H., Razal, J., Hall, L.J., Ferraris, J.P., Macdiarmid, A.G., Baughman, R.H.: Fuel-powered artificial muscles. Science **311**(5767), 1580–1583 (2006). (New York, N.Y.)
2. Guo, S., Nakamura, T., Fukuda, T., Ogura, K.: Development of the micropump using ICPF actuator. In: Proceedings of the 1997 IEEE International Conference on Robotics and Automation, vol. 1, pp. 266–271 (1997)
3. He, Z., Minteer, S.D., Angenent, L.T.: Electricity generation from artificial wastewater using an upflow microbial fuel cell. Environmental Science & Technology **39**(14), 5262–5267 (2005)
4. Ieropoulos, I., Anderson, I., Gisby, T., Wang, C.H., Rossiter, J.: Microbial-powered artificial muscles for autonomous robots. In: Taros Autonomous Robotic Systems, pp. 209–216, September 2008

5. Ieropoulos, I., Greenman, J., Melhuish, C., Horsfield, I.: EcoBot-III : a robot with guts. In: Proceedings of the Alife 7 Conference, pp. 733–740 (2010)
6. Ieropoulos, I., Winfield, J., Greenman, J.: Effects of flow-rate, inoculum and time on the internal resistance of microbial fuel cells. Bioresource Technology **101**(10), 3520–3525 (2010)
7. Kim, B., Kim, D.H., Jung, J., Park, J.O.: A biomimetic undulatory tadpole robot using ionic polymer–metal composite actuators. Smart Materials and Structures **14**(6), 1579–1585 (2005)
8. Prescott, L.M., Harley, J.P., Klein, D.A.: Microbiology, 3rd edn. Wm. C. Brown Publishers, Dubuque
9. Li, S.l., Kim, W.y., Cheng, T.h.: A helical ionic polymer–metal composite actuator for radius control of biomedical active stents **035008** (2011)
10. Liu, H., Wassersug, R., Kawachi, K.: The three-dimensional hydrodynamics of tadpole locomotion. The Journal of Experimental Biology **200**, 2807–2819 (1997)
11. Liu, J., Zhao, T., Liang, Z., Chen, R.: Effect of membrane thickness on the performance and efficiency of passive direct methanol fuel cells. Journal of Power Sources **153**(1), 61–67 (2006)
12. Onishi, K., Sewa, S., Asaka, K., Fujiwara, N., Oguro, K.: Biomimetic microactuators based on polymer electrolyte/gold composite driven by low voltage. In: Proceedings IEEE Thirteenth Annual International Conference on Micro Electro Mechanical Systems (Cat. No.00CH36308), pp. 386–390 (2000)
13. Shahinpoor, M.: Ionic polymer–conductor composites as biomimetic sensors, robotic actuators and artificial muscles—a review. Electrochimica Acta **48**(14–16), 2343–2353 (2003)
14. Walters, P., Ieropoulos, I., Mcgoran, D.: Digital fabrication of a novel bio-actuator for bio-robotic art and design. In: IS&T Digital Fabrication, pp. 6–9, October 2011
15. Winfield, J., Chambers, L.D., Stinchcombe, A., Rossiter, J.: The power of glove : Soft microbial fuel cell for low-power electronics. Journal of Power Sources **249**, 327–332 (2014)
16. Yeom, S.w., Oh, I.k.: A biomimetic jellyfish robot based on ionic polymer metal composite actuators. Smart Materials and Structures **085002**(18) (2009)

Multi-objective Optimization of Multi-level Models for Controlling Animal Collective Behavior with Robots

Leo Cazenille[1,2,3]($\boxtimes$), Nicolas Bredeche[2,3], and José Halloy[1]

[1] Univ Paris Diderot, Sorbonne Paris Cité, LIED, UMR 8236, F-75205 Paris, France
`{leo.cazenille,jose.halloy}@univ-paris-diderot.fr`
[2] Sorbonne Universités, UPMC Univ Paris 06, UMR 7222, ISIR, F-75005
Paris, France
[3] CNRS, UMR 7222, ISIR, F-75005 Paris, France
`nicolas.bredeche@upmc.fr`

Abstract. Group-living animals often exhibit complex collective behaviors that emerge through the non-linear dynamics of social interactions between individuals. Previous studies have shown that it is possible to influence the collective decision-making process of groups of insects by integrating them with autonomous multi-robot systems. However, generating robot controller models for this particular task can be challenging. The main difficulties lie in accommodating group collective dynamics (macroscopic level) and agent-based models implemented in every individual robot (microscopic level). In this study, we show how such systems can be appropriately modeled, and how to use them to modulate the collective decision-making of cockroaches in a shelter-selection problem. We address two questions in this paper: first, how to optimize a microscopic model of cockroach behavior to exhibit the same collective behavior as a macroscopic model from the literature, and second, how to optimize the model describing robot behavior to modulate the collective behavior of the group of cockroaches.

Keywords: Collective behavior · Decision-making · Multi-level modeling · Mixed-societies · Multi-objective optimization

1 Introduction

Groups of animals are able to reach consensus collectively, when presented with mutually exclusive alternatives. Previous studies have shown that it possible to influence the collective decision-making process of groups of insects by integrating them with autonomous multi-robot systems [12]. A mixed society is defined as a group of robots and animals able to integrate and cooperate: each robot is influenced by the animals, but can, in turn, influence the behavior of the animals and of other robots. Individuals, natural or artificial, are perceived as equivalent, and the collective decision process results from the interactions between natural and artificial agents [10–12].

S.P. Wilson et al. (Eds.): Living Machines 2015, LNAI 9222, pp. 379–390, 2015.
DOI: 10.1007/978-3-319-22979-9_38

A number of recent works in ethology have successfully used robots to investigate individual and collective animal behaviors, in particular by creating mixed robot-animals societies: robots are mixed with chicks in [10], cockroaches in [12,21], fruit flies in [22], honeybees in [16], guppies in [15] and zebrafish in [3–5,20].

In particular, Halloy *et al.* ([12]) demonstrates a system in which groups of robots are used to modulate the collective behavior of groups of animals (cockroaches *P. americana*). The same paper introduces a macroscopic Ordinary Differential Equations (ODE) model of the collective decision-making process of the mixed-society in a shelter-selection problem.

Macroscopic models can convincingly describe collective dynamics, but cannot be implemented directly into robotic controllers. Robot controllers are intricately microscopic, as they describe the behavior of individual agents. One of difficulties in experiments involving mixed-societies is to implement the dynamics described in a macroscopic model into robot controllers (microscopic models). In previous studies (including [12]), this process is often done empirically. Ways of handling different levels of descriptions is investigated in [17–19], but these studies do not address the issue of transitioning between models of different level of description automatically.

This paper introduces a novel methodology to navigate between models of different level of description by optimizing the whole range of parameter sets of models to get the same bifurcation diagram. This methodology is applied to the problem of modulating the collective behavior of a group of cockroaches with robots described in [12]. We take an agent-based modelling approach, and makes a number of assumptions: firstly, a model of the collective behavior of the animals already exists (the ODE model presented in [12]); secondly, robots can be attractive enough to the animals; and lastly, the number of robots is very small compared to the number of animals.

To describe the behavior of individual insects and robots, we use a Finite State Machine (FSM) agent-based microscopic model of cockroaches behavior. To test this FSM model in simulation, two sets of parameters are needed: one describing insect behavior, the other describing robot behavior. We address two questions: first, how to calibrate the FSM model describing insect behavior to exhibit the same collective behavior as the ODE macroscopic model, and second, how to optimize the FSM model describing robot behavior to modulate the collective behavior of the group of insects.

2 Multi-level Models

We use the same experimental setup as [12] (cf Fig. 1): a number of cockroaches (*P. americana*) are put in a circular arena with two identical shelters (resting sites). Cockroaches aggregate under the shelters. This setup is well adapted to study collective decision-making because it imply a trade-off between competition for resources with limited carrying capacity (the shelters) and cooperation (aggregation of the individuals).

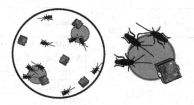

Fig. 1. Experimental setup used in [12] includes two identical shelters (150 mm) and both cockroaches (*P. americana*, approximate size: $\sim 4cm$, surface: $600mm^2$) and robots (surface: $1230mm^2$) in a circular arena (diameter: 1 m). The setup is symmetric.

2.1   Ordinary Differential Equation Model

A mathematical model describing the collective dynamics of mixed groups of robots and cockroaches was developed in [12] (based on [1]). In this model, robots and animals equivalently influence the collective decision-making process, and they exhibit homogeneous behavior. This model handles two populations (robots and animals) in setups with two shelters. The evolution of the number of individuals in each shelter (and outside) is represented by the following set of Ordinary Differential Equations (ODE):

$$\frac{dx_i}{dt} = x_e\mu_i\left(1 - \frac{x_i + \omega r_i}{S_i}\right) - x_i\frac{\theta_i}{1 + \rho\frac{x_i + \beta r_i}{S_i}^n} \tag{1}$$

$$\frac{dr_i}{dt} = r_e\mu_{ri}\left(1 - \frac{x_i + \omega r_i}{S_i}\right) - r_i\frac{\theta_{ri}}{1 + \rho_r\frac{\gamma x_i + \delta r_i}{S_i}^{n_r}} \tag{2}$$

$$C = x_e + x_1 + x_2, \quad M = r_e + r_1 + r_2, \quad N = M + C \tag{3}$$

Table 2 lists the parameters of the ODE model.

Because of crowding effects, the probability that an individual joins a shelter decrease with the level of occupation of this shelter.

We only consider the case where the two shelters have the same carrying capacity: $S = S_1 = S_2$. We define the measure $\sigma = S/N$ that corresponds to the carrying capacity as a multiple of the total population.

When only insects are considered, and no robots are present ($M = 0$), two different dynamics can be observed: When $0.4 \leq \sigma < 0.8$, only one configuration exists, corresponding of an equipartition of the individuals ($x_1/N = x_2/N = 1/2, x_e = 0$). In this case, the two shelters are saturated, with the remaining insects remaining outside. When $\sigma > 0.8$, two stable configurations exist, corresponding to all individuals in one of the shelter (either $x_1 \approx 0, x_2 \approx 1, x_e \approx 0$ or $x_1 \approx 1, x_2 \approx 0, x_e \approx 0$). These dynamics can be observed in Fig. 3, a bifurcation diagram of the occupation of the first shelter, as function of σ. Represented results are obtained by resolution of Eq. 1 using the Gillespie method [9]. A resolution using the Gillespie method allows to take into account experimental

Parameter for P. americana	Parameter for robots	Value for P. americana	Description
C	M	-	Total number of agents
x_i	r_i	-	Number of agents in shelter i
x_e	r_e	-	Number of agents outside the shelters
μ_i	μ_{ri}	$0.0027s^{-1}$	Maximal kinetic constant of entering a shelter
θ_i	θ_{ri}	$0.44s^{-1}$	Maximal rate of leaving a shelter
ρ, n	ρ_r, n_r	4193, 2.0	Influence of conspecifics

Parameter	Description
S_i	Carrying capacity of shelter i
ω	Surface of one robot as multiple of the surface of one animal
γ	Influence of animals on robots
β	Influence of robots on animals
δ	Influence of robots on robots

Fig. 2. Parameters list of the ODE model. Cockroaches (*P. americana*) parameter values are from [12]. We only consider the case where $N = 50$. In setups with two shelters, this model has 18 parameters. The influence of animals on animals is equal to 1, and is not considered in [12]: the assumption is made that this parameter is imposed by biology, and can't be changed in experiments.

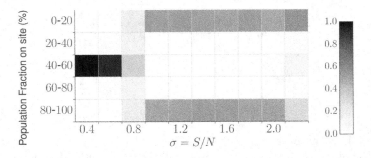

Fig. 3. Bifurcation diagram and distribution of $N = 50$ *P. americana* cockroaches in the first shelter, as function of σ. The bifurcation diagram is represented as bi-dimensional histograms of the results using 1000 solutions by parameter sets. The color of each bin of the histogram corresponds to the occurrence of experiments. The diagram is symmetric for all tested values of σ, so only one shelter is represented. When $0.4 \leq \sigma < 0.8$, only one configuration exists, corresponding of an equipartition of the individuals ($x_1/N = x_2/N = 1/2, x_e = 0$). When $\sigma > 0.8$, two stable configurations exist, corresponding to all individuals in one of the shelters (either $x_1 \approx 0, x_2 \approx 1, x_e \approx 0$ or $x_1 \approx 1, x_2 \approx 0, x_e \approx 0$). The bifurcation point is close to $\sigma = 0.8$.

fluctuations. Figure 3 only represents results with population of 50 cockroaches, but similar dynamics are observed with different population sizes.

Note that while models at the macroscopic level can easily describe the behavior of the dynamical system, in term of shelter selection, and offer a mathematical basis of description, they cannot explicit the behavior of individual agents, and cannot be implemented directly in actual robots.

2.2 Finite State Machine Model

We define a Finite State Machine as agent-based model of cockroaches and robots behavior. This model is very similar to the agent-based aggregation models introduced in [8,13] to describe the collective behavior of cockroaches in a similar setup.

Cockroaches tend to follow walls when close to the walls of the arena, and are gregarious during their resting period. We establish two zones in the arena: the peripheral zone, which is the ring that borders the walls of the arena, and the central zone, corresponding to the rest of the arena. In the central zone, agents exhibit a random-walk behavior, by following a recurring alternation of straight lines and rotations. In the peripheral zone, agents exhibit a wall-following behavior. Shelters are in the central zone. When an agent enters a shelter, it has a probability of stopping for a random duration before exiting the shelter. Similarly to [8], this probability depends on the number of present agents. Figure 4 provides a description of this model, with the relevant model parameters.

In our model (as opposed to [8,13]), the probability of stopping when reaching a shelter is not the same for both shelters. While it is not relevant when describing the behavior of cockroaches (the shelters in the setup are identical), it can be useful for describing robots that modulate the collective behavior of cockroaches.

3 Results

3.1 Numerical Computation

All results from the ODE model were obtained by resolving Eq. 1 and 2 using the Gillespie method ([9]). Results from the FSM model were obtained from simulations of 28800 time steps, of a setup similar to Fig. 1 (used in [12]): a circular arena (diameter $1m$) with two identical shelters (diameter $150mm$).

For both models, only populations of 50 individuals were considered.

3.2 Calibration of Models

In this section, we address the problem of finding parameter sets of cockroaches simulated using the FSM model that exhibit the same collective behavior as in the ODE model. FSM model parameters describing cockroach behavior can be derived (or 'Calibrated') from the ODE model.

As the ODE model is parameterized using experimental data, it allows the FSM model to be as close as possible to the behavior of cockroaches. This process is described in Fig. 6.

We optimize the parameter sets of the cockroaches individuals, for the FSM model. Instances of the FSM model using these parameter sets are simulated for different values of σ. The aim is to optimize parameter sets of the FSM model to obtain a similar bifurcation diagram as in Fig. 3.

As there is only few a-priori information about the parameter space, and as the parameter space has a relatively large dimensionality, we use the state-of-the-art CMA-ES evolutionary optimization method ([2], population size is 20, maximal number of generations is 500).

The fitness, minimized by CMA-ES, corresponds to a comparison between an optimized bifurcation diagram with the reference diagram from the ODE model. It is computed as follow:

$$\text{Fitness}_{\text{calibration}}(x) = D_{\text{Hellinger}}(B_{\text{optimized}}/N_u, B_{\text{reference}}/N_u) \qquad (4)$$

where x is the tested parameter set (genome), N_u is the number of considered values of σ in the bifurcation diagrams (10) and $B_{\text{optimized}}$ and $B_{\text{reference}}$ are one-dimensional histograms version of the bifurcation diagrams. The term N_u is used for normalization. The Hellinger distance ([7]) is defined by the equation:

$$D_{\text{Hellinger}}(P, Q) = \sqrt{2 \sum_{i=1}^{d} (\sqrt{P_i} - \sqrt{Q_i})^2} \qquad (5)$$

where P and Q are two histograms, and P_i, Q_i their i-th bins. The Hellinger distance is a divergence measure, similar to the Kullback-Leibler (KL) divergence. However, the Hellinger distance is symmetric and bounded, unlike the KL-divergence (and most other distance metrics). As such, it is adapted when comparing two histograms ([7]).

Figure 7 corresponds to the distribution of cockroaches in the two shelters, using parameters sets from the best-performing optimized individuals in 100 runs. All values of σ present in Fig. 3 are tested, and Fig. 7 shows typical results before and after the bifurcation point. Results before the bifurcation point ($\sigma < 0.8$) are similar to results at $\sigma = 0.4$, and results after the bifurcation point ($\sigma \geq 0.8$) are similar to results at $\sigma = 1.2$. Results show that it is possible to find parameters sets of the FSM model that exhibit the same collective choice that the ones from ODE. Similar results are obtained using the FSM model from [8] (results not shown).

3.3   Modulation of Collective Behavior by Robots

Our goal is to find sets of parameters of robots, capable of modulating the collective behavior of the group of cockroaches.

This process is described in Fig. 8. Populations of 50 individuals are considered, with a varying, but small, proportion of robots in the population.

The parameter set used for modeling cockroaches using the FSM model was taken from the best-performing optimized individuals during the calibration process described in 3.2.

An optimizer is used to generate the parameter sets of the robots modeled by the ODE and FSM models. Instances of the FSM and ODE models using

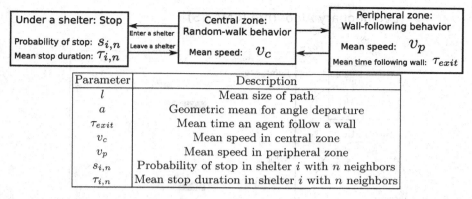

Parameter	Description
l	Mean size of path
a	Geometric mean for angle departure
τ_{exit}	Mean time an agent follow a wall
v_c	Mean speed in central zone
v_p	Mean speed in peripheral zone
$s_{i,n}$	Probability of stop in shelter i with n neighbors
$\tau_{i,n}$	Mean stop duration in shelter i with n neighbors

Fig. 4. Finite State Machine Model of cockroach individual behavior. The arena contains two zones: the peripheral zone (agents follow a wall-following behavior), and the central zone (agents follow a random-walk behavior). Shelters are in the central zone. When an agent enters a shelter, it has a probability of stopping for a random duration before exiting the shelter. The probability of stopping under shelter depends on the number of neighbors present in the shelter, and can be different for each shelter. Only 10 neighbors are considered in our experiments. In setups with two shelters, this model has 45 parameters per population.

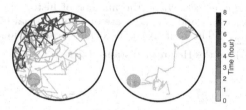

Fig. 5. Examples of the trajectory of an artificial insect, using the FSM model. The arena is circular and contains two shelters. Gray lines represents the trajectory of one agent. The brightness of the line reflects to simulation time. All experiments last 28800 time steps (corresponding to 8 hours). Note that the FSM model do not try to mimic the actual movement patterns of cockroaches. The arena contains two zones: the peripheral zone (agents follow a wall-following behavior), and the central zone (agents follow a random-walk behavior). Shelters are in the central zone. When an agent enters a shelter, it has a probability of stopping for a random duration before exiting the shelter.

these parameter sets are either simulated (FSM) or resolved using the Gillespie method (ODE), for specific values of σ.

There are two objectives to minimize:

$$\text{Fitness}_1 = D_{\text{Hellinger}}(\text{Hist}_{\text{optimized}}, \text{Hist}_{\text{reference}}) \qquad (6)$$

$$\text{Fitness}_2 = M/N \qquad (7)$$

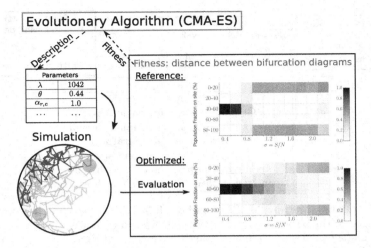

Fig. 6. Workflow of the Automated model calibration task by optimization. The optimized bifurcation diagram and the reference bifurcation diagram are both converted to one-dimensional histograms, by normalizing the sum of all bin values to 1.0. We use CMA-ES ([2]) as optimizer. The optimizer minimizes the fitness, which is computed by the formula: $\text{Fitness}_{\text{calibration}}(x) = D_{\text{Hellinger}}(B_{\text{optimized}}/N_u, B_{\text{reference}}/N_u)$ where x is the optimized parameter set, N_u is the number of histograms in the bifurcation diagrams (10) and $B_{\text{optimized}}$ and $B_{\text{reference}}$ are one-dimensional histograms version of the bifurcation diagrams. The term N_u is used for normalization. $D_{\text{Hellinger}}(P,Q) = \sqrt{2\sum_{i=1}^{d}(\sqrt{P_i} - \sqrt{Q_i})^2}$ is the Hellinger distance ([7])

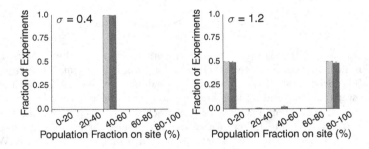

Fig. 7. Distribution of 50 cockroaches in the first shelter for chosen values of σ, using two different models: ODE (in dark grey) and FSM (in light grey). The parameter σ values are chosen before the bifurcation point ($\sigma = 0.4$), and just after the bifurcation point ($\sigma = 1.2$). Similar results are obtained for the range of values of σ present in Fig. 3. The best sets of optimized model parameters are used, after 100 runs of optimization. The diagram is symmetric for all tested values of σ, so only one shelter is represented. Calibrated versions of the FSM model behave similarly to the ODE model: (1) before the bifurcation point ($\sigma = 0.8$), only one configuration exists, corresponding of an equipartition of the individuals ($x_1/N = x_2/N = 1/2, x_e = 0$); (2) after the bifurcation point, two stable configurations exist, corresponding to all individuals in one of the shelters (either $x_1 \approx 0, x_2 \approx 1, x_e \approx 0$ or $x_1 \approx 1, x_2 \approx 0, x_e \approx 0$).

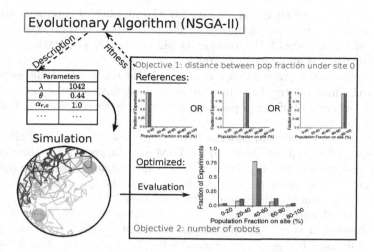

Fig. 8. Workflow of the process of modulating of the collective behavior of a group of cockroaches by robots. We use NSGA-II ([6]) as optimizer. The optimizer minimizes two objectives: (1) the difference between the optimized histogram and the reference histogram using the Hellinger distance $(D_{\mathrm{Hellinger}}(P,Q) = \sqrt{2\sum_{i=1}^{d}(\sqrt{P_i} - \sqrt{Q_i})^2}$ as described in [7]), (2) the portion of robots in the population. Three reference histograms are considered, resulting of three possible types of modulation.

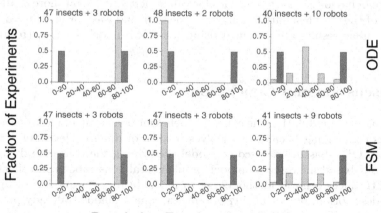

Fig. 9. Instances of results bio-hybrid group behavior when robots are optimized to change the reference behavior of cockroaches alone as much as possible (dark grey: reference animal-only model, light grey: optimized animals-and-robots models). σ values are chosen just after the bifurcation point ($\sigma = 1.2$). Results after the bifurcation point ($0.8 \leq \sigma \leq 2.2$) are similar. The three plots in the first line correspond to results obtained from the ODE model, the three plots in the second line are from the FSM model. These results are taken from the best-performing individuals in 30 runs.

with $D_{\text{Hellinger}}$ described in Eq. 5), and M the number of robots, from Eq. 3. We need a multi-objective optimizer to minimize these two objectives: we use the state-of-the-art NSGA-II evolutionary algorithm ([6], population size is 100, maximal number of generations is 1000).

Three reference histograms are considered: (1) where all of the population gather in the first shelter, (2) where all of the population gather in the second shelter, (3) where half of the population gather in the first shelter, and the other half in the second shelter.

Figure 9 shows several instances of interesting optimized individuals (on the Pareto Front), for both the ODE and the FSM models, and for the three different reference histograms. Small groups of robots are capable, using the optimized controllers, to modulate the collective behavior of the group of cockroaches to correspond to one of the three considered reference histograms.

When the objective is to force the cockroach population to select one of the two shelters, a very small portion of robots is required (typically 2 or 3). For the ODE model, this can be explained by the proportion of cockroaches to remain under shelter longer when a larger number of neighbors are presents. For the FSM model, the same behavior is evolved. This induces a progressive aggregation of the group of cockroaches toward the shelter occupied by the robots. If the objective is to force the cockroach population to occupy both shelters at the same time, it requires a larger portion of robots (10 robots). In this case, the robots have to occupy both shelters to lead the cockroaches into aggregating themselves in both shelters. Note that the modulation of the collective behavior of the cockroaches for values of $\sigma < 0.8$ is far more challenging because of the very fast saturation of the shelters, and was not considered in this study. Similar results are obtained using the FSM model from [8] (results not shown).

4   Discussion and Conclusion

The problem of modulating the collective behavior of a group of cockroaches with robots is challenging because it involves models of different levels of representation: an ODE-based macroscopic model (describing the collective dynamics), and a FSM-based microscopic model (implementable as robot controller). This paper introduces a novel methodology to navigate between models of different level of description, by optimizing parameter value of models already present in the literature. This approach makes three assumptions: firstly, a model of the collective behavior of the animals already exists ([1,12]); secondly, robots can be attractive enough to the animals; and lastly, the number of robots is very small compared to the number of animals.

The ODE model can describe the collective behavior of cockroaches, by using a parameter set obtained by experimentation with actual insects in [12]. A FSM model of cockroach behavior is introduced, with inspiration from [8,14]. This model is calibrated to exhibit the same collective dynamics as in the ODE model, using the CMA-ES evolutionary algorithm. FSM is a microscopic model that can be used as robot controller. The robot controller models are then optimized, using the NSGA-II multi-objective evolutionary algorithm, to modulate the collective behavior of the group of cockroaches, to match a user-defined reference.

Previous mixed-societies studies could only implement empirically the robot controllers used in experiments. The approach presented here is a first step toward generating them automatically, by deriving them from a validated macroscopic model of the animal collective behavior.

A subsequent study would include an application of this methodology to more complex setups, with more than two shelters and more than two population. Additionally, the calibration of models, and the modulation of collective behavior, could be performed in an online fashion, by using online evolutionary algorithms. The models investigated in this paper were only strictly macroscopic (ODE) or microscopic (FSM) – alternatively, a third kind of model could be defined, integrating both macroscopic and microscopic aspects.

Our methodology gives promising results, and could possibly be applied to model, calibrate, and modulate the collective behavior of other species (e.g. fishes, bees, or others).

Acknowledgments. This work has been funded by EU-ICT project 'ASSISIbf', no 601074.

References

1. Amé, J., Halloy, J., Rivault, C., Detrain, C., Deneubourg, J.: Collegial decision making based on social amplification leads to optimal group formation. Proceedings of the National Academy of Sciences **103**(15), 5835–5840 (2006)
2. Auger, A., Hansen, N.: A restart CMA evolution strategy with increasing population size. In: The 2005 IEEE Congress on Evolutionary Computation, vol. 2, pp. 1769–1776. IEEE (2005)
3. Bonnet, F., Retornaz, P., Halloy, J., Gribovskiy, A., Mondada, F.: Development of a mobile robot to study the collective behavior of zebrafish, pp. 437–442 (2012)
4. Butail, S., Polverino, G., Phamduy, P., Del Sette, F., Porfiri, M.: Fish-robot interactions in a free-swimming environment: effects of speed and configuration of robots on live fish. In: Proceedings of SPIE 9055, Bioinspiration, Biomimetics, and Bioreplication 2014, p. 90550I (2014)
5. Butail, S., Polverino, G., Phamduy, P., Del Sette, F., Porfiri, M.: Influence of robotic shoal size, configuration, and activity on zebrafish behavior in a free-swimming environment. Behavioural Brain Research **275**, 269–280 (2014)
6. Deb, K., Pratap, A., Agarwal, S., Meyarivan, T.: A fast and elitist multiobjective genetic algorithm: NSGA-II. IEEE Transactions on Evolutionary Computation **6**(2), 182–197 (2002)
7. Deza, M., Deza, E.: Dictionary of distances. Elsevier (2006)
8. Garnier, S., Gautrais, J., Asadpour, M., Jost, C., Theraulaz, G.: Self-organized aggregation triggers collective decision making in a group of cockroach-like robots. Adaptive Behavior **17**(2), 109–133 (2009)
9. Gillespie, D.: Exact stochastic simulation of coupled chemical reactions. The Journal of Physical Chemistry **81**(25), 2340–2361 (1977)

10. Gribovskiy, A., Halloy, J., Deneubourg, J., Bleuler, H., Mondada, F.: Towards mixed societies of chickens and robots. In: IROS (2010)

11. Halloy, J., Mondada, F., Kernbach, S., Schmickl, T.: Towards bio-hybrid systems made of social animals and robots. In: Lepora, N.F., Mura, A., Krapp, H.G., Verschure, P.F.M.J., Prescott, T.J. (eds.) Living Machines 2013. LNCS, vol. 8064, pp. 384–386. Springer, Heidelberg (2013)

12. Halloy, J., Sempo, G., Caprari, G., Rivault, C., Asadpour, M., Tâche, F., Said, I., Durier, V., Canonge, S., Amé, J.: Social integration of robots into groups of cockroaches to control self-organized choices. Science 318(5853), 1155–1158 (2007)

13. Jeanson, R., Blanco, S., Fournier, R., Deneubourg, J., Fourcassié, V., Theraulaz, G.: A model of animal movements in a bounded space. Journal of Theoretical Biology 225(4), 443–451 (2003)

14. Jeanson, R., Rivault, C., Deneubourg, J., Blanco, S., Fournier, R., Jost, C., Theraulaz, G.: Self-organized aggregation in cockroaches. Animal Behaviour 69(1), 169–180 (2005)

15. Landgraf, T., Nguyen, H., Schröer, J., Szengel, A., Clément, R.J.G., Bierbach, D., Krause, J.: Blending in with the shoal: robotic fish swarms for investigating strategies of group formation in guppies. In: Duff, A., Lepora, N.F., Mura, A., Prescott, T.J., Verschure, P.F.M.J. (eds.) Living Machines 2014. LNCS, vol. 8608, pp. 178–189. Springer, Heidelberg (2014)

16. Landgraf, T., Oertel, M., Rhiel, D., Rojas, R.: A biomimetic honeybee robot for the analysis of the honeybee dance communication system. In: IROS, pp. 3097–3102 (2010)

17. Mermoud, G., Brugger, J., Martinoli, A.: Towards multi-level modeling of self-assembling intelligent micro-systems. In: Proceedings of The 8th International Conference on Autonomous Agents and Multiagent Systems. International Foundation for Autonomous Agents and Multiagent Systems, pp. 89–96 (2009)

18. Mermoud, G., Matthey, L., Evans, W., Martinoli, A.: Aggregation-mediated collective perception and action in a group of miniature robots. In: Proceedings of the 9th International Conference on Autonomous Agents and Multiagent Systems. International Foundation for Autonomous Agents and Multiagent Systems, pp. 599–606 (2010)

19. Mermoud, G., Upadhyay, U., Evans, W.C., Martinoli, A.: Top-down vs. bottom-up model-based methodologies for distributed control: a comparative experimental study. In: Khatib, O., Kumar, V., Sukhatme, G. (eds.) Experimental Robotics. STAR, vol. 79, pp. 615–629. Springer, Heidelberg (2012)

20. Polverino, G., Porfiri, M.: Zebrafish (danio rerio) behavioural response to bioinspired robotic fish and mosquitofish (gambusia affinis). Bioinspiration & Biomimetics 8(4), 044001 (2013)

21. Sempo, G., Depickère, S., Amé, J.-M., Detrain, C., Halloy, J., Deneubourg, J.-L.: Integration of an autonomous artificial agent in an insect society: experimental validation. In: Nolfi, S., Baldassarre, G., Calabretta, R., Hallam, J.C.T., Marocco, D., Meyer, J.-A., Miglino, O., Parisi, D. (eds.) SAB 2006. LNCS (LNAI), vol. 4095, pp. 703–712. Springer, Heidelberg (2006)

22. Zabala, F., Polidoro, P., Robie, A., Branson, K., Perona, P., Dickinson, M.: A simple strategy for detecting moving objects during locomotion revealed by animal-robot interactions. Current Biology 22(14), 1344–1350 (2012)

Effects of the Robot's Role on Human-Robot Interaction in an Educational Scenario

Maria Blancas[1][(✉)], Vasiliki Vouloutsi[1], Klaudia Grechuta[1], and Paul F.M.J. Verschure[1,2]

[1] Synthetic Perceptive Emotive Cognitive Systems Group,
Universitat Pompeu Fabra, Barcelona, Spain
maria.blancas@upf.edu
[2] Center of Autonomous Systems and Neurorobotics - NRAS,
Catalan Institute of Advanced Studies - ICREA, Barcelona, Spain
paul.verschure@upf.edu

Abstract. In order for robots to be part of the education field, it is necessary to take into consideration the perception students have of them and of education in general. The aim of this study is to assess whether the role a robot plays in a classroom affects knowledge retrieval, subjective experience, and the perception of the learners. To investigate this, we developed an educational scenario and three questionnaires. The results show significant differences in the way the subjects perceived the robot as a tutor.

Keywords: Human-robot interaction · Robotic tutors · Education

1 Introduction

The increasing potential of robots that interact with people in daily life leads us to study the potential roles they will play in society. This makes us question different possibilities they might have in a number of relevant fields which include health care or education. Educational robots are being used in several instructional environments, like programming courses [1] or even language acquisition programs [2]. However, further research is needed to understand how to apply them as teaching assistants or learning companions as they seem to play a promising role [3]. With the advancement of technology, it is worthwhile to examine the diversity of the tools provided by the usage of robots as learning vehicles as well as their impact on students' development and intellectual growth.

Robots provide us with the ability to control, decompose and manipulate various behavioral cues such as gaze [4], as well as present the educational content in a more "socially present" manner [5,6] adapted to the needs of each individual [7].Furthermore, as perception plays an important role in experience, users perceptions should be taken into consideration to design suitable experiences with robots. Robots can assume the role of a teacher, a peer, or that of a teaching tool and behave accordingly; at present, it seems that preference is highly dependent on the age of the student [8].

© Springer International Publishing Switzerland 2015
S.P. Wilson et al. (Eds.): Living Machines 2015, LNAI 9222, pp. 391–402, 2015.
DOI: 10.1007/978-3-319-22979-9_39

The aim of the current study is to investigate how the role of an educational robot, either as a teacher or as a peer, can affect knowledge acquisition, the subjective experience of a user and their behavior towards the robot. In order to know how to design an appropriate educational scenario for a human and a robot, first we have to analyse which are the bases established by pedagogy in such an interaction between two humans. It is for that reason that we first provide information regarding the main educational approach we adopted for the current experimental protocol. Then, we explain the experimental setup devised to assess the role of the robot as a teacher and as a peer, and finally provide the obtained results.

1.1   Educational Paradigm

Some authors claim that the most valued feedback for students is being respected by their teachers as individuals and not treated as children who do not know much [9]. Therefore, the latest educational theories defend a change in the teacher's role, emphasizing the importance of moving from "telling" to "questioning". The figure of the teacher is no longer seen as someone who just gives a lecture, but someone who helps and guides the students to reason about the topic of the class by asking questions and making them perform tasks to improve their learning processes. This method of teaching balances the roles of students and teachers, decreasing the hierarchical differences and boundaries between them.

Such a change, from direct instruction to questioning, was already proposed by Constructivism, an educational perspective that defends the idea of learning through making and the use of technology-enhanced environments [10]. Tiberius and Billson [11] defined two types of teaching depending on the level of social supportive behavior offered by teachers: teaching can be seen as a mere knowledge transfer or as a social dialogue, with the latter case being where teachers guide students to achieve higher learning performance by positively motivating them. They also related this difference to the role which authority plays in education: although it can have immediate effects on students performance, learners may experience difficulties to engage with their teachers which can result in undesirable effects on their motivation.

In order to achieve these higher learning performances, they must cross the Zone of Proximal Development (ZPD), that is, the distance between a person's actual developmental level as determined through independent problem solving and their level of potential development as determined by problem solving under guidance or in collaboration with more capable peers [12].

Partnering and Peer Learning. Both Piaget's and Vygostky's theories have implications for the effects of peer learning in the educational mechanisms. On the one hand, [13], in his cognitive constructivism, proposes the construction of internal schemes for understanding the world. Peer learning is presented as a rather equal relationship where the power is distributed in a way that makes it more likely to be shared, contrarily to adult-peer relationships, where the child is placed apart so that his thought is isolated [14]. This allows a balanced assimilation, what eventually leads to cognitive structures more open to adaptation [15].

On the other hand, Vygotsky [12], in his social constructivism, remarked the importance of social interaction in order for children to scaffold each others learning. During the process of the ZPD theory mentioned before, the peer acts as co-learner, which minimises the excess of challenge. Similarly to Piagets theory, here, the relationship between peers is based on equality, facilitating self-discourse of ignorance and misconception and making diagnoses and corrections easier. Moreover, this equality leads to a sense of loyalty and accountability that might keep the learner motivated and on-task [16].

In order for the technology to be used successfully in classrooms, it must be combined with a new type of pedagogy: partnering. Although three levels of partnering can be defined, depending on the situation, the one recommended to start with is Basic partnering. This kind of educational action, often called inquiry-based learning, is based on providing the students with guiding questions and letting them work on their own, individually, or in groups to find the correct answers. Our approach is thus exploiting the idea of peer learning as a base for Human-Robot Interaction in educational environments.

1.2 Social Robots and Education

For a robot to be an effective tool in education, it needs to socially interact with humans during the teaching process. The term socially interactive robots defines robots with social characteristics, like expressing emotions (and sometimes, perceiving them, too) through the use of natural cues (like gaze or facial expressions) or establishing social relationships. These kinds of robots are highly relevant for the peer-to-peer interaction due to their social abilities and skills to solve specific tasks [17].

Sharing the same physical space with the user plays an important role in enhancing human-robot interaction [6]. When designing an educational technology tool, presence and anthropomorphism have to be considered as crucial characteristics for the setup, as an anthropomorphised body helps people to understand the robots gestures and establishes a common ground [2].

Robots in Education. Saerbeck et al. [6] examined the use of an expressive tutor, the iCat (Philips Research), to help children in a language-learning task. They set two conditions: a neutral one, where the focus was on a merely knowledge transfer, and a social, supportive one, where the focus was on active dialogue and positive social, supportive behavior, like nodding, shaking the head, using gaze to guide attention, and facial expression. The results showed a clear positive effect of the robot's behavior on the learning performance of the participants.

The use of robots as tutoring assistants has also been tested by Leyzberg et al. [18], who showed that a physically-present robot tutor produces better learning gains than on-screen or voice-only tutors. They also claimed that embodied agents are perceived with more authority and regarded more seriously than non-embodied ones. Embodiment, in contrast with an on-screen agent, produces a significantly greater sense of social acceptance.

Examining further the view of educational robots, the step goes towards crossing the bridge of telling to questioning as the Partnering theory claims. As some teachers could reject this change in hierarchy out of fear of losing their authority and students' respect, robots could play the role of this teacher-peer instead of the teachers themselves. Not having any previous role assigned, robots can take any role in education, contrarily to what happens with teachers.

The view of robots as peers was also researched by Tanaka et al. [19], who tested a social robot in a classroom of toddlers. After 45 days, the children not only did not lose the interest in the robot, but also treated it in a way more similar to a peer than to a toy. Based on their results, the authors claimed that robots can enrich the classroom environment, and consequently marked their potential in education as assisting teachers.

It is for that reason that we aim to test the difference in the role played by the robot in an educational scenario. The line between the role of the teacher and the role of the student is clearly defined in the current educational paradigm and, as some theories mentioned before, defend the use of peer learning in classes as highly recommended to achieve a deep knowledge retrieval. Knowing this, we want to test which role should a robot play if it was used in class - one more similar to a tutor or one more similar to a peer.

2   Methodology

The focus of this experiment was to test whether the differences between the role of a teacher and the role of a peer played by a robot could affect the acquisition of knowledge of a specific topic, the subjective experience of the user, and their behavior towards the robot. Our hypothesis is that the participants in the peer condition will have higher knowledge retrieval and a better subjective experience.

The robot performed the role of a professor, or a peer, depending on the condition, similar to the difference proposed by Tiberius and Billson [11]. Each role was characterized by differences between the posture, the degree of gestures used, and the use of colloquial expressions while speaking. In the Teacher condition (RT), the robot was always standing, with a minor use of gestures and its speech was formal. In the Peer condition (RP), the robot was seated for some time, used more and wider gestures, and was adding comments to the lesson. To assess whether knowledge acquisition is not due to the fact that the information was provided simply through speech, we also included a control condition where participants listened to the "Teacher" robot's voice through a speaker placed in front of a computer.

The two main conditions can be differentiated by the formality of the robot, the gestures, and its posture. The robot spoke in a more formal way in the RT condition than in the PR condition (never getting close to vulgarity in the PR condition, only making some comments about the topic). Regarding the gestures, the amount of gestures performed by the robot and their ampli-tude was different depending the condition: in the RT condition, the robot only used some small gestures to support the speech; in the RP condition, the robot used larger gestures, for example, waving or pointing (for example, the subject or the room). Finally, with reference to the robot's posture, it was always standing in the RT condition; and in the PR condition, the robot began the experiment standing and then sat on the table midway through.

2.1 Experimental Setup and Procedure

Given the fact that the experiment was conducted on the Poblenou Campus of the Pompeu Fabra University, the topic of the educational scenario chosen was the Industrial Revolution in Catalunya and its relation with the Campus. We used the humanoid robot Nao, developed by Alderaban Robotics. This robotic platform has 25 degrees of freedom and it is equipped with several sensors, including two cameras, RGB LEDs, and an inertial sensor. The robot's gaze was used during this experiment to follow the participant and maintain eye contact with them. A video camera was placed facing the participant to record the actions and behavior during the experiment, to be analysed later. For the first two conditions, the robot was placed in front of the participant and behaved according to its assumed role (teacher/peer), whereas in the control condition, the users listened to the robot's lecture from a speaker.

On arrival, the participants were presented a pre-knowledge questionnaire about the topic of the class and another one about their perception of the edu-cational system. The pre-knowledge questionnaire allows us to assess the par-ticipant's current knowledge regarding the examined topic, whereas the aim of the second questionnaire is to provide insights on the usage of technology in classrooms, assess the current educational approach (usage of examples, adapt-ability of the curriculum, interaction and practical learning) as well as the role of teachers (kindness, empathy, humour, authority etc).

When the participants completed the first two questionnaires, they were placed in front of the Nao and were informed that they were going to learn about history. Then, the robot delivered the "class", which followed a pre-programmed script. After the experiment, participants were asked to fill a subjective experi-ence questionnaire and the post-knowledge test. The post-knowledge question-naire serves as an assessment of knowledge acquisition. The subjective experience questionnaire consists of different parts, which provided information regarding the interaction, the perception of the robot and the quality of the received class. A video camera was placed in front of the participants to record their behavior.

Our sample consisted of 28 adults (age 27.04 ± 6.04; 16 male and 12 female) randomly distributed among three different conditions (RT=10, RP=10, C=8).

2.2 Data Collection

Data were collected from knowledge, subjective experience, and perception of education questionnaires, as well as the video recordings. In the knowledge

questionnaire, participants had to answer pre- and post-knowledge questionnaires related to the topic of the experiment: the Industrial Revolution in
Catalunya. The questionnaire consisted of seven multiple-choice questions,
ordered randomly.

Participants answered a questionnaire so that we could test the effect of the
robots social behavior. There were 32 questions and the answers were scored in
a Likert Scale from 1 to 5, based on their disagreement or agreement level. This
questionnaire was grounded on three different scales: the Basic Empathy Scale
[20], a self-reported measure of empathy, as much emotional as cognitive (in this
case, tailored to use with the robot); the Godspeed questionnaire [21], a 5-point
Likert scale created to test people's engagement with robots; and the Tripod
Survey [22], to test students' perception of their teachers and their engagement.
Moreover, there were 20 questions related to the personality of the subject and
the perceived personality of the robot, extracted from the Big Five personality
test.

The behaviours of the subject were video recorded to be analysed later using
the software ANVIL.

3   Results

A Wilcoxon signed-rank test showed that there was a significant improvement in
pre- and post- knowledge in all the conditions (Z = -4.303, p <0.001). There was
no significant difference among conditions (p = 0.629). Such results show that,
although there was an overall improvement in knowledge, there are not significant differences between the conditions in the amount of knowledge retrieved,
which suggests that the differences between the conditions may not be directly
related to the level of knowledge improvement.

A Kruskal-Wallis H test showed that there was a statistically significant
difference among conditions in the Tripod part of the Subjective Experience
questionnaire, $\&\Xi\chi^2(2)$ = 7.699, p = 0.021, with a mean rank score of 17.40
for the RT, 17.00 for the RP, and 7.75 for the control condition. The results of
the Mann-Whitney Tests showed significant differences between the RT (3.33 ±
0.75) and the Control (2.60 ± 0.34) conditions (p = 0.021); the RP (3.24 ± 0.49)
and the Control (2.60 ± 0.34) conditions (p = 0.009) (see Figure 1).

The difference among conditions for the Class part (regarding the expression
and the perceived authority of the robot) is close to significance (p = 0.053).
The results of the Mann-Whitney Tests showed significant differences between
the RT (4.23 ± 0.45) and the Control (3.48 ± 0.74) conditions (p = 0.025); but
not between the RP (4.20 ± 0.48) and the Control (3.48 ± 0.74) conditions (p =
0.070) (see Figure 2). This suggests that the robot (in both conditions) and the
audio are perceived differently in terms of their potential to transfer knowledge
and engage the learners.

Moreover, the results of these two conditions were significantly correlated
$r(27)$ = 0.471, p = 0.015 (see Figure 3); the perceived authority of the robot is
positively correlated with the clarity with which it expresses itself.

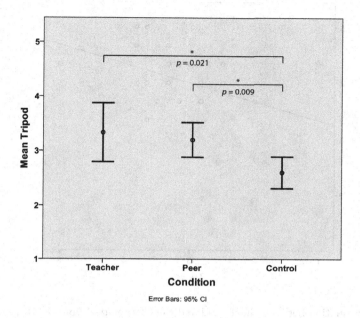

Fig. 1. Scores in the Tripod part from the Subjective Experience questionnaire

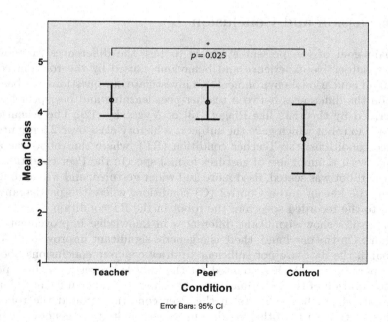

Fig. 2. Scores in the Class part from the Subjective Experience questionnaire

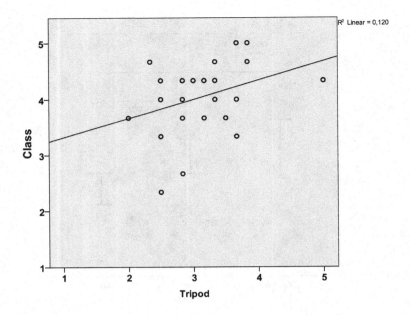

Fig. 3. Correlation between the Tripod and the Class parts from the Subjective Experience questionnaire

4   Discussion and Conclusion

The main goal of the present study is to test the differences in knowledge retrieval, subjective experience and behaviour caused by the role played by a robot in an educational environment. To investigate our questions, we based our design on the differences between teacher-peer learning and peer-peer learning, as described by theorists like Piaget [13] or Vygotsky [12]. The scenario contained a Nao robot which gave the subjects a history class over 2.5 minutes. We set three conditions: the Teacher condition (RT), where the robot was always standing, with a minor use of gestures formal speech; the Peer condition (PR), where the robot was seated, used more and wider gestures, and was adding comments to the lesson; and a Control (C) condition, where the participants only listened to the recorded speech of the robot in the RT condition.

Our results show significant differences in knowledge improvement among conditions. On the one hand, there is a general significant improvement. On the other hand, the data are not sufficient to draw concrete conclusions about the specific parameters which accounted for the knowledge retrieval. It is possible that the duration of the experiment was too short to find significant differences. Our results show that subjects in the Peer condition graded the robot as a tutor higher than in the other conditions. Such results were generally expected compared to the control condition where we do not introduce the embodiment. Other experiments about embodied robotic tutors support these results:

Leyzberg et al [23] found that physical presence may imbue the robot with more perceived authority than an on-screen agent, which causes the participants to take the advice of a robot more seriously, and accounts for a significantly greater sense of social acceptance; Saerbeck et al showed how the degree of sociability has an effect on students' motivation [24].

Nevertheless, as claimed by Leyzberg et al [18], novelty can lead the users to focus more on the robot itself than on the topic of the lesson. This was reported by some participants, who claimed to have been so engaged and enthusiastic that sometimes they were more focused on the robot than on what it was saying. A long-term study may be needed to test the evolution of knowledge retrieval where the subjective experience of the users after the novelty threshold diminishes.

Regardless of the statistical results, some subjects in the Peer condition reported that the robot was speaking to them, and one of the participants reported that the robot offered its hand to him, so he shook hands with the robot during the experiment (see Figure 4). This perspective can be related to *mindreading*, that is, estimating the mental states of the others by observing their behavior [25], and mainly the Intentionality Detector, a module that recognizes stimuli through visual, auditory and tactile stimuli and creates goal-desire relationships. The subject, seeing the hand of the robot and the way the robot was speaking to him, inferred it was offering its hand to him. This is important because, as Ono et al [25] claimed, *mindreading* is indispensable for reciprocal acts between humans and robots.

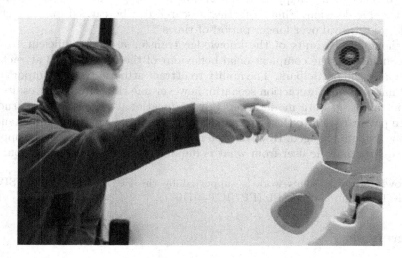

Fig. 4. A subject in the "Peer" condition shakes hands with the robot

Further steps in this research will include adding a new condition where the robot can recognize the speech of the subject. This allows for communication through dialog, one of the most relevant characteristics that the subjects needed in their classes. Another step will be to investigate how do long-term relationships evolve and whether the enthusiasm towards the robot is maintained or it was affected by the desire for communication and interactivity caused by the first impact. In any case, more experiments need to be performed, as the significance of the differences between conditions for the knowledge retrieval became higher as the number of subjects increased. As for our pilot study, further analysis of the behavioral data can provide insights regarding eye contact. The peer approach should be further investigated as it may provide more engagement compared to the tutor condition.

Due to technical issues, it was not possible to use the speech recognition of the robot; nevertheless, an approach where in the Peer Condition both the subject and the NAO listen to a lecture and then they fill the questionnaire together will be considered for further research. Our approach includes questions that are directly linked to the Industrial Revolution of Catalunya. This is fine up to a point, however knowledge acquisition in that scenario might reflect memorisation rather than clearly understanding the topic. In our next studies, we will add questions that were not part of the lecture but are directly linked to it (eg: Which other areas might be affected by the Industrial Revolution of Catalunya?). This way, we can ensure that knowledge acquisition is not the result of memorisation but of a deeper understanding of the topic.

One possibility following the same topic would be to present them conditions of a particular area of Spain, and then ask them to choose which one was most likely to reflect what happened in Catalunya, and provide evidence from the essay for this decision. Finally, further research can be carried out to test this knowledge retrieval over longer period of times.

Although the results of the knowledge transfer are not sufficient to draw clear conclusions, the complex social behaviour of the robot indeed attracts the attention of the participant. The ability to attract attention is very important in any Human-Robot Interaction scenario, however one thing from our results that is important is gesture moderation. In an educational scenario where students need to pay attention to the robot, attracting attention is of great importance to keep the student engaged. However, too many gestures seem to have the opposite effect, distracting the user from what is important: knowledge acquisition.

Acknowledgments. This work is supported by the EU FP7 project WYSIWYD (FP7-ICT-612139) and EASEL (FP7-ICT- 611971).

References

1. Mondada, F., Bonani, M., Raemy, X., Pugh, J., Cianci, C., Klaptocz, A., Zufferey, J.C., Floreano, D., Martinoli, A.: The e-puck, a Robot Designed for Education in Engineering. Robotics **1**, 59–65 (2006)
2. Kanda, T., Hirano, T., Eaton, D., Ishiguro, H.: Interactive robots as social partners and peer tutors for children: A field trial. Human-Computer Interaction **19**(1), 61–84 (2004)

3. Mubin, O., Stevens, C.J., Shahid, S., Al Mahmud, A., Dong, J.J.: A review of the applicability of robots in education. Journal of Technology in Education and Learning 1 (2013)
4. Lallée, S., Vouloutsi, V., Blancas Munoz, M., Grechuta, K., Puigbo Llobet, J.Y., Sarda, M., Verschure, P.F. Journal of Behavioral Robotics (in press)
5. Kanda, T., Sato, R., Saiwaki, N., Ishiguro, H.: A two-month field trial in an elementary school for long-term human-robot interaction. IEEE Transactions on Robotics 23(5), 962–971 (2007)
6. Saerbeck, M., Schut, T., Bartneck, C., Janse, M.D.: Expressive robots in education: varying the degree of social supportive behavior of a robotic tutor. In: Proceedings of the SIGCHI Conference on Human Factors in Computing Systems, pp. 1613–1622. ACM (2010)
7. Ramachandran, A., Scassellati, B.: Adapting difficulty levels in personalized robot-child tutoring interactions. In: Workshops at the Twenty-Eighth AAAI Conference on Artificial Intelligence (2014)
8. Shin, N., Kim, S.: Learning about, from, and with robots: students' perspectives. In: The 16th IEEE International Symposium on Robot and Human interactive Communication, RO-MAN 2007, pp. 1040–1045. IEEE (2007)
9. Prensky, M.R.: Teaching digital natives: Partnering for real learning. Corwin Press (2010)
10. Papert, S.: Mindstorms: Children, Computers, and Powerful Ideas. Basic Books Inc., New York (1980)
11. Tiberius, R.G., Billson, J.M.: The social context of teaching and learning. New Directions for Teaching and Learning 1991(45), 67–86 (1991)
12. Vygotsky, L.: Zone of proximal development. Mind in society: The development of higher psychological processes 5291 (1987)
13. Piaget, J.: The development of thought: Equilibration of cognitive structures. (Trans A. Rosin). Viking (1977)
14. Baines, E., Blatchford, P., Kutnick, P.: Changes in grouping practices over primary and secondary school. International Journal of Educational Research 39(1), 9–34 (2003)
15. De Lisi, R., Golbeck, S.L.: Implications of piagetian theory for peer learning (1999)
16. Thurston, A., Keere, K.V.D., Topping, K.J., Kosack, W., Gatt, S., Marchal, J., Mestdagh, N., Sidor, W., Donnert, K.: Peer learning in primary school science: Theoretical perspectives and implications for classroom practice 5(3), 477–496 (2007)
17. Fong, T., Nourbakhsh, I., Dautenhahn, K.: A survey of socially interactive robots. Robotics and Autonomous Systems 42(3), 143–166 (2003)
18. Leyzberg, D., Spaulding, S., Scassellati, B.: Personalizing robot tutors to individuals' learning differences. In: Proceedings of the 2014 ACM/IEEE International Conference on Human-robot Interaction, HRI 2014, pp. 423–430. ACM, New York (2014)
19. Tanaka, F., Cicourel, A., Movellan, J.R.: Socialization between toddlers and robots at an early childhood education center. Proceedings of the National Academy of Sciences of the United States of America 104(46), 17954–17958 (2007)
20. Jolliffe, D., Farrington, D.P.: Development and validation of the basic empathy scale. Journal of Adolescence 29(4), 589–611 (2006)
21. Bartneck, C., Kulić, D., Croft, E., Zoghbi, S.: Measurement instruments for the anthropomorphism, animacy, likeability, perceived intelligence, and perceived safety of robots. International Journal of Social Robotics 1(1), 71–81 (2009)

22. Ferguson, R.: The tripod project framework. The Tripod Project (2008)
23. Leyzberg, D., Spaulding, S., Toneva, M., Scassellati, B.: The Physical Presence of a Robot Tutor Increases Cognitive Learning Gains (1) (2012)
24. Saerbeck, M., Schut, T., Bartneck, C., Janse, M.D.: Expressive Robots in Education: Varying the Degree of Social Supportive Behavior of a Robotic Tutor, pp. 1613–1622 (2010)
25. Baron-Cohen, S.: Mindblindness: An essay on autism and theory of mind. MIT press (1997)

Adaptive Bio-inspired Signals for Better Object Characterisation

Mariia Dmitrieva$^{(\boxtimes)}$, Keith Brown, and David Lane

Heriot-Watt University, Edinburdh, UK
m.dmitrieva@hw.ac.uk

Abstract. Dolphins identify objects using their sonar, which works by emitting short acoustic pulses with high bandwidth and high intensity. These echolocation impulses have a double chirp structure. The complex signal structure allows animals to collect more information than simply the distance to the object. They can evaluate object's size, shape, and even the innards of the object, by processing the whole echo from the object.

The study of the dolphins' clicks inspired a simulation of the signals for echolocation purposes. They are already used for object characterisation.

In addition, dolphins' clicks are adaptive signals. Dolphins can change some parameters of the clicks during recognition process, which allows them to achieve better results for object characterisation.

This paper presents background and the main concept of the adaptive echolocation using bio-inspired signals. Implementation of adaptive echolocation is a new approach and can improve object characterisation and will help to achieve more accurate results.

1 Introduction

Dolphins inspect objects by transmitting sets of acoustic clicks. The clicks are acoustic echolocation pulses emitted in the object's direction. The animals acquire and process the reflected responses to recognize the object.

Dolphins demonstrate an adaptive control over the set of clicks by changing the interclick interval, the frequency components of the signal and other parameters. The efficiency of the dolphins' sonar has provided inspiration to simulate their signals for echolocation purposes. The bio-inspired sonar signals are a very powerful tool for object characterisation and recognition. The pulses with constant parameters values are already used for object characterisation purposes.

The paper demonstrates idea of adaptive bio-inspired echolocation. It is based on the work of Dorian Houser et al. [1], where a classification of the dolphins' clicks was proposed and the work of Chris Capus et al. [2] who presented a synthetic bio-inspired pulse set. These works describe a variety of dolphins' clicks and a way to simulate them. It is a good background for the adaptive bio-inspired echolocation. The idea of adaptive signalling was mentioned in the sources as

© Springer International Publishing Switzerland 2015
S.P. Wilson et al. (Eds.): Living Machines 2015, LNAI 9222, pp. 403–409, 2015.
DOI: 10.1007/978-3-319-22979-9_40

a possible evolution of the echolocation, but was not presented as a complete concept.

This paper describes basics of the bio-inspired echolocation and demonstrates a concept of the adaptive signals, as a next step in object characterisation using bio-inspired technology.

2   Dolphin's Adaptive Echolocation

The dolphins' clicks are changed according to task, distance to the object, surrounding environment and object characteristics. Capus et al. [2] describes three phases of the target detection and identification: initial search phase, transition phase and acquire phase. The initial search phase is characterized by wideband signal emission during the search process. When the target is found, the impulses are changed to double chirp structured signals. It is the transition phase between searching and acquire. Variation in the clicks appears during the acquire phase, when a dolphin discovers the target to recognise it.

A single click is like an impulse but with a composite structure, the simulated dolphin click is shown on fig. 1. The click can be represented as a combination of two linear chirps. One linear chirp is a signal, which frequency is changing with time. This type of the chirps is called the down-chirp where the frequency desraeases with time.

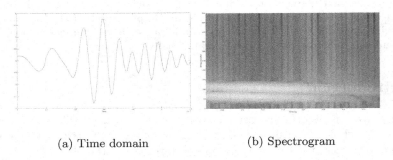

(a) Time domain   (b) Spectrogram

Fig. 1. Synthetic echolocation dolphin's click

Reflection of the click from the object creates a response, which the dolphin receives and processes. Depending on the result of the processing dolphins make a conclusion about the object.

As noted above, dolphins use different clicks. Houser et al. [1] studied the dolphins clicks recorded during object recognition tasks. The echolocation clicks were classified into groups based on their spectrum. They were labelled according to the number of distinctly bounded regions, number of peaks in the regions, peak frequency, secondary peak frequency (if it exists) and others properties. Houser et al. determines seven groups of the clicks.

The click's spectrum and other parameters can be changed from click to click during the target investigation process. It allows dolphins to "tune" the clicks for better object representation. The paper will consider a few parameters which have important influence in response processing and target recognition.

The interclick interval between the clicks does not stay constant during the object recognition task. It is strongly related to the distance between dolphin and target [3]. Also the neural processing time needs to be considered here, it is changed depending on task and animal capabilities. The interclick intervals can vary roughly between 19 and 34 ms, when dolphins emit one click after receiving and processing another [4]. Also they can emit set of clicks, called trains or packages. In these cases dolphins process the clicks all together. Even though, click parameters are changing inside of the click train.

Another important characteristic of the impulse is their spectrum, which is also changed from click to click. It can vary from unimodal to multimodal with peaks in low, high frequencies or both. The spectrum can be varied by changing the chirp characteristics of the click.

3   Bio-inspired Echolocation

Digital analysis of the dolphins clicks allows simulated impulses similar to the original dolphin ones. Based on the work of Houser et al. [1] six biosonar signals were synthesized by C. Capus et al. [2]. The impulses are formed by down-chirps with different frequency diapason. The chirp can be described as $chirp = cos(2\pi(at^2 + b_1 t))$, where a is the chirp rate and b_1 is the lowest frequency of the chirp. The High frequency chirp has a delay relatively to the Low frequency chirp.

Response is highly related to the transmitted pulse. Parameters of the pulse influence the way how the object interact with the acoustic wave and reflect it. Figure 2b presents an example of the response. The object is a buoy placed in a water tank and ensonified with a double chirp bio-inspired pulse. The recording was made in a tank with a 3 x 4 m perimeter and 2 m depth. The buoy was placed in front of the BioSonar.

The original recorded response contains reflection from the target (buoy), walls, bottom of the tank and other surfaces (fig. 2a). The reflection from the buoy presents useful information for the study. In the case, the second splash on the plot 2a is the response of interests.

The response can be presented in time domain, frequency domain and time-frequency domain (TFD). The time-frequency domain is more representative for the response processing, which shows changing of frequency components in time.

The buoy response was extracted from the full response manually based in the time domain representation of the signal. The Short-Time Fourier Transform and Gaussian window were used to present the response in the TFD.

Figure 3 presents Time-Frequency Representation (TFR) of two responses. Those responses present reflection from the same buoy, but the initial impulses are different: a single down-chirp impulse (fig. 3a) and double down-chirp impulse

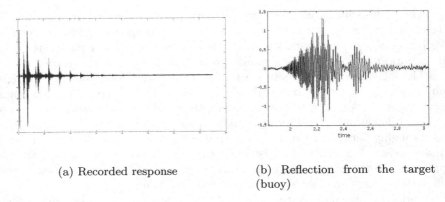

(a) Recorded response

(b) Reflection from the target (buoy)

Fig. 2. Response on bio-inspired signal

(fig. 3b). The TFR of the responses is very distinctive, one and double chirp structures can be easily recognized by the figures.

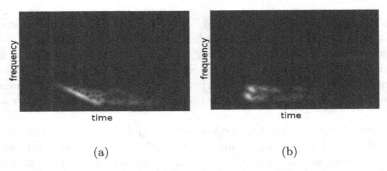

(a)   (b)

Fig. 3. Time-Frequency Representations of responses on bio-inspired synthetic signals: (a) one down-chirp (b) two down-chirp

The spectrum of the two chirp signals can vary depending on the chirp components of the impulse. It can be changed by the delay between LF and HF chirps in the click as well as frequency diapasons of both chirps. Some other parameters are also influence the response result, for example, the duration of the impulse and its intensity.

4   Adaptive Echolocation Concept

The idea of adaptive echolocation is based on the changing of the bio-inspired impulses for the current environment, object and task.

The response presents interaction of the initial impulse with the object during the reflection, that is why the parameters of the pulse are important to highlight

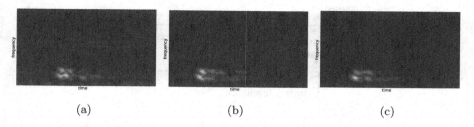

| (a) | (b) | (c) |

Fig. 4. Time-Frequency Representations of responses from a bio-inspired synthetic signal: (a) small buoy, (b) medium buoy,(c) big buoy

one or another characteristic of the object. Also different type of objects can demand different pulses.

For example, the spectrum of the echolocation pulse has a big influence on the result. Pulses with different spectrum interact differently with objects. Size, shape, density and other characteristics of objects reveal differently with different frequencies. Each object has a resonant frequency or frequencies, so the response will present more information about the object in the case that the click's spectrum excites the resonant frequency. The pulse spectrum can be set by chirp frequency diapasons. Figure 4 presents responses from different buoys. Some dependences can be seen from the image set. The high frequency components are changing with the target size. The changes appear only for the double chirp impulse with a particular frequency diapason. It allows us to assume that there is some influence of the pulse's spectrum into the result of object characterisation.

The interclick interval also can be varied proportionally with the complexity of processing and distance to the object. The interclick interval increases when processing time and distance to the object are increasing, and decreases otherwise.

The adaptive bio-inspired echolocation concept defines changes to be made based on the processing of already received responses (fig. 5). The adaptive module stores previous responses information and tunes the impulse generation module to adapt the next impulse for the particular environment and object. The tuning feedback is formed based on the learned knowledge about current target and environment. For example, if the distance to target is reduced, than the interclick interval can be reduced also.

The adaptive strategy first considers distance to the object to set the pulse, then size and shape of the object is taken into account. The final attribute to adapt to is innards of the object (material, structure, thickness of surface, etc).

Moreover, the impulse type can be changed according to the task. So the size recognition of the object can be made with one pulse type, while the shape recognition is studied by another pulse, which is more suitable for the object of the size.

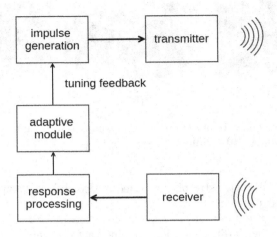

Fig. 5. Adaptive scheme

5   Conclusions

This work presents the concept of adaptive bio-inspired echolocation when the bio-inspired pulses can be changed during the echolocation process to adapt the signals for the specific environment, task and object type. The adaptive echolocation can enhance the object characterisation process and achieve better accuracy in recognition tasks. However there is a lot of research to do on these aspects to evolve the concept into a fully working model.

The implementation of the system demands better understanding of the signal's parameters and strategies which dolphins use to "tune" their clicks. The influence of pulse spectrum on the recognition process and manipulation of the spectrum by the chirps parameters are highly important in the study.

Future work will aim to study influence of impulse parameters into the target representation by running experiments and finding patterns in the responses for similar object attributes.

The first step is to investigate influence of target attributes on the response characteristics, it will lead to an advanced understanding of the signal-object interaction result. This result is represented in received signal. The knowledge about the dependency will allow the work to move forwards and discover relation between parameters of bio-inspired signal and target in received returns. This will inform the creation of rules to tune the transmitted bio-inspired pulses according to current environment, object and task.

References

1. Houser, D.S., Helweg, D., Moor, P.: Classification of dolphin echolocation clicks by energy and frequency distributions. The Journal of the Acoustical Society of America **106**, 1579–1585 (1999)
2. Capus, C., Pailhas, Y., Brown, K., Lane, D.M., Moor, P., Houser, D.: Bio-inspired wideband sonar signals based on observations of the bottlenose dolphin (tursiops truncatus). The Journal of the Acoustical Society of America, 594–604 (2007)
3. Au, W., Floyd, R., Penner, R., Murchison, A.E.: Measurement of echolocation signals of the atlantic bottlenose dolphin, tursiops truncatus montagu, in open water. The Journal of the Acoustical Society of America **56** (1974)
4. Herzing, D., dos Santos, M.E.: Functional aspects of echolocation in dolphins. University of Chicago Press, Chicago (2004)
5. Harris, F.J.: On the use of windows for harmonic analysis with the discrete fourier transform. Proceedings of the IEEE **66** (1978)
6. Au, W.: The sonar of dolphins. Springer-Verlag, New York (1993)

Mechanical Improvements to TCERA, a Tunable Compliant Energy Return Actuator

Ronald Leibach[✉], Victoria Webster, Richard Bachmann, and Roger Quinn

Department of Mechanical and Aerospace Engineering,
Case Western Reserve University, Cleveland, OH 44106-7222, USA
rjL32@case.edu

Abstract. TCERA (Tunable Compliance Energy Return Actuator) is a robotic emulation of a human femur-tibia system in a walking or running gait. TCERA features two parallel air springs to compliantly transmit or absorb torque about the knee. The radial positions of the springs as well as their relative lengths are adjustable, which effectively alter the torsional stiffness and equilibrium angle of the knee system, respectively. These functions are operable independently, and thus decouple the kinematic relationship and control of the torsional stiffness and femur-tibia angle. The actuator derives its inspiration from an analysis of the human gait, in which it can be seen that the torsional compliance about the knee is varied to specific stiffness values throughout the cycle. Special mechanical considerations have been taken into account to achieve swift changes in position without compromising the static integrity of the actuator.

Keywords: Biologically inspired · Tunable compliance · Energy return · Air springs · Legged robot

1   Introduction

Though muscles are by themselves rather inefficient (~25% efficiency) [1], many animals show higher efficiencies during locomotion. Humans, for example, have gait efficiencies up to 35% while walking [2], and up to 45% while running [1]. Additionally, individual muscle-tendon systems can have much higher efficiencies, with the Achilles system returning up to 93% of the energy stored during running. The ability of animals to achieve higher overall efficiencies than allowed by their individual actuators, stems from changing the natural dynamics of their legs during locomotion. The muscle-tendon systems in the leg absorb, store, and return energy with each step, reducing the overall energy input required. These compliant actuators also serve to reduce energy consumption by passively rejecting disturbances.

Locomotion efficiency is important to the development of untethered mobile robotic platforms. For legged robots, traditional actuator systems suffer from a lack of compliance, as well as low efficiencies. Pneumatic actuators, though compliant, can be up to 85% efficient during pressure cycling [3]. However they require a mechanical compressor (20% efficiency), which is often powered by a gearmotor

© Springer International Publishing Switzerland 2015
S.P. Wilson et al. (Eds.): Living Machines 2015, LNAI 9222, pp. 410–414, 2015.
DOI: 10.1007/978-3-319-22979-9_41

(70% efficiency). By combining a muscle-tendon inspired, compliant transmission, with traditional actuators. We can combine the energy return capabilities seen in biological systems, with the higher actuator efficiencies found in mechanical systems. We have previously reported on the design of a tunable compliance actuator, inspired by mammalian leg actuation [4]. However, the actuator was unable to provide the energy return desired. In addition, the transmission of the position actuator was unable to withstand the static torque induced by the segment's inertial loads. Here we present the mechanical design of a fully tunable compliance energy return actuator (TCERA).

2   Design Parameters

Biological inspiration and parameter optimization were two driving factors when it came to determining geometric and physical parameters. The lengths of the femur and tibia segments are those of a 50th percentile adult woman, which results in a femur length of 0.468 m and a tibia length of 0.375 m [5].

3   Mechanical Design

3.1   Mechanical Overview

The actuation scheme is derived from that found in the mammalian quadriceps/hamstring system. TCERA utilizes antagonistic co-contraction in order to control its stiffness, as well as taking advantage of the agonist/antagonist pairing to determine joint position via spring-system static equilibrium. Mounted on the most proximal end of the femur are two gearmotors. Each gearmotor is dedicated to actuating one of the independent features (i.e., stiffness or position). At the core of TCERA's mechanical capabilities is a device designated the "control bar", which can be seen protruding sagittally from the knee axis. The control bar not only houses the sliding mechanism that establishes the desired stiffness value, it also rotates as a whole, which alters the spring-system to resolve new joint angles.

The two parallel air springs (F1,F2) are attached at their distal end to their respective ankle pivots (G1,G2), while at the proximal end each spring is attached to a point (E1,E2) on the control bar (C), radially equidistant from the knee axis (D). To command a greater torsional stiffness, these insertion points travel across the control bar in the direction away from the knee axis. This effectively causes positive strain on the springs, which in turn incurs a greater symmetric recovery torsion to the control bar relative to the tibia. The angle of the control bar relative to the femur is controlled by the motor, while the angle of the control bar to the tibia resolves kinematically. Changes in the control bar-femur angle result in overall changes to the static system, which is used to control the femur-tibia joint angle.

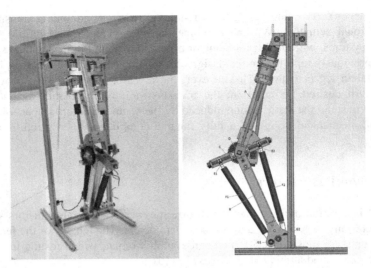

Fig. 1. The completed TCERA Assembly (left) and a labeled diagram of the core components of TCERA (right)

3.2 Stiffness Mechanism

To alter the stiffness, the insertion points move along the control bar, which is coincident with the knee axis, oriented in the sagittal plane at an angle to the coronal plane. The physical manifestation of each insertion point is a precision nut for an acme threaded lead-screw, which is housed along the interior of the control bar. Each lead screw must rotate in opposite directions in order to symmetrically position the insertion points. This is accomplished by using a series of bevel gears in an assembly that resembles a differential. Both lead screws are driven by "sun" bevel gears located at their interior ends. These "sun" bevel gears are coupled by a "planet" bevel gear that rotates about the knee axis. This "planet" gear is driven by the shaft to which it is attach, which in turn is driven by two addition bevel gears that exist to change the direction of transmission from the axis of rotation of the motor.

Due to the relatively large axial load upon the lead screw by the travel nut, the torque due to static friction is non-negligible. The torque required to overcome this friction can be calculated by drawing a single thread as an "unwrapped" two dimensional free body diagram. From there, the force (and thus torque) can be determined as a function of the axial load, friction coefficient, pitch diameter, and the lead. It was determined that for a conservatively stiff air spring with a spring constant of 16 kN/m, the frictional torque that must be overcome is 0.435 N-m.

The lead screws used are ½"-10 with 5-start threading. Multi-start threads were chosen to prioritize linear travel velocity over the greater holding torque that is inherent to 1:1 ratios of pitch:lead. This allows the insertion blocks to travel ½" per rotation of the lead screw. The full potential travel range for each insertion block is 2.5", or between 1.75" and 4.25" radially from the knee axis. With the motor at full speed (1450 RPM after the gearbox), the insertion block is capable of traversing the distance of the control bar in 0.21 seconds.

Fig. 2. Bevel Gear Assembly

3.3   Position Control

The femur-tibia joint angle can be controlled by manipulating the static equilibrium of the spring system composed of the air springs and the control bar. The full joint geometry is driven by the angle between the control bar and the femur. This feature is actuated by a large worm wheel which is rigidly connected to the control bar casing, which is connected to a driving worm gear. On the previous version of this robot, the position actuator was constructed with a similar bevel gear system as that used for the stiffness actuator. This system, coupled with a Maxon 241414 motor/transmission set, was unable to prevent the inertial forces from affecting the desired position. Thus on the previous version a friction clutch was implemented to enable position control, although only for low velocities and high femur-tibia angles. To improve upon this, a more robust transmission was designed with the following parameters.

The size and pitch of the worm wheel/worm gear combination were chosen for the unique difficulty in backdriving a worm gear transmission. Because of this, the reaction torque at the knee axis due to the inertial forces of the joint segments cannot affect the femur-control bar angle. In order to verify that a worm gear is inherently self-locking, the coefficient of friction must be less than the product of the cosine of the pressure angle and the tangent of the helix angle.

The worm wheel is a 3.33" diameter 40 tooth wheel, and is driven by a DeWalt 18V motor through a transmission at a 1:15 reduction. The total resulting gear reduction between the motor and the control bar is 1:600. At maximum angular velocity (1450 RPM), the control bar is capable of rotating through its 60 degree range in 0.28 seconds.

3.4   Electronics

TCERA is equipped with two potentiometers, two current sensors, and two voltage sensors (one of each sensor for each actuator). These are all fed into an Arduino Mega as analog inputs. The Arduino controls TCERA via serial communication to a Sabertooth 2x25, which can operate two DC motors simultaneously.

The voltage sensors and current sensors are both products of the Phidgets hardware platform, and produce 10 bit analog signals. The input ranges for these sensors are +/- 30V and +/- 30A, respectively. Due to different angular range requirements between the two actuators, each of the two potentiometers is a unique model. The position control actuator uses a VEX single turn potentiometer, which was selected for its ease of mounting and low requirements for rotational range. The stiffness control actuator must rotate five times in order for the insertion block to span the length of the lead screw, thus a Bourns 10 turn potentiometer was selected.

4   Future Work

In order to demonstrate energy return and the ability to continuously modify the passive dynamics of the system, the stationary device will be controlled in a simulated hopping regime. The device will be loaded and the control bar will be actuated briefly at regular intervals in order to add energy to the system, resulting in hopping-like behavior. The air spring system will store and return potential energy from both the actuator input and the potential change of the load. The total energy of the system will be recorded and compared to the theoretical energy demand for directly driving the load through the hopping motion. The passive dynamics, and thus natural frequency, of the system will be varied by changing the stiffness of the joint via the sliding mechanism. This will allow the gait frequency to be varied.

Once energy return has been demonstrated, the test stand will be redesigned to incorporate a movable hip, allowing forward locomotion, via hopping or walking, to be investigated. During forward locomotion inertial energy is stored at the beginning of stance and returned at the beginning of swing, rather than requiring additional input from the actuators as during the simulated hopping test. This will further improve efficiency.

Acknowledgements. The authors would like to thank Matt Klein for assistance with serial communication and Alexander Hunt for assistance in the mechanical testing of the air springs.

References

1. Alexander, R.M.: Principles of animal locomotion. Princeton University Press, Princton (2003)
2. Cavagna, G.A., Kaneko, M.: Mechanical work and efficiency in level walking and running. J. Physiol. **268**(2), 467–481 (1977)
3. Colbrunn, R.W., Nelson, G.M., Quinn, R.D.: Modeling of braided pneumatic actuators for robotic control. int. conf. on intelligent robots and systems. In: International Conference on Robots and Systems (IROS 2001), pp. 1964–1970 (2001)
4. Webster, V., Leibach, R., Hunt, A., Bachmann, R., Quinn, R.D.: Design and control of a tunable compliance actuator. In: Duff, A., Lepora, N.F., Mura, A., Prescott, T.J., Verschure, P.F.M.J. (eds.) Living Machines 2014. LNCS, vol. 8608, pp. 344–355. Springer, Heidelberg (2014)
5. N. S. D. A. of Japan, ST-E-1321 Japanese Female Body Size

Active Control for Object Perception and Exploration with a Robotic Hand

Uriel Martinez-Hernandez[1]([⊠]), Nathan F. Lepora[2], and Tony J. Prescott[1]

[1] Sheffield Robotics Laboratory and the Department of Psychology,
University of Sheffield, Sheffield, UK
`uriel.martinez@sheffield.ac.uk`
[2] Department of Engineering Mathematics, The University of Bristol and Bristol
Robotics Laboratory (BRL), Bristol, UK

Abstract. We present an investigation on active control for intelligent object exploration using touch with a robotic hand. First, uncertainty from the exploration is reduced by a probabilistic method based on the accumulation of evidence through the interaction with an object of interest. Second, an intrinsic motivation approach allows the robot hand to perform intelligent active control of movements to explore interesting locations of the object. Passive and active perception and exploration were implemented in simulated and real environments to compare their benefits in accuracy and reaction time. The validation of the proposed method were performed with an object recognition task, using a robotic platform composed by a three-fingered robotic hand and a robot table. The results demonstrate that our method permits the robotic hand to achieve high accuracy for object recognition with low impact on the reaction time required to perform the task. These benefits make our method suitable for perception and exploration in autonomous robotics.

Keywords: Tactile sensing · Active perception · Tactile exploration · Robotics

1 Introduction

The intelligent exploration of the environment performed by humans requires the use of exploratory procedures and intelligently controlled movements of their hands and fingers [1],[2]. The exploratory procedures are employed according with the information of interest from the environment, whilst active perception permits to decide where to move the hand and fingers to explore interesting locations and extract useful information [3],[4]. These are important features required for the development of intelligent robots capable to explore and interact with their environment in the presence of uncertainty.

In this work, we present a method for intelligent perception and exploration of objects using a three-fingered robotic hand. First, an exploratory procedure is developed to allow the robotic hand to explore various objects moving the hand and fingers around them, extracting both tactile and proprioceptive information.

© Springer International Publishing Switzerland 2015
S.P. Wilson et al. (Eds.): Living Machines 2015, LNAI 9222, pp. 415–428, 2015.
DOI: 10.1007/978-3-319-22979-9_42

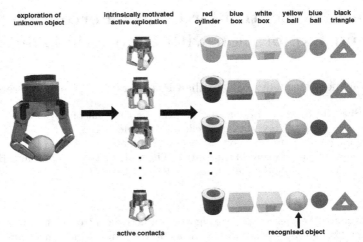

Fig. 1. Object recognition by the robotic hand using an intrinsically motivated active exploration approach. The robotic hand is actively moved towards interesting locations to improve perception. This process is repeated N times until a belief threshold is exceeded, permitting to make batter decisions about the object being explored.

Second, reduction of uncertainty is implemented with a Bayesian approach, which has been employed for object shape extraction [5],[6] and simultaneous object localisation and identification with a biomimetic fingertip sensor [7],[8]. The reduction of uncertainty is achieved by the accumulation of evidence based on the continuous interaction of the robotic hand with the environment.

An active exploration behaviour, similar to the one employed by humans, is performed by an intrinsic motivation approach, which permits the robotic hand to move towards interesting locations to extract useful information. This approach studied by psychology and cognitive sciences [9],[10], have found that intrinsic motivation is essential for cognitive development in humans, and also required for robust exploration and manipulation in robotics [11],[12],[13].

We implemented our methods in a sensorimotor architecture for the intelligent control of the exploration movements with a robotic hand for an object exploration and recognition task. Object recognition has been studied using tactile feedback with a simulated robotic arm showing accurate results [14]. The use of proprioceptive information from a five-fingered robotic hand has allowed to develop an object recognition task [15]. A fixed number of exploratory movements with a Self-Organising Map (SOM) approach was proposed for object recognition with a three-fingered robotic hand [16]. A drawback from these methods is that they are based on a single contact and passive exploration modality, where the robotic hand is not permitted to moved to interesting locations to reduce uncertainty. This contrasts with our method for active object exploration, which allows the robotic hand to intelligently move and improve perception from the object by the continuous interaction with the environment.

Our proposed methods were validated in simulation and real environments with an object exploration task. First, for the simulated environment we used the datasets collected from 6 test objects. Next, for the real environment, we used a robotic platform composed by a robotic hand and a positioning robotic table for exploration of various test objects. For both environments, the exploration was performed in passive and active modalities to compare their performance. Results demonstrate that our approach for active control of object perception and exploration permits to achieve higher perception accuracy over passive exploration modality, which offers a method suitable for autonomous robotics.

2   Methods

2.1   Robotic Platform

This study employs a robotic platform composed by a three-fingered robotic hand mounted on a positioning robotic table shown in Figure 2.

The three-fingered robotic hand from Barrett Hand has 4-DoF, with 1-DoF in each finger for its opening and closing, and 1-DoF for spreading the fingers around the palm of the hand (see Figure 2a). The robotic hand also is integrated with tactile and force sensors. Each finger is composed by 22 taxels (tactile elements), whilst the palm has 24 taxels of 12 bit resolution. The strain sensors are located in each finger which permit to detect when a tactile contact has exceeded a force threshold. Also, it is possible to obtain proprioceptive information from the robotic hand in real-time.

The positioning robotic table has 4-DoF that permit precise movements in x-, y-and-z axes, and rotations on *theta* (see Figure 2b). The three-fingered robotic hand is mounted on the positioning robotic table to allow a larger set of exploration movements: 1) opening and closing of fingers; 2) spreading of fingers around the palm; 3) rotation of the wrist (*theta*); and 4) displacements of the robotic hand (x-,y-and-z axes). This configuration permits the exploration of a large variety of objects by synchronising and controlling the robotic platform.

We developed a controller embedded in a microcontroller Arduino for the positioning robotic table. The data collection and exploratory movements performed by the robotic platform are controlled in real-time by tactile feedback and perceptions from the proposed methods. The synchronisation of the modules of software and hardware that compose the robotic platform is based on the YARP (Yet Another Robot Platform) middleware developed for robot control [17].

2.2   Data Collection

Our work is focused on object recognition with robotic hands using proprioceptive information. For this purpose, we collected information from the position and orientation of the fingers and hand in real-time for each contact performed on the set of test objects.

Figure 3 shows the sequence of movements performed by the robotic hand around two test objects. First, each finger moves independently towards the unknown object. They stop as soon as a contact is detected by exceeding the

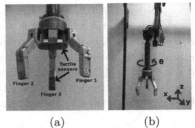

(a)   (b)

Fig. 2. Robotic platform used for data collection and validation of the proposed method. (a) Robotic hand with 4-DoF from Barrett Hand. (b) Positioning robotic table that provides mobility to the robotic hand.

tactile pressure and force threshold. The fingers keep in contact with the object for 1 sec, collecting 50 samples of proprioceptive information from the hand per contact. Second, the fingers are opened to a predefined home position, and then the wrist is rotated to collect data from a new orientation of the hand. The wrist is rotated in 12 degrees steps covering 360 degrees to explore the complete object. This process was repeated 5 times per object to have one dataset for training and four datasets for testing.

The data collected is stored in a 50×5 matrix per contact. The first three columns contain the positions of contacts detected by each finger, the fourth column contains the value of the spread motor, and the fifth column contains the angle orientation of the hand for each contact detected.

2.3 Bayesian Estimator

Robotics has made used of Bayesian methods to develop a variety of applications and estimate an state given the observations. Here, we use a Bayesian approach

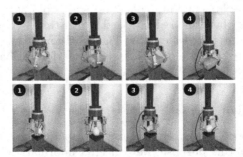

Fig. 3. Sequence of movements performed by the robotic hand around the test objects for data collection. For each contact, proprioceptive information was recorded. A total of 30 contacts were performed for each object, which was repeated five times, thus having one dataset for training and four datasets for testing. For visualization, here we only show a sequence of four contacts.

to estimate the most likely object been explored by using proprioceptive information from a robotic hand.

This probabilistic approach uses the Bayes' rule with a sequential analysis method, estimating the posterior probabilities recursively updated from the prior probabilities and likelihoods obtained from a measurement model. Then, the robotic hand makes a decision once the belief threshold about the object being explored is exceeded. This method has been tested for object shape extraction [5],[6] and simultaneous object localisation and identification [7],[8] using the fingertip sensors from the iCub humanoid robot [18].

Prior: an initial uniform prior probability is assumed for all the test objects to be explored. The initial prior probability for an object exploration process is define as follows:

$$P(c_n) = P(c_n|z_0) = \frac{1}{N} \tag{1}$$

where $c_n \in C$ is the perceptual class to be estimated, z_0 is the observation at time $t = 0$ and N is the number of objects used for exploration.

Measurement Model and Likelihood Estimation: each contact performed by the robotic hand during the object exploration task provides proprioceptive information from M motors: position and spread of the three fingers, and orientation of the hand. This information is used to construct the measurement model with a nonparametric estimation based on histograms. The histograms are used to evaluate a contact z_t performed by the robotic hand at time t, and estimating the likelihood of a perceptual class $c_n \in C$. The measurement model is obtained as follows:

$$P(s|c_n, m) = \frac{h(s, m)}{\sum_s h(s, m)} \tag{2}$$

where $h(s, m)$ is the number of observed values s in the histogram for motor m. The observed values are normalised by $\sum_s h(s, m)$ to have properly probabilities that sum to 1. Evaluating Equation (2) over all the motors, we obtained the likelihood of the contact z_t as follows:

$$\log P(z_t|c_n) = \sum_{m=1}^{M_{\text{motors}}} \sum_{s=1}^{S_{\text{samples}}} \frac{\log P(s|c_n, m)}{M_{\text{motors}} S_{\text{samples}}} \tag{3}$$

where $P(z_t|c_n)$ is the likelihood of a perceptual class c_n given the measurement z_t from M motors at time t.

Bayesian Update: the posterior probabilities $P(c_n|z_t)$ are updated by the recursive implementation of the Bayes' rule over the N perceptual classes c_n. The likelihood $P(z_t|c_n)$ at time t and the prior $P(c_n|z_{t-1})$ obtained from the posterior at time $t - 1$ are combined as follows:

$$P(c_n|z_t) = \frac{P(z_t|c_n)P(c_n|z_{t-1})}{P(z_t|z_{t-1})} \tag{4}$$

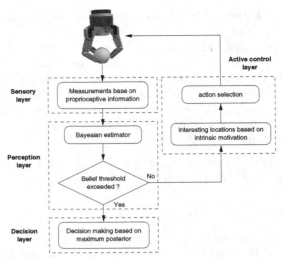

Fig. 4. Flow diagram with the steps required for the proposed intrinsically motivated active object exploration method. The robotic hand collects proprioceptive information from each contact performed. The robot is actively moved to interesting locations to improve perception based on an intrinsic motivations approach. Finally, a decision about the object being explored is made once the belief threshold is exceed.

Properly normalised values are obtained with the marginal probabilities conditioned from previous contact as follows:

$$P(z_t|z_{t-1}) = \sum_{n=1}^{N} P(z_t|c_n)P(c_n|z_{t-1}) \tag{5}$$

Stop Decision for Object Recognition: the accumulation of evidence with the Bayesian update process stops once a belief threshold is exceeded, making a decision about the object being explored. The object perceptual class is obtained using the *maximum a posteriori* (MAP) estimate as follows:

$$\text{if any } P(c_n|z_t) > \theta_{\text{threshold}} \text{ then}$$
$$c_{\text{decision}} = \arg\max_{c_n} P(c_n|z_t) \tag{6}$$

where the object estimated at time t is represented by $c_{decision}$. The belief threshold $\theta_{decision}$ permits to adjust the confidence level for the decision making process. Here, we have defined the belief threshold to the set of values {0.0, 0.05, ..., 0.999} to observe their effects on the object recognition accuracy.

2.4   Intrinsic Motivation for Active Exploration

Intelligent control of movements by an active exploration behaviour are achieved by the development of a computational method based on intrinsic motivation.

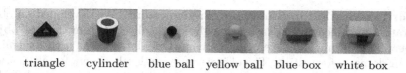

triangle   cylinder   blue ball   yellow ball   blue box   white box

Fig. 5. Test objects used for the experiments in simulated and real environments. The validation in simulated environment was performed using real data collected from these objects. For the validation in the real environment, the objects were placed and explored one at a time on a table.

It has been demonstrated by studies on cognitive development that intrinsic motivation is primordial to humans for engaging them to explore and manipulate their environment [9],[10].

In this work, we use a predictive novelty motivation model, where interesting locations for exploration are those for which prediction errors are higher [12]. This is defined as follows:

$$I(\text{SM}(t)) = E_I(t-1) \cdot E_I(t) \qquad (7)$$

where the interesting location I for the sensorimotor state SM is obtained by the prediction error $E_I(t)$ at time t multiplied by the prediction error $E_I(t-1)$ at time $t-1$.

We define the prediction error $E_I(t)$ as the distance between the MAP from the Bayesian approach and the belief threshold value for making a decision:

$$E_I(t) = \arg\max_{c_n} P(c_n|z_t) - \theta_{\text{threshold}} \qquad (8)$$

The active exploration performed by the robotic hand then is intelligently controlled by Equation 7, selecting the action for the highest SM state:

$$a = \arg\max_{\text{SM}} I(\text{SM}(t)) \qquad (9)$$

where a is the action selected by the robotic hand. Figure 4 shows the process described to perform object exploration. This process composed by the Bayesian method and intrinsic motivation is repeated until the belief threshold is exceeded to make a decision about the object being explored.

3   Results

In this section we present the results from the object exploration and recognition with passive and active modalities in simulated and real environments. Figure 5 shows the following objects used for validation: black triangle, red cylinder, blue ball, yellow ball, blue box and white box.

Object Exploration in Simulated Environment: We developed an object exploration and recognition task using the data collected from the test objects (see Section 2.2) in a simulated environment. One dataset was used for training and four datasets for testing. The objects were randomly drawn from the testing datasets with 10,000 iterations for each belief threshold in the set of values {0.0, 0.05, ... 0.999}.

Passive Object Exploration: First, the simulated robot moved the hand and fingers around the object to obtain an initial belief of the object being explored. Next, the hand and fingers were randomly moved, accumulating evidence from each contact and making a decision once the current belief threshold was exceeded. The perception accuracy and reaction time were evaluated for each belief threshold.

Figure 6a shows the results in perception accuracy for the object exploration process with passive perception (red curve). It is observed that the robotic hand achieved the minimum perception error of 60% for a belief threshold of 0.75. Similarly, the reaction time which refers to the number of contacts required for making a decision with passive perception (red curve) is shown in Figure 6b. The number of contacts increased for large belief thresholds, where a maximum of ~2 contacts were required to make a decision for a belief threshold of ~0.999. The results for perception accuracy and reaction time shown in Figure 6a and Figure 6b were obtained by averaging all perceptual classes over all trials for each belief threshold.

The confusion matrices (top panels) shown in Figure 7 permit to observe the performance of the classification accuracy with passive perception for each object and for different belief thresholds. These results show an slightly improvement of the classification accuracy with 68.28%, 71.77% and 76.18% for the belief

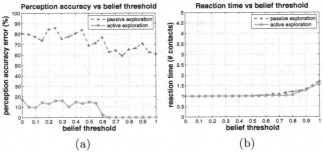

(a)   (b)

Fig. 6. Results from the passive and active object recognition in a simulated environment. Passive object recognition is presented by the red dotted-line. Active object recognition is presented by the green dotted-line. The experiment was performed for the set of belief threshold of {0.0, 0.05, ..., 0.999} with 10,000 iterations each. Results show the superiority of active over passive perception for object recognition with the robotic hand.

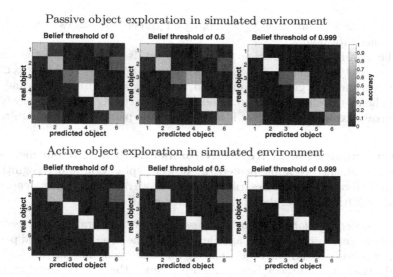

Fig. 7. Confusion matrices from the object recognition process with passive (top panels) and active (bottom panels) exploration modalities. The test objects used for the experiment are: 1) black triangle, 2) red cylinder, 3) blue ball, 4) yellow ball, 5) blue box and 6) white box. Results from passive perception show a small improvement in the object recognition for large belief thresholds. Results from active perception show higher perception accuracy over passive perception.

thresholds of 0.0, 0.5 and 0.999. These errors still can be improved by intelligent movements to interesting locations to reduce uncertainty.

Active object exploration: For the object recognition process with active perception, the robotic hand performed an exploration around the object to have an initial belief of the object being explored, similar to passive perception. Next, the robotic hand was actively moved, based on the proposed intrinsic motivation approach, towards interesting places around the object to improve perception. The active exploration process was repeated until the belief threshold was exceeded to make a decision. Similar to passive perception, the objects to be recognised were randomly drawn from the testing datasets with 10,000 iterations for each belief threshold in the set of values $\{0.0, 0.05, \ldots, 0.999\}$.

The perception accuracy results from active exploration are represented by the green curve in Figure 6a. It is clearly observed the improvement in accuracy by actively moving the robotic hand towards interesting locations for exploration, achieving an error of 0% for the belief thresholds of 0.65 to 0.999. This result validates our proposed method for active exploration, and also shows its superiority over passive perception. The reaction time required for making a decision with active perception also is represented by the green curve in Figure 6b. We observe that the reaction time increases for large belief thresholds, where $\sim$2 contacts are required for making a decision with a belief threshold of $\sim$0.999.

These results were obtained by averaging all perceptual classes over all trials for each belief threshold.

The classification accuracy for each object is presented by the confusion matrices (bottom panels) in Figure 7 for different belief thresholds. It is observed that the accuracy is gradually improved, achieving a 95.49%, 96.41% and 100.0% for the belief thresholds of 0.0, 0.5 and 0.999 respectively. The accuracy obtained by actively exploring an object is superior to the passive exploration process.

Object Exploration in Real Environment: To validate our methods in a real environment, we implemented the object exploration and recognition task with the robotic platform described in Section 2.1. For this experiment, we used the objects from the validation in simulated environment.

Passive Object Exploration: For the passive object exploration and recognition, the test objects were placed on a table one at a time. The robotic hand performed an exploration around the object through a fixed set of movements, building an initial belief of the object being explored. Next, the robotic hand started the random action selection of exploration movements, accumulating evidence to reduce uncertainty from the object being explored. The exploration process was repeated until the belief threshold was exceeded, making a decision about the current object.

Perception accuracy results are shown in Figure 8a for different belief thresholds. We observe that the error achieved for the object recognition process is improved with 26.66%, 16.66% and 10.0% for the belief threshold of 0.0, 0.5 and 0.999 respectively. The reaction time results required for making a decision are presented in Figure 8b. This result shows that for achieving the smallest error of 10% with passive perception, it was required ~15 contacts, whilst for the largest

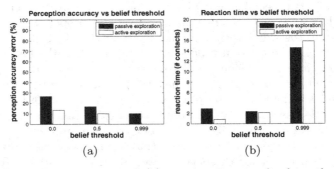

Fig. 8. (a) Perception accuracy and (b) reaction time results from the passive and active object recognition in a real environment. The experiment was performed with the belief thresholds of 0.0, 0.5 0.999. Passive perception was able to achieve the smallest error of 10% with 15 contacts for the belief threshold of 0.999. In contrast, active perception was able to achieve an error of 0% with 16 contacts for the belief threshold of 0.999. These results validate the performance of our proposed method for exploration.

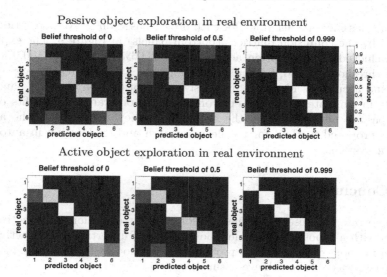

Fig. 9. Confusion matrices from the object recognition process with passive (top panels) and active (bottom panels) exploration modalities in real environment. The test objects used for the experiment are: 1) black triangle, 2) red cylinder, 3) blue ball, 4) yellow ball, 5) blue box and 6) white box. Results from passive perception show a small improvement in the object recognition for large belief thresholds, achieving an accuracy of 90% for the belief threshold of 0.999. Active perception shows higher perception accuracy of 100% for the belief threshold of 0.999.

error of 26.66% it was required ~3 contacts by the robotic hand. These results still can be improved by the use of our proposed method for exploration.

The classification accuracy for each object based on passive perception is presented by the confusion matrices (top panels) in Figure 9. The exploration task achieved the perception accuracies of 73.33%, 83.33% and 90.0% for the belief threshold of 0.0, 0.5 and 0.999 respectively.

Active Object Exploration: For the validation of the active exploration in a real environment, the test objects were placed on a table and explored by the robotic hand through a fixed set of movements. This step permitted to construct an initial belief of the object being explored. On the contrary to passive perception, here the robotic hand was able to selected the next action movement towards an interesting location around the object to improve perception. A decision about the object being explored was made once the evidence accumulated exceeded the belief threshold.

Figure 8a shows the perception accuracy results for the active exploration. We observe that the errors achieved for the object recognition process is improved with 13.33%, 10.0% and 0.0% for the belief thresholds of 0.0, 0.5 and 0.999 respectively. The reaction times required for making a decision are presented in Figure 8b. It is clearly observed that to achieve the best error of 0.0% it was required 16 contacts, whilst the error of 13.33% was obtained with 1 contact.

The classification accuracy for each object based on active perception is presented by the confusion matrices (bottom panels) in Figure 9. The exploration task achieved the perception accuracies of 86.66%, 90.0% and 100.0% for the belief thresholds of 0.0, 0.5 and 0.999 respectively. These results are improved over the accuracies obtained by passive perception. On the one hand, these results in simulated and real environments demonstrate the benefits of active over passive perception. On the other hand, they also validate the accuracy of our proposed method for tactile perception and exploration in autonomous robotic systems.

4   Conclusions

In this work we presented a method for object recognition using active exploration with a robotic hand under the presence of uncertainty. Our active exploration method, composed by a probabilistic method and an intrinsic motivation approach, was able to achieve accurate results.

We used a set of test objects for training and testing our methods in simulated and real environments. Tactile sensing was used for contact detection, whilst proprioceptive information composed by the position of the fingers and orientation of a robotic hand was used for object recognition. The robotic hand performed 30 contacts around each test object, which was repeated five times, to have one training dataset and four testing datasets.

A Bayesian method for uncertainty reduction through the interaction with an object was presented. This approach, together with a sequential analysis method, permitted the robotic hand to autonomously control the exploration and make a decision about the object being explored.

Active exploration behaviour was obtained with an intrinsic motivation approach by moving the robotic hand towards the more interesting locations for exploration. Interesting locations were represented as the locations with large variances, obtained from the distance between the posterior probability from the Bayesian approach and the belief threshold. The combination of Bayesian and intrinsic motivation approaches allowed to develop an active exploration behaviour, accumulating evidence and reducing uncertainty by exploring the most interesting locations of the object.

Our method was validated in simulated and real environments using passive and active exploration modalities. In simulated environment and active exploration the robotic hand achieved a perception error of 0% for belief thresholds from 0.65 to 0.99. These results contrast with the error of 60% for the belief threshold of 0.75 with passive exploration (Figure 6a). We did not observe large differences for the reaction time with both exploration modalities, where $\sim$2 contacts were required to make a decision for the smallest perception errors.

The validation in a real environment also shows the high accuracy achieved by the robotic hand using our proposed method. For active perception, the smallest error of 0% was achieved by the robotic hand with a belief threshold of 0.999 (Figure 8a). For passive perception, the smallest error of 10% was achieved for the belief threshold of 0.999. Similar to the validation in the simulated environment, the reaction time required to make a decision for the best accuracies did not present large differences, with 15 and 16 contacts for passive and active perception respectively. The validations from simulated and real environments show the benefits of our proposed method for object exploration.

Overall, we have observed how active movements performed by the robotic hand to explore interesting locations, improve the perception accuracy and decision making for an autonomous exploration task. For future work, we plan to extend our methods combining them with vision and implementing them with more complex robots to autonomously perceive and explore their environment.

Acknowledgments. This work was supported by EU Framework project WYSIWYD (FP7-ICT-2013-10). We thank to the technical staff and the equipment provided by the Department of Psychology and the Sheffield Robotics Laboratory at the University of Sheffield.

References

1. Lederman, S.J., Klatzky, R.L.: Hand movements: A window into haptic object recognition. Cognitive Psychology **19**(19), 342–368 (1987)
2. Bajcsy, R., Lederman, S.J., Klatzky, R.L.: Object exploration in one and two fingered robots, Number 3, pp. 1806–1810. Computer Society Press (1987)
3. Bajcsy, R.: Active perception. Proceedings of the IEEE **76**(8), 966–1005 (1988)
4. Prescott, T.J., Pearson, M.J., Mitchinson, B., Sullivan, J.C.W., Pipe, A.G.: Whisking with robots. IEEE Robotics Automation Magazine **16**(3), 42–50 (2009)
5. Martinez-Hernandez, U., Dodd, T.J., Prescott, T.J., Lepora, N.F.: Active bayesian perception for angle and position discrimination with a biomimetic fingertip. In: 2013 IEEE International Conference on (IROS), pp. 5968–5973 (2013)
6. Martinez-Hernandez, U., Dodd, T.J., Natale, L., Metta, G., Prescott, T.J., Lepora, N.F.: Active contour following to explore object shape with robot touch. In: World Haptics Conference (WHC), pp. 341–346 (2013)
7. Lepora, N.F., Martinez-Hernandez, U., Prescott, T.J.: A SOLID case for active bayesian perception in robot touch. In: Lepora, N.F., Mura, A., Krapp, H.G., Verschure, P.F.M.J., Prescott, T.J. (eds.) Living Machines 2013. LNCS, vol. 8064, pp. 154–166. Springer, Heidelberg (2013)
8. Lepora, N.F., Martinez-Hernandez, U., Prescott, T.J.: Active bayesian perception for simultaneous object localization and identification. In: Robotics: Science and Systems (2013)
9. Ryan, R.M., Deci, E.L.: Intrinsic and extrinsic motivations: Classic definitions and new directions. Contemporary Educational Psychology **25**(1), 54–67 (2000)
10. Barto, A.G., Singh, S., Chentanez, N.: Intrinsically motivated learning of hierarchical collections of skills. In: Proc. 3rd Int. Conf. Development Learn, pp. 112–119 (2004)
11. Baranès, A., Oudeyer, P.-Y.: R-iac: Robust intrinsically motivated exploration and active learning. IEEE Transactions on Autonomous Mental Development **1**(3), 155–169 (2009)
12. Oudeyer, P.-Y., Kaplan, F.: What is intrinsic motivation? a typology of computational approaches. Frontiers in Neurorobotics **1** (2007)

13. Stout, A., Konidaris, G.D., Barto, A.G.: Intrinsically motivated reinforcement learning: A promising framework for developmental robot learning. Technical report, DTIC Document (2005)

14. Pezzementi, Z., Plaku, E., Reyda, C., Hager, G.D.: Tactile-object recognition from appearance information. IEEE Transactions on Robotics **27**(3), 473–487 (2011)

15. Ratnasingam, S., McGinnity, T.M.: Object recognition based on tactile form perception. In: 2011 IEEE Workshop on Robotic Intelligence in Informationally Structured Space (RiiSS), pp. 26–31, April 2011

16. Johnsson, M., Balkenius, C.: Neural network models of haptic shape perception. Robotics and Autonomous Systems **55**(9), 720–727 (2007)

17. Fitzpatrick, P., Metta, G., Natale, L.: Yet another robot platform. http://eris.liralab.it/yarpdoc/index.html

18. Schmitz, A., Maiolino, P., Maggiali, M., Natale, L., Cannata, G., Metta, G.: Methods and technologies for the implementation of large-scale robot tactile sensors. IEEE Transactions on Robotics **27**(3), 389–400 (2011)

Fabrication of Electrocompacted Aligned Collagen Morphs for Cardiomyocyte Powered Living Machines

Victoria A. Webster[1]([⊠]), Emma L. Hawley[1], Ozan Akkus[1], Hillel J. Chiel[2],
and Roger D. Quinn[1]

[1] Department of Mechanical and Aerospace Engineering,
Case Western Reserve University, 10900 Euclid Ave., Cleveland, OH, USA
vaw4@case.edu
[2] Department of Biology, Case Western Reserve University,
10900 Euclid Ave., Cleveland, OH, USA

Abstract. Based on the need for small scale compliant devices for use
in medical robotics, there is increasing interest in the development of
muscle actuated biobots. Such biobots are traditionally fabricated using
nondegradable, synthetic polymers; however, such substrates require
micro-patterning and additional treatments in order to promote cellular
adhesion and induce cellular alignment. By using an organic substrate,
such steps can be eliminated, and degradable scaffolds can be produced.
Here we present a manufacturing process and culture conditions for fabrication of living machines using electrochemically compacted and aligned
collagen (ELAC) as a scaffold. Using collagen as a scaffold results in a
completely organic device. Milli-scale scaffolds were seeded with primary
cardiomyocytes isolated from chick embryos and electrically stimulated
to induce movement.

Keywords: Biohybrid robots · Biorobotics · Living machines · ELAC ·
Cardiomyocytes

1 Introduction

The development of milli and micro-scale actuators and devices is an area of
particular interest in the miniaturization of robotics, particularly for the development of medical technologies. In order to develop miniature medical devices,
milli to micro scale actuators are needed which can safely interact with living tissue while causing minimal damage. Unfortunately, many traditional small

V.A. Webster—This material is based upon work supported by the National Science Foundation Graduate Research Fellowship under Grant No. DGE-0951783.
This study was also funded in part by grants from the National Science Foundation (Grant Number DMR-1306665) and National Institute of Health (Grant Number R01 AR063701). Any opinion, findings, and conclusions or recommendations
expressed in this material are those of the authors and do not necessarily reflect the
views of the National Science Foundation.

© Springer International Publishing Switzerland 2015
S.P. Wilson et al. (Eds.): Living Machines 2015, LNAI 9222, pp. 429–440, 2015.
DOI: 10.1007/978-3-319-22979-9_43

robotic actuators, such as piezoelectrics, have small displacements and low compliance [1,2]. Others, such as shape memory alloys (SMAs), offer the possibility of compliant actuation, but the temperatures needed to elicit a shape change are often high relative to temperatures in organic systems (80°C +). Additionally, the environment must be conducive to the heating and cooling cycle needed to drive actuation. In both cases the actuators require an external power supply. Such characteristics do not lend themselves well to use in medical applications.

Recently there has been a growing body of research in the development of biohybrid actuators and devices. Such devices use synthetic substrates and are powered by muscle cells. Using such techniques, researchers have developed actuatable cantilevers [3] and micro-pillar arrays[4] as well as devices capable of crawling [5–7], and swimming [8,9]. Current biohybrid devices are typically fabricated using synthetic polymer substrates such as Polyethylene(glycol)Diacrylate (PEGDA) [5], Polydimethylsiloxane (PDMS) [8,10,11], or Poly(N-isopropylacrylamide) (PNIPAAm) films [12]. These polymers do not natively align cells and must be micro-patterned or micro-printed in order to promote both cellular attachment and alignment, thereby complicating manufacturing and requiring additional chemical processes.

Alternatively, tissue engineering scaffolds can be fabricated using collagen. Threads of 50-400 μm diameter or sheets can be manufactured by electrochemically compacting collagen [13,14]. Such electrochemically compacted and aligned collagen (ELAC) scaffolds can be fabricated as threads, which will align during compaction, or as sheets based on electrode configuration. While the threads are aligned during manufacturing, sheets are aligned by simple mechanical stretching. The resulting scaffolds have both stiffness and topological anisotropy, as well as natively promoting cellular attachment without micro-patterning. Additionally, in vivo studies have shown that ELAC fibers experience limited degradation over several months [15].

In this paper we present the first locomoting devices to be fabricated using electrochemically compacted collagen as a scaffold. As a result of using collagen, rather than a synthetic polymer base, these devices step out of the realm of biohybrid robots and towards that of functional, organic, living machines.

2   Methods

2.1   Scaffold Fabrication

Collagen was isolated and purified from lamb skin using the acid extraction method [16]. The resulting solution was dialyzed against ultra pure water at 4°C for 24 hours, with 3 water changes every two hours for the first 6 hours, in order to remove salts. This resulted in a viscous collagen solution (3.4 mg/ml). Electrochemically compacted and aligned collagen scaffolds were then fabricated in a manner similar to that described by Cheng et al. [13]. Briefly, the space between two electrodes is filled with dialyzed collagen. Application of a voltage across the electrodes compacts the collagen molecules onto the cathode. When using wire electrodes this results in an aligned collagen thread. The same process

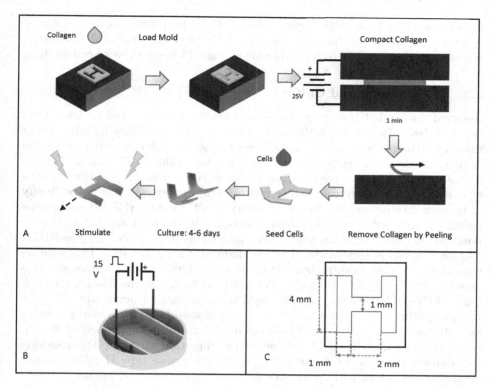

Fig. 1. Fabrication of a living machine using an ELAC scaffold. Collagen is injected into a mold and loaded between two electrodes where it is compacted by application of 25 V across the electrodes for 1 minute. The collagen is then removed by peeling the scaffold away from the cathode, which results in angled and aligned legs. Cells are then seeded on sterilized scaffolds and cultured for 4-6 days, after which they are electrically paced, resulting in forward locomotion.

can be completed by introducing the collagen solution into a mold placed between two plate electrodes. This results in a randomly aligned, compacted sheet in the shape of the mold. The scaffold can then be aligned by mechanical stretching. In this study a 1 mm thick mold between two plate electrodes has been used to compact the scaffold and collagen alignment has been induced by peeling the scaffold away from the cathode during removal. The scaffold, which is a symmetric H-shape (See Figure 1.C), is deposited on the cathode via compaction as previously described, treated with isopropanol for handling, and then gently peeled off of the electrode along the body axis (See Figure 1.A).

In order to demonstrate that the peeling process induces collagen alignment, a small number of samples were also removed from the electrode by ultra sonic vibration until the scaffolds released. The peeling process stretches the legs of the H and angles them back towards the body axis. Additionally, this angling gives the device a preferential locomotion direction. After manufacturing, the scaffolds are suspended in Phosphate Buffered Saline (PBS) for 6 hours at 37°C followed by overnight treatment in isopropanol at 4°C. After the overnight treatment, scaffolds are sterilized in 70% ethanol for at least 24 hours prior to cell seeding.

2.2 Cell Isolation and Culture

Assorted chicken hatching eggs (Meyer Hatchery) were incubated for 10-12 days in a Hovabator Incubator at 100° Fahrenheit and 40-60% humidity while being turned continuously by an automatic egg turner. In order to verify successful embryo development, eggs were candled two days prior to dissection. On the day of dissection, eggs were cut open under sterile conditions and the embryos were removed and sacrificed. Whole hearts were then excised and rinsed briefly in three successive ice cold Hank's Balanced Salt Solution (HBSS) wash steps before being stored in ice cold HBSS. Subsequently, the right and left atria were removed and the ventricles were finely minced in 0.1% (w/v) trypsin-EDTA (Sigma Aldrich) in supplemented M199 Medium (Sigma Aldrich). This solution was allowed to digest overnight at 4° C. Supplemented growth medium was prepared as 40% M199, 52% HBSS, 6% Fetal Bovine Serum (Sigma Aldrich #F2442, Lot 12J001), 1% Pen/strep, and 0.1 mM norepinephrine (Sigma Aldrich). The norepinephrine mixture was made in 30 mM of ascorbic acid in distilled water and filter sterilized. After 14-18 hours of digestion, warm supplemented growth medium was added in equal volume to the digestion mixture. The resulting supernatant was discarded and 0.1% (w/v) collagenase type II (Worthington Biochemicals), which had been sterile filtered, was added for 45 minutes at 37°C with gentle rocking. Again, warm supplemented medium was added in equal volume and the solution was passed through a 70 μm cell strainer. The remaining tissue was recovered from the cell strainer, suspended in HBSS and centrifuged for 5 min at 500 x g. The supernatant was discarded and the tissue was suspended in 0.25% trypsin-EDTA for 5 minutes at room temperature with gentle shaking. Subsequently, supplemented medium was added in equal volume to the trypsin and the solution was filtered through the 70 μm cell strainer. This process was repeated three times and resulted in complete digestion of the tissue. The resulting cell solution was centrifuged for 5 minutes at 500 x g and resuspended in supplemented growth medium. In order to reduce fibroblast contamination, cells were preplated at high density for 45 min at 37° C. During preplating, the scaffolds were removed from the 70% ethanol sterilization solution and rinsed three times in sterile HBSS, then allowed to soak for 30 minutes. The scaffolds were transferred to the culture dish and allowed to dry for 10-15 minutes prior to seeding, such that they remained lightly adhered to the dish. After the 45 minute preplate, unattached cells were recovered and counted. The solution was subsequently centrifuged for 5 minutes at 500 x g and resuspended to allow a seeding density of $1 x 10^6$ cells/cm^2. 10-12 hours after the initial seeding the medium was changed to remove non-adherent cells, and samples were allowed to culture overnight. After 24 hours in culture the samples were gently released from the culture dish and cultured with daily medium changes for 4-6 days prior to stimulation testing.

2.3 Stimulation Protocol

For paced testing, samples were transferred to a separate dish with two U-shaped platinum electrodes spaced 3 cm apart. The electrodes are supported by two polypropylene walls which span the chord of the dish at each location. Holes are regularly spaced along the bottom of each wall to allow fluid contact with the testing area while preventing electrolysis bubbles from interfering with the sample motion and imaging (See Figure 1.B). The electrodes are connected to a DC power supply via a L293NE H-bridge and paced at 15 V for 100 ms at a frequency of 1 Hz by an Arduino Uno.

2.4 Imaging

In order to verify collagen alignment, the scaffolds were imaged using a 43 mm polarizing filter (Prince) with a U-TP530 530nm gypsum plate on an Olympus BX51 System Microscope. Images were taken with a D5200 Nikon digital camera. Locomotion tests were recorded at 2 Hz with a Prosilica GX camera (Allied Vision) mounted vertically above the testing dish using a Macro 100 F2.8D lens (Tokina).

2.5 Scaffold Characterization

Scaffold height was measured using a FemtoTools FT-RS1002 Microrobotic System with a FT-S microforce sensing probe. Hydrated samples were placed in a glass dish for measurement. Using the force sensing probe the contact point with the glass dish was found and set as the reference height. Contact was then found with the scaffold and the height was recorded. Height measurements were made from seven samples with at least three measurements per sample.

The effect of compaction properties on ELAC thread geometries has been investigated with respect to compaction time and collagen stock dilution. For compaction time, collagen threads were fabricated from dialyzed collagen stock by compacting the solution between two wire electrodes for 30 seconds, 1 minute, or 2 minutes. For collagen dilution, the collagen stock was diluted to 25 %, 50 %, 75 %, or not diluted using ultra pure water. The resulting threads were imaged on an inverted microscope and the widths were measured using ImageJ [17]. For each group, at least two threads were fabricated and five measurements were taken from each thread. Significance was determined using one-way ANOVA, followed by two sample T-tests of individual pairs. Since multiple comparisons were made, a Bonferroni correction factor was applied in order to determine significance (Compaction Time: $p < 0.016$, Collagen Dilution: $p < 0.008$).

The effect of collagen crosslinking on scaffold stiffness has been investigated by crosslinking the scaffolds with EDC/NHS mixtures in 80 % Ethanol. Four crosslinker ratios (weight collagen:EDC:NHS) were tested: 1:100:250. 5:100:250, 10:100:250 and no crosslinking. The stiffness of the collagen samples was then tested using a FemtoTools FT-RS1002 Microrobotic System with a FT-S microforce sensing probe. Samples were rehydrated between measurements in order to

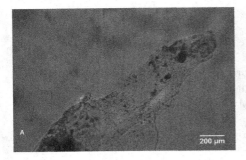 

Fig. 2. As a result of the peeling process the legs of the ELAC scaffolds are aligned, as revealed by polarized light microscopy (blue indicates collagen alignment along the axis of the leg). A) The leg of an ELAC scaffold removed from the cathode via peeling, with collagen alignment along the leg axis. B) The leg of an ELAC scaffold removed from the cathode by ultrasonic vibration, with no clear collagen alignment direction. The peeling process also eliminates damage caused by ultrasonic vibration.

ensure they did not dry during testing. Significance was determined using one-way ANOVA, followed by two sample T-tests of individual pairs, comparing the crosslinked stiffnesses to the uncrosslinked stiffness. Since multiple comparisons were made, a Bonferroni correction factor was applied in order to determine significance ($p < 0.016$).

3   Results

The fabrication techniques presented resulted in collagen scaffolds with a thickness of 203.8 ± 83.5 μm. These scaffolds showed collagen alignment along the leg axes of the device and were capable of locomotion when seeded with cardiomyocytes and electrically paced.

3.1   Effect of Fabrication Techniques on Collagen Alignment

Previous work has shown that compacting collagen between two plate electrodes results in randomly aligned collagen sheets. When working with large scaffolds, the resulting sheets can be easily fixtured and stretched. However, in this study the small scale of the scaffolds, combined with the leg geometries, prevents straightforward fixturing without milli-scale manipulators. Instead, collagen alignment is induced by slowly peeling the scaffolds off of the cathode on which they are deposited. Collagen scaffolds peeled off of the electrode showed strong alignment of the collagen along the axes of the legs (Figure 2.A), while scaffolds released from the electrode by ultrasonic vibration show a lack of structured collagen alignment (Figure 2.B). Additionally, ultrasonic removal appears to damage the collagen structure making it a nonviable method of scaffold removal.

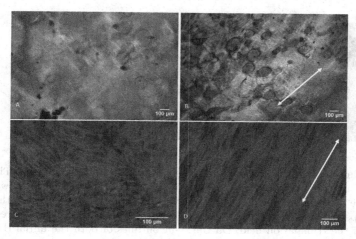

Fig. 3. Mechanically stretching electrochemically compacted collagen sheets results in alignment of the collagen molecules. Polarized Light Microscopy of the sheets reveals the collagen alignment, with blue coloration along the alignment axis (see double headed arrow in B). A) shows an unaligned sheet while B) shows a mechanically aligned sheet. Mechanical alignment of the collagen scaffold induces cellular alignment. C) F-actin staining of cells seeded on an unaligned sheet. D) F-actin staining of cells seeded on an aligned sheet with the arrow indicating the cellular alignment direction. (Cell and Collagen images from separate samples).

3.2   Effect of Collagen Alignment on Cell Alignment

Cardiomyocytes seeded on collagen sheets, aligned by mechanical stretching, show a strong cellular alignment along the direction of stretching. Figure 3 shows characteristic examples of polarized light microscopy of the unaligned (3.A) and aligned collagen sheets (3.B) and the effect each substrate alignment has on cell structure (3.C and 3.D, respectively). This cellular alignment helps to direct the actuation of the cells, producing predictable, coordinated actuation.

3.3   Living Machine Locomotion

For locomotion testing, samples were transferred to the stimulation dish in standard growth medium and paced as described above. Of 23 pacing trials, motion was observed in 13 trials. Some of the scaffolds were damaged during manipulation, or had deformed dramatically under the effect of cell stresses. Such devices often failed to translate but instead rotated. Figure 4 shows characteristic locomotion. Additionally, the relative rotation and global translation heading distributions of all moving devices, as well as a diagram giving the direction and magnitude of translation for each moving trial are presented in Figure 5. As a negative control, collagen scaffolds without cells have been tested with electrical stimulation and do not produce measurable motion.

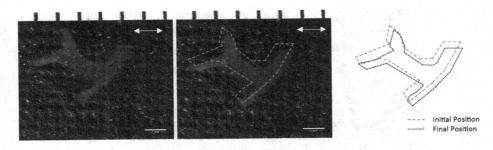

Fig. 4. Characteristic locomotion of an H-shaped ELAC living machine. The initial position of the device is shown on the left and the final position is shown in the middle. The blue dashed outline gives the original position of the device in the second image (Elapsed Time: 10 min). In both images the double headed arrow represents the direction of the stimulation field. On the right outlines of both the initial (blue large dashed lines) and final (red small dashed lines) are overlaid to show overall motion.

4   Discussion

Electrochemically compacted and aligned collagen natively promotes cellular attachment. Additionally, regional alignment of collagen (such as aligning the collagen along the leg axis of the living machines as presented here) results in alignment of cells seeded on the scaffold. This allows researchers to dictate the direction along which muscle cells will contract during actuation.

The devices presented here are slower than devices previously presented in the literature. Using the H-shaped scaffold presented, the maximum speed measured was 34 μm/min. However, the performance of the devices can be improved by optimizing the modulus and geometry of the scaffold. Initial results with alternative scaffold geometries indicate that higher velocities are achievable without modifying the scaffold fabrication techniques or culture conditions. In addition to modifying the shape of the scaffold, the thickness of the scaffold can be modified by varying the compaction time or by diluting the collagen solution used to fabricate the scaffolds (See Figure 6). Furthermore, the modulus of the scaffold can be varied by crosslinking. Simply crosslinking the scaffolds for 15 minutes in an EDC/NHS mixture with 80% ethanol results in a significantly higher stiffness (Figure 7). By varying the scaffold modulus via crosslinking and decreasing the thickness by varying compaction time or collagen stock dilution, the overall stiffness of the devices can be optimized to increase performance and ease of manufacturing. For example, much thinner scaffolds may be fabricated with light crosslinking that would have an overall lower stiffness than the devices presented. The specific material properties can be optimized using finite element analysis.

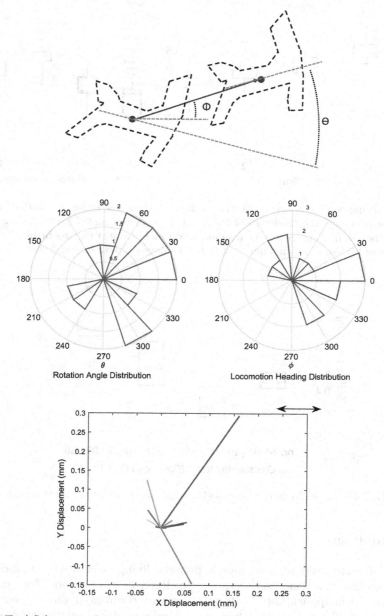

Fig. 5. Top) Schematic showing locomotion metrics investigated with the device moving from left to right. θ gives the relative angle between the initial and final headings of the device. ϕ gives the absolute translational heading of the device with respect to the horizontal. Middle) Distributions of the rotation magnitudes and translational locomotion headings for all trials in which the device moved. Bottom) The translational locomotion displacements and magnitudes for each trial in which the device moved. In the bottom image, the double headed arrow indicates the direction of the stimulation field.

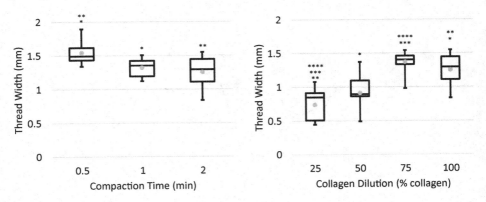

Fig. 6. Varying compaction time (left) and diluting the stock collagen solution (right) can be used to vary the thickness of the resulting collagen scaffold. Decreasing compaction time, or diluting the collagen solution results in thinner electrochemically compacted threads (Compaction Time: $p < 0.016$, Collagen Dilution: $p < 0.008$).

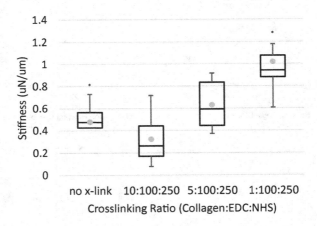

Fig. 7. Collagen scaffolds can be crosslinked in order to vary substrate stiffness

5   Conclusions

Functional, completely organic, muscle powered living machines were fabricated and locomotion was achieved in response to electrical stimulation. These devices demonstrate the applicability of electrochemically compacted and aligned collagen as a scaffold for compliant biobots. By using collagen, rather than a synthetic polymer, these devices provide the foundation for organic biobots which are intended to serve a temporary purpose inside an organism. Such applications include targeted drug delivery, vascular inspection, and self actuated stents. In order to improve control of living machines, neurons could be isolated and co-cultured with muscle cells. When combined with MEMs electrodes this could allow targeted stimulation of muscle regions, and thus, greater control of the device. While the devices presented are slower than devices previously developed using synthetic polymers, a number of techniques are available to allow optimization of substrate modulus and scaffold geometry in order to improve performance. In addition, the effect of stimulation parameters and methods on locomotion can be investigated.

Acknowledgments. The authors would like to thank Katherine Chapin for providing the collagen solution for scaffold fabrication.

References

1. King, T.G., Preston, M.E., Murphy, B.J.M., Cannellt, D.S.: Piezoelectirc ceramic actuators: A review of machinery applications. Precis. Eng. **12**(3), 131–136 (1990)
2. Uchino, K.: Materials issues in design and performance of piezoelectric actuators: an overview. Acta. Mater. **46**(11), 3745–3753 (1998)
3. Park, J., Ryu, J., Choi, S.K., Seo, E., Cha, J.M., Ryu, S., Kim, J., Kim, B., Lee, S.H.: Real-time measurement of the contractile forces of self-organized cardiomyocytes on hybrid biopolymer microcantilevers. Anal. Chem. **77**, 6571–6580 (2005)
4. Legant, W.R., Pathak, A., Yang, M.T., Deshpande, V.S., Mcmeeking, R.M., Chen, C.S.: Microfabricated tissue gauges to measure and manipulate forces from 3D microtissues. PNAS **106**(25), 10097–10102 (2009)
5. Chan, V., Park, K., Collens, M.B., Kong, H., Saif, T.A., Bashir, R.: Development of miniaturized walking biological machines. Sci. Rep. **2**, 857, January, 2012
6. Kim, J., Park, J., Yang, S., Baek, J., Kim, B., Lee, S.H., Yoon, E.S., Chun, K., Park, S.: Establishment of a fabrication method for a long-term actuated hybrid cell robot. Lab Chip **7**, 1504–1508 (2007)
7. Cvetkovic, C., Raman, R., Chan, V., Williams, B.J., Tolish, M., Bajaj, P., Sakar, M.S., Asada, H.H., Saif, M.T.A., Bashir, R.: Three-dimensionally printed biological machines powered by skeletal muscle. PNAS **111**(28), 10125–10130 (2014)
8. Nawroth, J.C., Lee, H., Feinberg, A.W., Ripplinger, C.M., McCain, M.L., Grosberg, A., Dabiri, J.O., Parker, K.K.: A tissue-engineered jellyfish with biomimetic propulsion. Nat. Biotechnol. **30**(8), 792–797 (2012)
9. Williams, B.J., Anand, S.V., Rajagopalan, J., Saif, M.T.a.: A self-propelled biohybrid swimmer at low Reynolds number. Nat. Commun. **5**, 3081, January, 2014
10. Feinberg, A.W., Feigel, A., Shevkoplyas, S.S., Sheehy, S., Whitesides, G.M., Parker, K.K.: Muscular thin films for building actuators and powering devices. Science (80-.) **317**, 1366–1370 (2007)
11. Kim, J., Park, J., Lee, J., Yoon, E., Park, J., Park, S.: Biohybrid Microsystems Actuated by Cardiomyocytes : Microcantilever, Microrobot, and Micropump. In: IEEE Int. Conf. Robot. Autom., Pasadena, CA, USA, pp. 880–885 (2008)
12. Xi, J., Schmidt, J.J., Montemagno, C.D.: Self-assembled microdevices driven by muscle. Nat. Mater. **4**, 180–184 (2005)
13. Cheng, X., Gurkan, U.A., Dehen, C.J., Tate, M.P., Hillhouse, H.W., Simpson, G.J., Akkus, O.: An electrochemical fabrication process for the assembly of anisotropically oriented collagen bundles. Biomaterials **29**(22), 3278–3288 (2008)

14. Gurkan, U.A., Cheng, X., Kishore, V., Uquillas, J.A., Akkus, O.: Comparision of Morphology, Orientation, and Migration of Tendon Derived Fibroblasts and Bone Marrow Stromal Cells on Electrochemically Aligned Collagen Constructs. J. Biomed. Mater. Res. A. **94**(4), 1070–1079 (2010)

15. Kishore, V., Uquillas, J.A., Dubikovsky, A., Alshehabat, M.A., Snyder, P.W., Breur, G.J., Akkus, O.: In vivo response to electrochemically aligned collagen bioscaffolds. J. Biomed. Mater. Res. B. Appl. Biomater. 400–408, December, 2011

16. Timpson, P., McGhee, E.J., Erami, Z., Nobis, M., Quinn, J.A., Edward, M., Anderson, K.I.: Organotypic collagen I assay: a malleable platform to assess cell behaviour in a 3-dimensional context. J. Vis. Exp. (56), e3089, January, 2011

17. Rasband, W.: ImageJ, U.S. National Institute of Health, Bethesda, Maryland, USA (2014). http://imagej.nih.gov/ij/

Extending a Hippocampal Model for Navigation Around a Maze Generated from Real-World Data

Luke W. Boorman[✉], Andreas C. Damianou,
Uriel Martinez-Hernandez, and Tony J. Prescott

Sheffield Robotics, University of Sheffield, Sheffield S10 2TN, UK
l.boorman@shef.ac.uk

Abstract. An essential component in the formation of understanding is the ability to use past experience to comprehend the here and now, and to aid selection of future action. Past experience is stored as memories which are then available for recall at very short notice, allowing for understanding of short and long term action. Autobiographical memory (ABM) is a form of temporally organised memory and is the organisation of episodes and contextual information from an individual's experience into a coherent narrative, which is key to a sense of self. Formation and recall of memories is essential for effective and adaptive behaviour in the world, providing contextual information necessary for planning actions and memory functions, such as event reconstruction. Here we tested and developed a previously defined computational memory model, based on hippocampal structure and function, as a first step towards developing a synthetic model of human ABM (SAM). The hippocampal model chosen has functions analogous to that of human ABM. We trained the model on real-world sensory data and demonstrate successful, biologically plausible memory formation and recall, in a navigational task. The hippocampal model will later be extended for application in a biologically inspired system for human-robot interaction.

Keywords: Autobiographical memory · Episodic memory · Hippocampus · Robotics · Temporally restricted boltzmann machine · Navigation

1   Introduction

For robots to interact with humans in a social manner they need to function in a flexible way. Greater flexibility can arise from robots making inference on future behaviour by recalling relevant past experience. This approach has parallels with human autobiographical memory (ABM), which is defined as the recollection of events from one's life. Human ABM is nonetheless very complex and its exact function is not completely understood. In contrast, navigational memory, especially in the rodent hippocampus has seen much research and is better characterised [1]. Navigational memory does not give insight into the higher level functionality required by human memory, such as comprehension, memory storage and memory recall through language, but instead our current understanding of rodent navigational memory offers a starting point for the development of a synthetic memory system.

© Springer International Publishing Switzerland 2015
S.P. Wilson et al. (Eds.): Living Machines 2015, LNAI 9222, pp. 441–452, 2015.
DOI: 10.1007/978-3-319-22979-9_44

A model for spatial navigation, based on a mapping between biological hippocampal function and a temporally restricted Boltzmann machine has been described previously [2, 3]. Following an extensive training phase the model was able to accurately predict the location of a virtual agent within a simple maze from exposure to location specific artificial features and images [4]. The hippocampal model has potential applicability for robot navigation, but also mimics functions of human autobiographical memory, such as memory compression, pattern separation and pattern completion [5].

However, the current model has only seen limited and highly constrained testing, for example, the original navigational task had an overly simplistic maze structure, with carefully placed sensory inputs. Thus, we set out to expand the generalisability and applicability of the model for real-world applications. The first step involved increasing the 'realness' of the original maze, by using real-world images from Google Street-view and removing the more 'artificial' sensory inputs. The new tests demonstrated that the model was able to accurately locate a virtual agent using novel real-world sensory image data, while retaining the original maze structure. The second step involved the construction of a larger and more complex maze, to test the ability of the model to accurately navigate a novel maze shape. Initial testing of the new maze exposed the constrained model design, as it could not accept the new maze topology. Thus, the software implementation of the original model was expanded beyond the original model design, to add new flexibility by allowing the automatic scaling of the number of hippocampal 'cells' used for encoding the agent position and the number of 'cells' used to store the memories which relate to each spatial location. Unfortunately, the testing of the new adaptable model showed it was unable to successfully navigate the new maze. This was most likely due to the software implementation of the original and modified models built in the Python programing language, having little documentation, multiple layers of functions, replicated function naming and recursive loops. We present our steps to making the model more generalizable and shown where we had difficulties in updating an existing model. We describe in detail the novel Google Street-view based maze generator in order to allow its use as a tool for testing alternative implementations of navigational models.

2   The Hippocampus: A Unitary Coherent Particle Filter Model

The rodent hippocampal system has been the basis for much invasive neuroscience research into memory function, especially following the discovery of the function of specific cells in the hippocampus, such as place [6] and grid cells [7]. The hippocampus is thought by many to play a central role in navigation and spatial reasoning [1]. However, it is not only thought to be involved in spatially related tasks, but has also been linked to a wide range of other non-spatial memory functions [8] and even beyond memory, with involvement, for example in decision making and emotion [9]. The hippocampus has been modelled extensively, with the majority of studies using tasks based on navigation to verify their models. The well characterised navigational abilities of the hippocampus, offer direct potential for localisation and mapping for autonomous robots similar to current simultaneous location and mapping (SLAM) methods [4]. However, the models based on hippocampal function can also offer features such as compression, pattern separation and pattern completion [5], which are also present in human autobiographical memory. A unitary coherent particle filter hippocampus (UCPF-HC) model [2] was previously tested using a navigational task [4], for applicability as a tool for SLAM.

Spatial navigation tasks generally require two phases, the first is to learn an environment and the second a test to demonstrate the quality of the learning of the environment, e.g. by traversing the environment to locate a food supply. Previous work has suggested that animals learning can be modelled as a particle filter [10], with sequential learning described using a machine learning approach such as a Temporally Restricted Boltzmann Machine (TRBM). A previous model [2] mapped hippocampal circuitry with a TRBM to produce a navigational learning system (fig. 1). The system was extended across a succession of papers, such as to include learning using a biomimetic sub-theta cycling [3] and to accept visual information using feature extraction [4]. The model is overhauled within this paper to add new 'realness' to the task, firstly by improving the sensory information within the virtual maze and secondly by building a model which can accept a flexible maze topology. This is a first step towards including non-spatial memory, namely to produce a more general memory model, for application in human robot interaction.

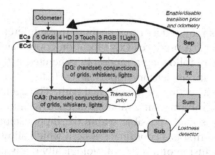

Fig. 1. The hippocampal model, adapted from Fox and Prescott (2010a)

3   The Hippocampal Model Implementation and Navigational Task

The ability of the existing hippocampal model to store and recall memories was previously tested using a navigational task, based on a simplistic maze shaped as a plus, which had 13 unique locations (Fig. 2). The details of the model implementation are summarised here, for a more complete description see Saul et al. [4].

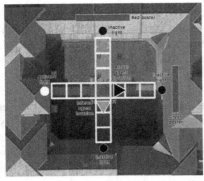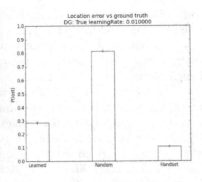

Fig. 2. Left: Plus-shaped maze of the courtyard of the Department of Computer Science, University of Sheffield, taken from Saul et al. [4]. The 13 tiles are marked as white boxes, with light sources at the end of each arm and coloured posters at the ends of two arms. The actual and inferred (by the hippocampal model) locations of the virtual agent are shown. Right: Assessment of the navigational accuracy of the original hippocampal model [4]. The bars show the proportion of steps in which the agent is lost over 3000 randomised steps, as a decimal percentage. 'Learned' used the hippocampal model with the weights set by training using 30,000 randomised steps over the maze. 'Random' used randomised weights in the model and 'Handset' used fixed GPS to include an exaction location of the agent during model training again over 30,000 steps.

The navigational task required a virtual agent to make inference about its current location within the maze, e.g., it produces (x,y) location coordinates. The virtual agent first learns the environment by following a randomly generated path around the maze, while being exposed to the various sensory inputs at each location. The agent is then sent around the maze, using a section of the original path. The agent then reports its inferred location, using sensory information available at each location and information on its previous location. This is made possible within the hippocampal model, by having two underlying systems. The first is a coordinate system which allows the agent to keep track of its location by the inclusion of a series of biologically equivalent cells (Fig. 1), including: 13 place cells, which encode for each maze location: a 3x2 grid of grid cells, which have a unique encoding for of (x,y); and 4 head direction cells, which encode direction (e.g., North, South, East, West). The second system is the sensory input and processing modality, which is used by the model to infer its current position within the maze. The 'plus' maze has been built to have sensory information available at each location which, when combined, is unique to that location, thus aiding inference (Fig 2. Left). The sensory inputs made available to the model included: photographs relating to specific directions at each location; coloured markers at the ends of two arms of the maze; touch/whisker sensors marking the location of walls; and an active light source on the west arm. The sensory inputs were processed as an equivalent to compression in ABM. For example, each of the images was converted to a 100-digit binary vector, by extracting Speeded-Up Robust Features (SURF) features [11] for each images and merging similar feature vectors using a k-means cluster approach.

The initial learning phase involved training the model (updating the weights within the TRBM), by having the virtual agent take 30,000 random steps around the maze. At each location a number of inputs were made available to the model, including the processed sensory inputs, grid cells marking the previous location, and head directions cells marking the current direction the agent was facing. Following the training phase, the virtual agent was subject to 3000 movements around the same 'plus' maze. The TRBM weights were fixed. At each step the agent then used the sensory inputs available, head direction cell firings and previous grid cell encoding, to make predictions about its current location, which could be read by decoding the firing of the grid cells. The location of the virtual agent was accurately inferred for around 72% of the steps (Fig. 2, right). A secondary loop was included within the model to allow for the detection and reaction to the agent being lost. This lostness system used the differences between the model's expected sensory inputs and its inferred position and the received sensory inputs; if the difference was too great then the priors would be reset (e.g., grid cells were cleared). The model's ability to store and recall memories in an unsupervised manner was compared to a control condition, named 'handset'. The same methods and inputs were used to train and test the model. However, during the training phase the model weights were set by using grid cells' firings based on the exact coordinates of each location (equivalent to using global position system; GPS), rather than using the previous location. The 'handset' control condition offers greater accuracy in predicting the location of the virtual agent for around 90% of the 3000 randomised steps within the maze. In addition to the 'Handset' condition an additional 'Random' condition used randomised weights in the model with the virtual agent's location again predicted for the 3000 randomised steps. Unsurprisingly the ability to detect the location of virtual agent was much lower than for the 'Learned' and 'Handset' conditions, at around 20%.

The hippocampal model demonstrated reliable memory storage and recall. The model was able to predict the current location of an agent using inference of past location and processed sensory data specific to that location. There are, however, a number of limitations in both the experimental design used to test the model and the implementation of the model. These not only limit the ability of the model to spatially navigate more 'real' environments, but also limit the applicability of the model for non-spatially specific tasks. We therefore extend the model to use more realistic sensory information and extend the model to navigate mazes with non-fixed topologies.

4   Extending the Hippocampal Model and Navigational Task

4.1   Making the Task More Realistic

To address the limitations imposed by the navigational task used in the original testing of the hippocampal model, a new experimental design has been developed and tested. These modifications are listed below with the ability of the model to navigate the maze assessed for each set of changes.

A Novel Path and a Randomised Start. The original navigation task used to test the hippocampal model used both a fixed start location and a section of the same path used in the learning phase. Thus, here the navigational ability of the trained model was tested using a novel path of 100 steps with a randomised start location (Fig. 3, left). The model was able to successfully infer the location of the agent approximately 65% of the time, which is only slightly less accurate than using a 3000 step segment of the original path, where the agent was accurately placed around 72% of the time.

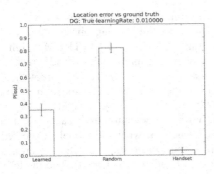 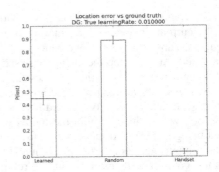

Fig. 3. Left: Assessment of the navigational accuracy of the original hippocampal model, using a novel test path (not a section taken from the training path), with a randomised start. The bars show the proportion of steps that the agent is lost over 100 randomised steps, as a decimal percentage. 'Learned' used the hippocampal model with the weights set by training using 30,000 randomised steps over the maze. 'Random' used randomised weights in the model and 'Handset' used fixed location coordinates (equivalent to using the global positioning system, GPS) to include an exaction location of the agent during model training, again over 30,000 steps. Right: Assessment of the navigational accuracy of the original hippocampal model, without the lights or colours placed at the end of each arm of the maze, again using a novel 100 step test path and randomised start.

Non-realistic Sensory Information Removed from the Maze. The visual sensory inputs to the original maze were deemed to be unrealistic, for example light sources and coloured panels were 'carefully' placed at the extremities of the arms of the plus maze. Thus, the coloured panels and light sources were disabled, as these were not considered to be authentic in a real-world maze environment. The colour and light inputs would also be difficult to place in a complex, non-uniform maze, especially when compared to the original 'plus' maze. The hippocampal model when trained (30,000 steps) and tested (100 steps) was able to reliably predict the location of the virtual agent for around 55% of the steps (Fig. 3, right), without the lights or colour inputs included.

'Real-World' Image Data. The model was tested using real-world sensory data from Google Street-view, while retaining the original 'plus' maze. The original images used by Saul et al. [4] for the sensory input within the 'plus' maze were taken in the Department of Computer Science building courtyard at the university of Sheffield, specifically for use within the model. The images were considered to be relatively limited both in their framing and content (constrained features available). Google Street-view offers 360° degree panoramic images of the roads around the world, taken with a car which has 360° cameras. Images were extracted manually from Google Street-view on-line. Images were collect at the intersection of Division Street and Carver Street (Fig. 4, left) in the centre of Sheffield, UK. The images were taken from 7 locations along both Division and Carver streets, which produced the identical structure to that of the original 'plus' maze. The hippocampal model was trained using the Google Street-view images with 30,000 randomised steps around the maze.

The model was then tested using 100 novel randomised steps, with a randomised start and was able to predict the virtual agent's location within the maze for around 72% of the steps (Fig. 4, right). This suggests the model is able to reliably store and recall memories based on real-world data.

 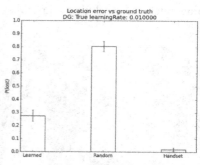

Fig. 4. Left: The locations from where images were manually extracted from Google Street-view and used to generate a 'plus' maze. Right: Assessing the ability of the hippocampal model to use real-world sensory data for navigation. The model was trained and tested using the Google Street-view images.

4.2   A Flexible, Reconfigurable Maze

A major limitation of the original experimental design is the fixed size and shape of the 'plus' maze, which also carries through into the implementation of the original model. The model has fixed numbers of encoded location tiles, as well as a fixed number of place and grid cells. This lack of flexibility severely limits the applicability of the model to a range of both navigational and non-spatial memory problems. Thus we generated a new larger and more complex maze using Google Street-view images for testing the hippocampal model.

Generating the Maze. Our previous development to the maze, taking 'real-world' images, relied on a manual online extraction of Google Street-view images. This approach was considered to be too repetitive, open to experimenter bias and would likely produce errors for the generation of a larger more complex maze. Google offers a free-to-use uniform resource locator (URL) based application interface (API), which allows the user to request and download Google Street-view images, for example: https://maps.googleapis.com/maps/api/streetview?size=480x480&location=53.37941 66,-1.4774962&heading=0&pitch=0&fov=90

The API allows the user to specify the location of the required image as longitude and latitude (Fig. 5, left). The service will not return the image of the exact location requested, but the image nearest to the requested location taken by the Google car when it drove past. The approximate image location makes the generation of the maze more difficult. For example, identical images are often returned for nearby locations and the identical images need to be detected and merged. In addition the non-exact location of the returned images means the physical spacing between images can vary, which adds complexity when attaching images to a fixed grid location. The maze generation module was built to standardise the spacing between images by using the minimum spacing between unique returned images. The Google Street-view API also allows the user to specify a series of parameters with each image request, such as heading, pitch and resolution (Fig 5).

Fig. 5. Left: Exemplar Google Street-view images extracted using the API, for a single location. The four images shown represent the four directions (left = North, middle = East, right = South and below = West). The images have 0° pitch, 90° field-of-view and a resolution of 640x640. Right: An aerial view of the centre of Sheffield, UK, from Google Maps is overlaid with the points marked in yellow that were used to predefine the maze. The blue line shows the included sections of road.

The automated generation of the maze uses pre-defined latitude and longitude location coordinates, in decimal degrees, of the start and end of each road and the intersections of one road with another (see yellow circles overlaid onto the map in Fig. 5, right), to be included. A minimum step spacing of 0.000125 decimal degrees was used as a stepping between requested image locations along each road. Returned blocks of four images (North, South, East, and West) were checked to make sure they were not identical to previous images. The image block was then named using a grid reference, with x (East positive, west negative) and y (South positive, north negative). A total of 708 images were requested and were used to build the maze (blue lines in Fig. 5, right).

A Flexible Hippocampal Model. The original model [4] was designed around the 'plus' maze, with each location 'hardcoded' within the software implementation of the model (written in Python). This prevented the model from navigating a virtual agent around alternative mazes. In order for the new model to navigate the new more complex maze, a substantial number of modifications were required to the old model. The first step towards a flexible hippocampal model implementation involved a maze generation module, which when provided with a folder of images extracted from Google Street-view, generated the maze from the image files available. This allowed for flexible maze generation, as different sets of images will produce different maze topologies in an automated manner. The increased complexity of the maze also called for a more straightforward approach to verify the consistency of the automatically generated maze. The checks allowed confirmation that the appropriate images were present at each location and were correct in terms of their location, direction and whether images at adjacent locations were consecutive. A graphical user interface (GUI) was developed to load the images and allowed a human user to navigate the maze using key presses, showing which direction steps were possible for each grid location. A map was automatically generated from the available images and included the current location and the previous path of the agent (Fig. 6).

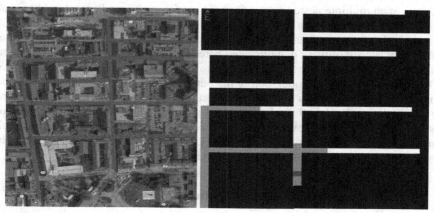

Fig. 6. Left: Google maps aerial image of Sheffield centre. Right: An autonomously generated top-down map of the maze. The map was generated from the locations of the images available in the given folder. The white boxes represent the locations where image blocks are available, the red box represents the current location of the virtual agent and the green boxes represent the previous path of the agent.

The sensory inputs at each location within the original plus maze were also 'hard-coded' into the model, for example, the model assigned sensory inputs from set tile locations to pre-defined place cells. Therefore each new sensory input had to be adapted for the autonomously generated maze and the libraries of sensory inputs built. Examples included:

1. Touch input – This gives details of the walls that are present [left, forward, right] e.g., dead end [1,1,1], intersection [0,0,0].
2. Available moves – Can the agent go; forwards, left, right or do a u-turn at each position.
3. Available grid locations – This gives the resulting location of the virtual agent following each type of move.
4. Grid cell encoding – The encoding of the x and y locations.
5. Head direction – The direction the agent is facing.

The GUI was found to be useful for testing the adjustments made to the model and was used throughout the different phases of training and testing. The debugging tools built-in to the maze visualisation GUI were used to verify that the processed location specific sensory information (e.g. Python dictionaries) was correct and was assigned to the correct tile location. The processed sensory information from the larger generated maze was found to be accurate across all the tile locations.

The heavily modified hippocampal model was trained on the larger maze data, but was unable to accurately infer the location of the agent within the maze, despite repeated modification, troubleshooting and re-testing. This failure to scale is not thought to have arisen from the biologically inspired design of the model or the selection of the TRBM, but instead arises from the existing practical software implementation of the model in 'Python'. In its current form the software is made up of multiple functions, with multiple nested subroutines. These functions send complex 'dictionaries' of encoded information between the different components of the model. The structure and current configuration of the model make it very difficult to modify and troubleshoot successfully, as demonstrated here. Troubleshooting is made all the more difficult with each software update having to rely on the time taken to train (e.g. 30,000 steps) and test the model, before the outcome of each update is known. Documentation of the software is also lacking. Further extensive modifications are therefore required to the update the software of the model to allow it to use the more complex and larger maze. However, we would recommend that the model is completely rebuilt using a structure based on the biological components (e.g. CA3, DG), with integrated scaling for different mazes and with more complete documentation.

5   Conclusions and Future Directions

This paper presents a number of steps in developing an existing hippocampal model [4] for initial application in a spatial navigation task and with the future aim of further developing the model for application to non-spatial memory. We firstly demonstrated the ability of the model to predict locations within a 'realistic' simple maze, which was built using data similar to that used by humans for spatial navigation (images from Google Street-view) and the pre-existing 'un-realistic' sensory streams were disabled. Secondly to assess whether the existing model could scale to a more complex, less structured task, the spatial size and complexity of the maze used within the navigational task was extended. This ability to scale is essential in producing a system which includes the functions similar to those of human ABM, where multiple streams of sensory information will need to be compressed and processed ready for storage and multiple predictions will need to be made through recall of multiple previous memories.

An automatic maze generation system was produced to generate the larger more complex maze. Initial testing with the larger maze revealed the model implementation was 'hardcoded' for the original 'plus-maze', in that it had an architecture with cell encoding (grid and place) fixed to the shape of the existing plus maze. The hippocampal model implementation was substantially updated to allow for learning of mazes with different topologies, using variable scaling, mapping and encoding of grid and place cells. An interactive visualisation tool was used to verify location specific compression of sensory information and to observe the agent moving around the maze using the paths generated for learning and testing. Testing showed the new version of the model was unable to predict the location of the virtual agent within the larger maze. A number of avenues will be explored to re-develop the model to learn and accurately navigate the larger maze, these include replacing the software implementation with a more simple structure, using deep Gaussian process techniques [12] instead of a TRBM and using active learning to improve the learning phase, for example by exploring the maze with the agent having a preference for novelty, rather than using a random exploration approach.

5.1   Towards a Model of Human Autobiographical Memory

The model and navigational task developed here are limited in that they replicate only spatial memory, however, they offer some insight into practical implementations of memory systems, which go beyond spatial memory, such as human ABM. However it should be noted this current system is in no way an equivalent for human ABM, but the testing of the model has demonstrated the model has some of the functionality associated with human ABM [5]:

— **Compression.** The model uses extrinsic compression with the extraction of SURF features from the Street-view images and the nearest neighbour clustering of features. The model also demonstrates intrinsic compression with the encoding of the sensory data within the TRBM during learning.
— **Pattern Separation.** The model was able to infer the location of a virtual agent within the Street-view plus-maze, using the provided sensory inputs available. The ability to recall unique locations demonstrates the ability of the model to store and recall information specific to each location, despite the need for compression. Thus, information is not merged where it is deemed to be essential.
— **Pattern Completion.** The initial model implementation used a series of well-placed sensory features, we removed these features, replacing them with real-world imagery. The model demonstrated pattern completion, as it was able to infer the location of the virtual agent within the plus-maze with only the limited sensory information available e.g. inferring location using only a set of image features for that location.

5.2   Future Objectives

The original model and experimental design have been developed to overcome the immediate limitations and has been shown to navigate with re-world image data from Google Street-view. However, the current implementation is unable to navigate different sized and shaped mazes, thus, as a next step we aim to build a complementary system that will use both spatial and non-spatial memory formation and recall, which will build on the biologically inspired design demonstrated here. This will be used with the iCub humanoid platform and will endow the robot with the ability to interact socially with humans, through recognition of humans, remembering actions and generation of specific actions and language. We will work towards the convergence of SAM systems based on human memory function [13, 14] with our current models which use well characterised biology.

Acknowledgements. The authors would like to thank the EU for funding this research (EU grant no. 612139 WYSIWYD - "What You Say Is What You Did"), our colleagues at Sheffield Robotics and Google for providing the Street-view images and API.

References

1. Moser, E.I., Kropff, E., Moser, M.-B.: Place Cells, Grid Cells, and the Brain's Spatial Representation System. Annual Review of Neuroscience **31**, 69–89 (2008)
2. Fox, C., Prescott, T.: Hippocampus as unitary coherent particle filter. In: The 2010 International Joint Conference on Neural Networks (IJCNN), pp. 1–8. IEEE (2010)
3. Fox, C., Prescott, T.: Learning in a unitary coherent hippocampus. In: Diamantaras, K., Duch, W., Iliadis, L.S. (eds.) ICANN 2010, Part I. LNCS, vol. 6352, pp. 388–394. Springer, Heidelberg (2010)
4. Saul, A., Prescott, T., Fox, C.: Scaling up a boltzmann machine model of hippocampus with visual features for mobile robots. In: 2011 IEEE International Conference on Robotics and Biomimetics (ROBIO), pp. 835–840. IEEE (2011)
5. Evans, M.H., Fox, C.W., Prescott, T.J.: Machines learning - towards a new synthetic autobiographical memory. In: Duff, A., Lepora, N.F., Mura, A., Prescott, T.J., Verschure, P.F. (eds.) Living Machines 2014. LNCS, vol. 8608, pp. 84–96. Springer, Heidelberg (2014)
6. O'Keefe, J.: Place units in the hippocampus of the freely moving rat. Experimental Neurology **51**, 78–109 (1976)
7. Hafting, T., Fyhn, M., Molden, S., Moser, M.-B., Moser, E.I.: Microstructure of a spatial map in the entorhinal cortex. Nature **436**, 801–806 (2005)
8. Eichenbaum, H.: Is the rodent hippocampus just for 'place'? Current Opinion in Neurobiology **6**, 187–195 (1996)
9. Cameron, H.A., Glover, L.R.: Adult Neurogenesis: Beyond Learning and Memory. Annual Review of Psychology **66**, 53–81 (2015)
10. Courville, N.D.D.A.C.: The pigeon as particle filter. Advances in Neural Information Processing Systems **20**, 369–376 (2008)
11. Bay, H., Tuytelaars, T., Van Gool, L.: SURF: speeded up robust features. In: Leonardis, A., Bischof, H., Pinz, A. (eds.) ECCV 2006, Part I. LNCS, vol. 3951, pp. 404–417. Springer, Heidelberg (2006)
12. Damianou, A.C., Lawrence, N.D.: Deep Gaussian Processes. arXiv preprint arXiv:1211.0358 (2012)
13. Pointeau, G., Petit, M., Dominey, P.F.: Embodied simulation based on autobiographical memory. In: Lepora, N.F., Mura, A., Krapp, H.G., Verschure, P.F., Prescott, T.J. (eds.) Living Machines 2013. LNCS, vol. 8064, pp. 240–250. Springer, Heidelberg (2013)
14. Pointeau, G., Petit, M., Dominey, P.F.: Successive Developmental Levels of Autobiographical Memory for Learning Through Social Interaction. Autonomous Mental Development, IEEE Transactions on **6**, 200–212 (2014)

Towards a Two-Phase Model of Sensor and Motor Learning

Jordi-Ysard Puigbò[1]([⊠]), Ivan Herreros[1], Clement Moulin-Frier[1], and Paul F.M.J. Verschure[1,2]

[1] Laboratory of Synthetic, Perceptive, Emotive and Cognitive Science (SPECS), DTIC, Universitat Pompeu Fabra (UPF), Barcelona, Spain
jordiysard.puigbo@upf.edu
[2] Catalan Research Institute and Advanced Studies (ICREA), Barcelona, Spain

Abstract. The cerebellum has an important role on motor learning. How sensory data arrives to the cerebellum is hardly understood. A two-phase model is proposed to understand how raw sensory data is processed to facilitate cerebellar predictive learning. Different candidates are presented for guiding the perceptual learning phase grounded on the role of the amygdala. A hebbian learning based computational model is presented with some preliminary results.

1 Introduction

The cerebellum is a region in the brain specialized in predictive learning. In a motor task, it allows to anticipate undesired stimuli eliciting motor actions, as in the paradigm of classical conditioning. Classical conditioning has also been related to the cerebellum [1] and will be explained in next section. Several models of the cerebellum exist and have been tested for motor control and avoidance learning in robots [2][3].

In the other side, current models of the cerebellum are mostly centred in the predictive learning part. Due to the high interconnectivity and multiple roles of the different areas of the brain, experimental setups to understand the cerebellum usually use simple, primitive sensory cues, in order to avoid the interference of higher brain areas. In the case of computational models, they hardly refer to the neural substrates that provide already processed sensory inputs to the cerebellum. Such models either replicate the biological setups or are tested in more complex experimentation where the data is provided as needed.

The aim of this work is converging to a model that overcomes the gap between raw sensory data and the preprocessed inputs to the cerebellum, as long as other brain areas. Such a model should reduce the dimensionality of the original sensory inputs. The neocortex is a plausible candidate for this processing. The neocortex is the broadest area in the brain, receiving most of its afferent connections from sensory areas. If one would want to describe the main cortical function in terms of data processing, it would be dual: to expand and replicate this data into several nodes and to recombine this data to produce more elaborated representations of the original raw inputs. Additionally, both the thalamus and pons

© Springer International Publishing Switzerland 2015
S.P. Wilson et al. (Eds.): Living Machines 2015, LNAI 9222, pp. 453–460, 2015.
DOI: 10.1007/978-3-319-22979-9_45

have been identified as relay areas, which mainly redistribute the data to and from the cortex, respectively. Which are the signal transformations happening within these two regions is still under debate. Literature about the thalamus points to its role in attentional processes, due to its tight recursivity with the cortical areas [4]. Literature about the pons is scarce, although it being the main source of afferents to the cerebellum.

Motor learning can't be achieved without sensory feedback and we propose a learning model in two phases to address this problem. The two-phase model must then be composed of:

- A cerebellar model of learning, that anticipates some *error signal* and elicits motor action from learnt predictive stimuli.
- A perceptual learning model, able to select relevant sensory features of the predictive stimuli.
- A nexus, that transforms the *error signal* into a learning signal, which enhances learning parameters of the perceptual system.

For the first point, an existing model [3] is used. For the second point, a model based in the synaptic plasticity of the cortex is suggested, being the main source of plasticity based on hebbian learning. Finally, a model based on the role of the amygdala in emotions is proposed for the third point.

The two-phase model of sensor and motor learning is presented in the following section 2. Next, section 3 presents the progress of this study towards the perception of stimuli. Section 4 shows some preliminary results and experimentation. Finally, section 5 wraps everything up and present the following lines of study.

2   A 2-Phase Model of Sensor and Motor Learning

Several studies have identified the cerebellum as the structure of the brain related to classical conditioning and fine tuning motor skills. In the classical conditioning paradigm, an animal is first presented with a neutral stimulus, the Conditioning Stimulus (CS). Short after the CS, a noxious Unconditioned Stimulus (US) is presented, eliciting a reflexive Unconditioned Response (UR). After a few sessions where the animal has faced the same CS-US-UR sequence, it learns to produce a Conditioned Response (CR) just before the US. This CR aims to avoid the noxious effect of the US. Existing models of the cerebellum [3] show how a learned motor action can be elicited in anticipation of a US after repeated exposure to CS-US input pairs.

On the other side, perceptual learning happens specially in the cortex of the brain, which receives afferent connections mainly from the thalamus, a relay structure connected to most of the primary sensory areas. The cortex is a layered region, which, despite having several functionally different areas (visual, auditory, somatosensory, prefrontal, etc.), shows an homogeneous anatomy over them. Nonetheless, hebbian learning (based on Hebb's learning rule [5]) has been significantly used for modelling local plasticity in the cortex, with encouraging results. Several different models have been proposed in the literature, specially for the visual cortex, which still only capture some of its most basic functionality.

The amount of data available through our senses is huge, but after being processed it becomes even more. The question that this study is trying to answer is how human brains are able to learn that a stimulus is relevant for a specific purpose. In the context of classical conditioning, the problem to solve is how does the brain *learn* which stimuli (CS) in the sensory stream are correlated with other noxious or rewarding stimuli (US).

In the cortex, rapid synaptic plasticity has been observed during various sensorimotor experiences, followed by the stabilization of the formed synaptic structures. This effect raises the question of which is the trigger of this episodes of rapid synaptic changes. Our hypothesis points to the amygdala, as a broadly connected structure in the limbic system.

The amygdala is known to be related to emotions, memory and decision-making, among other functions. In the context of classical conditioning, correlation between the amygdala and the detection of prediction errors and temporal expectancy [6] has been found. The amygdala could then be in charge of promoting synaptic plasticity in sensory cortical areas when emotional responses are present. If this hypothesis is correct, plasticity in the cortex would be enhanced during conditioning experiments that involved either fear or reward as in general motor learning, where the prediction of a stimulus would have errors till the motor action is learned. The two-phase model becomes then a combined theory of adaptation, as a goal oriented loop for learning action and perception through internal states, as detailed in figure 1.

3   Exploring Models of Perception

The second phase of the model, the cerebellum model, is a working implementation presented in [3] and tested in several occasions. The next step towards the two-phase model is then the definition of an appropriate model for perceptual learning: the first phase. The principal objective of this model, as introduced before, is reducing the dimensionality of sensory data, while providing a differential representation of the relevant stimuli. On this line, our proposal is the use of hebbian learning on multi-layered networks of neurons.

Hebbian learning has been presented on section 2 as a mechanism for learning correlations between neuron activity. This mechanism is unsupervised, in the sense that it doesn't need a target or error signal to compare with, but it learns the intrinsic characteristics of the presented data. Therefore, the external variation of synaptic plasticity should be able to bias the learning towards concurrent features. Current work is directed towards the identification of appropriate mechanisms for modifying this synaptic plasticity.

Present candidates are the modification of the learning rate or the selection of most active neurons. On the one hand, modifying the rate of learning would increase the connections between neurons active at the moment of learning. This effect is similar to the rapid synaptic plasticity episodes observed during

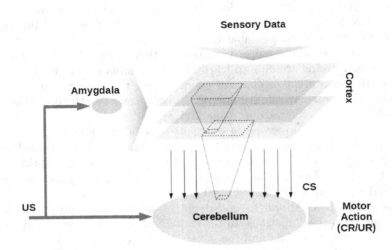

Fig. 1. Two-phase model Sensory data arrives to the cortex through several pathways. Inside the cortex, sensory data is condensed into elaborated perceptions. In the first learning phase, when a US is present, the amygdala promotes learning in the cortex based on emotional events, promoting the specialization of more neurons to a specific stimulus. Compact projections of this learned perceptions will become the CS for next trials. In the second phase, in the cerebellum, the US triggers a UR, while learning which inputs could be selected as CS. A correct learning of the predictive CS will elicit the CR in subsequent trials, avoiding the US. Having all the sensory data available in the cerebellum could overload the process or even make the discretization of a predictive CS impossible. This second phase of the model is using the working implementation of the cerebellum in [3].

sensorimotor experiences. This approach can be understood as a reward learning, and recalls some implementations of reinforcement learning. On the other hand, selecting a subset of neurons to be active and reducing the activity of the others would also increase the correlation between those neurons. De Almeida et al. proposed a model based on the selection of the neurons with more than the $E(\%)$ of the excitation, proposing it as a role for the synchronized activity produced by gamma oscillations [7]. In this specific case, the numerical value of this $E(\%)$ approximation was found to correspond to 95%. The use of an $E(\%)$ selection shows an interesting ability for noise reduction. This mechanism could be related to attentional processes, as there's some literature proposing the thalamus as the structure inducing gamma oscillations in cortical areas [8].

4   Preliminary Results

4.1   Inducing Plasticity by Changes in Learning Rate

The model has been tested on simulated data of a specific tone over background white noise. Simulated data has been generated in the form of a fast Fourier transform (FFT) with random noise on each frequency bin and peaks of intensity at specific frequencies and surrounding bins. A layer of 30 neurons was connected to this simulated input. Weights of these connections follow a Gaussian distribution along the neurons,

$$w_{ij} = \frac{1}{S\sqrt{2\pi}} e^{\frac{-(j-iM)^2}{S^2}} \tag{1}$$

being w_{ij} the weight of the connection between input i and neuron j, S the standard deviation of the Gaussian distribution and iM its mean, being M the ratio between the number of outputs and the number of inputs $\frac{N_{ou}}{N_{in}}$, each neuron becoming specially tuned to a specific frequency. Tonotopic organization of neurons is observed from the inner ear to the auditory cortex in mammal brain. The adaptation capacity of the organ of Corti to a continuous stimuli is modelled as a moving mean normalization as given by:

$$\mu[k + 1] = \mu[k](1 - \alpha) + x[k + 1]\alpha \tag{2}$$

$$\sigma[k + 1] = \sqrt{\sigma[k]^2(1 - \alpha) + (x[k + 1] - \mu[k + 1])^2\alpha} \tag{3}$$

$$x_n = \frac{x - \mu}{\sigma} \tag{4}$$

where the moving mean at iteration $k+1$ ($\mu[k+1]$) is iteratively calculated using a weighting factor $\alpha \leq 1$. α can be understood as a time constant, defining the adaptive dynamics of the input. Bigger α will then lead to faster adaptation. The normalized input x_n is then computed as x scaled over μ and σ.

Plasticity of those connections has been modelled as Hebbian Learning in the form of Oja's rule:

$$\Delta w_{ij} = \eta x_j(x_i - w_{ij}x_j) \tag{5}$$

Which updates the weight (w_{ij}) of the connection between an input neuron i and an output j in function of a learning rate (η) and the current activity rate in input and output neurons (x_i and x_j).

The experiment setup consisted of a training period of 8000 time steps and a testing period of 10000 time steps. During the training period, a new peak was generated at a random frequency out of 5 possible choices for a period of 30 time steps and learning rate was greater than 0.

During the test period, a stimuli, consisting of one of the 5 same frequencies, was presented 10 times for 50 time steps. Response to noise is measured for a

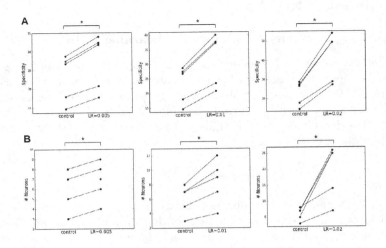

Fig. 2. Effects of induced plasticity Greater plasticity was induced in the non-control conditions by changing the learning rate in presence of a specific stimulus. a) In the STRF, specificity is defined as how different the activity of a neuron is when a specific stimuli is present compared to when it is not. A paired-samples t-test was performed to evaluate the increase in specificity of the most tuned neurons, with increasing LR. All plots have $p \leq 0.05$. b) A paired-samples t-test was performed to evaluate the increase in number of neurons recruited for the reinforced stimuli. The increase was significant with $p \leq 0.05$.

period of 500 time steps preceding the presentation of the stimulus. A Student-Fisher T-test is used used to evaluate whether there is significant difference between the response to noise and the activity within 10 time step bins, as well as up to 450 time steps after the stimulus ends. This allows to generate the Spectro Temporal Response Fields (STRF) of a neuron to an auditory stimulus.

The experiment consists of two conditions. A first control condition where the learning rate in eq. 5 is constant and a second reinforced condition where the learning rate increased each time a specific sitmulus is present and returned to normal for all other stimuli. Figure 2 show how both, the number of neurons tuned to the reinforced stimuli and their specificity increase significantly ($p < 0.05$) for different values of learning rate.

4.2 Noise Filtering with E(%)max

A single layer network has been built to show if hebbian learning can be tuned to a series of specific stimuli. Real sound samples have been used for the experiments. A dataset was constructed using real piano notes alternated with *silence*, always in presence of background noise. The input to our model was a normalized

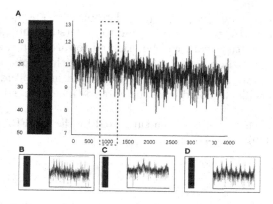

Fig. 3. Activity of a 50 neurons simulation Plot A, shows the activity of a network formed by 50 neurons, fully connected to the input as described in section 4.1 (left) and the current FFT (Fast-Fourier Transform) intensity on each frequency from 0 to 4096 Hz. The activity is presented after 5 minutes of training and using an $E(\%)$ selection of 95%. Plots B and C show the same results on the presence of different frequencies. Figure D shows that the activity with $E(\%) = 80\%$ is more sensitive to the harmonics of the played note and also more sensitive to noise.

FFT of those sounds, sampled at $8192Hz$ and sensitive to frequencies up to $4096Hz$. Figure 3 a, b and c suggest that neurons become tuned to specific frequencies over time. From d, tuning the value of $E(\%)$, exhibit more activity in frequencies far from the fundamental frequency of the played note for a lower value of $E(\%)$. This suggests a noise resistance capacity underlying high percentages of $E(\%)$ selection. This filtering ability promotes its use for dimensionality reduction, where higher neuronal activity in subsequent layers of the network would be due to *relevant* neurons instead of noisy ones.

Ongoing experiments aim to validate the effects of a varying $E(\%)$ selection on the tuning of neurons to specific stimuli.

5   Future Work

A two phase model of learning has been presented, with cortical areas for the role of perceptual learning, the cerebellum for motor learning and the amygdala as the learning control structure. The thalamus has been identified as an interesting area for attention to specific stimuli which serve to the goal of reducing dimensionality. Current work is being directed towards the development of an appropriate computational model of the cortex. Special effort is put on the identification of plasticity inducing mechanisms as those presented in section 3.

The presented model is of special interest for the developmental robotics community, as a model for the acquisition of perceptual and motor skill from experience. One of the main problems in state of art robotics is error evaluation and robust behaviour, combined with limited resources. The two-phase model addresses the sensorymotor problem by trying to understand how the brain deal with these situations. Future work will be oriented towards the validation of the presented learning model in a series of experiments with robots, such as sound localization and grasping. Additional efforts will be put in the study of attention related processes, specifically looking in the cortico-thalamo-cortical pathway.

Acknowledgments. This work has been supported by the EU FP7 project WYSI-WYD (FP7-ICT-612139).

References

1. Yeo, C., Hesslow, G.: Cerebellum and conditioned reflexes. Trends in Cognitive Sciences **2**(9) (1998)
2. Brandi, S., Herreros, I., Verschure, P.F.M.J.: Optimization of the anticipatory reflexes of a computational model of the cerebellum. In: Duff, A., Lepora, N.F., Mura, A., Prescott, T.J., Verschure, P.F.M.J. (eds.) Living Machines 2014. LNCS, vol. 8608, pp. 11–22. Springer, Heidelberg (2014)
3. Herreros, I., Paul, F.M.J.: Verschure, Nucleo-olivary inhibition balances the interaction between the reactive and adaptive layers in motor control. Neural Networks **47** (2013)
4. Guillery, R.W., Sherman, S.M.: Thalamic Relay Functions and Their Role in Corticocortical Communication: Generalizations from the Visual System. Neuron. **33**(2) (2002)
5. Donald, O.: Hebb, The Organization of Behavior: A Neuropsychological Theory. Wiley & Sons, New York (1949)
6. Daz-Mataix, L., Tallot, L., Doyre, V.: The amygdala: A potential player in timing CS-US intervals. Behavioural Processes **101** (2014)
7. De Licurgo, A., Idiart, M., Lisman, J.E.: A Second Function of Gamma-Frequency Oscillations: A E%-Max Winner-Take-All Mechanism Selects Which Cells Fire. The Journal of neuroscience **29**(2)3 (2009)
8. Saalmann, Y.B., Kastner, S.: Cognitive and Perceptual Functions of the Visual Thalamus. Neuron. **71**(2) (2011)

Telepresence: Immersion with the iCub Humanoid Robot and the Oculus Rift

Uriel Martinez-Hernandez[✉], Luke W. Boorman, and Tony J. Prescott

Sheffield Robotics Laboratory and the Department of Psychology,
University of Sheffield, Sheffield, UK
`uriel.martinez@sheffield.ac.uk`

Abstract. In this paper we present an architecture for the study of telepresence and human-robot interaction. The telepresence system uses the visual and gaze control systems of the iCub humanoid robot coupled with the Oculus Rift virtual reality system. The human is able to observe a remote location from the visual feedback displayed in the Oculus Rift. The exploration of the remote environment is achieved by controlling the eyes and head of the iCub humanoid robot with orientation information from human head movements. Our system was tested from various remote locations in a local network and through the internet, producing a smooth control of the robot. This provides a robust architecture for immersion of humans in a robotic system for remote observation and exploration of the environment.

Keywords: Telepresence · Human-robot interaction · Virtual reality

1 Introduction

Research on telepresence has grown in the last decades with the aim of providing humans with the sensation of being in a remote location. Driven by VR game industry this aim has motivated the development of sophisticated devices for the study and implementation of telepresence in domains such as virtual reality, office settings, education, aerospace and rehabilitation [1]. Despite advances in technology, robust telepresence systems that offer a friendly and natural human-robot interaction are still under development. This is mainly due to the vast number of human behaviours [2], and the required technological features to make the human feel physically present at the remote location, e.g. vision, tactile and proprioceptive information, depth perception, facial expressions, language and minimised time delays [3]. In this work, we present an architecture for telepresence, which allows the remote control of the head and eyes of the iCub humanoid robot. The observation and exploration of the remote location is achieved by orientation information and visual feedback with the Oculus Rift. Our approach offers a platform to provide the human with the capability to be physically immersed in a remote location.

© Springer International Publishing Switzerland 2015
S.P. Wilson et al. (Eds.): Living Machines 2015, LNAI 9222, pp. 461–464, 2015.
DOI: 10.1007/978-3-319-22979-9_46

2   Methods

2.1   iCub Humanoid

The iCub humanoid robot was chosen for the development of the telepresence system as it mirrors many human functions. The iCub humanoid is a robot that resembles a four year old child and has 53 degrees of freedom. The robot has integrated visual, vestibular, auditory and haptic sensory capabilities. It is one of the most advanced open systems suitable for the study of cognitive development, control and interaction with humans [4]. Its arms and hands allow dexterous, natural and robust movements, whilst its head and eyes are fully articulated. The iCub humanoid robot is also capable of producing facial expressions. The facial expression, e.g. sad, happy, are generated by LED (Light-Emitting Diode) matrix arrays located on the robots face, which can be controlled according to the feedback from the sensing modalities. Facial expressions are essential for providing a natural behaviour during telepresence and interaction with humans.

2.2   Oculus Rift

The Oculus Rift is a light-weight headset developed by Oculus VR that permits the immersion of humans in a Virtual Reality (VR) environment. The headset is composed by two adjustable lenses and was primarily developed for gaming, displaying virtual scenes. The stereo vision feature offered by the Oculus Rift provides the sensation of depth and more realistic immersion in a 3D world. The multi-axis head tracking capability integrated into the device allows humans to look around the virtual environment in a natural way as in the real world [5]. The Oculus Rift also includes a gyroscope, accelerometer and magnetometer, which together allow robust tracking of head position and orientation.

2.3   Control Architecture

Our goal is to immerse the human in a remote environment, through both the observation of the world through the eyes of the iCub humanoid robot and via the remote control the head of the iCub humanoid robot. This offers the human the ability to look around and explore the environment in a natural way as humans do. Thus, the first step involved the development of a module which capture the visual scene from both eyes of the iCub humanoid robot and display this in the Oculus Rift. Second, a module was developed to control the head movements of the robot, by reading the orientation of the human head with the Oculus Rift. This module relied upon a cartesian gaze controller developed previously for the iCub [6]. This module was required to both calibrate and transfer into a suitable format, the orientation information arriving from the human and being sent to the robot. Both modules were precisely controlled and synchronised, thus achieving a smooth and natural behaviour of the iCub humanoid robot. These features, important for telepresence, provide the human with a more real immersion into the world of the iCub, allowing a natural exploration and interaction

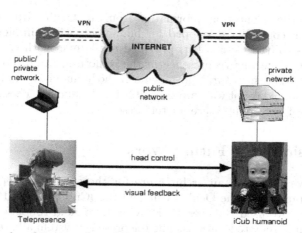

Fig. 1. Proposed architecture for telepresence with the iCub humanoid robot and the Oculus Rift via the internet. Visual feedback is provided by the iCub humanoid robot. Robot movements are controlled by information provided by the Oculus Rift.

with the remote environment. Figure 1 shows the proposed architecture with the connections and communication established between the human and the iCub humanoid robot for remote operation through the internet. All modules were developed using C++ language and the YARP library (Yet Another Robot Platform), developed for robust communication of robotic platforms [7].

3   Results

Our architecture for telepresence using the iCub humanoid robot and the Oculus Rift was tested in different locations at the University of Sheffield. The Oculus Rift and the iCub humanoid robot software was installed and executed on two different computer systems. To provide mobility and test the architecture for telepresence from different locations, the Oculus Rift was controlled by a mobile laptop with the following specifications: Core i5 Processor, 4 GB RAM, NVS 3100M Graphic processor, 512 MB for CUDA. The iCub humanoid robot was set up in the Sheffield Robotics Laboratory and controlled by a dedicated computer system with the following features: Xeon E5-1620 Processor, 16 GB RAM, Nvidia Quadro K2200 Graphic processor, 4GB RAM for CUDA. These systems have the computational power to minimise temporal delays in processing of the visual and orientation information that would reduce the immersive telepresence experience. A Virtual Private Network (VPN) was established to provide a secure and robust communication channel between the human and the iCub humanoid robot. The VPN also permitted access from different locations outside the University of Sheffield, to the local static IP address assigned to the iCub humanoid robot (Figure 1). The teleprescence system allowed a human participant to visually explore a remote environment through the eyes of the iCub humanoid robot. Robust eye and head movements were achieved with the robot based on the orientation information from the Oculus Rift and the open source Cartesian controller developed for humanoids which implements a PID controller [6]. Also, the Oculus Ritf provided the human with the sensation of visual depth and a feeling of presence in the remote location. For the current study, the delay between natural movements of the human head and the response of the iCub humanoid robot was imperceptible for the human. These results show the robustness of our architecture for telepresence.

4   Conclusions and Future Work

In this work we presented an architecture for the study of telepresence. The system was composed of the Oculus Rift and the iCub humanoid robot. As an initial approach, we used the eyes of the iCub humanoid robot for observation and exploration of remote locations. The telepresence system allowed the human to be successfully immersed in a remote environment by controlling the head and eyes of the iCub humanoid robot. Immersion using a telepresence system not only requires vision but also all the sensing modalities available with the chosen robotic system. Thus, for future work we plan to develop and integrate tactile feedback, speech capabilities and the control of the arms, hands and facial expressions. This will provide a complete telepresence system for robust and natural human-robot interaction.

Acknowledgments. This work was supported by the Cyberselves project. We thank the technical staff of the Sheffield Robotics Laboratory at the University of Sheffield.

References

1. Rae, I., Venolia, G., Tang, J.C., Molnar, D.: A framework for understanding and designing telepresence (2015)
2. Gibert, G., Petit, M., Lance, F., Pointeau, G., Dominey, P.F.: What makes human so different? analysis of human-humanoid robot interaction with a super wizard of oz platform. In: International Conference on Intelligent Robots and Systems (2013)
3. Stassen, H.G., Smets, G.: Telemanipulation and telepresence. Control Engineering Practice **5**(3), 363–374 (1997)
4. Metta, G., Natale, L., Nori, F., Sandini, G., Vernon, D., Fadiga, L., Von Hofsten, C., Rosander, K., Lopes, M., Santos-Victor, J., et al.: The icub humanoid robot: An open-systems platform for research in cognitive development. Neural Networks **23**(8), 1125–1134 (2010)
5. Desai, P.R., Desai, P.N., Ajmera, K.D., Mehta, K.: A review paper on oculus rift-a virtual reality headset (2014). arXiv preprint arXiv:1408.1173
6. Pattacini, U.: Modular cartesian controllers for humanoid robots: Design and implementation on the icub, Ph.D. dissertation, Ph. D. dissertation, RBCS, Italian Institute of Technology, Genova (2011)
7. Fitzpatrick, P., Metta, G., Natale, L.: Yet another robot platform. http://eris.liralab.it/yarpdoc/index.html

Biophilic Evolutionary Buildings that Restore the Experience of Animality in the City

Pablo Gil[1,2](✉), Claudio Rossi[3], and William Coral[3]

[1] The Barlett School of Architecture, London, UK
pablo.gil@uem.es
[2] Universidad Europea de Madrid, Madrid, Spain
[3] Centre for Automation and Robotics UPM-CSIC,
Universidad Politécnica de Madrid, Madrid, Spain

Abstract. In this paper, we present our work on the training of robotised architectural components of intelligent buildings, focusing on how architectural components can learn to behave animalistically, according to the judgment of human users. Our work aims at recovering the lost contact with animals in the urban context, taking advantage of biophilic empathy. The parameters governing the robotised elements we propose are mainly qualitative (emotions and aesthetical perception), which cannot easily be described by mathematical parameters. Additionally, due to their complexity, it is often impossible –or at least impractical, to hardcode suitable controllers for such structures. Thus, we propose the use of Artificial Intelligence learning techniques, concretely Evolutionary Algorithms, to allow the user to teach the robotised components how to behave in response to their resemblance to specific animal behaviors. This idea is tested on an intelligent façade that learns optimal configurations according to the perception of aggressiveness and calmness.

Keywords: Biophilia · Biomimicry · Wellbeing · Evolutionary robotics · Embodied evolution · Intelligent buildings

1 Introduction

There are many examples of bio-mimetic principles applied to the design of architecture by design professionals and researchers in the field of architecture [1], [12], [10], [7]. The approach to bio-mimicry is varied, as it is recognized by the analytical study of bio-mimetic design approaches developed in [11]. Among the different approaches, the most influential one is the attempt to develop bio-mimetic architectural technologies that seek performance optimization in comparison to established solutions in the design of structures, substructures, environmental systems or façades, components that are usually recipients of bio-mimetic design approaches. The approach of optimizing architectural solutions looking at mechanisms proper of living beings is one of the key fields that can have an impact in the improvement of the urban environment. But there is another aspect of living beings that has been less considered, but that holds an

© Springer International Publishing Switzerland 2015
S.P. Wilson et al. (Eds.): Living Machines 2015, LNAI 9222, pp. 465–472, 2015.
DOI: 10.1007/978-3-319-22979-9_47

intriguing potential: their ability to provoke an emotional impact on humans and the values and psychological consequences that are associated with it.

The importance of the connection between humans and the natural environment is recognized in the fields of psychology and neuroscience and by theories of designers that have discussed and experimented with this issue. The term biophilia has been coined to describe this relationship [17], [15], [11] and the evidence of the benefits of biophilic design has been linked to our hardwiring of the natural in our brain world as a consequence of the evolution, in relation to organisms and materials that configure our perceptive scenario, and that are needed for our well being. There is evidence that people feel less stressed, they are able to concentrate better and even heal faster physically and psychologically when connected to elements of the natural world [16], [5], [2].

The influence of the idea of biophilia has had a major impact on architecture, even before this term was coined, in the movement of organic architecture of the twentieth century [18]. The biophilic approach within this movement has focused on the organization of the plan of the architecture according to patterns of natural growth, the relationship with the context as if architecture would be a part of a larger ecosystem, the use of strategies of growth and structural support and the organization of parts as if belonging to an organic whole. These ideas serve as a counterpoint of the mainstream strategies of architectural design in the twentieth century, functionalism and rationalism, which dominate what is built today. The examples built and the more speculative designs that have aroused are mostly hereditary of forms and structures. Of these, most of them belong to plants. Yet, there is a lack of attention into the most biophilic range of materials of nature: animals.

According to [8], "[our] study shows that neurons in the human amygdala respond preferentially to pictures of animals, meaning that we saw the most amount of activity in cells when the patients looked at cats or snakes versus buildings or people". Animals are fascinating entities on many levels. They have a great impact in the formulation of moral principles as they represent examples of abstract ideas such as fierceness, ruthlessness, beauty, enigma or numinosity. They also have a great impact on the understanding of these ideas in the child. Animals show familiar and wild features representing an enigmatic ambivalence between closeness and otherness. Our fascination is also a consequence of the empathic relationships that humans are able to establish with animals, which experimental psychology has studied in numerous studies.

This fascination with animals is in contrast to our disconnection with them in our daily life in the city, especially with animals that are not domesticated or that have total liberty of action. City dwellers have lost any contact with the wild and enigmatic behavior of animals. This process has impoverished human experience and has had an impact in its emotional, moral and aesthetic development. But there is a way by which the art of architecture can present an alternative to this loss and rejoin us with the presence of the attributes associated with animals.

2   Behaviourally Biophilic Architecture

There are numerous mechanisms by which architecture can evoke or translate the attributes of animals into the city. Here, we focus on the possibility of implementing behavioral characteristics into buildings and the city in order to resemble animal behavior. The understanding of these potential characteristics by users is also related to other characteristics that are associated to architecture such as symbolic, artistic and expressive values, which depend on the variable understanding of these values in time and culture. This is often overlooked in the design of buildings. Within the design of biomimetic behaviorally based applications it seems important to attempt understanding the perspective of the user in the consideration of the object of experience as biomimetic or not.

2.1   Robotized Architectural Elements

To achieve this purpose of designing animalistic architectural elements that operate according to certain behaviour it is necessary to use dynamic systems able to perform or interact with the user. While in most cases dynamic and interactive architectural elements are meant as functional creations that deal with environmental issues, such as the changes in sun exposure, wind, or humidity, we attempt to charge them with emotional provocations in order to connect the users to their objects of experience, especially considering that the surroundings of the city become a close environment and sometimes a repetitive field of perception in the daily life of city dwellers. For this purpose, the architectural elements we consider are equipped with a set of sensors and actuating elements that shall operate according to some perception-control-action loop that also should in itself be understood as animalistic, biophilic and successfully biomimetic. Such elements can be considered robotic systems at all effects. Our aim is to provide them with suitable intelligent controllers, not hardcoded at design time, but rather learned by adaptive mechanisms, according to user-defined evaluation criteria, and thus enabling them to be instructed by a human instructor. In our current work, we are investigating the use of artificial intelligence and machine learning techniques, concretely Evolutionary Algorithms. The robotized elements we consider here have two distinguishing features w.r.t the mainstream research in Evolutionary Algorithms applied to robotics. First, from the mechatronics point of view, the high geometrical complexity of the structures, as well as the number of sensors and actuators involved. Second, our designs shall take into account factors which cannot be easily described by mathematical parameters, since they are related to social and/or aesthetic models and even to psychological states of the users that can not clearly measured. For this reason, we are applying human-in-the-loop learning strategies that easily allow non-quantitative and subjective parameters to be assessed and incorporated in the training. We consider that human-driven evolution is a key paradigm for dealing with intelligent structures, since it allows an interaction that brings closer the functioning of the machine to the necessities expressed by the user. In this case, the "feeling of animality", which could be related to qualities of wildness, fierceness, enigma and otherness in the object of

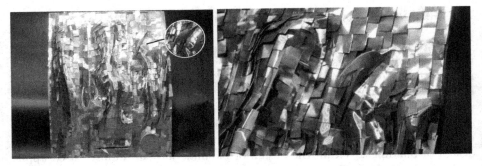

Fig 1. The Passive-aggressive façade prototype. Right: detail of a claw.

experience, can only be measured through the testimony of subjects of experience, the user of architecture.

2.2 Passive-Aggressive Façade

As an illustrative example of our work, we take into consideration the Passive-Aggressive Façade prototype, which has been robotised. The prototype (Fig. 1) is meant to interact with the user in either an inviting or aggressive way. The idea is that when the presence of people in a given range is detected, the surface of the door, which is built with a flexible steel mesh substructure, starts to fold and morph to express annoyance. Should undesired people get closer to the door, its reaction gets increasingly more violent, in an attempt to intimidate the intruder. As an ultimate attempt to avoid intrusion, three hidden steel claws are activated, coming out from the door surface.

Note that the Passive-Aggressive Façade shall be trained by a human operator, since the levels of aggressiveness require a human evaluator to be assessed. The Passive-Aggressive Façade, as vertebrate animals do, behaves expressing a phase of heightened awareness prior to fighting or fleeing conducts, communicating its state of anxiety and in preparation of the decision that the intruder might take. If a person is to be allowed in to the building, the façade must be taught to act in a calm way. This concept is meant to act also depending on the intentions of the persons that it perceives, although such cognitive feature is not taken into account here.

The Passive-Aggressive Façade prototype is built with a frame of Medium Density Fibreboard (MDF) of 19mm thickness and a flexible steel mesh that is covered with steel scales creating a double curved free form surface. The morphing movements are provided by several Shape Memory Alloy wires attached to various points of the structure. When powered, the wires contract, causing the flexible structure to morph. Commercial servo-motors are used to activate the claws. The door is also equipped with three proximity sensors to detect the presence of people. An Arduino UnoTM micro controller is used to drive the actuators and to receive the sensors readings. As actuation technology, we adopted FlexinolTM Shape Memory Alloys wires and commercial servomotors.

Fig 2. Detail of the mechatronics of the façade

Ultrasonic proximity sensors and light sensors are used to provide sensorial input to the robotized structures. Figure 2 illustrates the electronics of the prototype.

The intelligent controller of the prototype consists of a three layered Feed-Forward Neural Network. The network has three inputs (distance sensors) and six outputs (three morphing actuators plus three servo-controlled claws). The number of neurons of the middle layer has been empirically set to ten. The training process adjusts the connection weights.

3 Evolutionary Training

In this case at hand, supervised training techniques (e.g. backpropagation and its variants) cannot be used, due to the lack of pre-defined input-output training patterns. Thus, we adopted a reinforced learning scheme, implemented as a $(1 + 1)$- Evolutionary Strategy. Note that here we face a major constraint since we deal with a real, physical artefact of which one unique prototype exists. Other classes of population-based EAs commonly used to train neural networks cannot therefore be applied. As customary in evolutionary strategies, phenotypic traits (neural networks weights) are encoded in a $n + 1$-dimensional array of floating-point values, and a Gaussian perturbation mutation operator is employed. Gene $n + 1$ encodes a common mutation rate. Fig. 3 illustrates the training scheme we adopted. A training epoch consists of a series of stimuli to which the system reacts, taking a certain configuration according to its logic. The reaction of the structure to each of the stimuli is evaluated. At the end of the training epoch, the global behaviour of the robotised structure is evaluated, and if the instructor is satisfied with the behaviour of the structure the training ends. Otherwise, the network is modified. Note that the evaluation step (cf. Fig. 3) is performed against a reference value, which, depending on the application, can be performed either automatically for quantitative parameters or by a human instructor.

470   P. Gil et al.

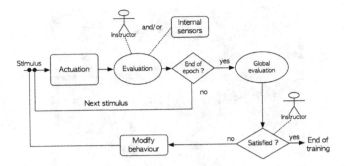

Fig 3. Training process. The external loop corresponds to one loop (generation) of the evolutionary algorithm.

Here, a subjective score is fed back to the training algorithm according to the feelings experienced by the trainer. In other prototypes (see, e.g. the *Armadillo Skyvent* [13]) a numeric fitness function has been adopted that takes into account internal sensors readings to assess the goodness of a behaviour.

3.1 Experimental Results

Using the prototype described in the previous section, we have performed pilot trials to test the system and the learning process. Experiments involved a set of three stimuli, corresponding to three distances of the person from the door: very close (distance less than 1 meter), close (between 1 and 2 meters) and far (over 2 meters), detected by any of the three range sensors. In case of o very small distance, the claw corresponding to the detecting sensor should de activated. Here, the reaction of the door is given a score from 1 to 10 by the human instructor, and training ends when the average score over the three stimuli is at least 8. Provided the evaluation of the behaviour of the system (fitness) is subjective and involves emotional response of the user, no statistical analysis was per-formed. Figure 4 shows an example training session results, where each point of the plot is the average over 5 training sessions. Ten to twenty iterations were normally sufficient for the trainer to be satisfied with the results.

3.2 Discussion

The experiments reported above have demonstrated that controllers for intelligent architectural elements can successfully be generated applying machine learning techniques. It must be pointed out that given the size prototype taken into account (number of inputs and outputs), from the point of view of artificial intelligence the learning task was expected to be easy to be learnt by the neural controller. Indeed, in these early trials the prototype was able to learn the task in few training sessions. For more complex structure more advanced EAs shall be employed, such as the CMA-ES (see, e.g., [4]), or more advanced neuroevolution techniques such as the Hyper-NEAT learning technique [3].

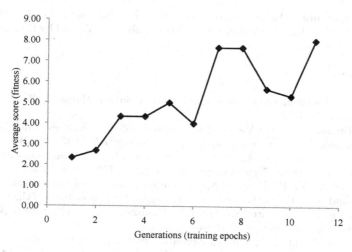

Fig 4. Results of the training of the prototype

4   Conclusion and Future Work

Robotised architectural components of intelligent buildings are elements provided with the capability of changing their morphology or geometrical configuration in order to dynamically adapt to environmental conditions to the preferences of the users. Such components are equipped with a set of sensors actuating elements, that are moved according to some perception-control-action loop. In our current work, we aim at providing such systems with some degree of autonomy, adaptivity and learning capabilities. From the Architectural point of view, considering intelligent buildings as autonomous embodied agents, allows us to apply the full arsenal of machine learning techniques, exploiting the potential of artificial evolution beyond structural design and optimisation. Current work is being devoted to exploring interaction with the user, either by direct communication through a user interface, or by a more sophisticated cognitive system capable of detecting automatically the users state and anticipate his/her needs and preferences. We envision intelligent buildings as complex dynamic systems whose behaviour will go beyond simple action-reaction functioning, where architectural components seemingly acquires a life of their own, where certain characteristics that were not programmed or considered at design time emerge as a property from the point of view of human perception. This is particularly relevant if we consider the range of opportunities to convert the city to become a more biophilic environment using not only formal resources, such as animal forms, or directly plant or animal species, but also, and more importantly considering action, movement and behavior. In conclusion, considering the building as an intelligent robotic system at all effects opens a new way of interaction between Architecture, Robotics and Artificial Intelligence which is full of potential both from the point of view of research and impact in peoples life.

Acknowledgments. The prototype used in this work have been designed by Guillermo Sanz, student of the course on "Digital Fabrication" held by the first author.

References

1. Benyus, J.: Biomimicry - Innovation Inspired by Nature. Harper Collins Publishers, New York (1997)
2. Faber Taylor, A., et al.: Views of nature and self discipline: evidence from inner city children. Journal of Environmental Psychology **22**, 49–63 (2002)
3. Gauci, J., Stanley, K.O.: Autonomous evolution of topographic regularities in artificial neural networks. Neural Computation **22**(7), 1860–1898 (2010)
4. Igel, C., Suttorp, T., Hansen, N.: A computational efficient co-variance matrix update and a (1+1)-cma for evolution strategies. In: Proceedings of the 8th Annual Conference on Genetic and Evolutionary Computation, GECCO 2006, New York, NY, USA, pp. 453–460. ACM (2006)
5. Lohr, V., et al.: Interior plants may improve worker productivity and reduce stress in a windowless environment. Journal of Environmental Horticulture **14** (1996)
6. Manrique, C.: Escrito en el Fuego. Edirca Editorial, Las Palmas de Gran Canaria (1988)
7. Mazzoleni, I.: Architecture Follows Nature-Biomimetic Principles for Innovative Design. CRC Press, Boca Raton (2013)
8. Mormann, F.: Captivated by critters: Humans are wired to respond to animals. ScienceDaily, September 9, 2011. California Institute of Technology (cited June 1, 2014). www.sciencedaily.com/releases/2011/09/110909091219.htm
9. Nagel, T.: What Is It Like to Be a Bat? The Philosophical Review **83**(4), 435–450 (1974)
10. Pawlyn, M.: Biomimicry in architecture. RIBA Publishing, London (2011)
11. Pedersen Zari, M.T.G.: Biomimetic approaches to architectural design for increased sustainability. In: The SB 2007 NZ Sustainable Building Conference, Auckland (2007)
12. Reed, B.: Shifting our mental model – sustainability to regeneration. In: Rethinking Sustainable Construction 2006: Next Generation Green Buildings, Sarasota, Florida (2006)
13. Rossi, C., Gil, P., Coral, W.: Evolutionary training of robotised architectural elements. In: Mora, A.M., Squillero, G. (eds.) EvoApplications 2015. LNCS, vol. 9028, pp. 819–830. Springer, Heidelberg (2015)
14. Stanley, K.O., Bryant, B.D., Miikkulainen, R.: Real-time neuroevolution in the nero video game. IEEE Trans. on Evolutionary Computation **6**(9), 653–668 (2005)
15. Storey, J.B., Zari, M.P.: Factor X - well being as a key component of next generation green buildings. In: Proc. of the Rethinking Sustainable Construction 2006 Conference, Sarasota, Florida, USA (2006)
16. Ulrich, R.: View through a window may influence recovery from surgery. Science **224**(4647) (1984)
17. Wilson, E.O.: Biophilia. The Human Bond with Other Species. Harvard University Press, Cambridge (1984)
18. Wright, F.L.L.: The natural house. Bramhall House, New York (1954)

Author Index